Cracking *The* Learning Code

The Science That is Transforming How The World Learns, Motivates And Changes Behaviours

JW WILSON

Executive Director, The Learning Code Institute

Testimonials

What market leaders are saying about the Learning Code Science:

I have helped to discover, and work with, some of the greatest authors of our time, and it is my opinion that this disruptive Learning Code research will have as profound an impact on the world as the works of Deepak Chopra and Wayne Dyer.

Daniel B. Levin, CEO, Mosaic, Founder, ZENsei
Former Director, Business Development Hay House Publishing

Why is it that of the multitudes of people exposed to new information from books, seminars and online programs, which could have a profound positive impact on their lives, only a small percentage of them take the information and actually put it into practical use in their own lives? There is a missing piece to this puzzle, and I believe The Learning Code is that missing piece.

Kim Kiyosaki, CEO and Cofounder, The Rich Dad Company

It is my opinion that every school district, and every homeschool parent in America, should be made aware of the innovative Learning Code science. This new science is so powerful that to try to educate without an understanding of it is a disservice, not only to our children, but to our world.

Nancy Monson, Learning Innovator, Founder Running River School

We are one of North America's leading seminar companies for market leaders like Tony Robbins and Sir Richard Branson. The Learning Code science is like the Intel Chip. Implant it into any transformational program and you will see astounding advances in the speed of learning and the depth of behavior change. It is a Disruptive Innovation that will transform our world

Ryan Mitchell, Owner-Director, Power of Success

We are among the leaders in the application of precision medicine for elite performers in extreme environments, ranging from spaceflight (NASA, commercial space), US Special Forces, NFL, NBA, US Olympic athletes, collegiate football, S.W.A.T., high altitude ascent, wilderness medicine, Lunar and Mars exploration, professional motorsports (Le Mans Series), and others. The Learning Code brings another layer of deep insight that promises to have a transformative influence on our field.

Michael A. Schmidt, PhD, CEO & Chief Scientific Officer, Sovaris Aerospace
Author, *Brain Fitness for Leaders*

The Learning Code is the holy grail in the field of education and behavioral change. This groundbreaking science will have a profound impact on the lives of millions of people around the world. The Learning Code is the missing piece and explains precisely why so many people are unable to achieve the life they deserve to live. As a leader in the financial industry for over twenty five years I can tell you firsthand, that the Learning Code science is a paradigm shift in how people live their lives, and its impact will be immense.

John MacGregor, CEO & Founder of ThrivePath: The Revolutionary Process to Financial Freedom and Peace of Mind, Best-Selling Author

The Learning Code is the holy grail in the field of education and behavioral change. This knowledge needs to be available worldwide

Dr. Jeanette Vos, National Institute for
Teaching Excellence, Coauthor, *The Learning Revolution*

The Learning Code is such a compelling approach to making learning meaningful. It doesn't just simplify and break things down-it teaches us the importance of connecting with other ideas, other people.

Danielle (Voirol) Ridgway, Grants Manager, Keep Indianapolis Beautiful
Former Senior Copy Editor, John Wiley & Sons
Editor of over 90 *For Dummies Books*

After integrating The Learning Code Science into our copy and website, we have seen a phenomenal 1000 percent increase in response rates for our social media services! This scientific approach to marketing has completely changed our mindset.

Ann Gray, CEO – Ann Gray Consulting
Selected Top 100 Influencers in Marketing

We specialize in creating peak performance for individuals and teams leading to profound culture shifts within the workplace. We've worked with a wide range of clients including: NASA, National Institutes of Health, The United Nations and the Department of Defense.

We are excited to say we are using elements of The Learning Code science to accelerate the creation of interpersonally safe and positive working environments that result in greater trust, engagement, and collaboration at work. Incorporating elements of the Learning Code enables us to develop training and coaching methodologies that support positive biological change in our clients, resulting in accelerated learning and real sustainable change, along with a culture that is good for the brain, good for the body and good for the bottom line.

Celeste Blackman, Co-Founder Green Zone Culture Group, greenzoneculture.com

The insightful Learning Code knowledge about how to create sustainable change in organizations by understanding the biology of change, has wholly transformed our perspective about how we implement change, and why our initiatives have experienced resistance in the past. This new insight is a quantum leap in comprehending "what we need to do next" when confronted with difficult change management issues.

Tiffany Goode, CEO, Goode Stuff Entertainment
Past Director, Workplace Initiatives, Turner Broadcasting

Most of the problems in our world start with our outdated and inefficient learning systems. The Learning Code opens our eyes to what went wrong and how to use the latest science to fix things. This research is a game changer."

Robert Kiyosaki, Best Selling Author Rich Dad Poor Dad

I have been around the genius of The Learning Code Institute and their investigations into the foundations of how we learn for over twenty years. The Cracking The Learning Code books and online programs will forever change the world for the better, of that I am certain. I am honored to have been able to see this disruptive approach to learning come to fruition.

Tom Chenault, One of the World's Top 25 Network Marketers
Host Network Marketing Leadership Radio Show, Founder ContactMapping.com

I believe the science of the Learning Code will revolutionize what we are doing with our school districts, individual schools, teacher groups as well as how we are homeschooling our children. I have been working in the educational space for fourteen years and, going forward, we will not consult, train or develop content without mapping it against the Learning Code.

Mack Weathers, Principal, School Ventures, COO, Sphere Capitol Group

The Learning Code is bringing about a sea of change. Universities and research centers around the world have focused on narrow aspects of learning and how to improve it, but no one had pulled all of that information together to build a powerful, comprehensive and breakthrough method for accelerating learning—until now.

Paul Kordis, Ph.D. International Expert in Creating Profound Change
Best-selling author 'Strategy of the Dolphin'

The Learning Code Science propelled us to a depth of understanding about our business that is beyond what we could have heretofore ever imagined. This cutting-edge scientific approach to marketing and business development has transformed our view of the world and how our company will operate in it going forward.

Nick Garvin, CEO, Stackup: The Leader in Learning Engagement

The Learning Code validates what all teachers have felt intuitively about what is right and what is wrong in education. A child is so much than their score on test. This science helps us see who each child is on a deep level. What a gift!

Eric Myhern, Educator, Coach, Mentor Founder, MetaFit+256

Today, the frontlines of learning research are found in Cracking The Learning Code. The impact of the Learning Code is transformative. This technology is the key to "active learning," where marketing and education intersect. For the savvy professional, owning the Code means owning the market.

Jon Ward, Founder & CEO, Braincat: Helping the World Think Better, Past Strategic Marketing Consultant: The Learning Company, The Rich Dad Organization, Del Web, and Zing Performance

I have been amazed (and very grateful) for The Learning Code. It has transformed how we produce our video projects. It is based in solid science, and has made all of my work, especially my interviews and studio work, much more effective and engaging. And it's changed my personal life as well!

Brian Doubleday, Producer Greenworks Video, Emmy Award-Winner

What's truly exciting is that The Learning Code has the potential to forever transform the ability of educators, businesspeople, homeschool parents, and the individual to thrive in this age of the information explosion. The Learning Code science is ushering in a Golden Age of Learning!

Neal Rohr, CEO, The Vermont Group, Co-Author *Self-Healing,* Producer The Legend of Spirit Dog Discovery Channel

Langford Learning focuses on improving learning organizations with Dr. W. Edwards Deming's System of Profound Knowledge. The Learning Code elements are now integrated into our Langford Learning theory and approach to improvement. Implementing The Learning Code Science has increased interaction with participants, created deeper intrinsic motivation, strengthened exponential learning, and developed new applications and applied knowledge for Langford Learning.

David P. Langford, CEO, Langford Learning

Acknowledgments

Over my thirty years of research, hundreds people as well as the works of experts from diverse fields have been instrumental, not only in helping our team uncover the scientific bases of learning, but also in keeping us inspired and sane during the difficult times. We are greatly indebted to all who have contributed in whatever small or large way. We apologize for any omissions; they were purely unintentional.

Gerald Edelman, Theodosius Dobzhansky, Donald O. Hebb, Anthony de Mello, Eckhart Tolle, Michael Singer, Robert Kiyosaki, Kim Kiyosaki, Daniel Levin, Celeste Blackman, Michael Schmidt, Nancy Monson, Marshal Thurber, Tom Rudmik, Gordon Penick, Trygve Myhren, Sheri Colberg, David P. Langford, Fernando Gonzales , Danielle Voirol, Ryan Mitchel, James T. Wilson, Neal Rohr, Chris Grout, George Moore, Eric Myhern, Sandy Draper, Jackie Bates, Cath Murray, Heather Pendley, Denise Wolk, David Bauer-Ray, Jerry Nihen, Ann Grey, Alex Spriggins, Ruth Penick, Catherine Jones Payne, Joel Laur, Dinny Wilson, Alan Pollock, Brian Doubleday, Gregg Lauer, Jake Rogers, Marykutty Mathew, Thomas Lehnert, Kirsten Kummert, Jean and Jerry Harrison, Claudia Sanchez, Mary Woods, Mark Reardon, Ronda Jaggers, Paul Kordis, Tim Artemenko, Steve Sorensen, Phil Garvin, Tom Nyiri, Keri Langford, Mark Jensen, Jacob L. Wilson, Rick Simms, Barbara Dishmon, Phil Streit, Paul Janssen, Rebekah Mitchel, Jon Ward, Tom Hart, Nick Garvin, Bob Blackman, Joe Palmer, Doug Curry, Chris Grout, Michell Umaki, Rick Alter, John MacGregor,

Tom Chenault, Lewis Pollard, Debby Wilson, Bernard Franklin, John Tomick, Dennis Maimen, Larry Niwa, Tommy Tucker, Arthur Victor, Tom Burns, Marc Cahn, Garret Sutton, Kim Pearson, Bob Browne, and special thanks to the staffs at Masters Academy, Calgary, Canada; Running River School, Boulder, Colorado; Gordon Parks School, Kansas City, Missouri; and Ingenium Schools, Los Angeles, California...

Published in the United States by Learning Code Publishing

ISBN: 978-1-953806-34-5

Library of Congress Number: 2021907027

Originally Published June 2021

www.thelearningcode.com

This book is dedicated to

all those who desire to

make learning

a more joyful and

meaningful process.

Contents

Introduction

Why This Book? and How to Read It

We believe this book will "break your thinking machine" about what learning is and how to create it. *Cracking the Learning Code* offers a more scientific perspective on how outdated learning strategies have created a world floundering in problems and why so many of us are not living up to our own expectations. It is based on thirty years of research into the science of learning, motivation, and behavioral change at The Learning Code Institute. We call this new perspective on how we naturally learn *The Learning Code*.

At The Learning Institute, we are committed to sharing what we know about learning so that everyone can improve their lives and those of their loved ones and also be a valuable resource for organizations and governments to make our world a more cooperative and fulfilling place. Our ultimate goal is to help you accelerate the speed and efficiency at which you learn and change any life-limiting behaviors, so you can live a more joyful, fulfilled, and meaningful life. And, if you so choose, to use this more scientific approach to make a greater impact on your work and in the world.

A Personal Mission

The Learning Code and Institute came about in the early 1980s when I was introduced to the works of two disruptive thinkers, management guru

Peter Drucker and visionary Buckminster Fuller, both of whom completely transformed how I looked at learning and the world.

Drucker[1] is given credit for putting forth the theory that we have moved from an industrial to a knowledge society. A knowledge society creates a brain-based economy where what makes you feel worthwhile and successful isn't dependent on your physical powers but on the size of your knowledge base and how effectively you apply it. Fuller[2] is given credit for presenting the concept that in modern times all the knowledge in the world is doubling at an exponential rate. For example, in the 1900s, it was estimated that the knowledge-doubling rate was once every 100 years. By the end of World War II, it was once every twenty-five years, and today it is estimated that knowledge is doubling at the phenomenal rate of once every eighteen months.

These two men helped me understand that if you want to live a fruitful, profitable, and joyful life in a knowledge society, your most precious asset is your ability to learn new knowledge quickly; it is the leverage point. But I recognized we had a severe problem. The methods we use today to help us acquire new knowledge were designed by the Ancient Greeks when knowledge was doubling once every 1,000 years. How could teaching strategies designed over 2,300 years ago, when knowledge was doubling at a snail's pace, work effectively in a world where information was doubling at a supersonic rate? The answer is… they don't. This realization turned into a passionate forty-year quest to understand the scientific underpinnings of how we learn, motivate ourselves, and transform our behaviors. My hope is that this book will take you on a journey of discovery, seeing what went wrong and how we can use a more scientific approach to make things right.

Peeling an Onion

Combining these profound scientific learning discoveries in one easy-to-read, comprehensive resource that would transform how the world perceives learning was a challenge. This is why this book has more than 500 pages and references over a thousand research studies and resources. But we took a unique approach and designed it so that you don't have to read every page of every chapter to get a grasp of the science of how you learn. After the first four chapters, you'll find that you can get a general understanding of the whole topic from reading the first few paragraphs of the chapter, then feel free to skip any part of the rest of it that doesn't match your immediate needs or resonate with you.

Starting in Chapter 5, you will begin to see these icons that alert you when a chapter will begin to go deeper into the science. Here you have a choice. You can either "fast forward" past that section if you feel you have enough understanding to satisfy your needs, or you can choose to peel another layer from the onion more and dive deeper. This approach allows you to come back and revisit any topic if and when you decide you want more depth. Our goal is to have all this learning science in one reference source to continue to refer to it as your needs demand.

or

While our approach is comprehensive, I often use metaphors and generalizations to give you a simple, quick, all-encompassing understanding. I want to apologize in advance to those in the scientific community, as in my effort to simplify, I sometimes leave out information researchers in a field would consider valuable.

Cracking the Learning Code contains more than 1,370 endnotes, so for easy reference, we've also listed them as hot links on The Learning Code Institute's website at www.thelearningcode.com, so you can follow-up and read the research for yourself if you choose. You will notice many of the general quotes I use to make a scientific point, while still relevant today, were made years ago. This is deliberate—we found that when the pioneers in a field first made an insight or discovery about learning, their excitement, passion, and energy was far livelier and more impactful than quotes from those who improved upon the research.

The following chapters cover the major scientific learning discoveries and neurochemical systems that need to activate in your brain to ensure learning has taken place from different angles. We use this method of looking at similar topics from different perspectives for two reasons: First, to provide you with a firmer foundation for understanding the science of learning; Second, in most cases, we have designed each chapter to stand alone so you can, if you choose, share that chapter with individuals or organizations that you believe could benefit from the specific knowledge presented without having them go back and read other chapters.

Chaos in the Making

And we can't leave the introduction to this book without addressing the chaos that Covid19 has created in our world. While coronavirus has been a disaster for our collective health, there is a silver lining. It has dramatically accelerated the speed at which we are being forced to realize that our existing online and classroom learning systems are flawed and have been for a long time. We hope that the pages of this book will enlighten the world's collective consciousness to what we have learned from researchers who focus on chaos and complexity science. That is, chaos provides an opportunity for systems to reach higher levels of organization; it is the fuel of transformation.

From the works of Noble Prize winner and complexity pioneer Ilya Prigogine and the renowned author of Chaos: Making a New Science, James Gleick, we find that when systems dissolve into chaos, that very chaos is the fertile ground from which reorganization to higher orders in chemistry, biology, nature, and society arise. Science informs us chaos is a gift. Our hope at The Learning Institute is that Cracking the Learning Code will be a guiding light in our collective effort to transform the chaos the world is experiencing at this time into a higher order so we can all experience greater levels of happiness, freedom, and fulfillment.

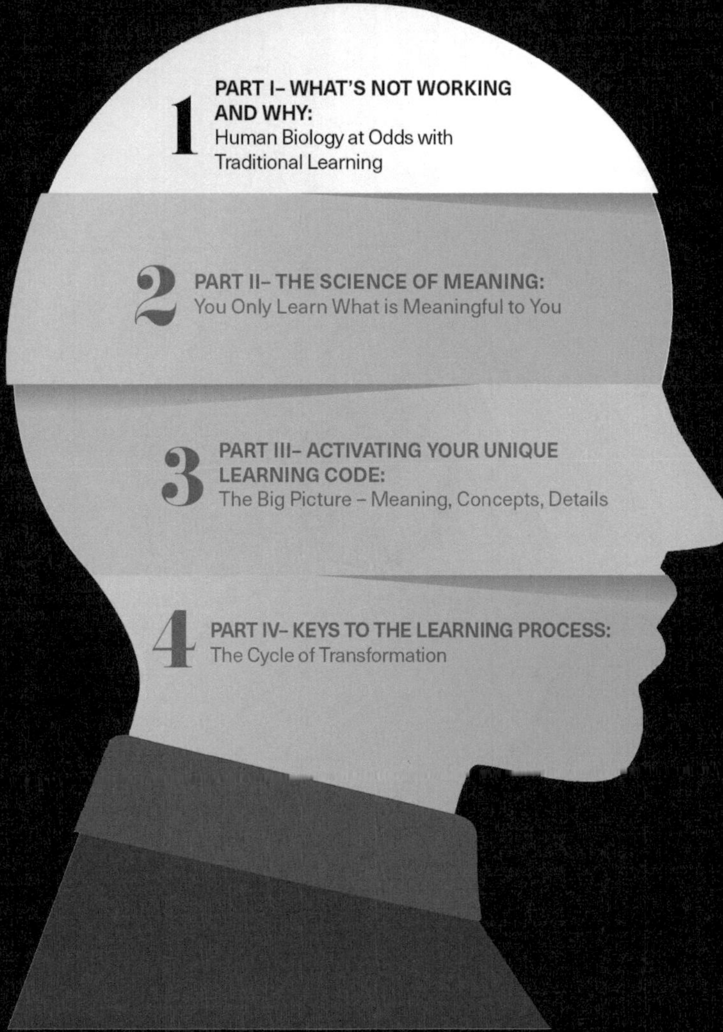
1
PART I– WHAT'S NOT WORKING AND WHY:
Human Biology at Odds with Traditional Learning
2
PART II– THE SCIENCE OF MEANING:
You Only Learn What is Meaningful to You
3
PART III– ACTIVATING YOUR UNIQUE LEARNING CODE:
The Big Picture – Meaning, Concepts, Details
4
PART IV– KEYS TO THE LEARNING PROCESS:
The Cycle of Transformation

1

"Houston, We Have a Problem!"

It ain't what you don't know that gets you into trouble. It's what you know for sure that just ain't so.[1]

—Commonly misattributed to Mark Twain

What happens when the traditional learning educational systems we've used over the last 100 years no longer work for you and our world? When "the way we've always done it" isn't just ineffective but actually keeps us from reaching our potential—as individuals and as a species? We see it on the news every day—a world spinning through one crisis after another.

As a society, we face poverty, violence, war, drug epidemics, extreme worker stress, and poor outcomes for our students. As individuals, we struggle with our health, finances, relationships, and often find ourselves feeling unfulfilled. As I described in the introduction, information now doubles at the rate of once every eighteen months. That's a lot of information but it hasn't given us the tools to effectively learn how to solve our world's or our personal problems.

For us to understand our current learning, motivation, and behavioral problems, it is vital to understand the history of how we got off track and what solutions we can apply to correct things. In this section you will get a sense of the depth and breadth of the problem, delve into how and why things have gone wrong, and get an overview of why science provides the solutions. This perspective is essential, so we don't try to fix today's problems with yesterday's solutions.

What is the Learning Code?

What is the Learning Code? In technical terms, it's the process by which your biology naturally selects knowledge into your brain, body, and genes, so you can use it to survive and thrive in your world. If you aren't accessing your Learning Code effectively, your ability to be successful —in a world where information is doubling at an exponential rate—will be severely compromised. We believe that the better each one of us understands the underlying biological processes of learning, motivation, and behavioral change, the greater our nation's, our world's, and our own chances of surviving and thriving will be.

The research that has led to the discovery of *The Learning Code* is complex and varied, it comes from the fields of neuroscience, genetics, epigenetics, molecular biology, evolutionary biology, quantum physics, complexity science, and chaos theory. Our goal at The Learning Code Institute has been to untangle all the research so we can find simple solutions that you

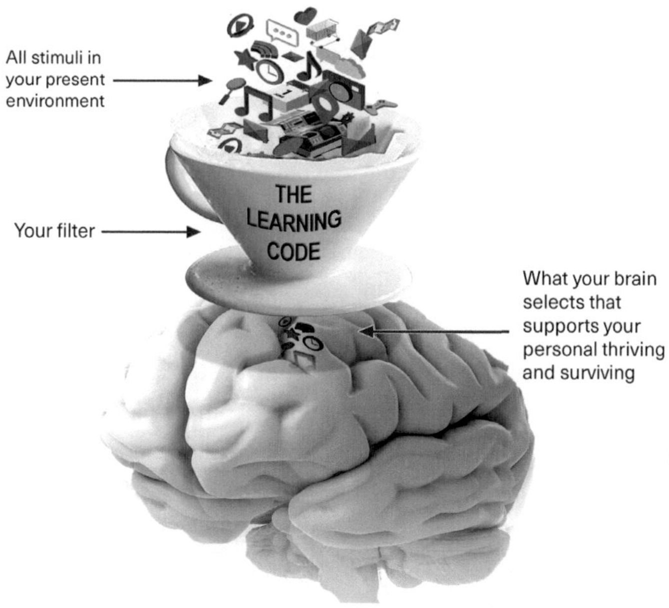

One way to look at the Learning Code is as your filter, ensuring that what you select into your neurobiological makeup directly supports your surviving and thriving in the world you're living in right now.

can quickly understand and apply. This untangling process was neither fast nor inexpensive. But don't worry—we will keep things simple. When you finish this book, your understanding of the science will dramatically support your personal success, and you'll be able to use it to help those whom you love and work with.

What You Are Not Aware of You Are a Slave To!

As I travel the world helping educational systems, governments, and corporations use the latest scientific research to accelerate the speed of learning, I'm continually struck by an interesting fact. Although the teachers, psychologists, trainers, social workers, coaches, administrators, and human resources departments in charge of creating learning, motivation, and behavioral change all know that the brain is the primary organ of learning, 98 percent of them have no idea how our biology actually produces the chemical and structural changes that lead to the very learning and behavioral change they so desperately want to create.

I often start my presentations with the quote "What you are not aware of you are a slave to. What you are aware of... you control." When I point this out to audiences, they immediately see the problem. How can they create profound transformation in themselves or others without understanding how the neurobiological process that directs that transformation works? The answer is they can't. And this is the root of our scholastic and corporate learning and behavioral change problems, that in turn cause our world and personal problems. As you will discover, while this book is about improvement, it is really more about transformation.

The Learning Code Institute's Goal: Keep It Simple

At The Learning Code Institute, we're now using a 30,000-foot view of the neurobiological research to dramatically accelerate the speed and depth of learning and social change in education, cultural transformation, motivation, and business development. The research draws from tens of thousands of peer-reviewed scientific papers and over thirty years of hands-on application. We've presented The Learning Code science to and worked with organizations worldwide, including Turner Broadcasting; Mayo Clinic; The Rich Dad Organization; seminar organizations for speakers like Bill Clinton and Tony Robbins; the Deming Institute; the group that conducts leadership training for NASA, the UN, and the National Institutes of Health; and presented to governments and educational institutions in Africa, Europe, Canada, Latin America, and Asia.

We have also taken the science and developed the first-of-its-kind online modules designed to transform how the world's educators and homeschool parents understand and implement the science of learning, and behavioral change. It is a disruptive scientific approach, designed not only to support the accelerated learning of students but also to make teachers' and parents' jobs less stressful, more fulfilling, and joyful. The prestigious Claremont Graduate University, which is part of The Claremont Colleges (a consortium that employs the Oxford-Cambridge model and includes Pomona, Claremont McKenna, Harvey Mudd, Scripps, and Pitzer colleges as well as Claremont Graduate University and Keck Graduate Institute), has made these disruptive online programs professional development courses that satisfy continuing education needs.[2] As an example of The Learning Code science in action, it is now being used at Master's Academy in Calgary, Alberta, to help create the first school system in the world based on neurobiology. Here, learning now takes 20 percent less time, standardized test scores are 400 percent higher than the province average in the "Excellent" category, and—because the learning systems match the students' biology and how they naturally learn—behavioral issues have been virtually eliminated.[3]

In addition, we're finding that The Learning Code science is transforming how companies market, position, and brand themselves. From a scientific viewpoint, a potential customer becomes a paying customer only when their brain shifts from one chemical state to another. In this context, we see that the main job of marketing and sales is to deliver information in a way that creates neurochemical change in the buyer's brain (and as you'll see later, also in their body). This approach is giving companies and individuals profound insights, allowing them to refine their marketing so it produces the biochemical changes in potential customers that can lead to substantial returns.

Furthermore, we're involved in two passions of mine, applying the latest scientific research to create systems that heal post-traumatic stress disorder (PTSD) and addiction (both of my parents and my son died from overdoses). We have shared this science with the first residential treatment center for Wounded Warriors and with treatment centers, schools, and governments worldwide.[4]

Learning Directs All Your Actions and Behaviors

From a neurobiological viewpoint, you behave the way you do because you've learned to behave that way. If your behaviors aren't getting you what you want in life, it's because you've learned to behave in a way that's counterproductive to fulfilling your dreams and goals.

Genuine learning literally changes your brain structures. These structures then dictate all your motivations, actions, and behaviors. Unfortunately, a lot of the learning the human species has done over the past few decades has led to behaviors that have produced a deeply troubled world, in which few of us are getting what we want in life when it comes to money, health, meaningful work, or peace of mind. In effect, we have a national crisis, a world crisis—and for many of us, a personal crisis—because we have a learning crisis.

When individuals understand the Learning Code science, they never look at how to create learning in themselves or others in the same way again. Because of the Learning Code science's ability to quickly transform the status quo, it has been called a *Disruptive Innovation.*[5] Coined by Harvard's Clayton Christensen, *Disruptive Innovations* like the arrowhead and cell phone displace previously held beliefs about what is possible. Our mission is to use the science so that never again does any child or adult have to experience feelings of pain or incompetency when trying to learn. One organization that implemented the science found it so disruptive to their learning and behavioral-change programs that they used the metaphor of the Intel chip to describe it.

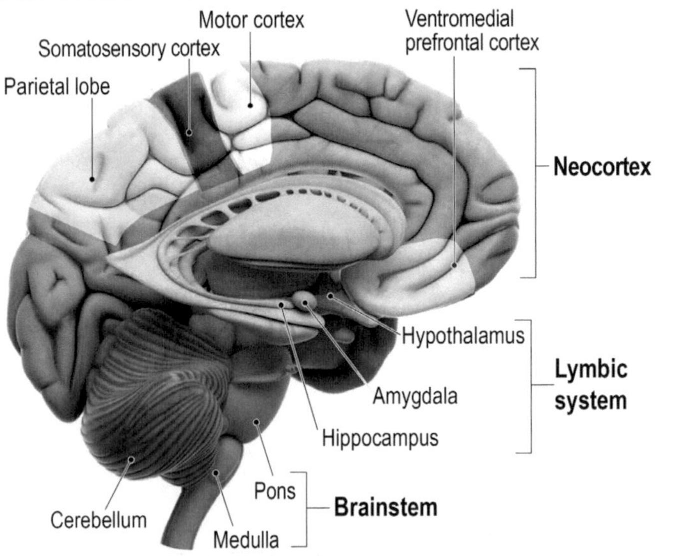

Genuine learning literally changes your brain structures, which then dictate all your motivations, actions, and behaviors.

The Learning Crisis Creates a Personal Crises

Are you overweight? Do you not exercise enough? Do you often find yourself depressed or anxious? Are you making less money than you would like? Are you failing to advance at the rate you desire in your profession? Are your personal relationships with your co-workers, your significant other, or your children not as deep as you would like them to be?

We often blame people's lack of success on not enough perseverance or grit, but a lack of success in health, finance, work, and relationships is more directly tied to what you've learned in life. What you've learned directs your behaviors, and sadly, what most of us have learned has produced behaviors that aren't getting us the "good" life we desire.

Consider our financial wellbeing:

- Seventy-six percent of all Americans are living paycheck to paycheck.[6]
- Fifty percent of American households report that they couldn't cover an unexpected $400 expense.[7]
- Fifty-nine percent of baby-boomer parents financially support their adult children.[8]
- Forty-five percent of baby boomers have no retirement savings. Only 55 percent of baby boomers have some retirement savings, and, of those, 28 percent have less than $100,000.[9]

Consider our health:

- Sixty-nine percent of all Americans are obese or overweight.[10]
- Eighty percent of Americans don't get the right amount of exercise.[11]
- Thirty-three percent of American have diabetes or prediabetes, and this number is projected to increase by 165 percent by 2050.[12]
- Nearly 70 percent of Americans are on at least one prescription drug, and more than half take two.[13]

And consider our joy, peace of mind, and sense of meaningful work:

- One-thousand percent more people suffer from major depression than in 1945.[14]
- The World Health Organization reports that suicide rates have increased 60 percent over the past fifty years.[15]
- Sixty-six percent of us don't feel "very happy" about our lives.[16]

Anxiety disorders are now the most common mental illness in the US. A Gallup poll found that 87 percent of workers worldwide "are emotionally disconnected from their workplaces."[17] Although some of these issues can be traced to other variables like the economy, the bulk of these startling statistics arise from what we've learned, like how not to manage our money, how not to eat right, how not to take care of our health, how not to find meaningful work, and how not to create peace and happiness in our lives.

The Learning Crisis is Creating National and World Crises

It could be said our world problems are also caused by our learning crisis. Just like many of us have learned how not to manage our lives effectively, so to our human collective conscious has learned how to build a world that is not working. What we have learned up to now has put Congress in gridlock and turned our medical systems into a mess. Corporations can't find enough skilled knowledge workers. We face a global financial meltdown as countries like Greece head toward bankruptcy. We are experiencing mass killings every two weeks. Countries such as Iran threaten with weapons of mass destruction, and terrorists and suicide bombers are destroying our trust in our daily safety. Climate change and energy disruption loom in our future. Also, consider that of the 7.5 billion of us on Earth, 50 percent of us try to exist on less than $2.50 a day.[18] Think about it this way, we've learned how to create our nation's and world's problems, but we haven't yet learned how to solve them.

Why Do We Have a Learning Crisis?

The learning crisis has reached this frenzied pitch for two main reasons. The first is that the world has changed, technology has changed, people have changed… yet we continue to rely on educational and learning methodologies designed for a time long since past. Clearly the "Stand and Deliver" memorization, cram-for-the-test learning strategies we've used since the 18th century aren't working in today's world.

When we expose the human brain to an educational system that for twelve to eighteen years primarily develops brain structures that support linear thinking and the storage of facts—at the expense of developing brain areas that support innovating, making order out of chaos, understanding complex systems, and amassing information from diverse sources—we produce brains that can't adequately deal with a society in which information is doubling at this exponential rate. The question we need to ask is, "Why are we using learning strategies that we relied on in agrarian and industrial ages, when knowledge was doubling once every 1,000 to twenty-five years?" In today's world, that approach is designed to fail.

The second reason we have a learning crisis is more disturbing: the science is revealing that the traditional memorization-and-testing learning strategies we rely on so heavily do not fit the way your biology naturally wants to learn. The traditional methods don't match your neurobiology; they don't encourage the changes in your brain and body that represent how learning and behavioral change can naturally and easily take place.

Because our existing learning systems aren't biologically friendly, they make the process much more complicated, stressful, boring, and inefficient than it needs to be. Think of your own experience. Of the tens of thousands of things you've memorized and been tested on in your life, how many do you remember right now? For most of us, the answer is *not very many at all.* In effect, you put in a lot of time and effort for little return.

A Guru's Take on the Learning Crisis

Over fifty years ago the Indian philosopher Jiddu Krishnamurti[19] had an interesting take on our learning problems, which I believe still hits the nail on the head today.

Here I paraphrase his thoughts:

Should education not be about understanding life? About gaining more freedom, fulfillment and joy? Should education not be about more than just being proficient in math, physics and chemistry? About more than just grades?

We do what education tells us to do, we pass exams, get employment, get married, have children. Yet we become more stressed, more anxious, less fulfilled… more and more like machines.

Is not life an extraordinary thing? The birds, bees, sunsets and flowers. Should not the function of education be about helping us to understand the whole process of life? Or is it merely to prepare us for a vocation, for the best job we can get?

Surely, education has no meaning unless it helps you understand the vast panorama of life, its subtleties . . . with all its extraordinary beauty, its joys and sorrows? You may get degrees, you may have a string of letters after your name, land a job; but then what? What is the point of it all, if in the process you become dull, weary and exhausted?[20]

Systems Dynamics Shines Light on the Problem

From the study of systems dynamics, we know that the harder you push an inefficient system, the greater the inefficiency and dysfunction you'll get. This is exactly what we're getting by pushing our outdated and biologically unfriendly scholastic and corporate learning systems harder and harder: greater waste and greater dysfunction.

Over the last decade, we have assembled the best educational minds in the world, spent billions of dollars, and committed millions of hours, but the solutions we're applying to today's educational and corporate learning problems aren't working. More money, longer school days, more homework, and more technology—none of it seems to be working the way we'd hoped. In many instances, it's making things much worse, and the consequences are devastating.

The research shows that indeed, the Emperor has no clothes. It might be hard to get our heads around the reports showing that the old learning standbys not only don't work but could be causing harm, but the data doesn't lie.

A depressing 65 percent of all eighth-grade youth in America are considered *not proficient* in math.[21] This means that after eight years of study, they lack the basic math skills to succeed in life. In some states, such as Louisiana, West Virginia, and Mississippi, a gloomy 80 percent of students lack math competency.[22] Fifty percent of all students in our major cities drop out of school before they graduate from high school, and an alarming number of those who do graduate are considered functionally illiterate.[23] The pain is also felt by teachers. Because they are so disillusioned and frustrated by the present system, huge numbers leave the profession after only five years. All this scholastic failure flows into corporate America, which can't find enough skilled workers. And because their in-house corporate learning programs are also not living up to expectations, corporations can't step in to produce the skilled, motivated workers the country and world demands. Unfortunately, not only are the old standbys not giving us what they promised, but our fixes to them—which just push the existing system harder—are also falling miserably.

$156 Billion Spending in Corporate Learning: An Outstanding Failure

A headline in the *Wall Street Journal* said it all, "So Much Training, So Little to Show for It." Ninety percent of all the knowledge communicated in the typical corporate training program is forgotten within one year. Some experts even suggest that 90 percent is lost a mere few days to a few weeks after the training program is finished.

Of the $156 billion organizations spent on employee learning in 2011, an estimated $140 billion in value was lost soon after the training was completed.[24] Such waste is destroying America's competitive advantage.

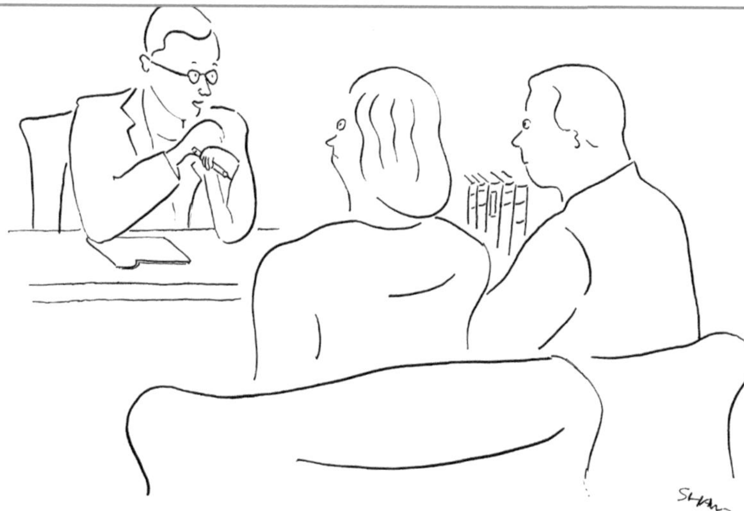

"The Feds have authorized me to leave your child behind."

No Child Left Behind Fails Dramatically

To improve classroom learning outcomes, the government spent an estimated $100 billion on the No Child Left Behind Act over ten years. It was a disaster. *The Washington Post* reported that in the years before the $100 billion was spent, the gains in math were actually larger.[25] Diane Ravitch, the former Assistant Secretary of Education, summed up the misadventure of these expensive initiatives nicely when she said, "Once they were tried, I was convinced that they didn't work and, in fact, not only were they failing, but they're ruining American education."[26] Because of the Act's emphases on testing, many educators jokingly renamed it the "No Child Left Untested Act"! It was replaced by the Every Student Succeeds Act in December of 2015, but children are faring no better under the new law.

The Gates Foundation Bombs: Spending Goes Up, Learning Goes Down

Okay, maybe the government can't fix learning, but what about private corporations? This didn't go well, either. Bill Gates assembled some of the best minds in the country, piloting smaller schools, closer relationships for students and teachers, and more flexibility and autonomy. Some schools were successful, but it wasn't sustainable and many "slid back" to their original ways. His foundation offered $2 billion in funding to create new initiatives to increase learning outcomes. And yet, he admits it was disappointing:

> *In the first four years of our work... most of the schools had achievement scores below district averages on reading and math assessments. In one set of schools we supported, graduation rates were no better than the statewide average, and reading and math scores were consistently below the average.*[27]

The Homework Myth

Let's look at a crystal-clear example of how pushing the existing system harder doesn't work: research is discrediting one of the most hated, feared, and yet relied-upon stalwarts of our educational systems—homework. In the last two decades, homework for elementary kids has doubled. The result? No appreciative increase in learning skills.

Mike Horsley and Richard Walker, authors of *Reforming Homework: Practices, Learning and Policies*,[28] did exhaustive evaluations of the homework research and concluded that most of it was for naught. They found that homework doesn't significantly impact achievement and can even be detrimental to the learning process. In Trends in International

"A lot of homework?"

Mathematics and Science Study, experts compared practices in more than fifty countries and came to a shocking conclusion:

> *Not only did we fail to find any positive relationships, [but] the overall correlations between national average student achievement and national averages in the frequency, total amount, and percentage of teachers who used homework in grading are all negative!*[29]

Is too much homework in American schools the reason so many other countries' educational systems are beating us? In many of the nations that outperform the US in student achievement, such as Finland, Denmark, and the Czech Republic, teachers tend to assign much less homework than American teachers (who assign an average of 3.5 hours a night, or 17.5 hours a week). On the other hand, instructors in low-scoring countries like Greece, Thailand, and Iran tend to give even more. Finland is consistently at the top of high-performing countries, yet its teachers tend to assign a total of only thirty minutes of homework a night.[30]

The past president of France, François Hollande, was so disturbed by the findings that in a speech at the Sorbonne, he proposed a law that abolishes homework for all primary and middle-school students in France.[31] After decades of studying the effect of homework on children, the leading expert in the world on this topic, Alfie Kohn, stated that homework "may be the single most reliable extinguisher of the flame of curiosity."[32]

Higher Education Isn't Delivering, Either

When we look deep into the studies, we see a nasty picture of higher education. Nearly 50 percent of all college students become disillusioned, bored, or discouraged and drop out before they graduate. Here's a statistic that no one envisioned: an astounding 78 percent of attendees at our private institutions drop out before they get a diploma.[33]

And it doesn't look like those who do graduate are getting their money's worth. The cost of college has gone up 600 percent since 1985—the average cost of four years' tuition is $40,000 at a state school and $135,000 at a private institution.[34] This forces the majority of students to take out loans to pay for their education. These loans now total over $2 trillion.[35] In fact, student debt is the US government's largest asset.[36] Yet much of that money probably won't be paid back. Why? Because almost 50 percent of twenty-five- to thirty-four-year-olds say they won't be able to pay off their debt because they're unemployed or underemployed.[37]

Clearly, these students aren't getting what they thought they were paying for, and neither is the country. Before we suggest that college education should be free, as some politicians are suggesting, we first better make sure it works.

"Welcome graduates, parents, guests, faculty, and home equity loan officers."

Is Online Learning the Answer? Another Dramatic Failure

Not long ago, online learning and MOOCs (massive open online courses) were all the rage. They were seen as the answer to our educational problems. The future looked very bright. These methods were to deliver information quickly and inexpensively to the masses. So far, research has shown online learning to be a dark hole, not at all living up to the potential once hoped for.

When compared to regular classroom learning, online students have lower learning outcomes and greater failure rates; an amazing 90 percent or more of the courses started online are never completed.[38] And prestigious names do nothing to improve online learning outcomes: When Princeton University presented the online course "A History of the World since

1300," only 0.08 percent of the 83,000 students who enrolled completed the course.[39] Obviously, technology, at least as we're using it now, isn't the answer.

How About Brain Games? Nope, They Aren't Working, Either

The brain game industry is booming; it's estimated that the industry generates $1 billion a year in revenue. But these games, although fun for some, aren't really working. Recently, sixty-nine respected neuroscientists at a Stanford University conference signed a petition stating that "brain games" are a major waste of time if you want to improve your general intelligence and memory. In another blow to brain games,[40] the Federal Trade Commission ordered Lumosity, which has thirty-five million members, to pay a $2 million fine for making deceptive claims that playing their games would boost performance at work and in the classroom and even delay dementia. Under the settlement, the company was also required to contact all its customers and offer them an easy way to cancel their subscriptions.[41]

Clearly, the Emperor is Naked!

This research about our failing learning systems comes from some of the most respected organizations in the world, including the National Assessment of Educational Progress, Harvard Graduate School of Education, the National Institutes of Health, the National Center for Education Statistics, the Trends in International Mathematics and Science Study, and Boston University. It's been reported in prestigious publications such as the *New York Times, Nature, Harvard Business Review*, and the *Wall Street Journal.*

For those with open minds, this kind of research is forcing us to reconsider much of what we believed about how to encourage profound learning. And the problems we've discussed are just the tip of the iceberg. As you go deeper into this book, you'll find more disturbing facts, including that our whole educational system is based on the concept that passing tests demonstrates we've learned. (It doesn't!) We've been told that high grades translate into high achievement in the real world. Research reveals that this just isn't true. Studies demonstrate that extrinsic motivators—that is, reward and punishment—actually inhibit the learning process. These are disturbing findings, as virtually every educational system in the world uses extrinsic motivation (grades) as the primary driver of learning.

Our obsession with grades may be creating a generation of drug-dependent youth. In the quest for better grades, 25 to 35 percent of our college students admit to illegally acquiring "smart drugs" like Ritalin and Adderall to help them study longer and harder.[42]

The dysfunction of the present educational system is so great that it's even driving teachers to commit crimes. In Atlanta, Houston, Columbus, Memphis, El Paso, and Washington, DC, teachers—who entered their profession with bright eyes, full of hope for their future and the future of our children—have been charged with fraud for altering test scores in a misguided effort to keep their jobs and protect their schools from being downgraded.

Teachers are not the only ones caught up in a dysfunctional system that is causing them to break the law, so too are parents. In highly publicized accounts actresses, business leaders and other wealthy individuals have been charged with paying up to $1.2 million to fraudulently get their kids into the right college. Parents either paid a college prep organizer to take entrance exam tests on behalf of their child, or they bribed college coaches to admit their kids as recruited athletes, even if they never played that sport in their life.[43] We have a learning crisis on our hands and the non-scientific solutions we are using to solve the crisis are making things worse.

We Have the Brain Capacity, But Something is Missing

Do we really doubt that the human species has the brain capacity to better solve our world's problems? There are more than 7.5 billion of us on this planet, and nature and genetic variation have given each of us a brain with a unique physical makeup, which in turn provides a unique way to survive and thrive in the world. Your genetic and neurological makeup gives you a brilliance that ensures you're a genius at something your neighbor is not. Through the lens of neurobiology, we can see that your individual brilliance was designed to combine with the world's other seven+ billion brilliances to create a human race with spectacular options for surviving and thriving. Each brain acts as a tiny droplet of "brilliance fuel" that the *global brain* uses to propel itself toward success. We're biologically wired with the capacity to transform our world and ourselves… but something is missing. We aren't maximizing the brilliance that nature has given us.

Why? Because our scholastic and corporate educational systems have generally ignored the biology of the learning process. They've been blind to the unique genius that each of us possesses. Instead, traditional learning systems treat us more like identical mechanistic parts on a conveyor belt than the individual droplets of brilliance fuel that we are. We get shuffled in and out of forty-five-minute, get-ready-for-the-test segments that do little to help us uncover or maximize our brilliance.

Although most good teachers and trainers do indeed see the brilliance in each student, the traditional methods they must implement force them to instruct us all as if our individual brilliance didn't exist. And if we don't learn in the way the traditional system wants us to learn, many of us are seen as not smart, not gifted, or just plain stupid. From the viewpoint of neurobiology, nothing could be further from the truth. The shout of traditional education is "Get good grades, or else!" When the shout should be "Let us help you unlock your brilliance so you can fulfil your dreams and help our world evolve into a peaceful, productive planet."

Cracking the Learning Code

What's the solution? We propose that it's understanding the science behind learning and behavioral change that we call "The Learning Code."

2

The Learning Solution We Need: Science

Do not confine your children to your own learning, for they were born in another time.

—Proverb

Embracing the latest neurological and genetic research allows you to perceive our existing learning systems from higher perspectives. This change to a more comprehensive view is vital to helping you understand why our present systems aren't working, and how we can transform them. In the 1960s, Thomas Kuhn became famous for naming this change in thinking a *paradigm shift*[1]—that is, a shift away from a previously fixed way of thinking to a more encompassing and valid way. Michael Walzer said, "Once the analogies and images that we have based our theories upon are called into question, it becomes impossible to 'think old thoughts,' the old paradigm dies."[2]

But it's important to understand that a paradigm shift doesn't wipe out what you've previously learned. Over the ages, knowledge moves in ever-expanding upward spirals. Each turn of the spiral allows us to view our previous knowledge from a broader perspective. New theories derived from new knowledge provide us with the opportunity to see that our view of the world is no longer inclusive enough to explain all that we see, hear, and feel. Einstein beautifully described how new theories like The Learning Code come about:

> *Creating a new theory is not like destroying an old barn and erecting a skyscraper in its place. It is rather like climbing a mountain, gaining new and wider views, and discovering unexpected connections between our starting point and its rich environment. But the point from which we started out still exists and can be seen, although it appears smaller and forms a tiny part of our broad view gained by the mastery of the obstacles on our adventurous way up.*[3]

With a more comprehensive view, we now have the perspective necessary to see the limits of our existing learning systems and construct more scientific and valid methods of producing learning, memory, and behavioral change.

The Learning Code Is a Trim Tab

Our experience at the Institute has shown that although the greatest successes are often made by transforming the existing system of learning completely, we can still make dramatic advances in accelerating the speed of learning and behavioral change by applying even a little scientific understanding to the system that's in place now. We call this the "trim tab effect."

Trim tabs are those tiny flaps you see at the ends of airplane wings. They're used to help a plane quickly stabilize or change direction by applying little force. When an airplane is going 700 miles an hour, and one of those flaps moves just a little bit, huge changes in the plane's direction are achieved. And when the plane is on a true course, it's easy to keep that course just by applying a small amount of pressure to the controls. The Learning Code science can be seen as a trim tab for learning systems and people's lives. Even by applying only parts of it, learning systems and people can quickly transform and maintain true direction with much less effort than would have been needed without the science.

As soon as we shift our paradigms about what actually causes learning, we can stop blaming teachers, administrators, parents, kids, trainers, corporations, and governments for our learning problems and focus on the real cause of our learning dysfunction: antiquated learning strategies that no longer fit a world where information is doubling at a phenomenal rate.

Your Genes Are Units of Learning

Your Learning Code is the process by which you log information into your biology to help you survive and thrive. This code dictates not only how you and the other members of our species learn and change behaviors, but also what you will remember. As you'll see, each species has a Learning Code that is specific to keeping its members alive, passed down from one generation to the next in their genetic makeup.

On this journey of discovery, let's first look at how your genes are involved in the learning process. A simple way to look at your genes is as vehicles in which past learning is sent to you from your ancestors. Nature has set it up so that for 3.5 billion years, ever since lightning hit the primordial ooze, information that has led to the survival of a previous generation is encoded

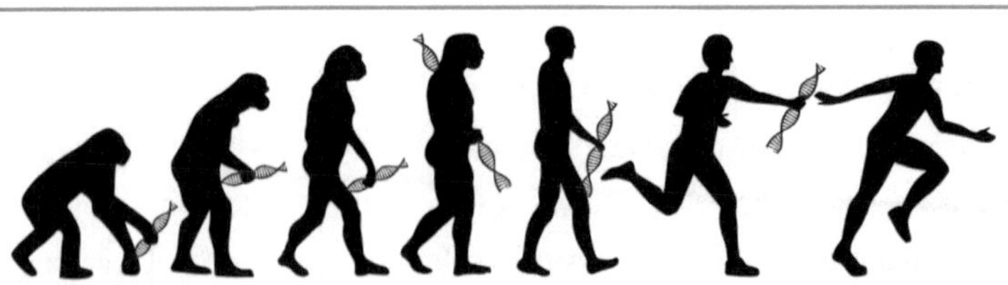

Genes are like units of learning passed down from one generation to the next.

into the genetic program of the next generation, and passed forward for the benefit of all future generations. Genes, then, are like units of learning passed down from one generation to the next. The genes that do not benefit the survival of a species are eventually bred out of the species altogether.

These genetic learning units are the reason you didn't have to be taught how to cry, suckle at your mother's breast, walk, duck when someone threw a rock at you, nest, hoard, deceive, or protect your things. All this instinctual learning that helped your ancestors survive in past environments—going back millions of years—has been passed down to you. This instinctual learning has been crammed into the 6000+ genes that control all your brain's functions. More than 2,300 years ago, Socrates maintained you remember things from a past life. He was right!

A Set of Learning Code Genes for the Here and Now

If the 6000+ genes that control your brain's function were encoded only with learning that helped our human ancestors survive in past environments, you would have little capacity to learn new information in the here and now, such as how to read, write, ride a bike, drive a car, operate a computer, create an app, or build a business. A code that enables you to quickly encode new learning from present-day environments was etched by evolution into the genes that control your brain's function, allowing humans to survive and thrive in the present.

Without this Learning Code for the here and now, your survival options would be as limited as a snake's. Reptiles like crocodiles, whose genetic plan affords them scant ability to encode new information, are trapped in a world where every reaction to an environmental stimulus resembles the reactions of their ancestors going back tens of millions of years. We have the greatest capacity to excel in the here and now because we have the most sophisticated and adaptable Learning Code of any species on the face of the Earth.

What Is Learning? The Short Story

No learning can take place unless there are chemical and structural changes in your brain. At its simplest level, the scientific explanation of how learning naturally occurs is that when your Learning Code is switched on *by meaningful stimuli*, a percentage of the 6,000+ genes that control your learning and behavior automatically express proteins and enzymes leading to changes in the structure of your brain tissue. If there's no change in your brain's chemistry and tissue, no information can enter your long-term memory. Trying to affect learning without accessing your biological Learning Code is like trying to open a bank vault with the wrong combination; it can't happen. This is why so many of us have been frustrated by our learning experiences in traditional education: our Learning Code wasn't turned on by it.

Change Your Biology and Your Behaviors Change

It's vital to understand that learning is not the only thing that occurs when your genetic Learning Code is activated, and neurological change takes place. Your present behaviors, which are dictated by the existing structure of your neural tissue, also change. You act the way you do because the configuration of your neural tissue prompts you to do so. If you change the existing structure of your neural tissue, your behaviors will automatically change. The only way to change the way you manage, play golf, make love, lose weight, or communicate with your boss, spouse, and kids is to change the configuration of your neural tissue.

Any learning system that wants individuals to change their behaviors must first cause brain-tissue to change. From this scientific viewpoint, we can see that one reason many psychological strategies fail is that they don't create enough neurological change to get people to substantially change their behaviors. While Gandhi had little understanding of neuroscience, he was right when he said, "Those people who want to change the world must first change themselves."[4]

In seminars, to graphically and physically demonstrate this, I often have plates of cheese handed out to the audience. Why? Because your brain is made up of fats and proteins, very much like soft mozzarella cheese. I then have the audience touch the sticky, gooey stuff and explain to them that this cheese is just like their neural tissue. And if they want to learn and transform their behaviors, they must "change their cheese." By the end of these seminars, people have quickly adopted this metaphor and joke to each other about how they can't wait to "change their cheese." As you learn more about The Learning Code, you will understand why this approach has such a long-lasting impact on audiences.

What is Learning? Going a Little Deeper

From this scientific view, you'll begin to understand that learning has little to do with passing a test and a whole lot more to do with your biology. If a learning program or strategy isn't causing the biological changes to switch on a percentage of the 6,000+ genes that alter your neurons, synapses, and the connections between brain cells (called *dendrites* and *axons*), learning and memory aren't taking place.

In effect, any learning program that tries to create memory formation or behavioral change without understanding that its goal is to cause neurobiological change will be a failure. This understanding frees us from the rigid confines of the "Stand and Deliver," lecture-memorization, cram-for-the-test-based classroom and allows us to adopt more science-based, creative, joyful, effective, and meaningful methods. When we design learning systems that match our biology, the pain and inefficiency of learning decreases, while the joy and efficiency of learning increases. Why? Because the scientific approach better fits the natural way your brain was designed to learn and make memories.

When you're learning things in ways that match how your biology was designed to make memories, not only are protein and enzymes expressed, but important chemicals like dopamine, serotonin, oxytocin, and

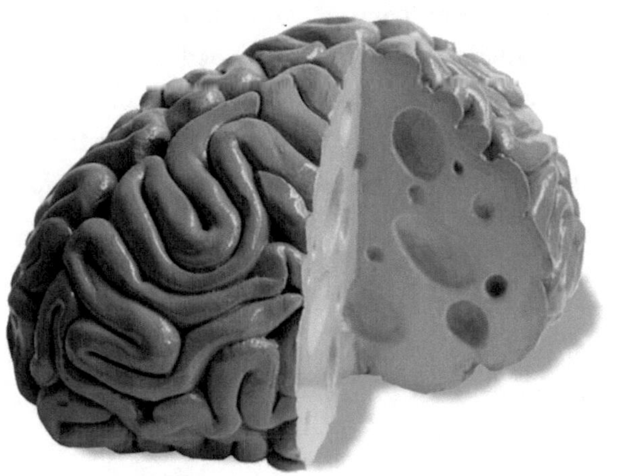

The brain is made up of fats (neurons) and proteins (synapses), very much like Swiss cheese. If you want to learn and transform your behaviors, you must "change your cheese."

norepinephrine also flood your brain. I'm simplifying greatly here, but as you'll see, these neurochemicals increase not only pleasure, excitement, and meaning, but also the efficiency of true memory formation and long-lasting behavioral change.

> **And that is the problem with our present scholastic and corporate educational systems; they don't align with human biology; they create high levels of pain and boredom and ultimately lead to inefficiency and a lack of meaning.**

Because our present scholastic and corporate educational systems don't align with our biology, they create high levels of pain and boredom and ultimately lead to a lack of meaning and inefficiency. We have educational problems *because* we're fighting Mother Nature. We're swimming against the biological stream instead of with it!

Your Learning Code Is an Acquisition Device

The idea that a Learning Code is preselected into our genetic makeup is not a new concept. In the late 1950s, renowned linguist Noam Chomsky kicked off the cognitive movement in psychology by asking how children so quickly and easily acquire language, with its many complex rules of grammar, with no instruction? Chomsky concluded that a child could learn to understand an amazing 20,000 to 30,000 words and the complex rules of grammar by the time he was only six years old only if the child came into the world with a "language acquisition device" prewired into his genes. Chomsky stated:

> *It seems that the child approaches the task of acquiring a language with a rich conceptual framework already in place and also with a rich system of assumptions about sound structure and the structure of more complex utterances. These constitute the parts of our knowledge that come 'from the original hand of nature,' in Hume's phrase. They constitute part of the human biological endowment, to be awakened by experience.*[5]

Since Chomsky first proposed an acquisition code for language, other researchers have theorized that genetically prewired acquisition devices enable children to quickly develop other traits and capacities, such as sight, sound, movement, math, the ability to distinguish between living and inanimate objects, and even the ability to grasp concepts such as circles and animals.

These experts point out that these preselected genetic acquisition devices are activated simply by introducing the right stimuli to the child, which prompts a high level of neurochemical activity in the brain, allowing for rapid learning. As researchers Peter Coveny and Roger Highfield point out in *Frontiers of Complexity*, "The DNA code 'programs' the brain architecture and learning process."[6]

A Universal Learning Code Active Throughout Your Life

These previous efforts to uncover our genetically encoded learning devices have focused mainly on how children acquire new information so rapidly, not on how adults learn. It's also been led by specialists in a particular field, and therefore they've tended to focus on specific traits like sight, sound, or movement. This book expands on these pioneering efforts that deal with specific traits, and proposes a universal Learning Code, which encompasses all forms of learning, a code that is active not only in childhood but throughout your adult life.

Why You Need a Learning Code: It's Your Filter

Not only does your genetically imprinted Learning Code provide the template for *how* you learn, but it also provides the template for *what* you learn. Without a Learning Code, your brain would have two primary options as to what it selected into its structures: it could encode everything it came in contact with, or it could select new information on a random basis. Neither of these options is attractive.

Because learning and memory are represented by neurological growth, if your brain selected all the information it came in contact with, your brain tissue would eventually expand until your head was as big as a blimp, limiting your ability to move!

Because your actions in the world are dictated by the learning that you've acquired, if your brain selected learning and memory into its structures in a random fashion, without rhyme or reason, the whole human species would behave senselessly.

Neither alternative is a solid survival strategy. One way to look at the Learning Code, then, is as a filter or limiting factor, ensuring that what you select into your neurobiological makeup directly supports your surviving and thriving in the world you're living in right now.

The idea that biological systems have limiting factors is not new to science. In the 1750s, one of the Founding Fathers of America, inventor Benjamin Franklin, speculated that limiting factors must exist for all species, or their populations would grow out of control. Reverend Thomas Malthus picked up on Franklin's idea in his famous "Essay on the Principles of Population," upon which Darwin heavily relied to develop his theory of evolution by natural selection.[7] All three men were clear: to successfully adapt to their world, all species needed to have limits on the growth of their populations.

More than 100 years after Franklin, Malthus, and Darwin studied the limiting factors on the population growth of mammal and insect species, scientists began to look at the influence of limiting factors on the growth of brain cells and neural connections. In 1959, one of the fathers of modern neuroscience, Jerome Lettvin, made a bold break from the ranks of his contemporaries, who maintained that the brain remembered everything it encountered. In his paper, *What the Frog's Eye Tells the Frog's Brain,* which became well-known to neuroscientists, Lettvin laid out a model for the mind and learning that showed the brain had definite limits as to what information it could select into its structures.[8]

Lettvin placed electrodes into the cells of the visual cortex of a frog's brain, which process signals sent from the frog's eyes. In his experiments, Lettvin found that these neurons didn't fire when just any object was presented to the frog's eyes. On the contrary, the frog's genetic code had programmed neurons in the visual cortex to fire only when presented with information that was instrumental to the frog's survival. The neurons fired wildly when objects such as lily pads, bugs, or frogs of the opposite sex were presented but failed to fire at the sight of things such as beer cans and pizza boxes.

Your Learning Code Imposes Limits on What You Can Learn

Lettvin and other scientists later extended their findings to humans, showing the human brain did indeed have limits on what information populations of neurons could select. Although there are similarities between the Learning Codes of each species, each species has its own genetically imposed limiting factors on what survival information its brain can select into its structures. Therefore, cats' brains can perceive ultraviolet light, bats' brains select ultrasonic sound, and bears' brains can recognize smells up to twenty miles from their source. Each species has variations in its Learning Code, which helps the individual select what information will help it best "fit" the environment in which it lives.

Although your brain can't select ultraviolet light, ultrasonic sound, or smells twenty miles away, humans do have the most complex and sophisticated Learning Code of any species on the planet. This high-level code allows us to select more information into our biology than any other species—which has enabled us to land men on the moon, build the internet, discover the cure for polio, and create the United Nations. We may not smell as well or see as well as some other species, but we sure can learn more and innovate better.

The sophistication of the human Learning Code also allows humans to be the most adaptable species on the planet. This is why we can successfully survive in environments as diverse as the frigid tundra of the Arctic and the scorching savanna of the Serengeti. Other species with less sophisticated Learning Codes have limits on their adaptability. Polar bears can't successfully survive in the Serengeti, nor can zebras successfully adapt to the Arctic.

Science Helps Us Build Learning Systems That Work

When we understand that our genetic code has built a human Learning Code that provides limits on how and what information can be selected into our long-term memory banks, we can begin to develop personal and institutional learning systems that match the way we most naturally, quickly, efficiently, and joyfully amass knowledge.

Unfortunately, our existing corporate and scholastic learning systems fail to create the levels of learning and behavioral change our society demands because they fail to effectively access our biological Learning Code. This is the scientific explanation of why programs like No Child Left Behind, Bill Gates's initiatives to fix America's schools, technology like massive online learning courses, games, and institutions like prisons are not living up to our expectations. It isn't from a lack of good intentions, money, or effort that they fail. They fail because their designers don't understand the biology of the learning process.

Later chapters will dive deep into what the Learning Code is but also how you can activate it to create profound neurological change—learning and behavioral—in yourself and others. These chapters also put into perspective how we've come to accept outdated learning theories, which prompt us to build painful *memorization factories* that try to affect learning in ways radically foreign to how your genes and neural tissue were designed to encode knowledge.

Cracking the Learning Code

Trying to affect learning without accessing our biological Learning Code is like trying to open a bank vault with the wrong combination, and this is why so many of us have been frustrated by our learning experiences in traditional education: your Learning Code wasn't turned on by it.

3

The Reason for Learning Failure Today: Ancient Learning Theory

If we teach today's students as we taught yesterday's, we rob them of tomorrow.

—John Dewey, American philosopher, psychologist and educational reformer[1]

In this chapter, we will go a little deeper into why our learning systems aren't producing the results we demand. We will look at how the knowledge we're applying to create learning and behavioral change today is based not on the modern sciences like genetics, neuroscience, epigenetics, evolutionary biology, molecular biology, quantum physics, and complexity science but instead on what we at the Institute call *Ancient Learning Theory*. This theory has its basis in the sciences and philosophies of Plato, Aristotle, St. Thomas Aquinas, Niccolo Machiavelli, Isaac Newton, John Locke, and the school of psychology called *behaviorism*. Although *Ancient Learning Theory* served us well from around 400 BC until the 20th century, it's failing miserably today.

Before we can transform our broken learning systems, we need to understand what is causing them to be broken in the first place. *Ancient Learning Theory* is that cause. It is *Ancient Learning Theory* that has prompted us to develop a scholastic educational system that relies almost exclusively on linguistic instruction, locking learners up in small boxes (50-by-50-foot classrooms), far from the real world, at a time when their brains are going through their most explosive growth period. It's a system that devalues your emotions and what is meaningful to you while treating you like machine parts on a conveyer belt, shuffling you in and out of 13,000 forty-five-minute "get ready for the test" segments over the course of your school life.

For twelve years, reward and punishment (grades) are the primary means of motivation used to get you to memorize 30,000 to 50,000 decontextualized, personally meaningless details, most of which are quickly forgotten. And then it follows you into adulthood—it also drives the "lock 'em in a box and lecture to 'em" training programs that most corporate learning systems and motivational experts rely upon. The main reason that online learning programs aren't living up to our expectations is that even though they use the latest computer technology, they still rely on *Ancient Learning Theory*.

Ancient Learning Theory depends heavily on what Ivan Barzakov of Optimal Learning called "the fallacy of the familiar."[2] That is, the theories of learning that we use today are virtually the same as those used by the generation before us, which are the same as those used by the generation before them, which are same as those used by the generation before them, going back 2,400 years to the time of the Greek authorities.

Ancient Learning Theory allows us to accept the bizarre concept that intelligence is represented not by how happy and successful you are in life, but instead by your score on pen-and-paper tests. And we now find these tests better predict how well you'll do in the classroom, not how well you'll do in the real world. A popular quote commonly attributed to Mark Twain sums up today's learning failures based on *Ancient Learning Theory* quite well: "It ain't what you don't know that gets you into trouble. It's what you know for sure that just ain't so."[3]

Why Is Ancient Learning Theory Obsolete?

The explosion of information growth in the 21st century is the main factor rendering *Ancient Learning Theory* obsolete. As I mentioned in the introduction, all the information that exists in the world is estimated to be doubling at the incredible rate of once every eighteen months. Consider that there are more than one trillion pages on the web now, with an amazing 82,000 new websites added every day;[4] Facebook users share more than 250,000,000 pieces of information each minute, and 5,300,000 Google searches are made every second.[5] And even the numbers of those old standbys—books—are exploding, with 7,000 new ones published every day.[6]

"We'll start out by speaking in simple declarative sentences."

This information explosion is putting many of us in what one researcher calls "information shock." That is, we're bombarded with more information than we could possibly assimilate. As an example, a few years ago at a well-known neurological conference, someone proposed that researchers stop publishing new research in order to give its members time to catch up with what had already been published![7] We are indeed experiencing an unprecedented explosion of information.

But new information hasn't always been generated at such a fast rate. Consider that, in the time of Jesus, it's estimated that all the knowledge in the world was doubling every 1,000 years; 1,250 years later in the Middle Ages, it was doubling once every 750 years; and in the 1700s, it was doubling once every 350 years. At the time of the US Civil War, information doubled only once every 100 years, and by 1950, it is estimated information was doubling at the rate of about once every twenty-five years.[8]

Ancient Learning Theory has its foundations in slower, more agrarian times when it was believed that one could memorize all the important data knowledge that existed in the world within the span of a little more than a decade.[9] It's been estimated that, to be considered a knowledgeable person in the 12th century, one needed to memorize only 3,000 bits of knowledge from the works of authorities such as Plato and Aristotle.[10]

Up until the start of the 18th century, all the information in the world was increasing at such a slow pace that it appeared that knowledge was static. This view allowed our schools to become what they are now, mainly memorization factories, where the primary job of the educational system is to pour a specific number of facts into the empty heads of students. Back in the early 1900s, only about 6 percent of the US population even graduated from high school.[11] This was because it was thought that most of what you needed to know to make a good living could be memorized by the time you left eighth grade. Even up until the 1960s, information was doubling so slowly that it was generally believed that after you left the confines of the educational system, you could be successful even if you stopped learning for the rest of your life!

It was during these unhurried and more knowledge-stable times—ranging from before 400 BC to the early 1960s—that the major philosophical schools of thought, upon which *Ancient Learning Theory* is based, had their genesis: such as idealism, scholasticism, rationalism, perennialism, pragmatism, authoritarianism, and behaviorism. It is vital for us to recognize that because *Ancient Learning Theory* was developed when information was doubling every 1,000 to twenty years, it can't possibly serve us in a world where information is doubling every eighteen months.

"Moore's Law" and "The Learning Law" Law

There's an observation in computer science called *Moore's Law.* It states that computer processing speed doubles every eighteen months.[12] At the Institute, we've made an observation in learning science that those associated with us have started calling *The Learning Law.* It states that the faster the rate of information in the world doubles, the less efficient the present educational systems will be. And because the rate of information doubling is increasing every eighteen months, our learning failures will only be magnified each year we rely on *Ancient Learning Theory.*

Is What You Learned Yesterday Adding Value Today?

With information doubling so quickly, what you learned three years ago about computers, apps, finance, marketing, customer service, medicine, taxes, telecommunications, sales, management, and distribution is no longer totally valid. If you had a brain tumor today, would you want a doctor to operate on you who had learned nothing new about tumors in the last three years?

In a knowledge society, the individuals who have learned the most relevant facts in a particular area become the most valuable assets to any

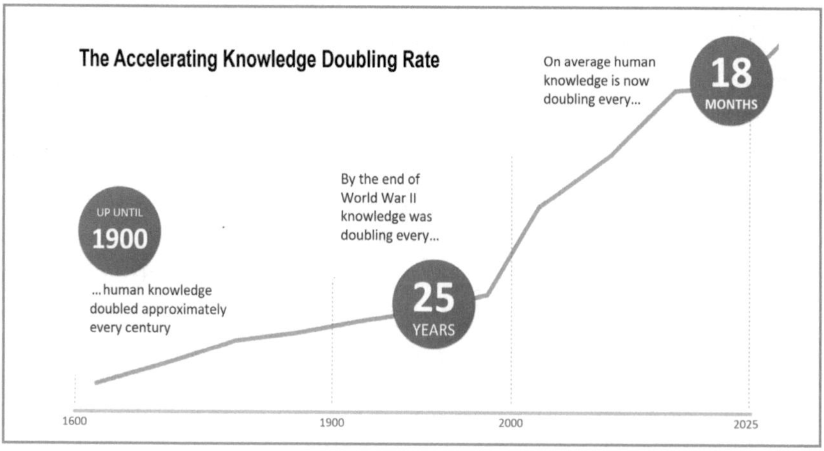

effort or organization. If you lack the ability to learn the right information quickly, you get left by the wayside.

As I touched on in the introduction, many decades ago, management guru Peter Drucker was one of the first to recognize this shift to a knowledge society. In his book *Managing for the Future,* he sums up this theory, *"From now on the key is knowledge. The world is becoming not labor intensive, not material intensive, not energy intensive, but knowledge intensive."*[13] Now your ability to have a successful life is directly tied to the speed at which you acquire new knowledge. Gold, silver, stocks, bonds, real estate, and business ownership are still valuable assets, but without the correct knowledge, your ability to understand and amass them is greatly reduced. The knowledge society creates a brain-based economy where it isn't the size of your biceps that makes you worthwhile, but the size of your knowledge base.

Make no mistake about it: our free-market economy makes us all entrepreneurs or free agents, with our personal value to any endeavor or organization directly tied to the relevant level of knowledge we've acquired. Never before in history has our ability to learn quickly been so important to individual and organizational success. Today, if you can't keep up your knowledge base, your job security has evaporated.

The Shift to Learning Organizations

Why do 88 percent of the companies that were listed in the Fortune 500 in the 1950s no longer exist? Because they failed to learn and adapt fast enough in our knowledge society—they either went out of business or got eaten up by the competition.[14]

Many leading corporations now recognize that knowledge is the most precious asset of the 21st century and are turning themselves into *learning organizations.* These more-aware organizations recognize that the key to their success and ultimate survival is directly linked to their employees' ability to learn relevant knowledge rapidly. Arie de Geus, past head of planning for Royal Dutch/Shell, pointed out, *"The ability to learn faster than your competitors may be the only substantial competitive advantage."*[15] History has shown that organizations that do not learn fast die.

Why Do We Continue to Embrace Ancient Learning Theory?

Faulty science has helped us hold onto *Ancient Learning Theory* long after it's outlived its usefulness. One of the main reasons we continue to rely on this theory is the myth that our brains are like digital recorders that somehow remember almost everything presented to them. This misconception has helped us maintain learning systems that have become little more than memorization factories. There would be no need to look for a Learning Code if the mind could remember almost everything it encountered.

Searching in the Wrong Place

A neighbor found Nasrudding on his hands and knees.

"What are you searching for, Mullah?"

"My key."

Both men got on their knees to search. After a while, the neighbor asked, "Where did you lose it?"

"At home."

"Good lord, why are you searching here then?"

"Because it's brighter here."

Why Memorization Factories and Not Learning Institutions?

As you'll see in more detail in later chapters, from the 1940s through the 1980s, the behaviorist school of psychology dominated popular thinking. By resurrecting the philosophy of John Locke, the behaviorists helped perpetuate the misconception that the mind remembers everything. In the 17th century, Locke theorized that the mind was like a *tabula rasa,* or blank slate, on which was imprinted everything it encountered.[16]

What gave Locke's misconception scientific validity was the work of renowned Canadian neurosurgeon Wilder Penfield. In the mid-20th century, Penfield performed pioneering neurosurgery on 1,132 patients with brain disease. With the permission of those patients, Penfield placed electrical probes in pinpointed locations in their brains. Because the brain contains no pain receptors, the patients received only local anesthesia to the scalp, thus letting them remain fully conscious while their brains were exposed. Penfield could then query them about the effects his electrical stimulation was having on them. From his efforts came the neurological maps of the motor and sensory strips that are still used in scientific textbooks today.[17]

The popular press of the day was fascinated by these operations and reported that Penfield had discovered that the brain was like a movie camera, recording everything it encountered. But years later, upon detailed examination of Penfield's records, scientists reached a very different conclusion. Those who pored over Penfield's work found that, of the 1,132 patients who'd had electrical stimulation applied to their brains, only forty had had stimulation-evoked experiences. The other 1,092 patients had had no evoked responses at all! There were other problems with what the press had reported. Of the forty patients who had reported that the electrical stimulation produced a response, many said it had had a dreamlike quality, more like a hallucination than a real memory, or that the memories were too generalized, like a dog chasing a cat, to be classified as a specific memory. Others flatly said the experiences provoked by the electrical stimulation had never happened to them. In the book *Conversations with Neil's Brain,* neuroscientists William Calvin and George Ojemann pointed out that Penfield's electric probes actually caused small seizures, which evoked nightmare-like hallucinations, not memories.[18] Now, at the start of the 21st century, neuroscientists are in agreement that your brain isn't like an empty vessel, nor does it record, consciously or unconsciously, all it comes in contact with.

Although the evidence that Penfield's conclusions were wrong is over sixty years old, the myth that the brain can remember almost everything persists. As you saw in the last chapter, the brain has limiting factors that restrict what you'll learn and remember. Why can't you pull something out of your mind that you studied in school? The false belief that persists is that the information is locked away in your subconscious, and you just can't figure out how to access it. This myth helps perpetuate the belief that memorization

> **From now on the key is knowledge. The world is becoming not labor intensive, not material intensive, not energy intensive, but knowledge intensive.**
>
> *Peter Drucker, Managing for the Future*

factories—that is, places where people are locked up in small classrooms, where the primary form of brain stimulation is linguistic and where thousands upon thousands of hours are spent getting ready for pen-and-paper tests—are the best way to create long-term memory.

As you'll see, your mind isn't at all like an empty vessel in which memory is created simply by pouring in information, as Locke and the behaviorists maintained (with Penfield's help). Instead, your brain is a biological instrument that must be stimulated by a very specific process before learning, long-term memory, and behavior change can be created in you. Without understanding that there's a Learning Code by which we gather new information, we'll continue to build learning systems that disappoint.

Fast Thinking Forces Us to "Recycle Ignorance"

Despite the respected research that undermines *Ancient Learning Theory*, we're resistant to change. It's much easier to push the existing system harder and harder, even though it causes even more dysfunction in the long term. Narcotics Anonymous observes, "Insanity is repeating the same mistakes and expecting different results." One of the answers to what is causing our collective educational insanity can be found in the work of 2002 Nobel Prize winner Daniel Kahneman, who brought to light the difference between fast and slow thinking.[19]

When we take a more contemplative approach to solving problems, we drop into *slow thinking* mode. When we react to our problems with little thought, we're in *fast thinking* mode. Kahneman is a psychologist, not a neuroscientist, so he didn't explore the neurological impacts of fast versus slow thinking. But when we look at his work through the eyes of neurobiology, we see that slow thinking employs our most frontal, developed, and flexible brain structures, which in turn allows us to engage in "big-picture" systems thinking.

When we use slow thinking, not only do we have the capacity to perceive how all the pieces of a system fit together, but we can also more accurately predict the consequences of our actions. Slow thinking is the kind of thinking that has produced profound solutions to problems and innovations like the printing press, $E = mc^2$, the antiretroviral therapy for HIV, and the mission to Mars. A review of history shows that virtually all the great innovations of humankind were achieved using a contemplative, slow thinking process.

On the other hand, fast thinking is what we do when we're under pressure to come to quick conclusions. As you'll learn in more detail in the upcoming chapters, stress and fast thinking compel you to employ our least flexible and least evolved brain structures, which in turn preclude you from systems thinking or accurately predicting the consequences of your actions.

Little creativity or innovation results when we're in fast thinking mode. We tend to default to the way we've always done things. This is because fast thinking forces us to engage brain structures that depend on what we already know. This form of thinking limits our ability to absorb and store new knowledge into our memory banks and, in the end, forces us to continually repeat what management guru Edward Deming called "recycling ignorance"[20] without being aware that we're doing it. Unfortunately, it's become the dominant form of thinking in a world where we feel we have too much to do and too little time to do it.

The cry of the fast thinker is, "We have to fix things now. Don't spend time thinking or looking at the big picture or importing new knowledge. Just get the information and get moving. We'll deal with the consequences later!" A wonderful example of how fast thinking makes us say and do stupid things is House Speaker Nancy Pelosi's infamous quote about the complex healthcare bill: "But we have to pass the bill so that you can find out what is in it."[21] Or take Senator Johnny Isakson, R-Georgia, who even though he had not had time to read the final of a gun control bill told CBS News, *"I think it deserves a vote up or down."*[22]

We know that education is important. We know that we're not living up to our values, and we're failing ourselves and the next generation. But it feels like we're powerless to stop it because we're pressured to find quick solutions that resonate with what we think we know about learning. This ends up sending us right back to *Ancient Learning Theory* and leaving us even more troubled and inefficient than we were when we started.

Systems Thinking Reveals What is Wrong and How to Fix It

In the effort to understand the causes of our present learning crisis and uncover the scientific bases of our Learning Code, we employed a form of analysis called *systems thinking*. Also called *systems dynamics*, systems thinking was discovered in the 1940s and popularized in the 1990s through the work

MIT's Peter Senge.[23] Unlike more traditional forms of analysis, which nphasize the breaking up of a system into smaller and smaller elements dissect the parts, systems thinking emphasizes a holistic approach, where e focus is on how the diverse patterns of interconnectivity between the arts ultimately dictate the behavior of the whole system. Systems analysis evealed that understanding the history of our current learning systems was ital to discovering solutions for it. It also made it clear that, along with the context of history, the new sciences—genetics, neuroscience, epigenetics, evolutionary biology, molecular biology, quantum physics, complexity science, and so on—needed to be considered before our Learning Code could be cracked.

Learning Systems Cause Their Own Failure

Systems thinking also brought with it a powerful benefit that illuminated the concept that any system, whether it be in business, society, or learning, is ultimately responsible for its own behavior. Systems analysis pioneer Edward Deming, whose work in systems behavior is so valued that *US News & World Report* called it one of the "nine hidden turning points in history," found that 94 percent of all system failures are attributable not to the individuals within the system but instead to the structure of the system itself.[24]

Because we fail to recognize that the malfunctions of our existing learning systems are caused by employing *Ancient Learning Theory*, we falsely blame the people in the system. I touched on this idea earlier—that is, because we fail to look at the structure of the system itself as the cause of its dysfunction, administrators, parents, trainers, workers, bosses, and students become the scapegoats.

We waste time, focus, and energy by blaming people and not the dysfunctional system itself. Just like the subjects of the emperor who wore no clothes, we become blind to the obvious: The system is broken. Systems analysis expert Donella Meadows pointed out that "a truly profound and different insight is the way you begin to see that the system causes its own behavior." We have a saying at our Institute that forces us to focus on the real cause of dysfunction: "Systems dictate behavior."[25]

Not recognizing that the system itself is causing our learning problems, we try to increase positive outcomes by pushing the existing inefficient system harder and harder. We try to improve the scholastic system by upping the homework load, increasing the length of the school day or year, adding more standardized testing, hiring private corporations to run public schools, initiating a voucher system, paying teachers for performance, and implementing programs like No Child Left Behind and Common Core. Corporations try innovative approaches such as internet learning, distance learning, and brain games, but our outcomes continue to disappoint.

As Senge pointed out, systems obey specific laws, and two of the most important are these: Today's problems come from yesterday's solutions, and the harder you push an ineffective system, the harder the system pushes back. In our scholastic and corporate educational efforts, we're doing both, applying yesterday's solutions and pushing the inefficient systems harder—of course things are getting worse![26]

As you'll see in the next chapter, to gain a powerful grasp on how to increase the speed, joy, and efficiency of learning, we must first define what intelligence is from a biological perspective. Without this perspective, we will continue to build learning systems and organizations that try to improve our ability to do well by focusing on passing tests instead of improving our ability to quickly and efficiently adapt to today's ever-changing complex information-rich world.

Cracking the Learning Code

To prosper, we need to search out learning which meets our biological and information requirements.

4

What Is Intelligence? Not What You Think!

For decades intelligence tests have been selling the public on the notion that a single score... reveals the single basic fact about people's intelligence... If we take a serious look around us, the entire fiction that the... test fully measures intelligence rapidly disintegrates.

—Robert Sternberg, The Triarchic Mind: A New Theory of Human Intelligence[1]

Ever since the fourth century, when St. Augustine declared, "The author and prime ruler of the universe is intelligence,"[2] humankind has been fixated on trying to define and measure intelligence. But *Ancient Learning Theory* means we tend to define intelligence primarily as an ability to do well in pen-and-paper tests—either standardized intelligence tests or academic exams. If you got high scores, you were considered more intelligent than those who got lower scores. At the turn of the century, with the development of their thirty-question IQ test, French educators Alfred Binet and his colleague Theodore Simon helped propel us toward another false belief: intelligence test scores could accurately predict future success.[3]

> **First of all, try to forget that you have ever heard of the concept of intelligence as a single property of the human mind, or of that instrument called the intelligence test, which purports to measure intelligence.**
>
> *Howard Gardner, Frames of Mind*

In fact, we have become so fixated on the power of the test to predict our success that it could be said that the goal of the educational system has been distilled to increasing a student's ability to score high grades. In 1923, as pen-and-paper testing was embedding itself into the fabric of the world's learning systems as the ultimate determinant of intelligence, Edwin Boring, a pragmatic scientist at Harvard, flatly proclaimed the view of intelligence that most of the world still wrongly holds today: that it is "the capacity to do well on an intelligence test. Intelligence is what the test tests."[4] Compare this to Peter Relic, past president of the National Association of Independent Schools, who made this heartbreaking statement:

> *Our fanatical emphasis on testing and academic performance sends the wrong message. A test score is more important to us than who a child is as a human being. That means an immense loss in terms of human potential because we have defined success too narrowly.*[5]

This chapter will delve into how we have gotten off track by mistakenly focusing on high grades, instead of focusing on how to change your brain structures and biology so that you may better adapt to your world and better succeed in life.

The Sinister "IQ Way of Thinking"

This misconception about intelligence is described by Harvard's Howard Gardner, one of the world's leading experts on intelligence and the creator of the concept of multiple intelligences, as the "IQ way of thinking."[6] That is, you believe you're either born smart or not, that your level of linguistic aptitude will determine your future success and that pen-and-paper tests can predict whether or not you'll do well in life. But under the light of research, these previously held beliefs about what high IQ/SAT scores, good grades, and class position really mean to your future success may begin to crumble. Because while studies indicate that these measures are very good at indicating whether you'll do well within the school system, they are poor indicators of whether you'll do well in the real world. Gardner has become so discouraged by the emphasis on test scores that it led him to write in the tenth-anniversary edition of his groundbreaking *Frames of Mind*:

> *First of all, try to forget that you have ever heard of the concept of intelligence as a single property of the human mind, or of that instrument called the intelligence test, which purports to measure intelligence."*[7]

Leading Soviet psychologist Lev Vygotsky agrees, saying, "IQ, schools, and tests fail to yield any indication of an individual's zone of potential."[8]

The idea that your intelligence test scores and grades are predictors of your future success is so engrained in society that it's important to take space here to note the experts in the field who have debunked this theory and explore the slightly sinister foundations of the "IQ way of thinking." I think you'll be shocked about how negatively these researchers view intelligence testing.

Let's start with Daniel Goleman, author of *Emotional Intelligence*, who summed up the ineffectiveness of intelligence tests, which are "based on a limited notion of intelligence, one out of touch with the true range of skill and abilities that matter for life over and beyond IQ."[9] He goes on to say:

One of psychology's open secrets is the relative inability of grades, IQ, or SAT scores, despite their popular success, to predict unerringly who will succeed in life.

We've been led to believe that the higher your IQ, the better off you'll be in life, but Leta Hollingworth, an expert on the study of gifted people, conducted seminal research that revealed a disturbing finding. She discovered that as IQ increased over a certain point, people's job performance and salary actually decreased![10] The Berkeley study on creativity also showed a lack of correlation between IQ and

A. independent thought and action
B. the ability to value or possess a good sense of humor, and
C. the ability to appreciate beauty, complexity, or novelty

Also, the study found that IQ didn't even closely correlate with the ability to be reasonable.[11] Intelligence researchers, Fred Fiedler and Thomas Link found that "Cognitive ability tests have been notoriously poor predictors of leadership performance" and that "Leader intelligence under certain conditions correlates negatively with performance."[12] Author Jensen Barclay, Professor of Education, who has written several books on intelligence, said of those with IQs in the top 2 percent of test-takers, which qualifies them to be part of an organization called MENSA:

"Shouldn't we hold off on artificial intelligence until we figure out actual intelligence?"

> *They are a very strange group. You run into people who have IQs of 160 or so and work as elevator operators or parking lot attendants. Their IQ is their only claim to distinction, and they make a lot of it.*[13]

The kind of mind that does well on tests and deals with miscellaneous facts doesn't necessarily do well in the real world. When asked why he hadn't received a promotion in eight years, one of the all-time money winners on *Jeopardy!* said flatly, "The *Jeopardy!* mind is not necessarily the most useful mind in the real world."[14] Many years ago, on a radio show and later a television show called *Quiz Kids,* the "brightest and the best" children with genius IQs of 140 to 200 were chosen to strut their stuff. But upon following these children through adulthood, Roth Feldman found that their lives were noticeably less distinguished than their IQs should have predicted. For the majority of the Quiz Kids, their highest achievement in life was their last appearance on the show.[15]

Susan L. Barrett, the author of *It's All in Your Heads,* says the test culture persists because "Old habits die hard [...]. They think that test scores can predict the students' future success—even though this has never been proved!"[16] After reviewing the research, Gordon Dryden and Jeannette Vos conclude in their bestselling book, *The Learning Revolution,* "Possibly the worst education invention of this century was the so-called intelligence test."[17]

Grades and Class Position: Poor Predictors of Life Success

Like high intelligence test scores, good grades and high class position are also poor predictors of future success in life. Consider a study that followed valedictorians and salutatorians from the 1981 graduating classes of Illinois high schools. It was found that although these students had the attributes to ensure success in school, these characteristics didn't necessarily translate into real-world success. By their late twenties, these superior students had reached only average levels of success in life. Only one-fourth were achieving at the highest levels in their chosen profession, and the rest were doing much less well. As one of the researchers, Karen Arnold, Professor of Education at Boston University, explains:

> *"I think we have discovered the dutiful people who know how to achieve in the system. But valedictorians struggled, as surely as we all do. To know*

> *that a person is a valedictorian is to only know that he or she is exceedingly good at achievement as measured by grades. It tells you nothing about how they react to the vicissitudes of life."*[18]

In another famous study involving grades, ninety-five Harvard students from the class of 1940 were followed into middle age. The men with the highest test scores in college weren't particularly successful in terms of salary, productivity, or status in their fields when compared to their low-scoring peers. The high-grade scorers also didn't have the highest levels of happiness, the strongest friendships, or superior family and romantic relationships.[19, 20]

To find out what made some scientists and engineers more productive than others, Janet Caplan and Robert Kelly did research and discovered a startling fact that they wrote about in the *Harvard Business Review:* "As it develops, academic type success was not a good predictor of on-job productivity nor was IQ discovered to be a factor."[21] Sternberg said:

> *"Many people who get good grades in school have come to be quite smug about their intelligence even in the face of repeated failure outside of the classroom."*[22]

In the fascinating book *The Millionaire's Mind,*[23] Thomas J. Stanley and Jon Robbin, a Harvard-trained mathematician, conducted in-depth statistical research to identify which variables caused people to become successful in business and super-wealthy. Their studies found that, contrary to popular belief, there was no significant statistical correlation between how successful these individuals were later in life and their grades in school, their class position, or their SAT scores. Another revealing study done in 1998 found that 15 percent of the individuals on the Forbes 400 list of wealthiest people either didn't start college or dropped out. Amazingly, these fifty-eight dropouts' average net worth wasn't less but more than that of their contemporaries that didn't drop out—and not by a little bit. Their average net worth was $4.8 billion, 167 percent higher than their college graduating peers, who averaged $1.5 billion. And when these individuals were compared to those who graduated from the prestigious Ivy League schools, such as Harvard, Yale, and Princeton, it was found that their net worth was 200 percent higher.

The founder of Kinko's, Paul Orfalea, who so poorly fit the school system that he was put in a special educational needs class, recounted that to bolster his feelings, his mother used to tell him, "The A students work for the B students, C students run the business, D students dedicate the buildings."[24] Consider that five of the most important minds of the 21st century—who have shaped the technology age—didn't finish college: Bill Gates and Paul Allen, the founders of Microsoft; and Steve Jobs, the founder of Apple; Michael Dell, the founder of Dell computers and Richard Branson, the founder of Virgin Galactic, who didn't even finish high school.

In simplest terms, academic performance and intelligence tests fail to predict real-world success because the knowledge and traits they emphasize are not necessarily what you need to survive and thrive in the real world. Unfortunately, abilities like fact memorization, exam technique, and essay writing often have little to do with attaining success in our chaotic, fast-paced, technology-driven world.

Biology Reveals What Intelligence Is

If the research is telling us that your intelligence isn't defined by school grades and test scores, then what is intelligence? From a biological perspective, intelligence is best defined as your ability to successfully adapt to the world around you. As the authors of *Cognitive Neuroscience: The Biology of the Mind* state, "Most practitioners of cognitive science and neuroscience don't yet fully appreciate the insight offered by an evolutionary perspective."[25] Gardner also points out our traditional linguistic-based educational system and intelligence testing "all ignore biology; all fail to come to grips with the higher levels of creativity, and all are insensitive to the range of roles highlighted in human society."[26] Because without a biological and evolutionary perspective, we'll continue to dupe ourselves into believing that our scores on a test paper can define intelligence and predict our success in the real world. As Derek Bickerton, author of *Language and Human Behavior,* said, "Let us look at intelligence in evolutionary terms; hopefully, it will soon be regarded as impossible to think about mind and intelligence in any other terms."[27]

With this more encompassing perspective of intelligence we can see that for 3.5 billion years, ever since life began, organisms have increased their intelligence by increasing their ability to adapt their behaviors to best "fit" their world. If you aren't good at adapting to the environments in which you live and work, your ability to survive and thrive in them is severely limited. Forget the test questions posed by authority figures. The most critical question you need to ask is this: "In my present environment, can I adapt my thoughts, behaviors, and actions well enough and quickly enough to survive and prosper?"

Until we recognize our ability to succeed in life isn't tied to a test score, we'll continue to build learning systems that better prepare us for an academic world than the real one. In the following chapters, we'll dive even deeper into how to define your intelligence, so you'll understand not only **what** you learn most easily but **how** you learn most easily.

Are We Being Bred to Best Fit School, Not the Real World?

Your brain is the most adaptable organ in your body and being locked up in a classroom setting for twelve to sixteen years—its most malleable period, when *super windows of learning opportunity* are at widest (more on super windows in Chapter 25)—forces your neurological structures to adapt their shape, and thus your behaviors, to fit this academic environment. Sadly, as you age, the neural networks shaped in childhood are resistant to change. Therefore, most of us exit school and enter early adulthood with large groups of neural networks somewhat hardwired to succeed in the isolated world of academia but not in the complex, teeming, sensory-rich, everyday world of real life. Gardner noted this failure of academia to shape minds to "fit" the real world in which "we evaluate everyone [...] according to whether they meet that narrow standard of success."[28]

This shaping of our young brain to best "fit" the school system may be one reason valedictorians don't necessarily achieve real-world success and why so many of us exit academia feeling unprepared to deal with life. Another reason some college and high school dropouts such as Gates, Allen, Jobs, Dell, and Branson may have done so well in the real world may be that their neural networks spent less time in academia and were, therefore, less shaped to fit the traditional academic setting.

Ugly History: Intelligence Tests First Used for Breeding

Unless you're part of academia, you may never have considered how we have become so obsessed with intelligence testing. When we take a deep dive into the roots of our "IQ way of thinking," we find the picture isn't a pretty one.

The genesis of officially ranking human beings based on their test scores started with Sir Francis Galton, who is recognized as the father of "intellectual measurement"—now known as *psychometrics*. Galton, a Cambridge-educated statistician, was the first scientist to develop methods of testing and ranking human intelligence.[29] Why did Galton want to test and rank humans? So he could breed a superior race. Sound a little scary?

Galton was a very insecure man who, to emulate his cousin's Charles Darwin's adventures, took an expedition to Africa, where he became convinced of the inferiority of any race that wasn't of European descent. His racial prejudice led him to take a particular interest in those who were born into the "right families" or possessed certain forms of genius. After reading Darwin's *On the Origin of Species* and his theories on "natural selection," Galton began to rework his cousin's theories to meet his own social and emotional needs. Darwin used the term "natural selection" to describe the process that leads to a species selecting favorable traits to enhance its survival chances. Those naturally selected positive traits are then passed down to successive generations, increasing the survival chances of the whole species.[30] Galton's idea, however, was to take Mother Nature completely out of the selection process and instead insert intelligence tests as the primary selector of who should procreate.

To accomplish his goals, Galton developed the philosophy of eugenics and founded the Eugenics Education Society of London in 1907.[31] *Eugenics* comes from the Greek *EU* + *genos,* which means "well-born" or "good birth," and with his concept of "heredity improvement," Galton believed the privileged classes should consider it their "paramount duty… to breed out feeble constitutions and petty and ignoble instincts and to breed in those who are vigorous, noble, and social." Intelligence testing was the management tool that eugenics would use to prevent European stock from deteriorating, while at the same time creating a superior strain of intelligent individuals who could rule over the less intelligent masses. In today's sound-bite terms, the eugenics movement could be called the "test 'em and breed 'em" movement.

In the early 1900s, eugenics quickly spread to the US, where some believed that immigration and intermarriage threatened the dominance of the European forefathers and the "country's vigor."[32] The lawyer Clarence Darrow expressed his fears, "The good old *Mayflower* stock is suffering the same unhappy fate as the good old pre-prohibition liquor. It has been mixed with all sorts of ailing and debilitating substances."[33] And George Bernard Shaw went as far as maintaining, "If we desire a certain type of civilization, we must exterminate the sort of people who do not fit in."[34]

Eugenics was no small movement. It raised money from the Carnegie Institute, the Harriman family, and John D. Rockefeller. The genetics laboratories of renowned schools such as Harvard, Johns Hopkins, Cornell, and Columbia were sympathetic to it.[35, 36] The former president of Stanford University, David Starr Jordan, chaired the America Breeders Association Committee on Eugenics, whose goal was to research heredity and to "emphasize the value of superior blood and the menace to society of inferior blood."[37] Eugenics laid the groundwork for Hitler, the Nazis, the Ku Klux Klan, and skinheads. Charles Lindbergh's famed sympathy for the Nazis came from his preoccupation with eugenics.[38] And in a sad chapter of American history, between 1907 and 1965, intelligence tests were legally used as criteria in thirty states to sterilize tens of thousands of individuals for such conditions as pauperism and "feeble-mindedness."[39]

The criteria that Galton and his followers first used were rather off-target, as Galton mistakenly equated intelligence with such things as an ability to hear high-pitched tones, sensitivity to the smell of roses, and the ability to discern the differences between the weights of gun cartridges, while other intelligence researchers included handshake pressure and pain tolerance.[40]

About the time eugenics was attracting a strong following, the French government passed a law mandating school for all children. Binet and Simon (in whose lab the developmental psychologist Jean Piaget worked in his youth) were commissioned to develop a test to separate those who would succeed in the school system from those who could not. Referring to Galton's intelligence tests based on sensory acuity as a "waste of time," Binet and Simon developed the world's first pen-and-paper IQ test.[41] It's interesting to note that their test of language and abstraction, upon which all of today's intelligence tests are built, was never intended to ascertain who would adapt or succeed in the real world, only those who would do well in an academic setting. In America, during the early part of the 20th century, immigrants just off the boat were given forms of these tests in

English to rank their intelligence, even though most of them couldn't speak the language. This misstep helped paint a picture of immigrants being ignorant, which in turn bolstered support of eugenics and sterilization.

Following Simon and Binet's lead, Carl Brigham, a psychology professor at Princeton, created an intelligence test for Army applicants and the next level of IQ testing, called the Scholastic Aptitude Test (SAT) for students wishing to go to college.[42] Unbeknownst to most of us, who had the fear of God put into us by this intelligence test, it was designed by Brigham to help identify, label, and select a preferred "natural aristocracy," creating a new frontier for the select few. In 1923, Brigham published his influential book *A Study of American Intelligence,* which clearly laid out his philosophy where he stated his tests:

> *... had proven beyond any scientific doubt that, like the American Negroes, the Italians and the Jews were genetically ineducable. It would be a waste of good money even to attempt to try to give these born morons and imbeciles a good Anglo-Saxon education, let alone admit them into our fine medical, law, and engineering graduate schools.*[43]

In his essay "The Racist Origins of the SAT," Gil Troy recognizes Bingham as a "Pilgrim-pedigreed, eugenics-blinded bigot." The roots of intelligence testing are not pretty at all.[44]

Are You Born with a Fixed Intelligence?

The persistent concept that your intelligence is fixed at birth can also be traced back to Galton and the eugenics movement. Galton believed that intelligence was inherited, like brown eyes or your height. He even experimented with blood transfusions to see whether it altered intelligence traits.[45] The idea that intelligence is passed down from your parents continues to dominate today's "IQ way of thinking."

During the first thirty years of the 20th century, researcher Sir Cyril Burt conducted research bolstering the concept that intelligence was inherited and fixed. Burt claimed to have performed painstaking studies that conclusively proved that the lion's share of an individual's intelligence was inherited.[46] But analysis of Burt's work in the mid-1970s by Leon Kamin uncovered irregularities. Kamin found that not only had Burt falsified his data, but had also made wild claims that even his own falsified research didn't substantiate and the associates from whom Burt said he had collected his data didn't exist.[47] Nevertheless, Burt's work continues to dominate the general "IQ way of thinking," and as brain-based educators, Dryden and Vos commented, "It is impossible to say how many millions of people have had their lives ruined by this appalling myth."[48] Daniel Goleman, the former editor of *Psychology Today*, noted that: "At best IQ contributes to about 20 percent of the factors that determine life successes, which leaves 80 percent to other forces."[49]

Complexity Science and Quantum Physics Challenges Test Predictability

The discoveries of the 20th and 21st centuries involving quantum physics and complexity science provide an enhanced perspective of why IQ and standardized test scores can't accurately predict success. These new sciences reveal a multifaceted world in which predictability is decimated. Newtonian physics maintained that humans were like machines: if we had the right measurements, we could make exact predictions.

The classical science view held the whisper of "Ye shall be God." But quantum physics and complexity science have revealed a world so chock-full of variables exerting influences on every one of us, from so many different

"I dreamed I was being chased by a giant standardized test."

directions, that predicting accurate futures with present information becomes a frustrating, self-defeating endeavor. If our academic vehicles of prediction worked, young educational misfits such as Darwin, Bell, Edison, Einstein, and Churchill would never have amounted to anything, least of all become some of the most significant contributors of our modern age.

Do Tests Limit Your Possibilities?

Considering the research, it's worth asking yourself: "Why should I let a grade on a test determine my future?" Although the evidence clearly shows that academic tests have very little predictive power of who you'll become in the future, many of us continue to wrongly identify who we are today, and who we will be tomorrow, based on our past academic achievement. From the strange world of quantum physics, we get a startling image, called the *collapse of the wave field*.[50] This scientific concept helps us more deeply understand the damage we inflict on ourselves when we allow our life potential to be limited by such false beliefs.

Quantum physics deals with atoms and their subatomic particles: protons, neutrons, and electrons. Everything in the universe is made of atoms: you, me, birds, bees, DNA, brains, rocks, plants, asteroids, everything. Understanding how atoms and their subatomic particles behave gives us a greater understanding of how our world behaves. But in analyzing and studying atomic matter, we enter a world that shakes our hold on the conventional view of reality. Quantum physics presents a shocking truth where Aristotle, Descartes, and Newton's classical logical, linear, and predictable machine-like paradigms are no longer adequate.

More than seven decades of the most intense and rigorous scientific research clearly shows that when an atom is observed and measured by a physicist, it takes on definite properties. It looks and acts like a real object—one of the tiny tangible building blocks upon which the entire physical world is built. But the second the physicist stops his measurement, he's presented with an extraordinary dilemma: that is, the atom can no longer be described as a particle. Amazingly, the atom gives up all its definite properties and takes on those of a wave. Unlike a wave in a pond, which has specific limits, it's called a *possibility wave* because it spreads itself out over the entire universe all at once! This state where an atom is everywhere at once is frequently called the *field of all possibilities*. I've often thought this field of all possibilities must be a very joyful place for an atom to hang out.

What jerks an atom out of its field of all possibilities and reverts it back to becoming a particle? The simple act of measurement. The critical point for our discussion is that the mere act of measuring *collapses the atom's wave function* forcing it to abandon its joyful, unbounded dance in the field of all possibilities, allowing only one "reality" to be singled out. In his award-winning overview of the new physics, *The Dancing Wu Li Masters,*[51] Gary Zukav speaks of the power of measurement upon the atomic matter:

> *The effect of perception is immediate and dramatic. All of the wave function representing the observed system collapses, except one part, which actualizes into reality.*

If we can snatch atoms out of the field of all possibilities by merely measuring them, could the same also be true of us After all, we are made up of atoms. And isn't it so often true that test scores—and judgments by others—collapse our field of all possibilities, limiting who we may become?

A famous study conducted in the 1970s showed we may do just that. In this study, students were given IQ tests and then split into two groups, labeled high-IQ group, and low-IQ group. The two groups were then sent to teachers, who were informed which group they were to instruct. As the teachers expected, the group labeled high-IQ did well and outperformed the low-IQ group. But the researchers who conducted this study had tricked the teachers and reversed the two groups. In both groups, it could be said that the students' perceived intelligence test scores and the teachers' judgments collapsed the students' field of possibilities down to one reality, which was the reality that the teachers most expected, good and bad.[52]

Over the last forty years, large numbers of studies have confirmed this IQ study, showing that we perform based on how we are prejudged. In their book, *The Quantum Society,* Danah Zohar, and Ian Marshall point out the role of judgment in the collapse of the wave function: "In the quantum realm, the observer plays a crucial role in bringing about the very situation that he observes. His presence and his expectations physically alter what he sees."[53] As the philosopher Nietzsche said, more than forty years before quantum mechanics was articulated, "We cannot see around our own corner."[54]

It's important to recognize that collapsing the wavefield isn't in itself a negative event. Quantum physics tells us there must be a collapse for matter to have a definite shape. What I'm addressing is collapsing the wave function with inaccurate forms of prediction like IQ tests, SATs, and grades—judgments that fail to consider our true potential. All that intelligence test scores and grades tell us is how well "authority" thinks we'll do in academic settings. It says little if anything of our ability to succeed in the real world of infinite possibilities.

If your wave function must be collapsed for you to take shape in the world, shouldn't you accept that power for yourself? Imagine the vast numbers of individuals who would have been the next Einsteins, Edisons or Churchills but, in taking their academic performance as an accurate indicator of their potential, collapsed their wave functions, and accepted a mediocre future. Shouldn't you collapse your wave function around what you define as successful, meaningful, and worthwhile? Shouldn't you define our own success quotient—that is, figure out what will make you happy, successful, safe, and fulfilled—and then collapse your wave function around those personally chosen possibilities? When Henry Ford said, "Whether you believe you can do a thing or not, you are right."[55] Unbeknownst to Henry (who knew little if anything of quantum physics), he was talking about collapsing your wave fields around your personally chosen reality.

Who Is Collapsing Your World of All Possibilities?

It begins to get scary when we look a little deeper at the tests that may be snatching us out of life's field of all possibilities. Have you ever taken a standardized, intelligence, SAT-type, or other high-stakes test and wondered, "How could they come up with a question that's so hard to understand it makes no sense to me? Am I stupid?"

No, you're not stupid.

Ask yourself these unasked questions: "Who are these people designing the tests that I'm judging myself and others with? What are their lives like? Are they successful? Happy? Balanced? Who would want to sit around all day in a tiny cubical and come up with complicated questions often designed just to trick the test-taker? Why am I giving my power to people whose life's work is to develop test questions that have nothing to do with the real world?"

Of course, everyone deserves to make a living with their core competency. The people who make up IQ and standardized tests have a core competency with which they're making a living, and this is good for them. Most of them probably fit very well into academic settings and got good grades. As you'll see in more detail later in this book, our species has over 7.5 billion different brain plans, but only a tiny number effectively "fit" the academic environment. The creators of IQ-type tests aren't bad or wrong for making up challenging and often obtuse test questions—it's just the way their brains work. But as a society, the mistake we're making is allowing our test results to collapse the field of all possibilities for ourselves and our children, who may have very different brain plans from the test creators.

Our mistake is we haven't taken a holistic system's view of the situation, and in the process, many of us have given up our power to the wrong group of people. This, in turn, forces us to develop educational systems that focus more heavily on getting us to answer test questions than preparing us to adapt successfully to the real world.

The following are real test questions from various grade levels, which are somewhat shocking in their lack of clarity—or connection to the real world. As you read them, you can almost feel the designers trying to make something simple more complex, either because their brains work that way or because they need to do it that way to keep their jobs. Yes, we did pick some of the worst, but all you have to do is use Google to find a bunch more.

Example 1 [56]

Add 26 + 17 by breaking apart numbers to make 10
Use a number that adds with the 6 in 26 to make a 10
Since 6 + 4 = 10, use 4
Think: 17 = 4 + 13
Add 26 + 4 = 30
Add 30 + 13 = 43
So, 26 + 17 = 43

Example 2[57]

Juanita wants to give bags of stickers to her friends. She wants to give the same number of stickers to each friend. She's not sure if she needs four bags or six bags of stickers. How many stickers could she buy, so there are no stickers left over?

Example 3

Madison's father was being ________ when he refused to buy her a Range Rover for her birthday?

A. pragmatic
A. frugal
B. recalcitrant
C. unfair
D. proprietary

The point I'm making is don't judge your capacity to succeed in life, or your children's prospects, on the ability to answer such questions designed by those who have a very different brain plan than you. Test scores are not the sum total of your intelligence. Your intelligence is so much more than an answer to a test question. So, if your intelligence isn't based on a score on a test, what exactly is it, and how do you increase it? This is what you'll discover in the following chapters.

Cracking the Learning Code

To understand what intelligence is for you as an individual, it's essential to understand what intelligence is for the whole human species. Your *Adaptability/Intelligence Factor* then acts as a drop of unique fuel that helps propel your personal success.

5

How Humankind Learns: Your Adaptability/ Intelligence Factor

When you know better, you do better.

—Dr. Maya Angelou, African American poet and scholar[1]

To give you a firm grasp on the fundamentals of the Learning Code science, this chapter takes a multifaceted look at how a whole species learns, and how it affects what and how you yourself learn. Before you take that dive, I need to make an important distinction in the way "intelligence" is presented and discussed in the rest of this book. Because, as pointed out in the previous chapter, instead of thinking about intelligence as a good grade on a test or a high IQ, you need to shift to a more biological way of thinking.

To help you do this, I use the term *Adaptability/Intelligence Factor* when referring to the attributes, skills, and behaviors that allow our species to better adapt, survive, and thrive in the real world. This term embodies a more natural, encompassing, and scientific understanding of what's necessary for our species to flourish. Learning is basically the process by which you and our species adapt to constantly changing environments, so I use *Species Adaptability/Intelligence Factor* when referring to the whole species' ability to learn and survive and *Individual Adaptability/Intelligence Factor* when referring to your personal ability to learn and succeed in the real world.

Understanding our *Species Adaptability/Intelligence Factor* is important because it's been hardwired into your genes over millions of years and supports how and what you learn and remember today. Each section of this chapter and the next goes deeper and deeper into the evolutionary development and genetics of how our *Species Adaptability/Intelligence Factor* and *Individual Adaptability/Intelligence Factor* came into being. Taking time to absorb this information is vital to understanding yourself as a learner. If you don't understand the history of how your "learning machine" came to be, you'll have a much harder time creating powerful, efficient learning and behavioral-change systems for yourself or others.

A Shared Experience: Species Learning

Let me ask you a few questions. Why is it that when you see sweet fruit, you salivate? Or a charging bear, you run? A good-looking man or woman, you gape? Mere moments after you exit the womb, you searched out your mother's face and breast, but who taught you to do this? Why do you hoard more stuff than you need? Who taught you how to flirt? When fashions change, why do most of us feel compelled to imitate the current style? Why is it that when you get caught doing something wrong, you often feel the urge to lie? When you find yourself in a group, why do you unconsciously search out your place in the pecking order?

All those mentioned above are instinctual behaviors that arise from *species learning.* This is learning that has proven so successful to our ancestors' survival in past environments that it has been passed down to each successive human generation in our genes. Without species learning, your survival chances and the survival chances of the human species would be quite limited. An infant who fails to suckle or an adult who fails to run from a charging bear is a member of our species who lacks good instinctual survival advantages and is less likely to pass on their genes to the next generation.

Though the eyes of biology, when we talk about species learning, we're referring to how our species adapts. The definition of adaptation is "the adjustment or changes in behavior, physiology, and structure of an organism to become more suited to the environment." Therefore, it can be said that your instincts are adaptations that your ancestors acquired to help them better survive and thrive in days gone by, going back millions of years. Whether you like it or not, you rely on these genetic adaptions to survive and thrive. They play a powerful role in what you learn and how you behave in the here and now.

Before we can fully appreciate intelligence, we need to understand what intelligence is from a biological perspective. Science reveals that what you learn today is directed by "units of learning" in your DNA, passed down to you from your ancestors. Without comprehending this, the genetic stream that flows to you from your forebears, we end up building learning systems that make us a species of great test-takers but not a species great at developing deeply meaningful, successful, and joyful lives.

Your DNA: A River of Learning

Your ancestors going back millennia possessed the successful genetic changes that were eventually selected into the whole human species and provide our innate survival advantages today. Like a river of learning, our DNA flows from one generation to the next. Modern science has a deep appreciation that change at the genetic level is the source of species learning, but this hasn't always been so. When considered on the timeline of written history, the idea that change has occurred within the biology of a species is a very new concept. This is because for 2,100 years, up until the 19th century, the ancient sciences painted a picture where nothing in nature changed—ever!

Prior to the publication of Charles Darwin's *On the Origin of Species* in the 19th century, the prevailing beliefs about species change were those put forth by Plato and Aristotle in the third and fourth centuries BC, which held that all living things were created separately, spontaneously, and precisely. The Greeks maintained that all beings that God put on earth were placed here in perfect and unchangeable form, which the Christian Church embraced, making it part of its doctrine. To say otherwise was blasphemy, as it would suggest that God had made "mistakes" and that his creations had somehow been imperfect in the first place.

This "no change" doctrine remained unshakable until the early 1800s when paleontologists began to discover remains of species that no longer existed and which appeared to be between one species and another, which gave rise to some fantastical, theories. It wasn't until 1859, when Charles Darwin published *On the Origin of Species,* that the no-change doctrine was obliterated along with all the other theories of how change might take place in nature.[2] Darwin's research proved to be the greatest paradigm-buster since Copernicus discovered that the Earth orbited the sun.

Darwin's investigations allowed him to clearly see that survival advantages went to those species that had a vast reservoir of variation among its members. When environments changed, variations in body plans and behaviors could be drawn upon to help a species better fit the environmental changes presented to it. After reading Thomas Malthus's work on how populations of a species have limits on their ability to grow unchecked,[3] Darwin came

Going back through time, our DNA is like a river of learning flowing from one generation to the next.

up with the theory of natural selection.

The theory of natural selection became the first general rule of modern biology.[4] It provided a unified answer to what scientists had found in the fossil records. It laid out the process by which existing species changed body shapes and behaviors (selected variations), how new species evolved (successfully selected variations allowed them to diverge greatly from the species' original forms), and why extinct species no longer existed (they lacked enough of the right variations to adapt to changing environments).

Although the theory of natural selection did provide a scientific explanation for species learning, it failed to address two important points:

- How were these successful variations passed from generation to generation?
- What was the source of variation?

FAST FORWARD *or* DEEP DIVE

The rest of this chapter goes more deeply into the history and science of how we came to understand that the genes passed to us from our ancestors form the basis of how and what we learn and remember in the here and now, and how the biological variation between each member of our species is designed to ensure our group and individual survival. The following sections and chapter also help us grasp the genetic basis of why we don't all learn the same way. If you would like a more profound understanding of how we discovered this process and how it is impacting your learning right now, please read on. If what you have learned up to this point gives you enough background for your needs, please jump to the next chapter.

Genes are a Transmission Device for Learning

Today we recognize genes are transmission devices, and they were discovered while Darwin was laboring away on his theory of natural selection at Down House in the UK. Some 800 or so miles away in Brunn, Austria, an Augustinian monk named Gregor Mendel (who, like Darwin, didn't excel in school) came up with what many consider "the biggest idea in genetics." After spending years meticulously researching and breeding pea plants, Mendel proposed that all organisms contained hidden "atoms of inheritance" which provided instructions, or intelligences, that could be passed from parent to offspring.[5] Mendel's work was initially ignored, but in 1900 three scientists independently confirmed his efforts, and this set off the race to understand these "atoms of inheritance," as Mendel described them.

In the early days of genetics research, scientists had three major questions:

- Where were they located?
- What did they look like?
- What were they made of?

Today we know there are six feet of DNA packed into every one of your 100 trillion cells, and if these DNA strands were strung out end to end, they would stretch approximately sixty-seven billion miles. That's 150,000 round trips to the moon! However, each strand of DNA only has a tiny circumference, and can shrink up to a trillionth of an inch, and is invisible to the naked eye. The size of a single gene in a whole length of DNA is so small that it's been compared to finding an ant on Mount Everest. To complicate matters, when DNA splits to replicate itself, it spins at about 8,000rpm. Getting a handle on this tiny spinning stuff was no easy task.

Although a daunting task, this didn't stop modern scientists from setting out to solve the puzzle. In the 1920s, Columbia University Professor Thomas Hunt Morgan found the location of genetic material while studying the favorite subject of geneticists, the fruit fly. Morgan discovered that genetic material was arranged in tiny worm-like bodies in the nucleus of a cell.[6] The makeup of this genetic material was revealed in the 1940s by Oswald Avery,[7] who established that genetic material is deoxyribonucleic acid, or DNA, which consists of sugar, phosphate, and four nucleotide bases, called adenine, thymine, cytosine, and guanine, which scientists abbreviate to A, T, C, and G.[8]

However, what slowed the advancement of genetic discoveries was that the structure of DNA hadn't been worked out yet. Then, in 1944, Erwin Schrodinger, one of the fathers of quantum physics, wrote an intriguing and profound book, *What Is Life?* linking genetics and quantum physics.[9] Schrodinger's observations had a profound impact on many geneticists, allowing them to recognize that if the structure of the atom could be worked out without ever seeing one, then the structure of the material that held species learning might also be worked out.

Inspired by Schrodinger's insight, American James Watson and Briton Francis Crick, drawing on research by Rosalind Franklin and Maurice Wilkins, made the most well-known discovery in genetics in 1952.[10] Using LEGO-type models, Crick and Watson worked out the complicated structure of DNA to be a double-helix structure with the bases A, T, C, and G separated into pairs and placed like rungs on a ladder between the sides of the helix.

Two years later, the man who had put forward the Big Bang theory of the creation of the universe made a gigantic contribution to the understanding of how species learning is passed from one generation to the next. Focusing on Crick and Watson's discovery, George Gamow determined that the sequence of the four bases strung out on the DNA strand acted like an alphabet. When putting together in the right order, these bases produced the words and sentences of a genetic language.[11]

Just as an alphabet allows us to transmit information from generation to generation, so Gamow found that genetic material has a four-letter alphabet of nucleotide bases and, when arranged in the right order on the double helix, the bases A, T, C, and G code for the learning that previous generations selected as necessary to create and maintain life. (A, T, C, and G stand for the chemicals adenine, thymine, guanine, and cytosine.) Gamow's work, showing how species learning was passed from ancestor to heir in a

GENES ARE A TRANSMISSION DEVICE FOR LEARNING. Just as our twenty-six-letter alphabet allows us to transmit written information from generation to generation, your DNA has a four-letter alphabet of nucleolide bases (AGTC) that allows our species to biologically transmit survival information to the next generation.

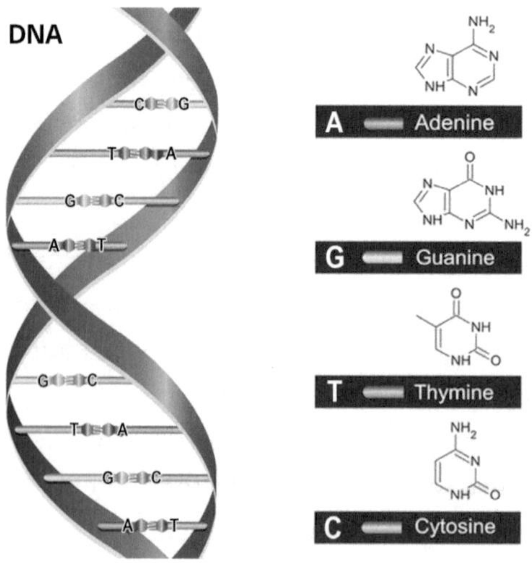

genetic language, delivered a severe blow to the then-popular behaviorist school of psychology, which maintained that all learning occurred only from environmental interactions in our present lifetime.

Since Gamow's discovery, we now know that the human genome has an amazing three billion base pairs strung out on the DNA strands in each of your cells. There are approximately 20,000 distinct segments, called genes, of these bases that code for specific information.[12] As I mentioned earlier, about one-third (or around 6,000) of these genes are involved in how the brain processes information.

Your genes average 27,000 base pairs apiece, but some have up to 2.4 million base pairs.[13] Gene segments code for all of a species' pre-learned characteristics and behaviors. You're able to suckle after you exit the womb, walk by age two, speak fluently at eight, and duck when somebody chucks a rock at you, all without anyone ever teaching you how, because this information is written in detailed segments of bases spelled out something like TACGGCGTTAGACAAGTGCGTGA (continuing for millions of letters, depending on the information they code for).

The Source of Genetic Learning is Variation

At the same time as the quest to understand the physical makeup of DNA, researchers were also trying to discover the source of the variations that drive our ability to adapt to an ever-changing environment. The first clues about where they came from were presented by one of the founding fathers of modern genetics, the Russian-born Theodosius Dobzhansky, who in 1937 put forth the "modern synthesis" of genetics in his esteemed *Genetics and the Origin of Species*.[14]

Dobzhansky firmly believed that the most effective genetic research should be conducted in the field. This real-world approach allowed him to establish that discrete variations in each individual's genetic makeup were the "raw material" upon which Darwin's natural selection worked. Dobzhansky's insights, that variations in genetic material are the source of species ability to learn how to adapt to changing environments, are greatly admired, especially considering that they came a decade and a half before the double helix structure of DNA was worked out!

Not only did Dobzhansky find that variations in genes were the source of species—the *Species Adaptability/Intelligence Factor*—but he also took the radical position that genes could manufacture their own diversity.[15] In his 1962 work, *Mankind Evolving*, he wrote: "To maintain itself in harmony with a changing environment, the organism must not only be adapted but also adaptable."[16] And just nineteen years later, microbiologist Julian Adams showed Dobzhansky's theory to be correct, finding variation so valuable to species learning and survival that genes are preprogrammed to produce it.[17] In his experiment to test his hypothesis, Adams brewed a batch of genetically identical bacteria. He kept these little clones in a uniform climate where they all received the same nutrient bath for 400 generations.

You might expect that because the 400 generations all came from clones with the same genetic composition and lived in the same environment, all the offspring would be the same. But the scientific community was surprised to find that later generations of these bugs had generated new strains by spontaneously mutating their DNA. The most fascinating finding of Adams' experiments was that these bacteria created variations in their genetic material without any environmental cause at all. Adams' work showed that variation is so important to species survival and learning that the genetic code is designed to produce it in every species on Earth.

Simply stated, every reptile, animal, and human on Earth has preprogrammed differences in the way their segments of A, T, G, and C are arranged on DNA's double helix. As an example, if we take two humans and compare their three billion base pairs, we will find that because of preprogrammed variation, there are approximately three million differences between how their A, T, G, and C are configured. Some experts estimate that the chance of any two humans ever having the same genetic makeup is 10 to the power of 1,200. These variations are the wellspring we can draw on to enhance our collective ability to learn and change behaviors to better "fit" ever-changing environments. This is the *Species Adaptability/Intelligence Factor* in action. The variations that code for successful survival body and brain changes are then selected and spread throughout the whole species.

How Species Learn from Each Other

The human species learned from animals, reptiles, and even algae. The idea that humans' genetic makeup is chock-full of successful learning might be hard to get your mind around, but science shows it can be no other way: species learn from each other. In fact, as as evolutionary biologist Richard Dawkins suggested, DNA is like a river of learning that flows directly to us from our ancestors.[18] What is often difficult to comprehend is that the source of this genetic code goes all the way back 3.5 billion years ago to when lightning struck a small pool of primordial soup, causing methane, ammonia, hydrogen, and carbon vapors to produce the nucleotides that are the building blocks of the DNA in your cells today.

Darwin was the first scientist of note to grasp the strong connection between humans and animals. Although his idea that selected variations were the cause of species change was a somewhat easy concept for the Victorian populace to digest, the idea that humans evolved from lower life forms was a bitter pill to swallow. Darwin's theory of evolution shocked the world by maintaining that not only did man carry in his body the indelible stamp of his lowly origins,[19] but human thoughts and behaviors were also driven by urges we inherited from ancestors we shared with our animal cousins.

A decade after publishing *On the Origin of Species,* Darwin produced two subsequent works, *The Descent of Man* and *The Expression of Emotions in Man and Animals,*[20] in which he left little doubt of humankind's connections to our lowly ancestors. In these efforts, Darwin pointed out numerous links between the urges of animals and man, including urinating and defecating in the face of fear and erect body hair and goosebumps in dangerous situations. He also quoted Dr. Maudsley, who specialized in what was called "human insanity," as questioning where the savage snarl and wild howl of a deranged person came from, "unless he has the brute nature within him."

Darwin's efforts produced a public outcry and, as one distinguished scientist of the day was alleged to have written: "An embarrassing discovery and the less said about it, the better." However, further genetic investigations showed Darwin was on the right track, and nature uses a sort of "if it ain't broke, don't fix it" rule. Shapes and sizes differ, but Mother Nature conserves the basic plan, modifying it to create a new species. The successful learning that served a species on one branch of the evolutionary tree is retained and passed to those species on another. As Erich Jantsch pointed out, a biological system "keeps the memory of its evolutionary path."[21]

Under similar circumstances, very different animals behave in very similar ways, and as evolutionary psychologist Robert Plutchik put it, we "all have the common function of separating an organism from a threat to its survival."[22] This means the most successful survival learning is coded into DNA strands, which are conserved across species. An amoeba wiggles away from threat just as surely as a human runs from it; monkeys and humans both salivate at sweet fruit; and all species that reproduce have some form of mating behavior because as Howard Gardner commented, "Everything that we have ever achieved has been coded into our DNA."[23] The stretches of DNA that code for our human instincts were so successful in helping our amoeba, reptile, and mammal relatives survive that they easily made it through the sieve of evolution, which allowed them to be selected into the genetic material of modern humans.

Deeper Interconnections

DNA is amazingly uniform in each species; the difference between an amoeba and a human can be reduced to how the nucleotide bases A, T, C, and G are strung out upon the double helix. Eric Davidson of the California Institute of Technology said, "All genes that a liver or a brain uses are already there in worms."[24] And a revealing example of how close we are to other species is represented by looking at a protein called cytochrome c, whose code is 339 base letters (A, T, C, and G) long. Only one letter change in cytochrome c separates monkeys from humans; a twelve-letter change separates us from horses, and only a forty-five-letter change separates us from lowly yeast!

It's been found that more than 140 human genes are the result of horizontal transfer from bacteria. Some human genes are so similar to worm genes that scientists have inserted human genes into worms and watched as the implants worked perfectly. In the end, humans are really only slightly remodeled chimpanzees, as we share 99.4 percent of functionally significant genes.

Your Brain: Built Bottom-Up by Nature

In the 1970s, decades before efforts like the Human Genome Project allowed science to recognize that gene segments were transferred between the species, Dr. Paul MacLean, Director of the Laboratory for Brain

Evolution and Behaviors at the National Institute of Mental Health,[25] worked out which brain regions hold our ancestral learning. Not only do genes transfer learning from one generation to the next, but they also build the brain parts that store and implement this learning. Back then, to find out which brain regions held genetic learning, MacLean performed what MIT Harvard University cognitive psychologist Steven Pinker described as "reverse engineering"—that is, taking the brain apart to see what evolution was up to when the thing was put together in the first place.[26]

MacLean saw that evolution used a bottom-up approach. The least-evolved and lowest structure of the human brain he called the "reptilian brain,"[27] which first appeared 350 million years ago. The reptilian brain contains the brain stem, pons, and medulla, which regulate and maintain vital body functions such as digestion, reproduction, circulation, and breathing. MacLean pointed out that this part of the human brain appeared to contain the same structures as the complete brains of turtles, alligators, and lizards.

The next piece of the MacLean evolutionary puzzle is filled by the limbic system. MacLean called this area the *paleomammalian brain* because this area corresponded to the arrival of nonplacental mammals. This brain area appeared 250 million years ago on the evolutionary scale, at about the same time dinosaurs dominated the Earth. MacLean noted that we share this brain structure with other mammals such as rats, rabbits, cats, kangaroos, and horses. These structures are housed in the midbrain and include the

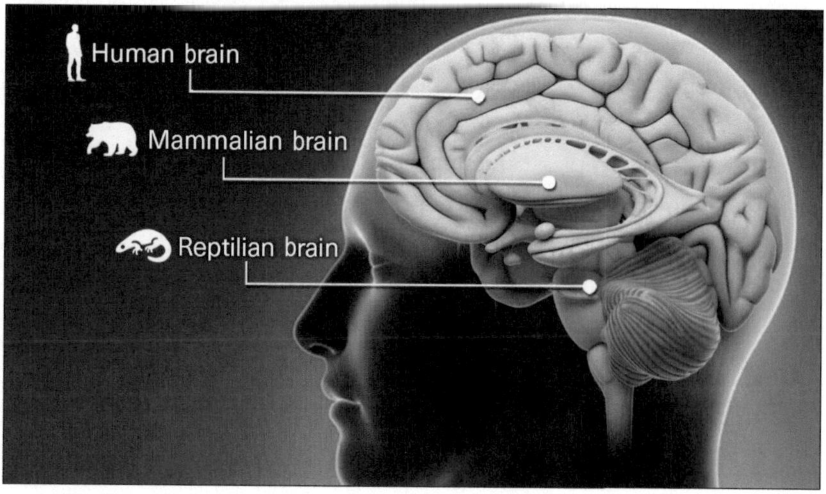

hippocampus, mammillary bodies, amygdala, and the cingulate cortex. The limbic brain is responsible for the production of base emotions such as fear, anger, sadness, and joy. As author Daniel Goleman said, "When we are in the grip of craving or fearing, head over heels in love or recoiling in dread, it is limbic that has us in its grip."[28]

The instincts held in the limbic and reptilian brain areas are very resistant to change—as well they should be, because the information coded here over millennia ensured the survival of literally trillions of our ancestors. A joke among neuroscientists is that these two brain areas house the four *F*s of survival: fighting, fleeing, feeding, and f— (sexual reproduction). Although more recent brain research has revealed some flaws in MacLean's initial claims, his overall general theory has stood the test of time.

As I'll emphasize numerous times, you learn and remember most of what you value. What is personally meaningful to us today and what we find easiest to remember has its roots in our *Species Adaptability/Intelligence Factor*. Today, we value instincts like bonding to our parents and friends; eating; acquiring, marking, and protecting territory; copulating; hoarding more than we need; moving away from pain and toward pleasure; imitating styles of the day, and locating our place in the hierarchy—all because this information was instrumental in keeping each and every one of our ancestors alive. What you remember most easily in life has its foundations in this species learning.

We don't often see humans running around kissing spiders, eating mud, or copulating with trees, because our instincts preclude us from engaging in acts that hinder the survival or propagation of our species. Instincts also act as limiting factors on the human ability to encode new information. As we discussed earlier, if the brain didn't limit what information it selected into its structures, your head would be as big as a blimp. As you journey into the following chapter, you'll get an understanding of the type of biological change that takes place at the level of your neural tissue. This change increases your *Individual Adaptability/Intelligence Factor*, allowing you to more successfully survive and thrive here and now. In short, genes learn slowly; brains learn quickly. Understanding this change process is instrumental in developing learning and behavioral change programs that complement, rather than impede our biological processes.

Cracking the Learning Code

Your genes are units of learning that have been passed on to you from your ancestors going back millions of years. These inherited genes form the basis of how and what you will learn and remember today. Failing to understand how this river of learning affects us in the here and now, means we will never be able to develop effective learning and behavioral-change systems.

6

How Do You Increase Your Individual Adaptability/Intelligence Factor? Learn Quickly

Intelligence is the ability to adapt to change.

—Dr Stephen Hawking, theoretical physicist[1]

In the last chapter, you saw how the *Adaptability/Intelligence Factor* of a whole species increases through biological change at a genetic level. The type of biological change that puts learning into the genetic makeup of a whole species has many drawbacks. The biggest problem is that it's a glacially slow process; genes just aren't capable of split-second adaptations. In general, it takes hundreds to thousands of generations to select successful variations in the structure of DNA (deoxyribonucleic acid) into the genome of a whole species. Changes in body shape or behavior that can increase a *Species Adaptability/Intelligence Factor* take a phenomenally long time. It took three billion years of genetic learning before organisms bonded strongly with their young and 550 million years of genetic adaptations to create two-legged locomotion, which freed the front appendages from supporting the upper body, allowing them to be used for efficient grasping.

Another problem with the genetic survival information packed into every one of your cells is that it isn't easily changed. Genetic learning is highly inflexible and resistant to change,[2] as well as it should be. Evolution saw it was a good idea to keep the memories of drinking water when your body needed H_2O, escaping from dangerous creatures, and passing your genes to your heirs through sex. But this indelible nature of genes limits their speedy new learning in the here and now. The titanium-hard quality of the information in our DNA ensures that, at the genetic level, split-second adaptations cannot take place. Another predicament with genetic learning is that it can't deal with the incredible flood of complexity that the world sends at us. Your relatively limited number of genes—estimated at around 20,000—couldn't possibly deal with the changing array of the thousands upon thousands of micro- and macro-environmental juxtapositions presented to you each day of your life.

Genes Change Slowly, Your Brain Changes Quickly

To create organisms that could substantially and quickly increase their *Individual Adaptability/Intelligence Factor* in a complex world, Nature had to come up with learning devices that had greater speed, flexibility, and capacity than genes. Once these cells and their connections were created, organisms could then make split-second adaptations in response to complex and rapidly changing environmental stimuli. Writing on nervous-system learning in *River Out of Eden*, Richard Dawkins pointed out, "It is important because now action can be taken on a timescale much faster than the one that genes… can achieve directly."[3]

Your nervous system has the wonderful capability to improve on information that's been coded into your genes. Your genetic makeup not only codes for instinctual learning—such as fighting, fleeing, hoarding, and marking territory—but also provides the code for the nervous system to improve on this pre-existing survival data. Genes lay down the general plan for survival, and our cells and connections refine that information. Genes tell you to run from big furry things, while your nervous system and brain cells allow you to learn to make the distinction between a bear and your unshaven Uncle Harry. What's more, the nervous system also has almost unlimited learning capacity.[4]

When compared to the titanium-hard, inflexible quality of genes, our nerve cells and their connections are extremely flexible learning devices. The effervescent quality of your nervous system allows it to quickly reshape itself over and over again in response to changing environmental stimuli. In a fascinating experiment, scientists covered chrome pellets with a bitter substance, a taste that represented poison to a chicken's taste buds and nervous system. They then mixed the chrome pellets with the chicks' usual food. Soon after the chicks pecked at the bitter pellets—for only a short period—the researchers found key neurons that recognize bad food sprouted up to 60 percent more connections in their brains. These new "instant" connections ensured that the chicks quickly learned to never peck at the perceived poison again.[5] Similarly, nerve and brain cells can change their structure within seconds,[6] which is why a child who burns their hand on a hot stove immediately learns never to touch it again.

Your nerve and brain cells have another distinct learning advantage over genes: neurons can improve on existing information in their structures without having to rely on chance variations.[7]Unlike your genes, which follow a strict Darwinian process and can change their structure only by choosing variations that best fit environmental stimuli, cellular learning devices like the brain can select new information through use.[8]Genetic learning is a slow and laborious process that takes eons to accomplish as random variations are tried and discarded over and over again until the ones that best fit the environment are selected into the whole species. However, by eliminating variation as the only criterion by which change can take place, the nervous system and brain can adapt their shape and structure in seconds. Not only can your brain quickly improve on genetic information passed down to us from our ancestors, but it can also quickly improve on the information we acquire during our lifetime.

Unlike Genes, The Brain Learns Through Use

Fifty years before Darwin detailed his theory of change through the selection of the variations of characteristics that best "fit" the environment, the French master botanist Jean-Baptiste Lamarck had put forth a theory of change through *use.*[9] For over 125 years after the publication of Darwin's *On the Origin of Species*, Lamarck's idea of change through use fell into disfavor and was ridiculed by the scientific community. If Lamarckism worked, you could pass on the information you learned today about how to run the Windows operating system, or how to drive a car, to your heirs through your genes.[10]

Today, neuroscientists are taking a fresh look at Lamarckism as a model for learning not in your genes but in your nervous system and brain. Research shows that, unlike genetic changes, the changes in your brain structures which lead to learning and memory, do indeed occur through use. And these changes occur on the time scale of seconds to days, not centuries. In your brain, a new generation of connections can be created each time the structure of an existing generation of connections is improved through use. Improvements in the structure and function of first generation brain connections are what create second-generation connections, and changes in second-generation connections are what creates third-generation connections—and so on, ad infinitum. This improvement process through use is how you learn how to improve your golf swing, dance the tango, manage people, operate an iPhone, or love others. It can be said that for the brain, "no use, no improvement."[11]

In his excellent book, *The Cerebral Code*, neuroscientist William Calvin points out that the brain is an appropriate home for Lamarckism:

> *In the cortex… there are generations of lateral copying attempts, and their improved versions can feedback into the connectivity changes to influence what happens tomorrow when another copying attempt is made.*[12]

Renowned neuroscientist and author Antonio Damasio echoes Calvin: "the genome… helps with a general rather than a precise arrangement of a system of circuits… under the influence of environmental circumstances."[13]

In the end, what allows your brain to learn in seconds is that life experiences can create new generations of nerve tissue immediately. In Chapter 7, we'll take an in-depth look at how your brain selects new information based on information already laid down by your genes and your life experiences. For the rest of this chapter, we'll explore the evolution of your *Individual Adaptability/Intelligence Factor* and how it became so efficient.

FAST FORWARD *or* DEEP DIVE

If you feel you now know enough about the general concept that genes learn slowly and brain tissue learns quickly, feel free to skip the rest of this chapter. If you would like to peel the onion more about the history of how human learning became so much more refined and efficient than animal learning, and why and how you can learn so easily, please read on.

The Evolution of Your Learning System: More Complexity More Learning

Before we look at the specific biological change that must happen in your neural tissue so that learning and long-term memory can take place quickly, it's worthwhile understanding how the modern human brain—the most intelligent and adaptable instrument on the face of the Earth—evolved from its earliest form.

The nervous system is the rich soil from which faster learning, greater adaptability, and higher intelligence take root. The more complex the nervous system of a species is, the higher its *Species Adaptability/Intelligence Factor* and the higher the *Individual Adaptability/Intelligence Factor* will be. Over millennia, more thresholds of learning speed, capacity, and flexibility have been crossed as evolution built more and more complex nervous systems.

Over a billion years ago, multicellular organisms with tiny sensory receptors located on their surfaces evolved. These small receptors allowed information about things like heat, light, and acidity to pass from an organism's exterior to its interior, which increased the speed at which an individual organism could adapt to its environment. Eventually, more complexity was added when receptors evolved that not only received messages but also sent messages back from inside the organism to its surface. This simple two-way system helped these minuscule creatures more successfully wiggle away from pain and toward pleasure. This two-way system is considered the genesis of the nervous system.[14]

Around 750 million years ago, another threshold of learning speed and capacity was crossed when segmented worms developed nervous systems with specific front and rear ends. The sense organs and associated nerve connections gathered at the front end. As soon as nature got clear on which end would be the head and which would be the tail, it was able to increase the speed of adaptation by adding more and more nerve cells at the end that mattered. Over the millennia, more nerve cells were added to the head until the first small brain appeared 500 million years ago in primitive fish, during a period of rapid evolution called the Cambrian explosion.

Another threshold of complexity was crossed 350 million years ago, when reptiles emerged with a simple brain structure that we see reflected

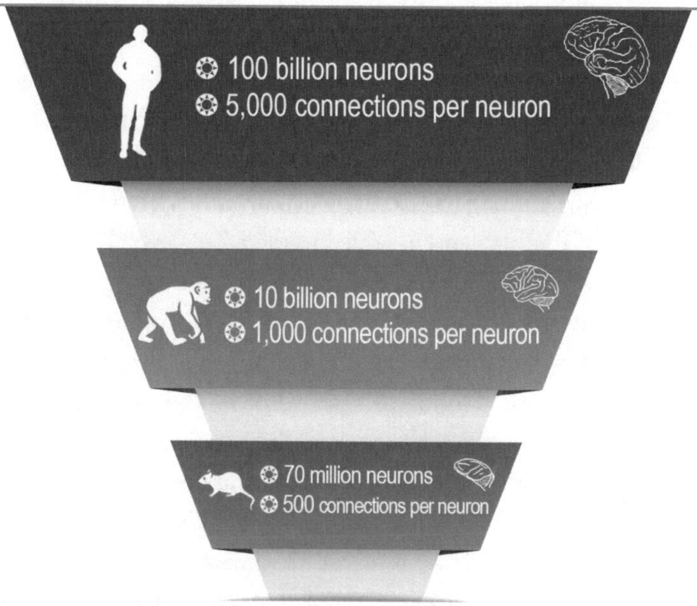

Humans have the highest *Species Intelligence/Adaptability Factor* because we have the most neurons and connections.

in the human brain stem. As Dr. Paul MacLean found, this reptilian brain not only held *species learning* that provided for vital body functions such as digestion, reproduction, circulation, and breathing—which was much more sophisticated at processing environmental stimuli than the fish brain—but also, at this time the first signs of a cortex appeared: a tiny one-cell-deep layer of neurons that sits atop this reptilian brain. Though minute, this ancient cortex amplified the reptile's ability to store small bits of survival information—such as the location of the closest watering hole or the hottest rocks—which allowed the reptile to suck up the sun's energy. This layer of cells is so small that up until recently, many neuroscientists denied its existence.[15]

Two hundred million years ago, nonplacental mammals made their appearance. With these non-womb-bearing mammals, evolution upped brain complexity by adding a group of structures we now call the limbic system (MacLean's *paleomammalian brain*). The limbic system provided not only more intricate processing of external stimuli but also the structures that allowed for animal emotions. Emotions increased the *Individual Adaptability/Intelligence Factor* of these early mammals dramatically because they now could make refined distinctions between what gave them the most pain and the most pleasure.

The importance of emotions in accelerating the learning process cannot be overestimated. They provide powerful criteria for the brain to decide which information should be stored and which should be ignored (more on emotion's role in learning in Chapter 14). These ancient mammals also had a second layer of flexible neurons added to the cortex, which allowed them to store many more bits of survival data at faster rates than reptiles.

Around sixty-five million years ago, with the appearance of the first mammals who carried their young in the womb, another important learning threshold was crossed. During this period of evolution, four new layers of neurons were added to the ancient cortex, dramatically increasing complexity, the speed of information processing, and the amount of emotionally charged information that could be stored for later use. With a total of six layers of malleable neurons, which we now call the neocortex, animals could make even more refined approach-and-avoid decisions. The difference between the speed and storage of the ancient two-layer cortex and the modern six-layer neocortex is profound, like comparing an ancient Chinese abacus and a modern supercomputer.

One reason humans are considered to have the highest *Species Adaptability/ Intelligence Factor* of any species is that our brain evolved to have the most complex configuration. Consider the brain of mice, apes, and humans: The neocortex of a mouse has seventy million neurons, each averaging 500 connections per neuron; our nearest cousin, the ape, has approximately ten billion neurons, averaging 1,000 connections each; and a human has 100 billion neurons, averaging 5,000 connections per neuron. It's been estimated that this complexity makes the human brain 40,000 percent more powerful than a mouse's brain and 4,000 percent more powerful than an ape's.[16]

Prefrontal Lobe Increasing Our Individual Adaptability/ Intelligence Factor

Humans' advantage over animals in learning isn't limited to having a dramatically more complex neocortex. Evolution's march toward a more complex brain took a quantum leap around 50,000 to 100,000 years ago when a large bulge appeared not in our biceps but at the front of our heads.

In 1868, railway workers cutting through a hillside in southern France

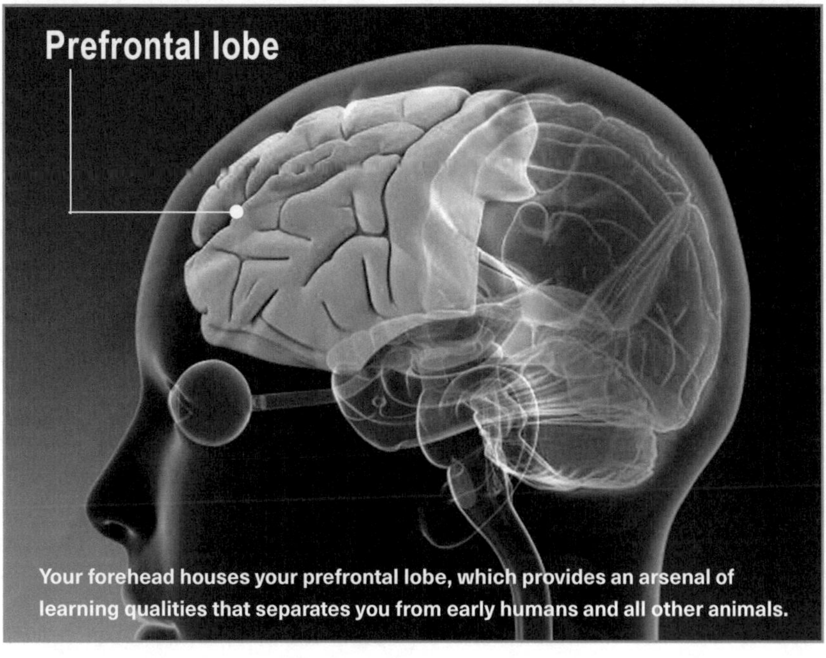

Your forehead houses your prefrontal lobe, which provides an arsenal of learning qualities that separates you from early humans and all other animals.

came upon four prehistoric bodies, which looked, unlike any other previously found prehistoric human remains.[17] What fascinated scientists most about the Cro-Magnon (named for where they were discovered) was the location of the brain in the skull. It had elevated in the head and expanded in the area above the eyes. Cro-Magnon's change in brain shape had forced the front of the skull outward, while all other prehistoric humans—Australopithecus, Homo habilis, and Homo erectus—all possessed foreheads that sloped away from the eyes toward the backs of their heads.[18]

This bulge in the modern human brain, which we call the prefrontal lobe, provided modern humans with an arsenal of learning qualities that would separate us from early humans and all other animals. The collection of abilities includes insight, abstraction, self-awareness, sympathy, empathy, altruism, anticipation, association, metaphor, judgment, sophisticated future planning, analysis, sequencing, large-scale cooperation, creative thinking, intuition, enhanced social skills, expanded working memory, big-picture thinking, speech, and the mediation of our basic animal emotions. With underpowered prefrontal areas, many animals have so little self-awareness that they can't recognize themselves in a mirror and so little creativity that they can't even untangle themselves from a leash.

Steven Mithen, in his respected book *The Pre-history of the Mind*, pointed to a lack of communication between mental structures as the major block to rapid social and technological advances in animals and early humans.[19] Modern human's complex prefrontal lobes removed this block. With rich feed-forward and feedback loops with virtually every part of the brain, these upfront structures can act as a control and coordination center. Because of this executive function, the prefrontal lobes are identified as the seat of human consciousness.[20]

Changing Our Environment to "Fit" Us

Before the appearance of the frontal lobe, the highest level of adaptability/intelligence most animals could achieve was to alter its behavior to fit a specific environmental challenge. With the more complex human frontal lobe, not only are we able to fit ever-changing environments effectively and rapidly, but we can also manipulate our environments to fit us. The frontal lobe has allowed humans to develop innovations such as the plow, the city, medicine, the automobile, the computer, the internet, space travel, and so on. The shift from merely adapting our behaviors to fit changing environments to manipulating the environment to fit us is directed by our more complex prefrontal lobe.[21]

The test that anthropologists use to judge the intelligence of our human ancestors is tool use. The more sophisticated the tool use, the higher our prehistoric relatives' *Adaptability/Intelligence Factor* was judged to be. To get an idea of how much the prefrontal lobe advanced humans' *Adaptability/ Intelligence Factor*, it's enlightening to look at tool usage before and after the appearance of the frontal lobe.

While the human branch on the evolutionary tree is estimated to have started six million years, the first sophisticated tools for cutting didn't make a widespread appearance until 2.5 million years ago with Homo habilis (literally, "handyman").[22]It took Homo habilis about fifteen blows with hard rock to shape a primitive cutting tool. A million years later, Homo erectus was using twenty-seven blows to make a sharper version of the same tool. And 1.4 million years later, early Homo sapiens were making a chopping tool with 111 whacks.

During this time, while humans were learning faster than other species, our brain expanded from $387cm^2$ to $1,200\ cm^2$—but in this case, a size increase was not enough. Without a prefrontal lobe, humans were learning how to add a chip to sharpen a tool at the unbelievably slow rate of one chip every 21,000 years—definitely not a learning society. Mary Leakey, the matriarch of archeology, said she got sick of working on tools from Homo erectus because, for a million years, each generation kept making the same hand ax over and over again. She states that pre-prefrontal man

must have been "a very dimwitted fellow."[23]

Then, between 100,000 and 35,000 years ago, up sprang the modern prefrontal lobe, and learning, intelligence, adaptability, and environmental manipulation really took off. As an example in only 5,000 years—that is, in only 25 percent of the time it took our ancestors to learn that one more whack with a rock would make a sharper cutting tool—humans learned how to create the wheel, the plow, the Agricultural Revolution, the Iron Age, the printing press, the Industrial Age, space travel, and the internet. In effect, what the highly complex human neocortex and expanded prefrontal lobes have allowed humans to do is make the leap from unconscious to conscious adaptation.[24] This leap gives us the highest *Individual Adaptability/Intelligence Factor* of any species on Earth and is driving the phenomenal doubling rate of information that we are now experiencing. This is one reason why to increase our individual ability to adapt to the real world and successfully and safely manipulate our environment, our learning systems must focus on developing the function of our prefrontal lobe.

Unfortunately, our emphasis on testing has helped push the most important values made possible by our expanded frontal brain areas—such as compassion, empathy, spirituality, big-picture thinking, intuition, insight, innovation, and personal meaning—out of our learning systems.[25] Our memorization and test-dominant educational systems primarily focus on the parts of the brain that store miscellaneous facts for very short periods, overstimulating these areas and under-stimulating our most-evolved frontal brain structures.

Now that we have grasped a little of the history and big picture of how our *Species Adaptability/Intelligence Factor* and our *Individual Adaptability/Intelligence Factor* influences our learning, in the next chapter let's go a little deeper and look at the changes that must take place in your brain before accelerated learning and behavioral change can actually take place.

Cracking the Learning Code

Your genes learn slowly, your brain cells learn quickly. Genes aren't capable of split-second adaptations. The genetic changes that can increase a *Species Adaptability/Intelligence Factor* take a phenomenally long time to be selected into your genetic makeup. Fortunately, unlike your genes, your brain cells can learn and adapt by changing their structures in very short time spans that range from seconds to days.

7

What Changes Must Take Place in Your Brain Before You Learn?

Memory itself is not a thing. It is a process of neural network activation.

—J. Allan Hobson, Harvard neuroscientist[1]

So, if humans have the most evolved *Individual Adaptability/Intelligence Factor*, what specific biological changes must take place in your brain before your *Individual Adaptability/Intelligence Factor* increases?

This question is crucial because it allows us to look inside your brain and get a visual representation of what the goal of any personal or institutional learning or behavioral change system must be. Without an understanding of what specifically creates learning and long-term memory in your own brain, you have little hope of developing learning systems that effectively increase the *Individual Adaptability/Intelligence Factor* in yourself or others.

The "Grandmother" Theory and the Hebbian Rule

Over fifty years ago, in what has become known as the *grandmother* cell theory of learning, scientists theorized that a single isolated neuron in the brain held a single isolated memory.[2] That is, one neuron held the memory of your grandmother; another, of your pet; another, of your house; another, the date Columbus sailed the ocean blue; and so forth. Under this theory of learning, brain cells were believed to be like tiny empty cups, each waiting to be filled with a specific bit of information. Not much change needed to happen in your brain for learning and memory to take place—a pre-existing neuron just needed to be empty.

This theory of learning helped fortify the belief, first put forth by John Locke in the 1600s and reenergized by the behaviorist school of psychology in the 1950s, that your brain is an empty vessel that can be filled up by memorizing of thousands of isolated details. Unfortunately, our memorization-dominant scholastic and corporate educational systems have also unconsciously embraced this outdated grandmother theory of learning.

Today, we know that learning and long-term memory don't take place at the level of just a single cell. In 1949, the beloved Canadian scientist Donald O. Hebb, in his immensely influential book *The Organization of Behavior: A Neuropsychological Theory*, put forth a theory of brain organization that led to the demise of the single-cell theory of learning.[3] His theory, now called the Hebbian Learning Rule, has become the basis for the neuroscientific, 21st-century view of how learning takes place in your brain. Hebb's theory says that, for learning and memory to occur, large numbers of neurons in your brain must all be activated at the same time, and this simultaneous activation produces changes in neurons, which creates a long-term memory. As Francis Crick, co-discoverer of the structure of DNA, said, "An individual neuron is, in fact, rather dumb. It is the intricate interaction of the many of them together that can do such marvelous things."[4]

Neuroscience now knows that the simultaneous activation of large groups of neurons, which Hebb called "cell assemblies," triggers an electrochemical process that produces a change in the existing architecture of the brain.[5] It is this physical change that increases your *Individual Intelligence/Adaptability Factor* and indicates that learning and memory have taken place.

The Biological Change That Creates Learning: The Short Story

At its simplest level, long-term memory formation process can be reduced to the simultaneous firing of large groups of neurons, which causes magnesium plugs in special receptors called NMDA receptors to be blown out, which in turn produces physical changes in the strength and shape of your neurons. This change in strength and shape signifies that learning and long-term memory formation has taken place in your brain. Unfortunately, traditional classroom-based learning systems, which are devoid of personal meaning and real-world experience, often have a limited impact because

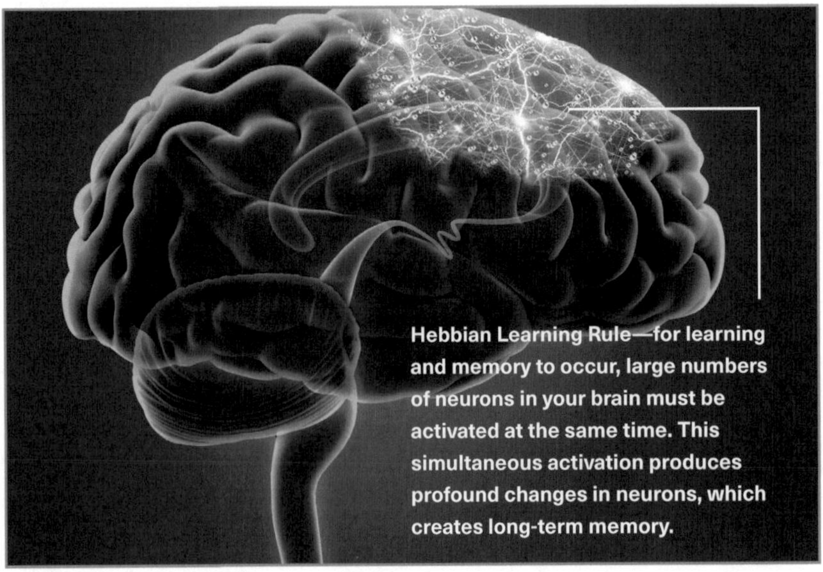

they fail to create the massive neural firing necessary for profound memory to be cemented. ***It is estimated that our stand-and-deliver learning formats fire only 5 percent of your neural structures, mainly in the linguistic areas of your brain.***[6] This is not nearly enough to ensure efficient long-term memory formation.

Accelerated learning pioneer Tony Buzan, in his groundbreaking book, *Use Both Sides of Your Brain*, equates *Ancient Learning Theory* methods with the old Chinese custom of binding girls' feet to keep them "dainty."[7] To Buzan, an educational system that focuses on limited circuitry of the brain damages your brain's ability to develop naturally. He writes, "Our minds, like the women's feet, have been 'bound' by the way we have misjudged and mistreated them, and they are therefore not naturally developed." Brain-based educators Renate and Geoffrey Caine point out how most educators believe that learning occurs due to memorization. In their book, *Making Connections*, they point out, "This is like looking at the moon and believing we have understood the solar system."[8]

FAST FORWARD *or* DEEP DIVE

What follows next is the long story of how learning takes place in your brain. I've done my best to keep this essential complex science as simple as possible, but if you only want the short story of how learning occurs, as outlined above, feel free to skip ahead to the next chapter. If you want a more robust understanding of science, then I suggest you skim or dive into the following sections. They will greatly enhance your ability to use this science for yourself and others.

The Long Story: Ancient Learning Theory Fires Limited Neurons

You've already seen that long-term memory is created when a meaningful environmental stimulus causes large groups of neurons to fire simultaneously. When this happens, glutamate, your brain's most abundant excitatory neurotransmitter, floods your brain and binds to a special receptor on neurons, called the NMDA receptor. Usually, an NMDA receptor on an individual neuron is blocked by a magnesium plug. But if glutamate attaches to the NMDA receptor at the same time the neuron is being stimulated by other neurons at other receptor sites, glutamate can blow the magnesium cork out of its NMDA socket. With magnesium no longer blocking the NMDA receptor, calcium, which has been waiting outside the cell, can rush into the neuron. This sets off a complex neurochemical process, which prompts the release of the CREB molecule; this, in turn, causes genes to express the proteins and enzymes necessary to cause the growth and strengthening of the synaptic receptor buds, which are the physical foundations of long-term memory.

One way to look at this process is to think of the magnesium plugs as boosters; they hold their power in reserve until an extraordinary response is required to emphasize special, important, and meaningful occasions and knowledge that needs to be remembered. You have a hard time remembering meaningless or boring information because it lacks the power to cause massive simultaneous neural firing. One of the main goals of any learning or behavioral change system, then, is to create a meaningful environment for the learner that causes massive neural firing, leading to large numbers of

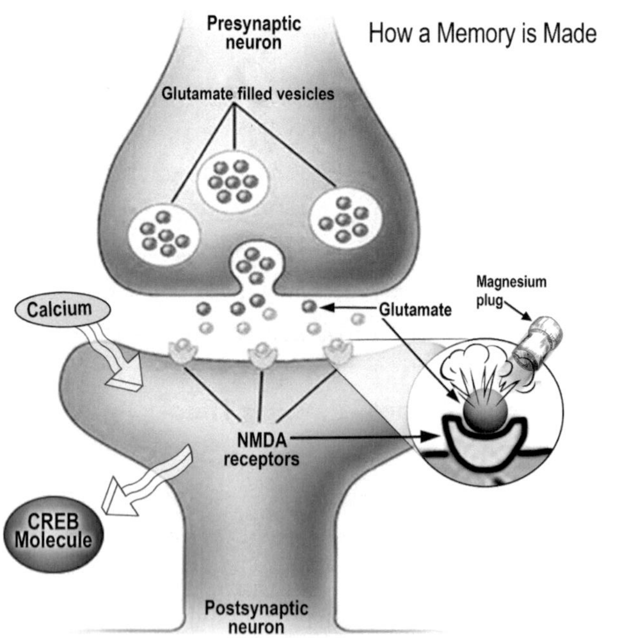

Profound learning takes place only when magnesium plugs are blown out of NMDA receptor sites.

magnesium plugs being blown out of NMDA receptors so that the neurological change, which is the basis of increasing your *Individual Adaptability/Intelligence Factor,* can more easily take place.

Over the last several years, we've learned a lot about the power of the NMDA receptor and the CREB molecule in the process of long-term memory formation. Years ago, researchers started to call the NMDA receptor, the "Doogie Howser receptor site."[9] *Doogie Howser, M.D.*, was a popular TV show that ran from 1989 to 1993. It featured the genius hero Doogie (Neil Patrick Harris), who graduated from college at the age of twelve and became a noted doctor by fifteen. Princeton's Joe Tsien genetically altered mice, which he called "Doogie mice," in a way that increased the numbers of NMDA receptors on each neuron.[10] Tsien found that, with more NMDA receptors available for glutamate to bind to, these mice became tiny, four-legged geniuses that had dramatically increased abilities to learn and remember.

We now know that if NMDA receptors aren't functioning correctly, memories can't be efficiently formed. What is the reason we often have poor memory or even blackouts after a night on the town? The excess consumption of alcohol interferes with the magnesium plug's ability to be blown out of the NMDA receptors. This is the reason alcoholics often can't remember the fight they got into in a bar or how they got home. New research also indicates that NMDA receptors in Alzheimer's patients are hypoactive—that is, their performance is slow and sluggish, thus inhibiting memory formation.[11]

On the other hand, in an epileptic seizure, NMDA channels are wedged open too long, which makes long-term memory of the pattern of the seizure quickly hardwired.[12] This jamming open of the NMDA receptors allows the seizure to be recreated over and over again with less and less stimulation. In other words, the brain has learned how to have seizures more easily.

CREB Molecule: Shifting Short Term to Long Term Memory

Research has also shown when large numbers of neurons fire, CREB proteins are expressed, which help shift short-term memory into long-term memory through the creation, growth, and strengthening of synapses.[13] Eric Kandel, professor of physiology and psychiatry at Columbia University, and director of the Kavli Institute for Brain Science, said, "Remembering something in the short-term uses proteins already there, but to shift memory into the knowledge base you draw from every day… you need a new set of synapses."[14]

Research at Cold Spring Harbor Laboratory in New York has shown that by increasing CREB protein in a fly's visual neurons, the visual information goes directly into the fly's long-term memory, effectively giving it a photographic memory.[15] Tim Tully, a geneticist at Cold Spring, said, "CREB functions as the gate that determines when an experience will be transferred into long-term memory."[16]

CREB is a transcription factor—meaning a protein that causes some genes to spring into action while inhibiting others from acting at all. There are two types of CREB molecules involved in the learning process: the CREB activator molecule, which turns on genes that express proteins and enzymes that cause neurological strengthening and growth; and the CREB repressor molecule, which stops genes from expressing the proteins and enzymes necessary to encode memory. Geneticists believe that the CREB repressor molecule remains dominant when information lacks meaning or is uninteresting. As Dr. Alcino Silva, a neuroscientist who led much of the CREB research at Cold Spring, said, "Memory is not about storing information; it is about storing useful information."[17]

You can look at the CREB repressor as a memory filter. It dominates until something meaningful is presented to your brain. As soon as your brain receives information that it judges will increase your ability to survive and thrive, the magnesium plug is blown out of your NMDA receptor, and calcium rushes into the cell, which in turn prompts the CREB activator. CREB activator molecules then remove the CREB repressor allowing genes to express the proteins that build and strengthens synapses, which in turn helps cement long-term memory.[18] No wonder we forget so much of the information we tried to memorize in school—it failed to stimulate the CREB activator molecule.

Massive Neural Firing Births Your Neurons

Extensive simultaneous neural firing caused by meaningful stimuli can also prompt the growth of new neurons. The hippocampus (which you'll become very familiar with as this book unfolds) is located deep in the center of your brain. It's instrumental in shifting information from your working memory into long-term memory and in placing specific incoming information into the correct concept networks. The hippocampus ensures that when you encounter a new make or style of car, that information is placed in the specific neural networks that deal with cars, not the networks that deal with fruit or dogs.

Up until about thirty years ago, neuroscientific dogma dictated that the brain couldn't produce new neurons. Research at various research centers now confirms massive neural firing causes the formation of new cells in the hippocampus and other brain areas.

Neurons that have been stimulated at multiple receptors by other neurons also produce growth tonics called *Brain-Derived Neurotrophic Factors* (BDNF). When massive brain cell firing takes place, these special growth-producing factors are sent back to the neurons that initially produced them. This causes the strengthening of existing synapses and sprouting of new ones. The release of this elixir of synaptic growth not only strengthens the bonds between your sending and receiving neurons but also supports neurogenesis—the growth of new brain cells—and their longevity.

BDNF is very much like a protein fertilizer for brain tissue. Anything you do to stimulate it will accelerate the growth of neural tissue. Not only does growth in new brain cells and connections help make you a more efficient learner, but it also provides power to make you more proficient at adapting to an ever-changing world.

Education Starts to Embrace Physical Movement

A growing body of research indicates that learning and memory increase if you move when you're engaged in the learning process. Why? Because movement stimulates the activation of BDNF. Because of this, some forward-thinking schools have installed stationary bikes, exercise balls, and elliptical machines in the classroom. For example, a study with young adults in Ireland showed that exercise increased the function of the medial temporal lobe, which enhanced success on face/name-matching tasks.[19] Additionally, in a study at the University of British Columbia, scientists found that exercise increases the size of the hippocampus, indicating learning is taking place.[20] Another small controlled study found that grades rose 16 percent for students who did very low exertion movement while learning (versus those who did not move while learning).[21, 22]

What often isn't well explained is the evolutionary roots of *why* the BDNF fertilizer is released when we move. Consider that for six million years, your ancestors, from whom you got your present genetic makeup, did virtually all their learning not in a classroom sitting stationary, but while moving toward prey or away from predators, moving away from pain or toward pleasure. Because our genes are simply units of past learning sent forward (see Chapter 4), the human genome is exquisitely coded to learn most effectively when you're experiencing physical movement. It's your genetic heritage.

Traditional education is starting to embrace more of the science about learning and behavioral change. However, creating natural, profound, joyful learning environments is so much more intricate than just sticking some exercise machines in the classroom or standing up when you use your computer. These are steps in the right direction, but they're only a tiny fraction of where we need to focus our attention to build natural learning and behavioral change systems that dramatically, quickly, and easily increase our *Individual Adaptability/Intelligence Factor.*

Cracking the Learning Code

The simple rule is... the more brain cells that fire, the more effective your learning. Comprehending this rule is essential to increasing the speed and efficiency of learning and behavioral change. Learning systems that fail to create massive neural firing in learners' brains, are systems that are designed to fail.

8

There Are 7.5 Billion Types of Intelligence—You Have A Special One!

But once we realize that people have very different kinds of minds, different kinds of strengths—some people are good in thinking spatially, some in thinking language, others are very logical, other people need to be hands on and explore actively and try things out—then education, which treats everybody the same way, is actually the most unfair education.

—**Howard Gardner**, *Harvard professor, author, and intelligence researcher*[1]

By relying on *Ancient Learning Theory*, modern scholastic and corporate educational institutions design learning systems as if there is no variation in the way our brains process information. Therefore, they proceed as if all brains learn in precisely the same manner. Science is proving this belief as a myth. Your brain is not like your neighbor's. It is different, and that variation is so vital to our species and individual survival that no two humans—not even identical twins—share the same genetic makeup or the same brain plan. Therefore, no two humans can learn in exactly the same way. The variations we see among the 7.5 billion humans on Earth have their genesis in that none of us share the same configuration of the three billion base pairs strung out on the double helix of our DNA.

This genetic variation ensures that you're brilliant at something your neighbor is not. Genetic variation builds unique brains that have unique gifts and allows each of us to have a distinctive way of surviving and thriving in the world. Each of our unique *Individual Adaptability/Intelligence Factors*—that is, gifts—then combine to create a species that has a spectacular reservoir of surviving and thriving options. Although most good teachers and trainers do indeed see the brilliance in each of their students, traditional teaching methods instruct us as if we have the exact same biology. And if you don't learn like the way the system wants you to learn, you may be seen as not "smart" or not gifted. Nothing could be further from the truth.

Your Unique Genius Represented by Your Unique DNA

Each of us has slight-to-great genetic differences in our DNA—the arrangement of the nucleotide bases of adenine, thymine, cytosine, and guanine (A, T, C, and G). These unique genetic variations build unique brains, which in turn ensures that all of us will think, act, and learn in our own ways. This is why we say from the biological perspective there's no one form of intelligence, as the "IQ way of thinking" maintains, but 7.5 billion forms of intelligence, one for every person on Earth—including you.

Genetic and neurological variation ensures that each of us will process information in different ways. This variation is the source of your unique genius. Therefore, designing and implementing learning systems as if we all learned in the same way, will continue to produce less-than-successful outcomes. Institutions that try to create learning as if everyone had identical brain networks fail to account for the biological diversity among learners—no wonder traditional learning systems are more painful and inefficient than need be.

Genetic variation is the stuff of life, and modern biologists and geneticists have a deep appreciation for it. Darwin was one of the first scientists to understand that variation is the fuel of life. Today, scientists universally recognize that without variation, there can be no species change; without species change, there can be no adaptation, and without adaptation, there can be no life on Earth.

Lack of variation within a species is the kiss of death, which is why only about 1 percent of all the species that ever made an appearance on Earth still exist today. The major reason that species like trilobites and saber-toothed cats failed to make it through the sieve of evolution is that they lacked enough useful variations in the language of their genes (A, T, C, and G—strung out on the double helix of their DNA) to adapt to fit into new environments. We now know that in changing environments the species with the most useful genetic variations in body structures, immune systems, nervous systems, and behaviors have the greatest chances of survival.

Slight-to-great genetic variation in the genome and brain of every member of a species ensures that, when confronted with similar environmental pressures, not all members of that species will adapt in the same manner. This is a wonderful survival advantage. If humans didn't have genetic variations in their immune systems, diseases like the bubonic plague or scarlet fever would have completely wiped us all out. This genetic variation also ensures

Lack of variation within a species is the kiss of death, which is why only about 1 percent of all the species that ever made an appearance on Earth still exist today.

unique neural circuits within each member of the human species, which in turn provide for unique strategies of behavioral adaptation and learning in each of us.

If the first small group of our human ancestors all had the exact same brain structures and their water source ran dry, they might all have taken the same route and not spread out to find a new water hole. If they all took the same route and that water hole was also dry, that could have been the end of our species! This wouldn't have been a great survival strategy at all. We only need to look around our offices, our neighborhoods, or even our immediate families to confirm that unique brain plans ensure that none of us have the exact same form of intelligence or the exact same strategy of adaptation.

Variation: The Gift That Supports Thriving

Our variation of brain plans not only ensures our collective survival but also our collective thriving. The solutions to the monumental problems that have confronted humankind are solved by many people, taking many different routes to resolve the same dilemma. Lots of brains processing information in divergent ways have allowed us to put men on the moon, create the internet, and produce treatments for HIV. It's our diversity of thinking that allows humans to produce such wonderful things. The real value of teamwork comes not from the similarities of how our brains process information but from the differences. In the end, it's the variations in neural processing of our 7.5 billion human brains that are driving today's incredible growth rate of knowledge.

One way to look at each of the 7.5 billion unique brains on Earth is as tiny droplets of variation fuel that the global brain of our human species uses to propel itself toward success. In the end, it can be said that the human species has not one form of intelligence to help it adapt to changing environments, but 7.5 billion different types of intelligence. Any learning or behavioral change system that hopes to add value to our world must embrace, not ignore, the variation among human brains, which is the key to our species' survival.

Neurological variation also creates a gift of unique meaning to each of us. Distinct brain differences dictate that we all find different things meaningful. And even while many of us derive meaning from the same sources—whether it is music, cooking, business, sports, etc.—our individual genetic and neurological divergences ensure that each of us will pursue a somewhat different path within that interest.

The brain is a dramatically complex system, where everything is either directly or indirectly connected to everything else. In this complex environment, even the minutest differences in a few of our 100 billion neurons or some of our 10^{27} neural connections can create vast differences in brain processing and what we find meaningful in life.

As complexity science informs us, the small flap of a butterfly's wing in Tokyo can set off a tornado in Kansas. So too, small differences in brain structure can cause dramatically different outcomes in what we find meaningful. But research shows that, in general, our neurological differences are not minute. Total brain size in individuals can vary by up to 50 percent.[2] Individuals with normal vision can have up to a 300 percent difference in the number of neurons in the visual cortex.[3] And in a fascinating discovery revealed by PET and MRI scans, even identical twins have vastly different neurological compositions[4] (see page 114).

These differences in our neurological structures ensure we will all adapt to our world and find meaning in our lives in very different ways. No two businesspersons, psychologists, computer programmers, or bricklayers pursue their lives or their crafts in exactly the same way. This means that no one will ever do exactly what you do… in exactly the way you do it. This is a biological gift not only to you but to our world!

EVERYONE IS A GENIUS
No two brains are the same; it's this uniqueness that gives each one of us a special genius that no one else possesses.

Education Ignores Biological Diversity to Its Detriment

It can be said that the idea that we all learn and remember in the same manner has been part of the fabric of education since Plato's Academy and Aristotle's Lyceum were influential over 350 years before Christ. In America in 1892, the concept of learning sameness was formalized by the National Education Association when it created the powerful Committee of Ten, a group of educational leaders who laid down a strict framework for American schools. The committee did not mince words: "Every subject, which is taught... should be taught in the same way to the same extent to every pupil so long as he pursues it, no matter what the probable destination of the pupil."[5] This policy of treating every brain in the same way still reverberates within the halls of education today, yet from a biological standpoint, it is completely wrong.

Embracing Neurological Diversity

To create successful learning and behavioral change systems, we must embrace three things regarding brain diversity.

1. Your Individual Variation Must Be Nurtured

First and foremost, it's the variation in our individual adaptations to similar environmental pressures that affords our species such a high *Adaptability/Intelligence Factor*. The goal of any learning system must be to nurture, not ignore, the neurological diversity that has led to humankind's technological, spiritual, and social advances. Any effort that limits variation in the way our species responds to changing environments, whether it be a conscious effort like eugenics or unconscious like traditional educational systems, inhibits our overall *Species Adaptability/Intelligence Factor* and thus the individual's ability to survive and flourish.

2. Unique Brains Create Unique Differences in What We Find Meaningful

Your unique neural circuits provide the basis of what you find personally meaningful in life (as you'll see in more detail in Chapters 12–14). Your unique brain plan becomes the primary foundation upon which new information is selected into your long-term memory bank. By ignoring the fact that the networks that hold meaning are unique to each of us, we inhibit the speed and efficiency of the learning process for us all. Without stimulating your unique *Meaning Network*, essential neurotransmitters, which are vital for creating joy in your life and are necessary for forming long-term memory, aren't activated. This is why you have forgotten most of the stuff you memorized in school that was not meaningful to you.

3. Our Educational System Can Quash Personal Meaning

Traditional learning systems that ignore the unique structural organization of each brain and proceed as if all brains learned in the same manner, may kill your ability to find meaning in your life. This is an unintended consequence, which I'll delve into in later chapters, but for now, consider this. The brain receives about twenty watts of power in the form of glucose and oxygen via the carotid and vertebral arteries.[6] During our first twelve to sixteen years, when the brain is at its most malleable and developing at its fastest rate, traditional educational systems primarily keep this energy flow directed at the circuits that process detailed linguistic information. If the unique neural networks that code for what you find personally meaningful in life aren't in these linguistic areas, the brain regions that house your unique "gifts" can be deprived of the vital energy flow they need to develop fully, and thus their growth may be stunted.

In effect, learning systems that treat all brains the same are in danger of limiting your ability to develop and strengthen the very networks upon which personal meaning is built. This redirection of energy away from your unique neural networks may be one reason that when many of us leave learning institutions, we are not energized and confident but exhausted and confused.

How Your Genes Built Your Unique Brain and Individual Adaptability/Intelligence *Factor*

These next few sections go a little deeper into the science, delivering a more profound understanding of why we must stop building educational, training, and behavioral change systems as if all brains were the same. As we have been discussing, research shows that the human genome is pre-programmed to automatically vary the three billion base pairs that are strung out on the double helix of your DNA, which builds unique brains, which in turn gives each of us unique gifts and unique ways to adapt to our world.[7] I hope you're beginning to see that because of genetic and neurological variation, you have more in common with those mutant comic book heroes from Marvel like the X-Men and Black Widow than you might have imagined. So how do those variations in your DNA, also called mutations, arise?

Let's Start with Sex!

One of the reasons we're programmed to desire, get pleasure from, and constantly think about sex is that it's a major source of the dramatic genetic variation that supports life itself. There is an amazing amount of genetic scrambling that goes on during the making of sex cells. During sperm and egg production, called meiosis, large segments of A, T, C, and G bases—some short, some long—get randomly snipped from one part of your DNA's double helix and randomly reinserted into another part.[8] This random snipping and inserting are like thoroughly shuffling and reshuffling a deck of three billion cards.

As if this genetic shuffling during the creation of the sperm and egg wasn't enough, another level of shuffling occurs at conception. The three billion scrambled base pairs in your father's sperm are mixed with the three billion scrambled base pairs in your mother's egg, producing a zygote, which is the single-cell stage before the multicellular embryo is formed. The result is a distinctive combination of three billion base pairs, unique from either one of your parents. Meiosis and fertilization create so much variation that an amazing 70 trillion combinations of bases are possible in the resulting offspring.[9] With this high level of variation for each conception, is it any wonder that none of the 7.5 billion of us, even brothers and sisters, will ever have a chance of acting, thinking, or learning like any of the other 7.5 billion?

FAST FORWARD *or* DEEP DIVE

If you feel like you have a good general understanding of the value of variation in increasing our *Species and Individual Intelligence/ Adaptability Factors*, you can skip the next few short sections on Jumping Genes and on how mutations in your DNA are opportunities, not mistakes, and go directly to the section titled "Danger: Focusing on Similarities, Not Uniqueness."

Mutations: Opportunities Not Mistakes

The definition of a mutation is the unexpected and chance rearranging of genetic material caused by the adding, dropping, or rearranging of one (point mutations) or more (chromosome mutations) base pairs on a sequence of DNA, which in turn changes the meaning of the information conveyed.[10] The word *mutation* is often taken to mean "mistake." But when an evolutionary biologist hears that word, they think "opportunity." Although mutations can cause diseases such as Huntington's and schizophrenia as well as birth defects, they are also the source of new characteristics and behaviors that support individual and species options for surviving and thriving.

The mutations in base pairs are what drive evolution by providing the small to large genetic differences that allow a species to try out completely unique physical and behavioral characteristics. As Darwin suggested, the variations that aid survival are selected into the species, and those that harm survival are selected out.[11] Physical characteristics such as the eyeball, the thumb, and upright locomotion, as well as behaviors such as bonding, mating, and imitating, were initially random mutations. These successful mutations or opportunities were eventually selected into the whole species because they helped us better fit our world. Research now indicates that to ensure the survival of our species, an amazing 40 percent of your genetic makeup has been mutated from its original form.[12]

Small Changes: Big Variations

Although large-segment mutations can cause alterations in genetic information, even a single-base mutation can cause profound changes in meaning. The genetic alphabet of A, T, C, and G form three-letter words called codons. A codon is a sequence of three DNA nucleotides that corresponds with a specific amino acid. A one-letter change in one three-letter word codon can have weighty consequences in the meaning of a genetic sentence, which in turn can cause impressive variations in the way your genes express and the way your brain processes information. Just as only one letter makes a huge transformation in the meaning of the sentences "Get me a pet" and "Get me a jet," so it is with genes.

Jumping Genes: Your Evolutionary Toolbox

It's been clear for a long time that genetic mutations were causing changes in the genome that drove evolution, but the exact genesis of most of these mutations is only now being uncovered. Recent studies at Johns Hopkins and other leading research institutions are shedding light on transposable elements, or "jumping genes."[13] These are intriguing segments of bases that create mutations in stretches of DNA by doing incredible things. They can spontaneously jump from one segment of the double helix and reinsert themselves into another, or they can jump from their initial position and totally knock out another section. They can even jump up, spin around 180 degrees, and reinsert themselves backward. Jumping genes provide for a profound source of genetic mutations on a scale unimagined before Barbara McClintock uncovered their existence over sixty years ago.

In *Wisdom of the Genes*, Christopher Wills calls the mobile genome an "evolutionary toolbox" that facilitates the evolutionary process itself.[14] Jumping genes validate Dobzhansky's claim in the 1960s that genes produce their own variation,[15] and they appear to be the reason that Adams' bacteria clones we saw in Chapter 5 were able to mutate so drastically from their original form.

Genetic differences between individuals ensure that neurons and their connections will be like fingerprints: none will ever be the same in any two of us on the face of the Earth. Gerald Edelman, Nobel Prize winner and developer of the theory of neural group selection, says even if we "were to number the branches of one neuron and to number in corresponding manner neurons it touched, the numbers would not correspond exactly in any two individuals of a species—not even in identical twins or in genetically identical animals."[16]

Danger: Focusing on Similarities, Not Uniqueness

Your brain is one of the most complex systems in the world, with more than 100 billion neurons that have over 100 trillion connections. These connections have a relationship with another estimated 85 billion special support cells called *glial cells*.[17] This creates an environment in which the complexity is almost unimaginable to the very instrument that houses it! Sorting out exactly which neurons and their connections do what is complicated by their microscopic size: 60,000 neurons possessing thirty-four miles of DNA and 300 million connections to other neurons can fit on the head of a pin![18] This complex world is made all the more difficult to examine by the fact that working with the brain is like working on a wet blob of Jell-O.

Figuring out which brain structures do what, in this highly complex, gooey, sticky environment, has proven to be a very drawn-out and laborious process—and it isn't nearly complete. Since the dawn of the 21st century, hundreds of new discoveries about the structure and function of different brain areas have been made every year. To facilitate understanding, the anatomical charts, PET scans, and MRIs that you see in the popular press and even in neuroscience textbooks emphasize the similarities between brains, giving the wrong impression that all brains have the same configuration. As Michael Posner and Marcus Raichle point in their award-winning book, *Images of Mind*, "When human brains are normalized into standard coordinates, we learn a lot about common areas of activation, but little about detail differences among people."[19] When budding neuroscientists perform their first brain autopsies, they're often taken aback to find that the brains they're dissecting are very different from the anatomical charts and images they studied in the classroom.

The Sad History of Identifying Brain Uniqueness

No one is clearer about the differences in brains that than those who cut them open in the effort to heal. It was during the 1940s and 1950s, as surgical procedures improved to the point where greater numbers of brain surgeries could be performed that neurosurgeons found that brains were too different to rely on the "one chart fits all" mentality to guide their hands during surgery. In these pioneering days, neurosurgeons found they needed to focus on the differences between brains, not similarities, or risk maiming or killing their patients.[20] Surgeons eventually started to develop techniques to ensure that, during surgery, they weren't going to cut out something that should be left in!

One of the first sad indications that all brains were unique came from brain surgery that inadvertently damaged patients' abilities to speak. In the early 20th century, neurosurgeons believed that speech capacity was controlled by the left side of the brain. By mid-century, advanced surgical techniques allowed more brain operations to be performed by surgeons following anatomical charts. But often after the operation, surgeons would be shocked to find they had accidentally wiped out the patient's ability to verbally communicate. Their charts were wrong!

In response to this problem, Japanese Canadian neurologist Juhn Wada devised a test to determine which side of the brain-controlled speech.[21] The brain's main energy supply of glucose and oxygen comes from two carotid arteries. The right carotid artery supplies blood to the left side of the brain, and the left artery, to the right side. Wada's procedure consisted of injecting a short-acting anesthetic into one of the two arteries to see which one would affect speech. The Wada Test, as it's now called, revealed that the anatomical charts were indeed wrong and that 8 percent of the population had their speech capacity on the right side of the brain, not the left!

In another example, in the early years of modern neurosurgery, it was found that removing parts of the temporal lobe could cure some epileptic seizures.[22] In performing these surgeries, surgeons tried to avoid damaging the sensory, motor, language, and memory areas, which surround the area slated for removal. After the first few surgeries, it was clear that no two patients had the exact same brain structure. In one patient, a surgeon could cut out a small piece of cortex with little or no consequence, while cutting out what looked like the exact same section in another patient's brain would cause tremendous damage. After a successful surgery that helped solve their epilepsy problems, some patients would lose specific motor functions, experience a lack of feeling in certain body parts, have speech capacity damaged, or even lose complete sections of memory for categories such as tools, faces, and vegetables.

To solve the problem of how to operate on brains that were so different, surgeons started to map each individual brain before they operated. In a process called "opening," a hand-sized piece of bone was extracted, revealing the location of where the "temporal lobectomy" would be performed.[23] The patient received local anesthesia and was awake during the procedure. The surgeon then stimulated parts of the cortex with a mild, painless electric probe, which interrupted a specific brain function. With feedback from the patient, the surgeon could then map the areas of the brain to be avoided during the lobectomy. As more of these operations were performed and surgeons compared neurological data, they were amazed at the spectacular variation in human brain structures.

Brain Imaging Revelations

In the late 1980s and 1990s, the ability to see the differences between brains was accelerated by brain-imaging processes called PET (positron emission tomography) and MRI (magnetic resonance imaging). These scanning processes allowed brains to be mapped without actually cutting into them. Subjects could be asked to perform a specific function, and the parts of the brain that lit up on the scans could be mapped.

The unique way different brains process information is highlighted by studies conducted by Julie Fiez and her colleagues at Washington University in St. Louis. Fiez gave subjects the simple task of remembering four nonsense words while she took PET scans.[24] The scans revealed that each subject used different parts of their brain to perform the same task. In-depth neuroscience research points to differences in thinking processes among twins, caused by the brain map variations. Autopsies, operations, and scans now clearly reveal that no two brains, even those of identical twins, look or process information in exactly the same manner.

Microscopic differences in lower brain structures, such as the substantia nigra, locus coeruleus, raphe nuclei, and pituitary, which control important neurotransmitter and hormone production, can have a remarkable impact on the way we learn and interact with our

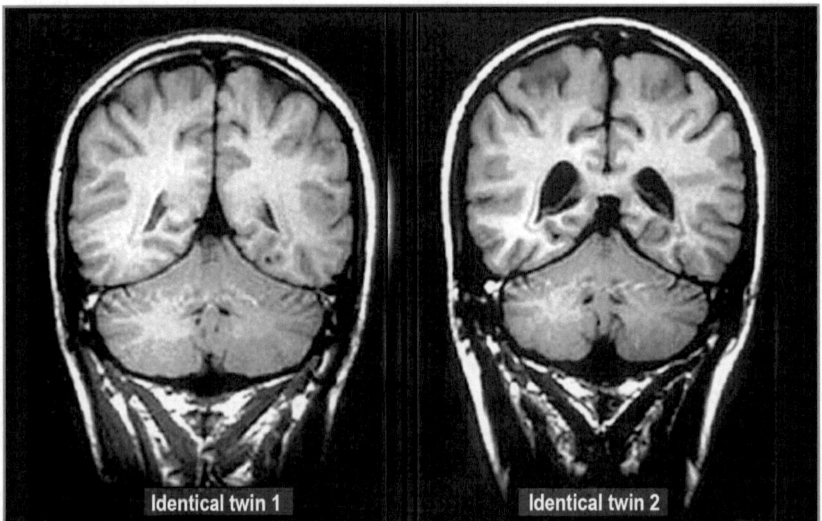

All brains are unique. as these brain scans of identical twins show: they are totally different from each other.

environment.[25] In a classroom setting, the person next to you may be able to sit completely still and focus on the blackboard for forty-five minutes, while you may not. One isn't smarter or a better student than the other. This difference occurs in most cases simply because of differences in the brain structures that produce neurochemicals that control meaning, focus, and stress—like dopamine, serotonin, norepinephrine, and cortisol. Thus, many people who've been diagnosed with attention deficit and hyperactive disorders merely have small variations in the structures that produce these neurochemicals. Most of the individuals who have difficulty sitting still in a classroom for hours simply require a learning environment that doesn't force them to sit motionless for hours on end and focus on someone talking. And it's this kind of research on our individual genetic and neurological variations that allow us to understand more deeply that each of us has a unique *Individual Adaptability/Intelligence Factor* and that each one of us, therefore, possesses a unique way of processing information and learning.

Cracking the Learning Code

There are 7.5 billion humans on Earth. Genetic variation ensures that none of us will share the exact same brain plan. Your unique brain plan is like a biological gift that makes you brilliant at something your neighbor is not. Unfortunately, our current corporate and scholastic learning programs proceed as if this genetic and brain variation did not exist and all brains learn in exactly the same way. This misconception causes undue harm.

9

Understanding Your Biological Preferences

Every student can learn, just not on the same day, or the same way.

—George Evans, cartoonist[1]

As we just saw, each of the 7.5 billion of us on Earth has a unique makeup of genes and brain tissue that give us a unique *Individual Adaptability/Intelligence Factor* that no one else on the planet possess. At The Institute, we have found that to help educators, business leaders, students, and parents more effectively comprehend and apply this concept of brain variation, an understanding of what we call the *Biological Preferences* is advantageous.

We debated long and hard before including this and the next chapters. The main advantage of understanding the different *Biological Preferences* presented here is not to encourage breaking students into groups that are taught in what their primary preference might be, like visual, auditory, spatial, and kinesthetic, as some suggest. This just makes the whole learning process way too mechanistic and is the main reason the psychometric research on teaching to a preference is so disappointing.

Ever since Harvard's Howard Gardner published *Frames of Mind, The Theory of Multiple Intelligence*[2] in 1983, scholastic and corporate educational systems have been trying to figure out ways to imbed the theory of multiple intelligences into the learning process in order to accelerate it. While the anecdotal evidence is very positive, the research proving that the theory works is very weak. The teachers who understand it and apply it rave about it. Yet, in general, the researchers who study its effect with scientific rigor find the evidence does not support the rave reviews at all. Why this discrepancy?

At The Institute, we believe this discrepancy exists because the simple yet disruptive idea that each one of us has a unique way to learn is vital to understanding why our existing learning systems aren't working and what to do about it. It helps educators more simply and easily appreciate the diverse biology of their charges. Those educators and trainers who gravitate to the *Biological Preferences* way of thinking see it as a simpler way to understand why the existing system cannot possibly account for our individual uniqueness, as discussed in the last chapter. And this alone makes a dramatic difference in how teachers and business leaders connect with learners. They no longer see people as similar machine parts, all to be molded in the same way, instead they begin to see them as unique souls that need unique treatment in order for them to thrive and prosper in the real world.

In the end, we decided that exploring our *Biological Preferences* is helpful because it will give you a different perspective, and clearly shows that we do all learn differently. Please understand that to provide learning environments that naturally create profound learning, knowing that each of us learns in a different manner, is only a tiny, tiny piece of the puzzle. It is not "the answer." The rest of the pieces of the puzzle are presented in the other chapters of this book that lay out the other elements of The Learning Code science.

Evolution Is Like a Sieve

So how does knowing your *Biological Preferences* fit into helping you understand learning and behavioral change? We can look too evolutionary biology for the answer. Evolution is like a sieve: over the millennia, nature poured in billions upon of billions of variations of body types and nervous systems in at the top, and only the 8.7 million species that best "fit" today's world made it through.[3] It is estimated that less than 1 percent of all the basic nervous systems and body plans that were tried out different species over the last three billion years survive today. But within each element of the general design of each species hides a profound level of diversity.

Take humans, for example. There is a general design for fingers, ears, and eyes, but within those general designs lies massive diversity. We all have fingers, but every fingerprint on earth is unique, we all have ear lobes, but all ear lobes are unique, and each of our eyes has an iris, yet each iris in the world is even more unique than our fingerprints. As we have seen, it is the same for genes and brains.

As anthropologist Stephen Jay Gould explained, the "surprising fact of life's history [is] marked decrease in disparity followed by an outstanding increase in diversity within the few surviving designs."[4] These general plans provide each of you with a very general modality in which you prefer to learn and adapt to your world. You see these general brain plans creating preferences in everyone you know. These preferences are not the only way you learn, but they make you predisposed for things and information that resonate with your preference. Some of your friends might have a musical preference and are drawn to music; others for numbers and are attracted to math; others for words and are attracted to literature. At The Institute, we've used the latest scientific research on the general brain plans to define the eleven main *Biological Preferences*, which you'll find in Chapter 12.

The Medical View of Structure-Function

The theory that humans have *Biological Preferences* that help us adapt to our world comes from many disciplines, including not only evolutionary biology but also psychology, education, genetics, medicine, and neuroscience. At The Institute, medical research has had the greatest impact on our uncovering and understanding *Biological Preferences.*

The structure of an organ dictates its function, so hands grasp, legs run, hearts pump blood, and lungs exchange carbon dioxide for oxygen. Because the brain is such a complex organ, it is only in the last fifty years that the major (macro) structure-function brain areas have been uncovered. On the microscopic level, because of advanced detection techniques, many more minor areas of structure-function are now discovered every year.

Research on these biological predispositions is in its infancy. But we've found that discussing the *Biological Preferences* and understanding which one yours might be helps bring greater clarity to those of us who want to

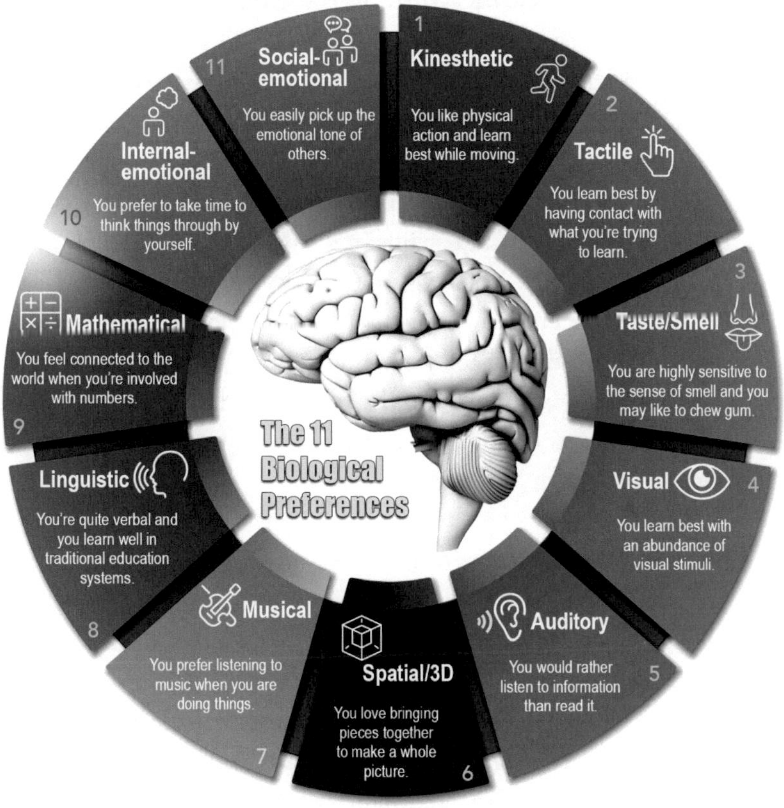

create profound learning and behavioral change in ourselves and others. Please note we are not suggesting that the following is a definitive list of *Biological Preferences*. This list is based on our research to date and what has been shown to work to accelerate and deepen the understanding of the learning process. Remember, because of genetic variation, there are really 7.5 billion types of intelligence.

It's essential to recognize that no single general brain design is better than any other. The survival of our species depends on our different individual *Biological Preferences* working together as one global brain to support the survival of the whole human species. As we saw in the last chapter, the different preferences are the fuel our species uses to enhance our group survival and success. But we can only reach our maximum species potential when we learn to work with, rather than against, others in our species with different *Biological Preferences*.

Your Brain Has Variations within Variations

It is important to note that while we tend to have a primary *Biological Preference*, that preference will vary slightly to greatly among everyone who possesses the same preference. This is because of the biological diversity among all 7.5 billion of us that we discussed in the last chapters. Therefore, even if you have a spatial, musical, or linguistic preference, it will hook up to all your other brain areas differently than anyone else on Earth. This makes every butcher, baker, or candlestick maker unique in the way he or she applies the craft and interacts with the world. As an example, while Mozart and Beethoven both had musical preferences, the individual variation within their preference and the variations in how their preference connected up to the rest of their brain, ensured that they would never create the exact same music.

The Source of Your Biological Preference

The only reason the brain came into existence and acquired its particular structures and designs is that these features contributed to the survival of the individual and the species. Evolutionary biologists present what has come to be called the "Swiss Army knife view" of how the human brain developed.

Over twenty-five years ago, in his influential book *The Modularity of Mind*,[5] Jerry Fodor proposed that the brain is a collection of domains, such as

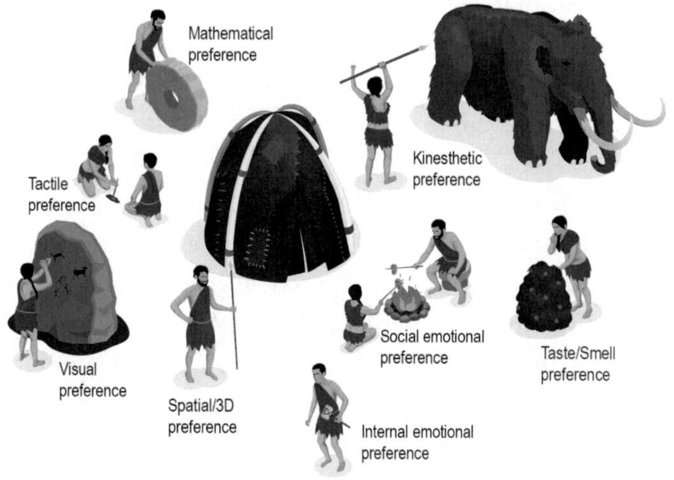

Our ancient ancestors lived primarily in small groups of twenty-five to fifty and for the welfare of the tribe it was vital for each member to process information a little differently from each other.

touch, hearing, and sight, with each modality selected over millions of years of evolution to help a species better adapt to ever-changing environments. Under Fodor's theory, evolution started off with touch and smell as two of the first blades of the knife and selected in more and more designs, such as social, emotional, and mathematical, for even more efficient surviving and thriving advantages.

For six million years, until the advent of farming 10,000 years ago, humankind and our direct ancestors lived primarily in small groups of twenty-five to fifty, in which it was advantageous for the whole tribe if each member processed information a little differently from everyone else.[6]

Individuals with a good sense of taste and smell would help decide which food and water sources were healthy and which were poisonous, and those with the kinesthetic preference were the first up in the morning, looking for the woolly mammoth to kill for dinner. If the individuals with the kinesthetic preference found no food, those with the spatial/3D preference could map out where the tribe last killed a mammoth three months ago so that they could go back and get some woolly mammoth jerky. The job of those who excelled socially was to keep the tribe together and limit confrontation, so they could survive and thrive with as much harmony as possible.

In this manner, an individual's preferred mode of processing information aided not only individual survival but also the success of the whole tribe. Just like the brains of our ancestors, each of the 7.5 billion brains on Earth today tends to process information in our *Biological Preference*.[7]

Genes Build Your Brain's Preference

As advances in uncovering the human genome are made every week, we're discovering more and more genes that code for our *Biological Preferences*. At MIT, Susumu Tonegawa and Matthew A. Wilson found a gene that controls spatial map building. Also, at MIT, Myrna Gopnik, working with a British family who had "specific language impairment," found that a single gene was primarily responsible for the grammatically incorrect speech.[8] At Harvard Medical School, researchers have discovered nearly all of the more than 200 genes responsible for vision, and the PAX3 and MITF genes have been identified as coding to produce our hearing.[9] Scientists at Columbia University uncovered eighteen genes necessary for building the sensory neurons that register a gentle touch.[10] A team at the University of

An autopsy on the brain of the greatest spatial thinker of our age, Albert Einstein, reported that his parietal lobe (which supports spatial/3D thinking) was 15 percent wider than normal.

California at San Diego identified genes specifically geared to the sense of taste.[11] At the rate the mapping of the human genome is progressing, very shortly we should have a complete list of which genes code the eleven *Biological Preferences* we refer to in this chapter.

How is your *Biological Preference* represented in your brain? As discussed, evidence from disease research, brain imaging, and autopsies indicate that the brain areas that house our individual *Biological Preferences* differ from one person to another. These differences provide mass, power, and the neurochemicals of learning. They appear to exhibit themselves in three main ways:

- More neurons and/or connections
- Receive more of the brain's energy in the form of glucose and oxygen
- When activated, they produce more of the neurochemicals of joy and learning

Research on the brains of individuals with outstanding abilities in a particular preference shows they have more neurons and synaptic connections in the brain areas responsible for these abilities. For example, visual artists' brains tend to have larger and/or denser visual cortexes, and athletes tend to have larger and/or denser sensory-motor areas. An autopsy on the brain of the greatest spatial thinker of our age, Albert Einstein, was reported to show he had a parietal lobe (which supports spatial/3D thinking) that was 15 percent wider than normal.[12]

Your *Biological Preference*, with a substantial amount of neural tissue laid down by your genetic program, provides a solid structural landing pad for new learning. An important point to understand is that newly learned information can't float about freely in your brain like so many loose butterflies but must attach itself to existing neural networks in the form of new synapses. Your existing neurological connections act like hooks, trapping information that resonates with the information that already exists in those brain areas. This is why people with kinesthetic preferences can more easily learn a new sport, those with taste and smell preferences can more easily remember so much about recipes or wines, those with a mathematical preference can more easily remember equations, and individuals with a linguistic preference can more easily memorize information in word form.

Initial research also demonstrates that the brain areas holding your *Biological Preference* naturally receive a substantial percentage of your brain's energy. As discussed, your brain receives about twenty watts of power in the form of glucose and oxygen via the carotid and vertebral arteries. The more energy the brain area receives, the longer that area remains activated, and the more effective its functioning. When subjects are hooked up to brain imaging devices and are given facts to learn or problems to solve, the parts of their brain that correspond to their preference immediately light up, showing that those brain areas are demanding a large portion of the brain's energy to process the information.

The research also indicates that when your *Biological Preference* is activated, the flow of the neurochemicals of learning and behavioral change are enhanced. When you engage in activities that stimulate your *Biological Preference*, dopamine, serotonin, norepinephrine, endorphins, and other neurochemicals of joy, fulfillment, and learning are expressed. This is one main reason you tend to pursue activities that stimulate your *Biological Preference*.

Activating your *Biological Preference* helps you learn something because it is like igniting a blasting cap, triggering the explosion of massive neural activation; this increases energy flow and neurochemical activity to that area, which in turn helps to produce engagement and long-term memory formation. And because your brain is an associative instrument, once you've used your *Biological Preference* to understand a concept, there's much less need to present further information about that concept in your primary preference. It can be presented in new and different ways, yet still be quickly understood.

One way to look at this process is like priming a pump. After a learner who has a tactile, kinesthetic, or spatial/3D preference is taught the concept of addition using stones or marbles, it's easier to transfer that understanding to the symbolic form of 2 + 2 = 4 and 2 + 3 = 5 written on a blackboard or spoken in the classroom. Researchers Dryden and Vos comment: "And if the current authors had to choose any one step needed to transform the world's high-school systems in particular, it would be this: find out each student's combination of learning styles and talents—and cater to it."[13]

Meaning and Your Biological Preference

One of humankind's most often-asked questions is, "Where does meaning come from?" *Biological Preferences* give us a strong clue. Substantial neural material, increased energy flow and boosted neurochemical activity in your primary *Biological Preference* help build the foundation for what you find meaningful.

As soon as your genetic program lays down the neural structures that code for your preference, you start selecting information that matches it. Soon after toddlerhood, it becomes apparent that you value things that match your primary preference. Children with a social preference find group projects most meaningful. Those with a linguistic preference are drawn to language (this helps explain early readers). Musical dominants are drawn to music, and kinesthetic dominants naturally seek out activities that engage them in physical movement. But because of the vast neurological variation among us all, no two of us sharing a primary *Biological Preference* could ever find the exact same things in the same category meaningful in exactly the same way. This is why, although many of us may take similar paths in life, genetic and neurological variation ensures that our routes of adapting to the world will never be identical.

Teachers Brains vs. Student Brains: The Unfortunate Mismatch

Unfortunately, one of the major failings of the *Ancient Learning Theory*-based learning systems is that they tend to focus on only two of the *Biological Preferences*: linguistic and mathematical. From our analysis of educators, we estimate 80 to 90 percent of all professors and teachers have a preference in the linguistic and or mathematical areas. Yet an estimated 85 to 95 percent of their students do *not* possess a primary preference in these two brain areas.

One reason that most educators go into the educational system in the first place is that their brain plan fits the system. They get a positive neurological blast of neurochemicals (discussed in more detail in Chapter 17) because the educational system stimulates their *Biological Preference*. But if we hope to increase learning and behavioral change outcomes, it's vital to understand that, in general, the people designing and implementing our learning systems have a very different *Biological Preference* from the majority of those they're designing the system for. This creates an unfortunate gap in understanding.

What we end up with is an educational system that is a disadvantage for the majority of the population, for whom linguistic delivery and finite math are not the primary methods creating profound memory by effectively blowing magnesium plugs out of NMDA receptors. From a neurobiological point of view, then, it can be said that the linguistic-math-based educational system misses its mark by neglecting to stimulate the primary centers of preference for up to 6.4 billion humans on the face of the Earth whose primary preference is not linguistic or mathematical. No wonder things aren't working. As Paul Newman said in the classic movie *Cool Hand Luke,* "What we've got here is a failure to communicate." Partly because of this mismatch problem and the pain it causes those of us in the system, many educators are discouraged; this is one reason why 50 percent of teachers are reported to leave the profession after only five years.[14]

Understanding this mismatch also helps parents of non-linguistic preference children relax into the reality that their kids aren't lazy or unmotivated but instead trapped in a system that doesn't fit their preference and therefore causes them pain. This realization, when communicated to the child, greatly enhances the parent-child relationship. As Harvard's Gardner muses: "… Individuals have distinctly different kinds of minds.… Unfortunately, this fact has not been seriously confronted in schools."[15]

I always find it interesting that when I speak to groups of teachers and ask how many got good grades in school, about 90 percent raise their hand. When I speak to entrepreneurs and ask the same question, only about 7 to 10 percent raise their hands. There is clearly a mismatch that, unfortunately, has not been communicated to the general population.

Breeding Out Your Gift

This mismatch between most learners' brains and the traditional educational system may eventually negatively affect the *Biological Preference* you were born with. Just as muscles can fail to develop from lack of use, brain areas can also become underdeveloped from the lack of proper stimulation. Remember that the brain receives twenty watts of power. When someone is in the educational system for twelve to sixteen years, and their *Biological Preference* doesn't match the system, most of this energy is redirected away from that individual's *Biological Preference* during the brain's most extraordinary developmental period. This may stunt the development of an individual's *Biological Preference* and affect how they interact with the world. Writing in the *UCLA Educator* on this tendency to over-focus on one brain area, Joseph Bogen says, "The current emphasis in education on the acquisition of verbal skills… is starving one-half of the brain and ignoring its contribution to the whole person."[16] Another educator, speaking about failing to stimulate a learner's primary gift, says, "His brain is being systematically damaged. In many ways, he is de-educated."[17]

If, as a youth, Mozart had been locked up in small rooms for twelve years and bombarded with primarily linguistic and math data, redirecting his twenty watts of neural power away from his musical preference, would he ever have been capable of producing Symphony No. 40 or Violin Concerto No. 5? Beethoven had learning circuits so unmatched to the way the educational system taught math that biographers report he never learned his multiplication tables. If, at age seven, instead of creating music, Beethoven had been forced to adapt his brain circuits to the linear world of traditional education, would his natural spatial/3D and music circuits ever have developed to the point where he could have created his famous, fulfilling music?

As mentioned earlier, what you find personally meaningful in life has its foundations in your *Biological Preference*. How many of us today have difficulty finding deep meaning in our lives because our primary preference was ignored during our youth? How many of us are latent Mozarts or Beethovens who have never reached our potential because our gift of interacting with the world has been ignored, devalued, and atrophied by the traditional scholastic system? We can create the opposite problem too. Overdeveloping a few brain areas, such as *Broca's* and *Wernicke's areas* (where some of the main circuits for language and math reside), at the expense of other brain areas may push us to adapt to the real world using only limited brain areas. Not a good survival strategy. Suppose you already have a linguistic preference that fits the school system, and these circuits are overstimulated for years on end. Is overstimulating them at the expense of your other circuits doing your future a disservice? This may be one reason that the research on valedictorians' success in the real world is so disappointing (see Chapter 4).

Overstimulating Your Preference: Not a Good Thing

Although we need to fully develop our primary intelligence to realize our potential, we also need the balanced development of our other brain areas to secure the greatest adaptive advantages. Relying on only one modality doesn't serve us, our primary modality of learning wasn't designed to stand alone. We could never really learn what a lemon was if we only saw it; we need to experience the fragrant smell, the sour taste, the oily touch, the squirting sound of the juice being squeezed, and the emotions we feel when we bite into one to really know it. To be well-rounded, we need to effectively develop all our brain modalities to successfully navigate the world.

As we've seen, the brain adapts its shape to fit the world presented to it. Natural, holistic brain development cannot occur when environmental stimulation is aimed at only one or two brain areas. Our brains were designed by 3.5 billion years of evolution to develop most effectively in the real world, where all brain areas are barraged by billions of bits of multisensory information each second. Real life automatically activates most of our brain areas. You can't operate in complex environments without stimulating visual, spatial, auditory, social, tactile, and emotional circuits (more on how the real world accelerates learning in Chapters 22–23).

The classroom isn't the real world, and sitting still for hours on end, listening to lectures, memorizing facts and taking tests is not the way most people succeed in real life (unless you work in the school system). Peter Russell, the author of *The Brain Book,* notes:

> *If only the verbal-analytic side is being educated, the student is effectively being cut off in many ways in which he could directly experience the world around him. And without direct experience, education can become dry, meaningless, and boring.*[18]

In his book *Evolution's End,* educator Joseph Chilton Pearce minces no words in criticizing an educational system that focuses on a limited locus of neurons: "The massive evidence presented… shows that our schools have collapsed not from 'bad" educational systems, which have always plagued us, but from a majority of our children have been neurologically damaged past the point of educability."[19] As Howard Gardner says:

> *... We subject everyone to an education where, if you succeed, you will be best suited to be a college professor. And we evaluate everyone along the way according to whether they meet this narrow standard of success. We should spend less time ranking children and more time helping them to identify their natural competencies and gifts, and cultivate those.*[20]

Savants: Over Learning in One Modality

It's believed that those individuals whom we used to call *idiot savants*, because of their astounding abilities, perceive the world and process information almost exclusively in one *Biological Preference*. Dustin Hoffman's character in the Academy Award-winning 1988 movie *Rain Man* is a perfect example of an individual who processed information primarily through one Preference. This ability allowed him to count and remember thousands of cards and become a phenomenal blackjack player. The inability of those with "savant syndrome" to effectively access their other brain modalities limits their capacity to successfully adapt to life. It's also believed that the conflicting relationship between extremely high IQ scores and real-world success has its root in some individuals' over-focusing on the language or math preferences, which excludes their ability to effectively process other important stimuli.

More on the Mismatch

When you're forced to learn new information in a modality outside of your preference, not only can learning become more difficult, but the effort can raise your stress hormone (cortisol) levels, triggering not joy in learning but anxiety. One theory says that many of us are labeled slow, dyslexic, deficient in attention, or learning disabled because we're simply overstressed by traditional educational settings that deliver information all day long to us in a way that doesn't consider our *Biological Preference*. As you'll see in Chapters 37–38, excess stress is the greatest inhibitor of learning because it shifts our neural energy away from our most flexible brain structures down to our least flexible brain structures, which inhibits efficient memory formation. Learning pioneer Tony Buzan, in *The Mind Map Book*, notes that:

> *The members of this race are squeezing their intelligence out only through the incredibly narrow restrictive channel of language. As a result, many of them experience actual nausea at the mere prospect of learning, and in the millions of learning institutions dotted around the planet, most of the students are either sleeping or trying to get out!*[21]

As Gardner says in *The Unschooled Mind*, "Students with strength in the spatial, musical, or personal spheres may find school far more demanding than students who happen to possess the 'text-friendly' blends of linguistic and logical intelligence."[22] To get compliance, we often end up giving children Ritalin or some other form of pharmaceutical, when simply creating an educational environment that considers their biology would be a much more rational and effective approach.

This mismatch is also the likely cause of why so many of our world's greatest minds had difficulty in school, because their *Biological Preference* didn't match the linguistic-based school system. The man who came up with the equation $E = mc^2$ had a spatial preference; he complained, "The words or the language… do not seem to play any role in my mechanism of thought!"[23] Like anyone who possesses neurological circuits that don't match the educational system, Einstein found his school days painful and upsetting. After finally graduating university, he commented, "This coercion had such a deterring effect that after I had passed the final examination, I found the consideration of any scientific problems distasteful to me for an entire year."[24]

- Sir Winston Churchill—one of the greatest military leaders of the 20th century—had a learning modality so different from the one employed by the school system that he failed his military school exams not once but twice![25]
- Yale's Robert Sternberg, the author of forty books and one of the most respected intelligence researchers in the world, was labeled a "dumb-dumb" and "loser" because he did so poorly on linguistic-based intelligence tests.[26]
- Robert Sylwester, professor of education at the University of Oregon and a vocal champion of science-based learning, flatly states, "Teachers should adapt their instruction to the built-in biases and limitations of their students'… mechanisms."[27]

MOTOR SENSORY MAP OF THE BRAIN

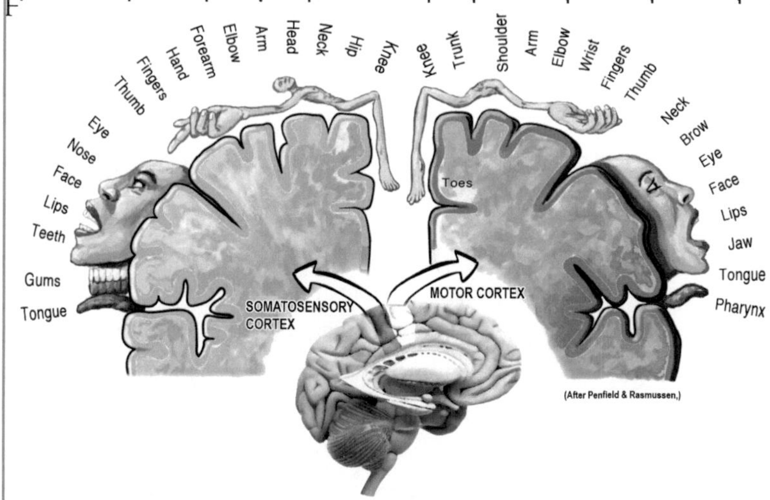

In the 1940s, Canadian neuroscientist Wilder Penfield mapped the motor sensory areas of the brain; these maps are still used in most neuroscientific textbooks today.

students' preferences have seen exceptional results. Years ago, Brightwood Elementary School, located in a socioeconomically disadvantaged area of Greensboro, North Carolina, had math and reading scores in the thirtieth percentile. Three years after implementing a method of identifying students' primary preferences, it reached the eighty-fifth percentile. It was reported that the only thing teachers did differently was to become aware of their students' learning preferences.[28]

The Origins of Your Biological Preference

The first sustained effort at identifying which brain areas produced specific functions was undertaken in the late 1700s by the German anatomist Franz Joseph Gall. As a schoolboy, Gall observed a relationship between certain mental characteristics of his schoolmates, such as compassion, self-esteem, hope, and reverence, and the shapes of their heads. From these observations, Gall later developed the discipline of phrenology, which maintained that there were thirty-five different "intellectual faculties" and that the dominance of any one of these intelligences could be read by observing the bumps on a person's head.[29] In recent times, phrenology has rightly come to be ridiculed, yet Gall must be commended as the first modern scientist to stress that different parts of the brain mediated different brain functions.

Amazingly, there are still cults today that hold firmly to Gall's ideas.

In the 1860s and 1870s, French surgeon Pierre Paul Broca and the German neuroanatomist Carl Wernicke made sound scientific discoveries, earthshaking at the time, that identified specific structures on the left side of the brain that coded for language production and word comprehension. Broca and Wernicke found that when these areas became damaged, an individual's ability to process language would be severely disrupted.[30]

In the 20th century, the development of more modern medical techniques enhanced our ability to learn more about brain function and location. In 1909, the neuroanatomist Korbinian Brodmann analyzed the cellular organization of the cerebral cortex and separated it into fifty-two distinct regions.[31] In the 1940s, Canadian neuroscientist Wilder Penfield used electrical probes to stimulate portions of over 1,000 patients' brains.[32] His yeoman-like efforts, while misrepresenting what the brain remembered (as seen in Chapter 3), helped map out the motor and sensory areas of the brain; these maps are still used in most neuroscientific textbooks today (see image above). Now, at the start of the 21st century, neuroscientists are using microscopic techniques and brain scans to identify hundreds of new areas of distinct structure and function.

Starting in the 19th century, research in the fields of psychology, evolutionary biology, and education began to parallel the discoveries in medicine. About the same time, Broca and Wernicke were using patients with trauma and disease to make their neuroscientific findings, the founder of the infamous eugenics movement, Darwin's cousin Sir Francis Galton, was studying men of position and power. In 1874, in what many consider the first formal study of how different people process information in different modalities, he surveyed men of science and found that some relied primarily on the vision to remember information, while others relied on words.[33]

Nearly fifty years after Galton's efforts, Swiss psychiatrist Carl Jung undertook what is believed to be the first major psychological effort to understand the mental process of preference. Jung's work led him to classify people as feelers, thinkers, sensors, or "initiators." It was Jung who popularized the concept that we interact with the world as either extroverts or introverts.[34] Jung can be considered the founder of what is now called *emotional intelligence.*

The next major psychological effort toward identifying preferred modes of mental processing was undertaken fifty-nine years after Jung's studies. In the 1970s, psychologists Richard Bandler and John Grinder wrote *The Structure of Magic I: A Book about Language and Therapy* and the small book *Frogs into Princes,* which helped launch the field now called NLP or *neuro-linguistic programming*. NLP maintains that people have three main ways of processing information: auditorily, visually, and kinesthetically. Bandler and Grinder found that a patient suffering from psychological issues could be cured much more quickly if the therapist delivered information in the patient's preferred learning style.[35] Soon after these books were published, NLP attracted a strong following, which still exists today, mainly through the work of motivational speaker, author, and infomercial guru Anthony Robbins.

In 1983, the same year that evolutionary biologist Jerry Fodor came out with his "Swiss Army knife " view of the mind, Harvard professor and intelligence researcher Howard Gardner proposed what could be called the first educational view of brain dominance. In his groundbreaking book *Frames of Mind: The Theory of Multiple Intelligences,* Gardner utilized psychological research on savants, prodigies, and people from diverse cultures to propose that humans are blessed with a set of limited intelligences; he initially focused on seven of them.[36] Gardner was one of the first respected researchers to recognize that, because there are many forms of intelligence, an educational system that focused on only the linguistic style of learning would be doomed to failure.

In the 1990s, Jonathan P. Niednagel became well-known in the world of sports for developing what he calls *brain typing*.[37] He expanded on Carl Jung's efforts from the 1920s and came up with sixteen brain types, including logician, strategizer, inventor, and investigator. Niednagel's work has helped sports teams, as well as corporations, support those in their employ.

Over the years, many other models put forth by individuals like Walter Burke Barbe, David Kolb, and Anthony Gregor have helped us reach the conclusion that we all have a *Biological Preference*, but it's the medical structure-function research that gives us the most clarity. That said, we want to emphasize again that though there are a limited number of *Biological Preferences*, we know from neurobiology that each individual has a unique genetic and neurological makeup that gives each of us a unique form of intelligence—limiting the view of intelligence to a finite number distorts reality.

It needs to be repeated: knowing that each of us learns differently is only a tiny, tiny piece of the puzzle. It is not "the answer." The rest of the pieces of the puzzle are presented in the other chapters of this book that lay out the other elements of the Learning Code science. In the following chapter, we'll go into more depth into each of the unique learning preferences.

Cracking the Learning Code

Each of us has a particular *Biological Preference* for learning. This understanding allows us to create more natural and engaging learning environments. The mistake that most educational, training, and motivational programs make is in trying to get people transform without understanding their primary mode of learning.

10

The Eleven Biological Preferences: Which Is Yours?

Enlightenment is man's leaving his self-caused immaturity... it is not caused by lack of intelligence, but by lack of determination and courage to use one's intelligence without being guided by another. Sapere Aude! Have the courage to use your own intelligence! is therefore the motto of the enlightenment.

—Immanuel Kant, 18th-century German philosopher[1]

In the previous chapter, we explored the theory and development of the eleven *Biological Preferences*, and in this chapter, we'll look at the attributes of each one. Of course, numerous other intelligences, such as natural and intuitive, have also been proposed by fields outside of neuroscience, but for the time being, we'll only focus on those *Biological Preferences* which have scientific and medical criteria upon which to draw conclusions.

There's also much speculation among researchers as to what percentage of the population has a particular preference. Professors Ken and Rita Dunn from St. John's University suggest that 15 percent of us learn tactually, and another 15 percent learn kinesthetically.[2] Other NLP experts suggest 20 percent of us will be "visual only" learners. To date, the field of identifying what percentage of the population has what particular intelligence is hampered by the lack of brain imaging on large numbers of individuals while they perform diverse learning tasks.

The Eleven Biological Preferences

Again, it is important to understand that I am not suggesting we create educational and training systems that cater directly to any of these preferences. The learning process is more complex than that. We're only suggesting the importance of recognizing that you (and the rest of us) have a predisposition to process information in one or several of these modalities. As you read about these *Biological Preferences*, see which ones resonate with you.

FAST FORWARD *or* DEEP DIVE

If you only need a cursory understanding of each preference, we suggest you only read the first several and the last few paragraphs of each one. For those wanting a deeper scientific understanding, such as the main structures and neurochemical systems that must be activated, we have placed this more detailed information in the middle paragraphs.

1. Kinesthetic Preference: Keep Moving!

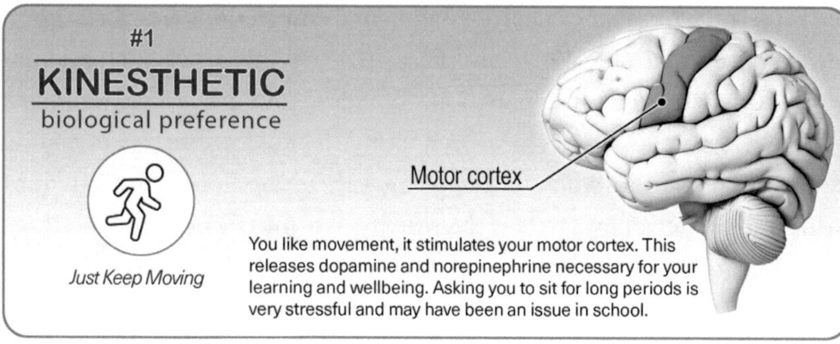

With a few exceptions, the ability to move separates the plant and animal kingdoms. As the ability to control physical movement has progressed up the evolutionary ladder from single-celled bacteria to fish to reptiles to mammals to humans, the ability to improve survival options by physically moving away from pain and toward pleasure has become more and more sophisticated. For 3.5 billion years—until about 50,000 years ago, when new brain areas brought with them the capacity for complex language—all learning by our ancestors was done on the fly. Before the advent of speech and writing, if you weren't moving, you weren't learning.[3]

In prehistoric times, individuals with a kinesthetic preference are likely to have supported the group's survival by being the first up in the morning, taking off in search of the day's prey. To be effective at their job, they had to have not only physical prowess but also the ability to continually shift their focus, picking up clues to help them escape predators and find meaty prey. For our ancestors who operated in a dangerous world, fixating on any one stimulus for too long could lead to death. Today, experts see a connection between many individuals labeled ADHD and the kinesthetic *Biological Preference*.

Those with a kinesthetic preference like to learn while moving because physical action not only stimulates brain areas that support learning but also prompts the release of neurochemicals like dopamine and norepinephrine necessary for learning. Kinesthetic learners experience discomfort and inefficiency when they try to learn without movement, for two main reasons.

First, when kinesthetic types are made to sit still for hours at a time, the motor cortex, which occupies a wide strip behind the prefrontal lobe and runs from temple to temple, doesn't receive enough neural energy to fire up the substantial number of neurons and neural connections that act as hooks for new information. When any networks that hold personal meaning remain unstimulated for long periods, boredom and discomfort set in.

Second, kinesthetic types have been found to have low levels of dopamine and norepinephrine when not moving—these are focus and attention neurotransmitters. Low levels of these neurotransmitters prompt kinesthetic types to continually want to move and/or shift their focus of attention to increase dopamine and norepinephrine. Although this may be a great survival advantage, in the real world, it causes discomfort and learning difficulties when they need to sit still and focus on a single stimulus such as a teacher in the classroom. Forcing someone with a kinesthetic preference to sit for long periods is like asking an auditory-dominated person to listen to fingernails scratching on a blackboard—very painful.

As you'll see in Chapter 17, no matter what your primary intelligence, any time you move, your levels of dopamine and norepinephrine increase. This is why it's so much easier to focus on and remember things you did while you were moving through the real world. But getting up and moving is frowned upon in our traditional learning systems. Therefore, to keep kinesthetic types in their seats, we often drug them by prescribing stimulants like Ritalin, which artificially increases dopamine and norepinephrine.[4] Although stimulant drugs can be of great benefit to individuals properly diagnosed with attention deficit disorder, we may be causing great harm by prescribing

stimulants at a very young age to the 15 percent of the population whose primary *Biological Preference* is kinesthetic. While there needs to be more research in this area, many experts believe these drugs can set up children for a lifetime of addiction.[5]

Examples of those with a kinesthetic preference are Tom Brady, LeBron James, Tiger Woods, and Serena Williams. They've all used their kinesthetic dominance to learn not only how to play but also later how to leverage their preference—and what they've learned employing it in the sports world—in business.

2. Tactile Preference: Get Dirty

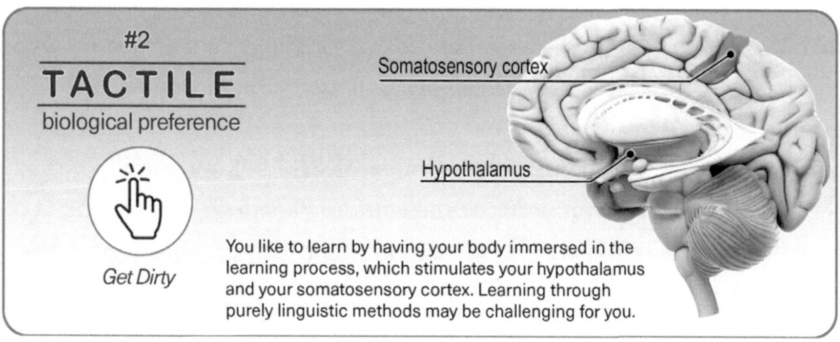

Tactile feedback (perceived by touch) from our contact with the environment is how everything from a slug to a human knows how to regulate itself. Tactile feedback is recorded in our somatosensory cortex, located just behind the motor strip in the brain (*soma* is the Greek word for the body). The tactile strip works in tandem with the hypothalamus in the limbic system, which regulates the neurotransmitter and hormonal responses to all stimuli you receive through your skin.

Those with a tactile preference like to learn by having their bodies immersed in the learning process, which stimulates both the somatosensory strip and the hypothalamus. Having a primary tactile preference was valuable to the tribe during ancient times for several reasons. Their fine sensory abilities made them more skilled in manipulating common materials available to the tribe. It is believed these individuals could determine through touch and feel which soils would best grow crops, which clay would be best for making pots, and which stones would be best for shaping arrowheads. It was probably the tactile types who figured out which stones could make a spark to start a fire.

Many believe those with a tactile preference were also the first weather forecasters. Their preference would have made them extra sensitive to changes in humidity, temperature, and air pressure, which indicated changing weather conditions. They could then alert the tribe to early winter or fast-approaching weather change so that the community could take appropriate action.

Those with a tactile preference learn best by having some sort of contact with what they're trying to learn, which makes strictly linguistic instruction very difficult for them. Primarily tactile learners are strongly affected by room temperature, how the furniture touches their skin, and their proximity to other students. Authors Gordon Dryden and Jeannette Vos point out that "tactile learners are the main candidates for failure in traditional school classrooms."[6] Learning style researcher Michael Grinder reports that this group, which is made up of only 7 percent or so of the population, is responsible for a disproportionate number of school dropouts.[7]

One of the greatest minds of the millennium, Charles Darwin, appears to have been tactile dominant. He relied on his constant physical contact with plants, insects, and animals to allow him to work out what many consider the most important idea ever discovered: the theory of evolution by natural selection. Darwin needed to touch and feel so much that he said he found lectures "intolerably dull," and after seven years of effort at classical teaching, he concluded that his mind was "simply a blank."[8]

Geneticists Theodosius Dobzhansky, whose work on variation helped put the first coffin nail in the eugenics movement, and Barbara McClintock, who discovered jumping genes, are two scientists who, like Darwin, learned best in this way. They both strongly believed that genetics could not be learned in a classroom, and they encouraged their students to get hands-on learning in the wild. Tactile preference vocations include massage therapy, sculpture, surgery, carpentry, gardening, and farming.[9]

3. Taste/Smell Preference: Loving the Food Channel

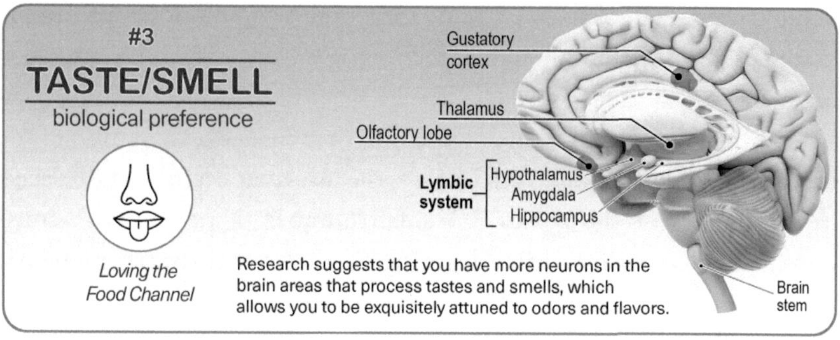

Taste and smell sensations from our noses and mouths alert us about what to approach and what to avoid. Taste and smell helped our ancestors decide which food and water were palatable and which was poisonous. Every living entity—be it predator or prey, nutritious or poisonous—carries a distinctive odor. Even the sexes give off different smells, as we've learned from pheromone research.

Modern bears can smell prey twenty miles away,[10] and in primitive times, our command of taste/smell was paramount to the continuation of life; the species that weren't skilled in this modality no longer exist. Today, with bottled water and packaged foods, taste/smell preference plays a lesser role in human survival, and it's estimated that only about 5 percent of people possess this as a primary preference.

Although the neural systems for taste and smell are distinct from one another, we've combined them for two reasons. First, the sensations of flavors and aromas are so closely intertwined that it's impossible to deal with one without dealing with the other. Second, taste and smell use the same processing mechanisms. The process begins when molecules detach from substances that float into our noses or are put into our mouths. These molecules must then dissolve in watery mucus to bind to and stimulate special receptor cells, which transmit messages to the brain centers where we perceive odors and tastes.

Where do odor signals go after they activate the sensory neurons in the nose? They're sent to the olfactory bulbs, one on either side of the bottom surface of the brain. In our ancestors, these bulbs were huge, but in modern humans, they've shrunk to small, poorly defined nubbins.[11] From the olfactory bulbs, smell signals go by way of the olfactory tract to the limbic system, which houses raw emotions. In the early 20th century, neuroscientists felt that smell was so important to emotional processing that they called the limbic system the *rhinencephalon,* which means "smell brain."[12]

The olfactory tract also sends signals to the sensory relay station of the brain called the *thalamus,* which passes the information to the associative memory areas of the cortex.[13] A small number of people with smell dominance have a hard time learning whenever a distracting smell is present. Some of these people are so sensitive that learning can't take place at all in the presence of just a few of the molecules responsible for an obnoxious smell. At seminars, I've heard people complain that the smell of a person's perfume next to them was so strong that they couldn't learn. Some people hypothesize that the field of aromatherapy has the greatest effect on those with a smell dominance.

The taste buds on your tongue send signals to the brain stem, limbic system, and thalamus, which in turn pass this information on to the gustatory area of the somatosensory cortex.[14] Research shows that taste-dominant individuals have an overabundance of taste receptor sites on their tongues, which prompt more neural activity in the gustatory brain centers.[15] Dominant tasters do not have to eat while processing information, but many smack their lips and tongue in imitation of tasting while learning. Often, those with a taste/smell preference suck on things, such as pens and pencils, while trying to sort out a problem. Chewing gum often helps them

"It's not icky– it's tactile. That's my ***biological preference****."*

to process information.

Those with primary taste/smell preference often gravitate toward jobs that include being chefs, sommeliers, perfumers, and aromatherapists. Julia Child and Robert Mondavi are examples of famous individuals in the fields of cooking and wine because of their taste/smell preference.

4. Visual Preference: In the Eye of the Beholder

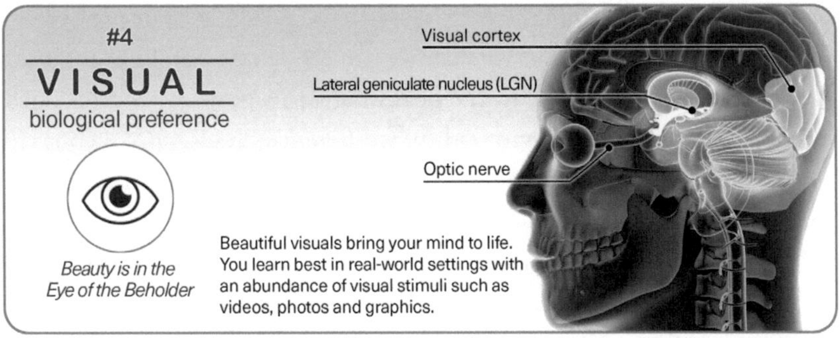

Vision became a dominant survival characteristic 550 million years ago during the Cambrian explosion, a period in which nature tried out billions of new survival characteristics, of which only a few million survive today.[16] Vision was a tremendous survival advantage for all organisms. As soon as organisms could shift away from relying solely on smell, taste, and touch to avoid pain and approach pleasure, their survival chances were enhanced dramatically. Now organisms could see what they wanted to have sex with, eat, or flee from. Some researchers estimate that up to 30 percent of us rely primarily on this preference to survive and thrive in our world.

The value of vision to the human species is represented by the neural connections in the brain allocated to this modality—one expert has estimated that as much as 50 percent of neural tissue is involved with vision.[17] Vision begins when light and color-sensitive photoreceptor cells in the retinas pick up signals. This information leaves the eye by way of the optic nerve, which sends information to a section of the thalamus called the *lateral geniculate nucleus* (LGN); from here, axon connections fan out to the visual cortex, which takes up the whole back of the brain, where connections spread out to the rest of the brain.

To someone with a visual preference, rich visual sensory input is necessary to fire up a large number of neural connections that prompt the NMDA receptors and CREB molecules to action. Sitting in a small classroom for hours with only a teacher to look at is the fastest way to put a visual dominant to sleep or drive him or her out of the traditional learning system altogether. These learners encode knowledge best in real-world settings with an abundance of visual stimuli or with videos, pictures, and infographics. The most obvious jobs that attract those with primary visual intelligence include photography, painting, fashion, film, video, and structural and interior design.

It must be noted that visual preference can reside in color, line, angle, or some combination of these. In the field of art, this gives our world a rich pool of wonderful painters with radically different styles, such as Leonardo da Vinci, Monet, Frida Kahlo, and Andy Warhol; their distinct works are thought to be driven by different genetic variations.

Many in fields far from what could be considered art have also relied on visual intelligence to accomplish great things. Mathematician Benoit Mandelbrot, a leader in the field of fractal geometry, said that to understand complex concepts, he needed to think in images.[18] Physicist Richard Feynman finds that he can really understand an equation only if he looks at the illustration that goes with it.[19] And geneticists Watson and Crick could not solve the puzzle of the structure of DNA with chemical equations alone; it was only after they drew and constructed visual models that the pieces fell into place.[20]

5. Auditory Preference: Let Me Hear You

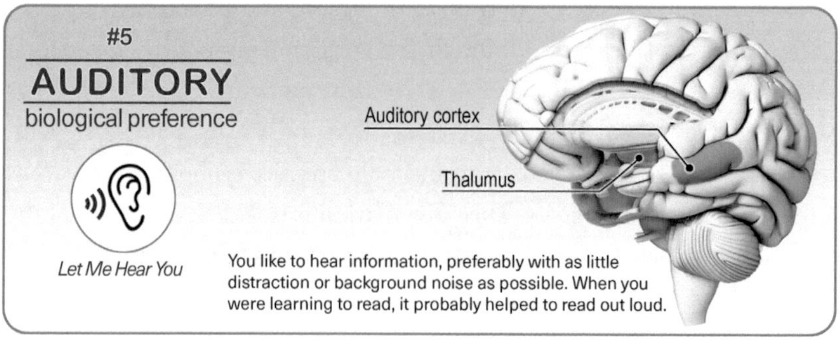

As organisms marched up the evolutionary ladder, the physical mechanisms that could pick up vibrations and translate them into survival information became more and more sophisticated. Like smell, the auditory modality was important because it allowed our ancestors to locate pleasure and danger in their environment without having to expose themselves to potential prey, predators, or sexual partners. This allowed them precious time to stay out of the line of sight while they developed effective strategies of action and decided whether they should attack, run, or get randy.

In modern humans, the auditory modality is controlled by a very complex process. Essentially, hearing starts when sound waves vibrate the eardrum at the end of the ear canal. These vibrations are sent to the middle ear, where three small bones refine the initial vibrations and send them on to the inner ear structure called the *cochlea.* Thousands of tiny hairs in the cochlea have receptor cells, which process auditory information in the range of fifty to 20,000Hz and pass this information to the switchboard of the brain, the *thalamus;* the thalamus, in turn, sends this information on to the auditory cortex located in the temporal lobe, where the sound is identified.

The auditory preference is closely connected to the linguistic and musical modalities because they share many of the same structures that translate vibrations into sounds. Individuals with an auditory preference prefer to hear information, often with as little other sensory distraction or background noise as possible. The main difference between auditory and linguistic dominance is that auditory-dominant types would rather listen to information than read it. Many beginning readers with primary auditory intelligence prefer to read out loud to themselves and often continue to move their lips while reading as adults.

It shocks many people to realize that reading silently is a new process for the brain. It is reported that people initially had to say the words they were reading out loud to comprehend them.[21] History tells us that it was only 600 years ago that St. Augustine noticed that St. Ambrose was doing something that had been previously unheard of—reading without a sound!

Auditory-dominant types can often learn more from listening to audiobooks than from reading. When these folks are wrestling with a problem, they often talk out loud to themselves to stimulate the parts of the brain that allow them to process information more effectively.

Those with an auditory preference are good listeners and often take jobs as social workers or psychologists or in radio stations or human resource departments. I have an auditory-dominant business partner who gets almost no information from written reports. Instead, to process information, he has to hear what employees have to say about a project to grasp significant points. What business is he in? Radio, of course.

6. Spatial/3D Preference: Engineering Your Way to Joy

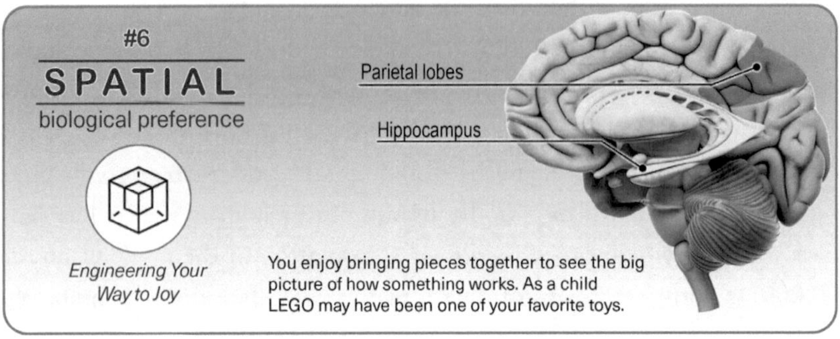

The more efficiently our brains map our world, the more effective our survival actions can be. Just getting sensory feedback from the environment isn't enough. Sight, sound, touch, smell, and taste must be put together, so we have accurate 3D maps of our world from which to take appropriate action. Spatial intelligence controls the map-building function of the brain.

Spatial intelligence is located in the parietal lobes just behind the somatosensory cortex and is supported by "place cells" located in the hippocampus in the limbic system. Any time you try to find your glasses at home or look for your car in the parking lot, you use the map-building capabilities of your spatial/3D intelligence. People in professions that require good mapping capabilities, such as taxi drivers, have increased activity in the brain areas where spatial intelligence is housed. Howard Gardner found that individuals whose survival depended upon finding their way in vast territories, such as the desert, Arctic, and oceans, have highly developed forms of spatial/3D intelligence. He notes that in Solomon Islanders, the parts of the brain that deal with spatial information were highly developed.[22]

Math concepts such as geometry, which need to be understood on spatial/3D planes, stimulate these brain areas. Individuals with a primary spatial/3D preference love bringing pieces together to make a whole picture. These people include architects, engineers, explorers, typographers, computer programmers, and print and video editors. At this time in human history, those with a spatial/3D intelligence are being handsomely rewarded. In a technical world, the mind that can see how to fit the world on a computer chip or manipulate the web is highly valued. However, our traditional educational system has proved inefficient at satiating these types that many of those who have been most influential in developing our information age. For example, Bill Gates and Paul Allen of Microsoft, Steve Jobs, and Steve Wozniak of Apple, Evan Williams, founder of Twitter, and Michael Dell of Dell Computers, all dropped out of high school or college to stimulate their spatial/3D preference on their own.

7. Musical Preference: Widespread Activation

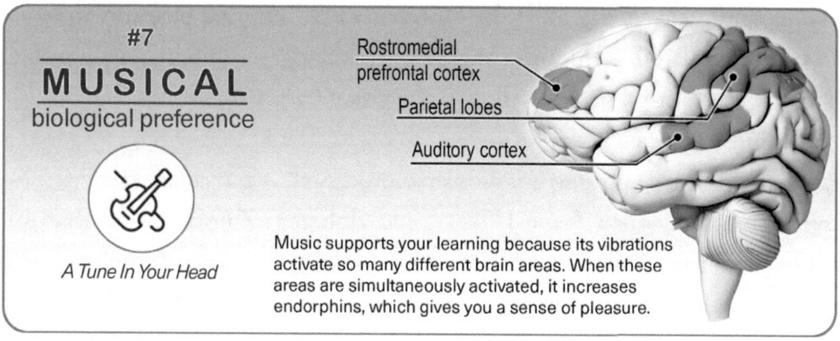

Music could be called humankind's first language because until we developed complex speech and language around 50,000 years ago, rhythmic vibrations were our primary method of communication—from the chirping of crickets to the grunts and whines of gorillas to the more complex combinations of pitch and tone that songbirds and whales employ to communicate. Over the last decade, a substantial amount of research has gone into locating the circuits that hold our musical intelligence. Evidence suggests that musical processing overlaps with other brain areas and that our musical networks are more spread out within the brain than any other modality is.[23]

Musical vibrations picked up by the ears are sent to the auditory cortex in the left temporal lobe for processing. One area located here, called the *planum temporale,* is called on to recognize not only pitch but also words[24] Damage to these temporal areas causes *amusia,* a condition that inhibits people from recognizing melodies. Brain scans also reveal that these temporal areas connect to networks in the right temporal lobe that can differentiate changes in note duration and separation. Lawrence Parsons and his colleagues at the University of Texas in San Antonio found that an area in the right temporal lobe that interprets written musical notes and passages corresponds to an area in the left temporal lobe known to interpret written sentences.[25]

It's also been found that the parietal lobe, which houses our spatial/3D modality, is involved in producing music.[26] A decade ago, researchers at Dartmouth College discovered that an area of the prefrontal cortex called the *rostromedial*, which processes emotional feedback, is also necessary for processing the harmonic relationships of music.[27] Amazingly, research shows that even the visual cortex gets into the act. Hervé Platel, Jean-Claude Baron, and their colleagues at the University of Caen used positron emission tomography (PET) to monitor the effects of changes in pitch. Much to their surprise, they found distinct areas in the visual cortex that lit up.[28]

But music goes even deeper than the cortex. The limbic system, where our raw base instincts and raw emotions are processed, is activated when we hear music. Neuroscientists Anne J. Blood and Robert J. Zatorre found that this is one reason we feel good when we hear our favorite music.[29] Their research uncovered that the same limbic pleasure centers that are activated when we engage in sex, eating, and taking mind-altering drugs are also activated when we listen to music—so, yes, it looks like we can become addicted to our favorite tunes! Dr. Mark Jude Tramo, the neurobiologist at Harvard University Medical School, says music is so ingrained in our brain structures that it "is biologically part of human life."[30]

Plato once said that for education, "musical training is a more potent instrument than any other." It appears that he was right. One study at the University of California has shown that piano lessons given to preschoolers helped them score 34 percent higher on math and science tests than a control group who did not take piano lessons. Evidence shows that high school students with a music background do better than their peers on college entrance exams.[31] Other studies have shown that both SAT and IQ scores can be temporarily driven up by exposing students to music before the tests (the lowering of stress hormones may be the reason for this).[32] Research indicates that many students learn better with music playing in the background as they study, even if adults consider that same music distracting.[33] One researcher found that, for many students, the hard rock and rap that their parents hate is just what they need to integrate information.[34]

My elder son has such a strong musical dominance that, in third grade, he couldn't process new information in the classroom without humming or singing. This kept his young butt in trouble with his linguistic-dominant teacher, who couldn't fathom why he needed to hum all the time. We eventually had to move him to a school that had a greater tolerance for different *Biological Preferences.*

It appears that music supports learning because its vibrations activate so many brain areas. As you saw in Chapter 7, the more neurons that are activated during the learning process, the greater the chance that the magnesium plug will be blown out of the NMDA receptor so that memory can form. Research shows that electroencephalogram (EEG) waveforms in the left and right hemispheres synchronize when we listen to certain types of music. Scientists believe this synchronicity may help the flow of information from one side of the brain to the other.[35]

No discussion of the musical modality could be complete without reference to spirituality and health. Before the Industrial Revolution, music was seen as the way to align one's body with the laws of nature and one's soul with the laws of God. Music was the path to salvation.[36] One of the greatest musicians of our times, Beethoven, said that music is the soul in which the spirit lives. The Christian, Islamic, Buddhist, and Hindu chants that have soothed souls for centuries offer examples of humans using music to connect themselves to the higher realms.[37] Music is now even being used to heal. Several studies reveal that playing specific music to premature babies and those with diseases speeds healing.[38]

Musicians—from composers such as Mozart and Gershwin to monks who perform Gregorian chants to Beyoncé and Jay Z—have a musical preference.

8. Linguistic Preference: Words, Words, and More Words

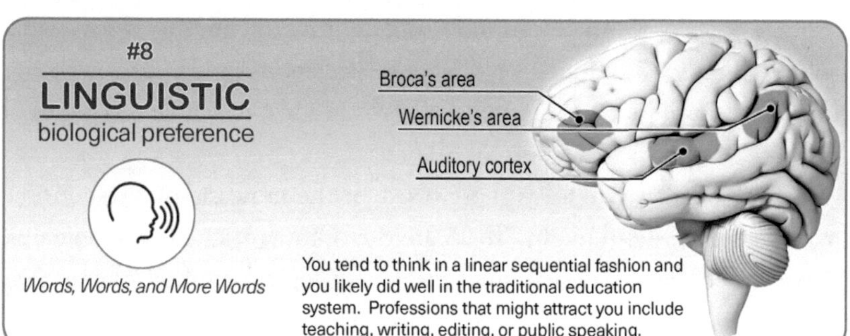

Language is the new kid on the evolutionary block. Our linguistic intelligence developed quite suddenly in evolutionary terms, about 100,000 to 50,000 years ago, and coincided with the expansion of the frontal brain areas and the elongation of the palate.[39] Before these changes, humans had neither the mental capacity for complex language nor the physical apparatus by which to form words. Our primate cousins who lack these evolutionary adaptations have a mind and mouth that can produce enough sounds, but they lack the brain wiring to produce speech.[40]

Language allowed humans a new, fast, efficient, and safe way to learn survival information without having to personally experience a dangerous world. No longer did an individual have to encounter a snarling saber-toothed cat to learn that the animal could be dangerous. Sitting by the fire in their cozy cave, one tribal member could tell the whole tribe to stay clear of these nasty cats. This ability to share complex surviving and thriving information is one of the humans' greatest advantages over our animal and ape cousins.

Humans are genetically wired for language. We come out of the womb with the ability to utter every sound of every language on the face of the Earth. Noam Chomsky, linguistics researcher, and professor at MIT, maintains that we're also born with a genetic blueprint for grammatical rules, which allow for an incredibly complex speech by the age of six without any formal instruction.[41]

In the left hemisphere of the brain, our genes build specific language-processing networks in our *Broca's area*, which is located at the rear of the frontal lobe. Language comprehension resides in our *Wernicke's area* in the temporal lobe behind the ear. These two language areas are connected by a tract of fibers called the *arcuate fasciculus*. All these structures share many connections with the auditory cortex. Research has shown that, within the language centers, the location for different word concepts is somewhat separate. That's why trauma or disease in specific locations can wipe out the ability to produce or comprehend verbs, nouns, or whole categories of things such as vegetables, tools, or animals.[42]

These linguistic brain areas share many of the same circuits responsible for linear sequential thought. These are the dominant circuits of Socrates, Aristotle, and Plato—designers of the Greek logical method of learning, now used throughout the educational world.

Professions that attract those with a primary linguistic dominance include teachers, professors, authors, editors, copywriters, seminar presenters, and many word-based comedians.

9. Mathematical Preference: The Joy of Counting

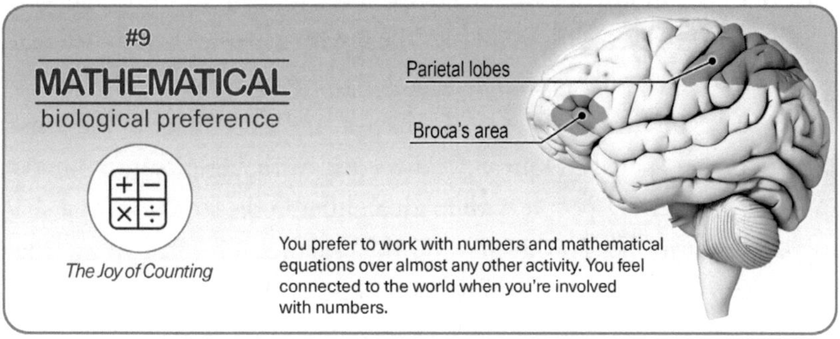

Evidence suggests that the origins of our mathematical modality are genetic. Many of those who study the genesis of our math abilities embrace a view similar to Noam Chomsky's theory on grammar, maintaining that math is hard-wired into the genetic plan. Dr. Brian Butterworth, author of *What Counts: How Every Brain Is Hardwired for Math* and *The Mathematical Brain,* points out that newborns have been shown to have concepts of numbers and arithmetic because all humans are "born with a start-up kit for numbers."[43] One of the things that make humans so unique from animals is that our ability to process mathematical concepts is so advanced.

Research at MIT and other institutions has demonstrated that we have two sets of neural networks that deal with math.[44] The oldest network, in evolutionary terms, is located in the parietal lobes, the same structure where most of our spatial/3D abilities are housed. Circuits here supply us with the ability to make estimates about which groups of things are larger and which are smaller. These structures also provide the foundation for the understanding of general math concepts like geometry. Newer evolutionary networks, which handle linear, exact calculations and rote memorization such as multiplication tables, share many of the same structures as our linguistic intelligence.

The theory is that many of the same networks that decode language are also necessary for decoding finite math. A study led by cognitive neuroscientist Stanislas Dehaene of INSERM used functional brain imaging to find that circuits in and around Broca's, in the left frontal lobe, known to identify and make associations between words, also lights up when we try to make exact computations.[45]

Less than 5 percent of the population appears to have a mathematical preference. These individuals prefer to work with numbers and mathematical equations over almost any other activity. They feel connected to the world when they're involved with numbers. Those with a mathematical preference do well in fields like business, banking, accounting, stock analysis, engineering, and physics. People who have a large percentage of their math abilities in the spatial/3D areas tend to deal with big-picture numerical concepts. Conversely, people with math abilities in the structures that share language tend to focus on details and the exactness of calculations. These people also do well on the math portion of IQ and SAT tests.

Famous math-dominant individuals include Euclid, Isaac Newton, René Descartes, John von Neumann, and Alan Turing.

The Two Emotional Preferences: I Feel Therefore I am

Just like the first nine *Biological Preferences*, emotion is an evolutionary contribution that has been selected into our species because it aids our surviving and thriving. Emotions mark the information in your world that you should pay attention to; they're the sole criterion for evaluating the correctness of our perceptions. Once those calculations have been made, our emotions prompt us to act. Our emotions are also primarily responsible for tagging specific information for selection into our long-term memory banks (see Chapter 7).

Without emotions, your world would be a gray, bland, and unmemorable place in which there would be little reason to seek pleasure or avoid pain. Humans have the most advanced emotional networks in the animal kingdom, which give us an advantage in the quest to survive and thrive. Anger, rage, joy, sadness, grief, curiosity, love, jealousy, greed, surprise, embarrassment, disgust, and pride are not intangible or elusive. They're in-your-face motivators that help you pay attention, calculate, remember, and make refined decisions about what to approach and what to avoid in your life.

The structures that produce and mediate your emotions are also part of the *Meaning Network* (the all-important group of structures we'll look at in detail in Chapter 14), which must be stimulated before we can create meaning in our lives. Joseph LeDoux (author of *The Emotional Brain*),

Antonio Damasio *(Descartes' Error),* and Daniel Goleman *(Emotional Intelligence)* have done yeoman's work to bring the concept of emotional intelligence into the consciousness of the general public.

Around 50,000 years ago, the brain area behind the human forehead exploded in size and sophistication. This dramatic increase in brain complexity created not only a set of networks that allowed for language acquisition but also structures in this prefrontal cortex that allowed humans to refine and mediate their raw, base, instinctual emotions. These prefrontal areas provide the capacity for the modern human brain to take the impulses served up by the limbic/stem areas and permit us to think before we act. Our animal cousins with a less complex prefrontal area live in a world where they must respond to every environmental encounter with the same reaction and intensity as their ancestors. It's the analysis and refinement of limbic/stem urges by the brain's forward areas, especially the ventromedial area (located where ancient Eastern traditions place the third eye), that produce the human abilities of insight, abstraction, altruism, anticipation, association, metaphor, judgment, future planning, sequencing, creative thinking, big-picture analysis, self-awareness, and empathy.

The psychological evidence suggests that there are two types of emotional intelligences: one that helps us be responsive to our own internal emotions and another that helps us be sensitive to the emotions of others.[46] Although both forms of emotional intelligence rely on the same brain structures, individuals with a preference toward one or the other have very different ways of interacting with the world. Estimates are that 20 percent of the population has dominance in one of these two types.

10. Internal-Emotional Preference: "I Want to Be Alone!"

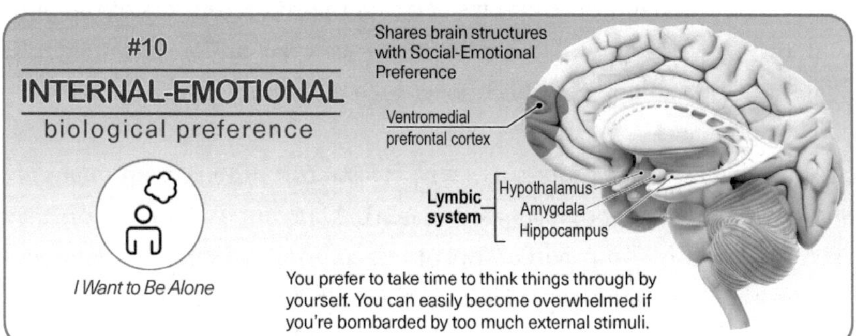

While your brain perceives billions of bits of sensory information each second from the skin, tongue, nose, ears, and eyes, it's your emotions that tag which of these bits to pay attention to and which to ignore. There are members of our species who are much more sensitive to these bits of information than others. It can be said these people *feel* more than the rest of us, and because of this exquisite sensory acuity, they expend a good percentage of their neural energy emotionally processing sensory information so they can figure out exactly what is meaningful to them and what is merely noise.

Individuals with this internal-emotional preference learn best by first taking in sensory information and then emotionally processing this information in private. Like Greta Garbo, the famous Swedish movie star of the 20s and 30s, said when she wanted to think, "I want to be alone." Those with internal emotional-preference learning mechanisms cannot be hurried; they need time to sort through the complex array of emotional feed-forward and feedback loops that billions of bits of stimuli produce. Internal emotional processors are easily overwhelmed if their brains are bombarded by too many environmental stimuli. Over ninety years ago, Carl Jung would have labeled those with primary internal emotional processes as introverts. Psychologist Elaine Aron has written a very useful book for parents with children who have an internal emotional preference called *The Highly Sensitive Child*.[47]

Internal emotional processors are the "thinkers" of our world. Our society relies on these internally sensitive individuals to solve problems and create art. Some of the most innovative and creative people in our society have an internal-emotional preference. Many inventors, scientists, investors, researchers, philosophers, poets, and painters are primarily internal emotional processors. It's believed that the inventors of the plow and the wheel were internal processors, as were philosophers such as Plato and Immanuel Kant and poets such as Walt Whitman and Emily Dickinson.

Note that internal processors aren't necessarily proficient in analyzing their own emotions. Some are, some aren't. Many internal processors may be unconsciously competent at using their emotions to solve problems and be creative, but there's a paradox, as many of these individuals lack self-awareness. They don't have to be aware of their feelings for their brains to effectively use their emotions on a subconscious level to make order out of the chaos in the world and solve problems.

11. Social-Emotional Preference: "I Know What You're Feeling"

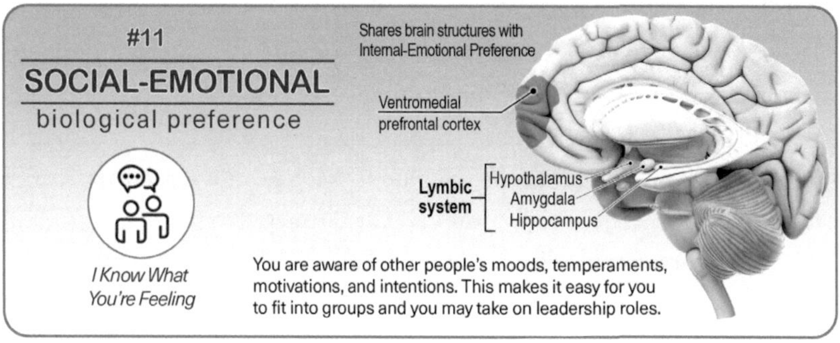

The second type of biological, emotional preference is a social-emotional preference. Being able to read others' emotions allows us to be empathetic, which increases smooth interactions with others and our success in life.[48] Whether you're climbing the corporate ladder, chasing a political dream, pursuing romance, or just trying to get out of the checkout line of the grocery store, being able to read others' feelings makes it all easier. Howard Gardner describes this modality as "the ability to notice and make distinctions among other individuals and, in particular, among their moods, temperaments, motivations, and intentions."[49]

Tests with more than 7,000 people show the benefits of reading others' feelings, including being popular, outgoing, social, and well-adjusted,[50] meaning our social skills are disrupted when the prefrontal lobes are injured by trauma or disease. The radical prefrontal lobotomies that were carried out on psychiatric patients until the 1960s wiped out these individuals' ability to emotionally connect to themselves or others.

Englishman Nicholas Humphrey, who introduced evolutionary psychology over forty years ago, pointed out that those individuals with the ability to intuit the behavior of others not only had the highest capacity to enter into diverse cooperative relationships but also achieved the greatest reproductive success. In *Consciousness Regained*, Humphrey says our social skills go "hand in hand with intellect."[51]

Some anthropologists maintain that social interaction, not tool use, is not only the primary marker of evolutionary advancement but the reason that the human brain increased in size and complexity. This progressively greater capacity for understanding each other's emotions and intentions has allowed us to move from living in small groups of twenty-five or fewer millions of years ago to living in groups of a million or more today in cities like Tokyo and New York.[52]

Individuals with the social-emotional preference as their primary modality have the exquisite ability to judge the emotional tone of others, which allows them to easily fit in with or lead groups. It's the socially dominant among us who act as the glue to keep groups operating at high levels of efficiency. The socially dominant are attracted to roles such as many politicians, corporate heads, mediators, social workers, salespersons, nurses, and psychologists. Mahatma Gandhi, Lyndon Johnson, and Bill Clinton are examples of individuals with primary social intelligence. The fellow who was my youngest son's godfather is a networking marketing guru who has used his emotional-social *Biological Preference* to amass a downline of over a million people.

Whereas internal-emotional processors need to be alone to learn, those with the social-emotional preference desperately need interactions with others to stimulate the neural firing necessary to create a memory. They learn best in groups. Next to the kinesthetic and tactile dominances, they have the most difficulty with traditional Stand and Deliver instruction. The teacher is always yelling at them to be quiet. For these individuals, the lack of social stimulation while they learn impedes logging information into their long-term memory. While those with a social-emotional preference are great at intuitively reading others' emotions, because their neural energy is directed outward, we have found many have a difficult time connecting to their own emotions.

Findings show that serotonin, the main neurotransmitter that regulates mood, is highly concentrated in the brain centers that modulate social intelligence. Research on primates reveals that individuals whose behavior is socially well-tuned have an extremely high concentration of serotonin receptors in these emotional centersÏ[53]

Conversely, low amounts of serotonin have been implicated in depression. Being depressed dramatically limits our ability to interact with others. Depression demands that we deal with this affliction in solitude; we just aren't social when we're depressed. The antidepressant Prozac is reported to act by blocking the uptake of serotonin in the brain, making sure that there will be more of it available to the brain, thus lessening depression and increasing the capacity for social interaction.

Identifying Primary Preferences

The key to remember is that no matter what your primary intelligence, it will hook up to all your other brain areas in unique ways, making you a genius at something your neighbor is not. However, a word of caution about this approach of identifying your preferences. For many of us, our school experience has redirected our twenty watts of neural energy away from our primary intelligence for so long that the result is our "neurological gift" atrophies. The result is we are unable to easily access what is truly meaningful in our lives. I'll explain in more detail how meaning is created in your brain and life in Chapters 12–14 on your *Meaning Network* and motivation.

Cracking the Learning Code

Understanding what your primary *Biological Preference* is provides deeper insights into why you are compelled to learn some things and not others. In addition, being able to recognize the *Biological Preferences* of others allows you to communicate at a deeper level with them.

11

You Learn Through Selection, Not Instruction!

Looking back into the history of biology, it appears that wherever a phenomenon resembles learning, an instructive theory was first proposed to account for the underlying mechanisms. In every case, this was later replaced by a selective theory.

—Niels Jerne, Nobel Prize Winner Physiology[1]

If you passed through the educational system, the title of this chapter might sound outrageous. You could respond, "Of course, we learn through instruction. We spend the first eighteen years of our lives being instructed by our parents and teachers and the rest of our lives being instructed by our bosses, mates, seminars, and corporate trainings. Instruction is how we learn!" While it may appear as if you acquire learning and long-term memory through instruction, we're now discovering that under the microscope of science that this is not true at all. From the fields of neuroscience, genetics, and molecular biology, what we're uncovering is very disruptive to a world that has relied primarily on instructive processes to create long term memory and behavior change for so long. We now know learning occurs from a selective process, not an instructive one. And this is no small discovery!

After years of working with, studying, and researching learning systems and the biology of memory formation, motivation, and behavioral change, I've adopted the sincere belief that the concept that we learn through selection, not instruction, will have a more profound impact on accelerating learning and increasing the joy of behavioral change than any previous learning advancement. Failure to understand this fundamental biological concept is one of the major reasons we are trapped in a world where painful and inefficient learning and motivational systems continue to dominate.

Throughout history, every time science has looked at how learning and change occur in biological systems—such as genes, bacteria, antibodies, enzymes, or brains—an instructive hypothesis has first been presented. In virtually every case, with the acquisition of further knowledge, the instructive model was discarded and replaced by a selective one. Neuroscientist Massimo Piattelli-Palmarini nicely sums up the new scientific view of how we learn and change: "There is no known process, either in biology or in cognition, that literally amounts to learning in the traditional 'instructive' sense… All the mechanisms of acquisition… are due to a process of internal selection."[2] This statement is very disturbing, especially considering all education is based on instructive methods, and in the corporate world instructional design programs account for over $168 billion.

Instruction versus Selection: A Disruptive Difference

There are substantial differences between the biological term *selection* and the academic term *instruction*. The academic term implies a method in which learning takes place through directives from a central authority. Instructive methods view the brain as an empty container to be filled using the tools of force with which every learner is familiar: the reward and punishment of good and bad grades. The biological term *selection* defines the natural process of physiological change that takes place within an organism when it selects new information into its biology that helps it more effectively adapt to its world. Your human brain can only learn through a selective process. This is because, as we will see in great detail, all new information you learn must, in some way, physically attach itself to knowledge that has been previously selected into your brain to be remembered.

It is important to understand that pre-existing information has been encoded into your neurological structures either by your genetic program or your previous encounters with the environment. In selective systems, it is the individual—not some authority like a teacher or boss—that has the power, because it is the individual's brain that is unconsciously selecting new information that resonates with previously acquired knowledge.

Traditional efforts to develop effective learning systems that produce dramatic neurological change and thus dramatic learning have been hampered because we've failed to understand that learning occurs through the biological process of selection, not the academic process of instruction. In simplest terms, instructionists say learning can take place without accessing information that already exists in your brain, while selectionists maintain that all new learning is dependent upon what already exists in your brain.

Any learning or behavioral change system that wants to profoundly impact your biology must rely primarily on selective methods of knowledge acquisition, not instructive methods. This is not a subtle difference! It is the key to understanding what we must do to transform our outdated learning systems. So, here we'll dive deeply into this topic. It's our belief that once you grasp the history and science behind the difference between instruction and selective systems, you'll more clearly see how to accelerate your own and others' learning, motivation, and behavioral-change processes.

A Little History: Instructivism Gets Us Off Track

Ever since the introduction of formalized learning through stand-and-deliver verbal instruction by the Greeks over 2,400 years ago, instructive methods have been the mainstay of creating learning and change. Instructive forms of learning embrace the philosophy of the 17th-century Englishman John Locke, who in his work *An Essay Concerning Human Understanding,*[3] introduced the idea that the brain came into the world with no prior knowledge; it was like a *tabula rasa,* a blank tablet upon which new information could be written simply through instruction. This was very wrong.

From the 1950s through the 1990s, the behaviorist school put a psychological spin on Locke's philosophy by declaring that the brain was like an empty black box into which information could be easily forced with the right leverage. The right form of instruction and force, in the form of reward and punishment, was all that was required. This instruction-plus-force philosophy of the behaviorists is clearly illuminated in the famous statement of one of its founding fathers, an advertising executive turned psychologist John B. Watson, who proclaimed:

"Give me a baby… And I will make him climb and use its hands… And I will make him a thief, a gunman or a dope fiend. The possibilities of shaping in any direction are almost endless."[4]

The mid- to late-20th century was a heady time for educational and corporate systems with behaviorist leanings, for they firmly believed that with the right external motivators and proper instructive methods, they could completely change and control any human-machine. Even today, virtually every learning system that exists has its foundations in the instructional ideas of the Greeks, Locke, and the behaviorists.

Selectionism Struggles to See the Light of Day

The evidence that we learn and change through selection has flown beneath the radar of our corporate and educational systems, as well as the popular press. But ever since the fifth century BC, when the Sophists (Greek for "wise men") first proposed that you could recognize a specific idea only if that idea already existed in your mind, courageous philosophers, psychologists, and scientists have slowly been unveiling the truth about how learning and change take place through a selective process.

In the 18th century, the Prussian philosopher Immanuel Kant proposed that acquiring new knowledge of the world wasn't possible without prior (a priori) possession of certain types of information.[5] But with no understanding of genes and their ability to code for learning passed to us from our ancestors, Kant was unable to speculate as to the origins of this prior knowledge. As we have discussed, the idea that change and learning occurred through a selective process was first given scientific relevance through the work of Charles Darwin. In his profound work *On the Origin of Species*,[6] Darwin presented the idea that all biological change that occurs in animals and humans to help them better "fit" their world is the result of a selective process that he called "natural selection." But as Jonathan Weiner writes in *The Beak of the Finch*: "After Darwin's death, many biologists found it easy to accept evolution and impossible to accept Darwin's chief explanation for it. Evolution, yes, selection, no!"[7] In 1913, William Bateson, a founder of modern genetics, wrote in *A Eulogy for Darwin* that selection was "so inapplicable to the facts that we could only marvel... at the want of

penetration displayed by the advocate of such a proposition."

Few people realize that Darwin was also the first scientist to understand that human consciousness and mental capacities emerged from the process of selection.[8] His books *The Descent of Man* [9] and *The Expression of the Emotions in Man and Animals* [10] were dedicated to the idea that selection gave rise to the basic features underlying human thought processes.

Very soon after Darwin put forth his theory, a small band of psychologists interested in cognition began to look at selection as the basis for learning, motivation, and change in the mind. Only twelve years after the publication of *On the Origin of Species,* psychologist William James was writing letters to friends about the concept that thought followed a process of selection.[11] James was so convinced that the mind used the selectionist model that in his book *Principles of Psychology,* he maintained, "Selection is the very keel on which our mental ship is built."[12] Darwin, James, and the followers of a selectionist representation of the mind were right in their appraisal of how we acquire mental processes, but these early selectionist models were seen as too radical and lacking in evidence to be credible, so they fell out of favor.

"Survival of the Fittest" Getting Us More Off Track!

In addition to the history above, one of the most significant things that prevented us from understanding that we learn through a selective process was the phrase "survival of the fittest." In our quest to understand the biological bases of learning, it is one of the most influential and wrongly interpreted phrases in popular culture.

It wasn't Darwin who introduced this phrase. In fact, it wasn't until five years after Darwin first published *On the Origin of Species* that the most famous sociologist of the day, Herbert Spencer, formally presented the phrase "survival of the fittest."[13] Here, Spencer was speaking not of survival in the animal kingdom but of survival in the world of economic and social life. Because Darwin's writings were somewhat abstruse, Spencer, a wordsmith whose skills had made him an editor of the prestigious *Economist,* appointed himself the unofficial translator of Darwin's natural selection theory of change. Spencer is also given credit for transforming Darwin's phrase "descent with modification" into the catchier term we use today, "evolution." But not being a skilled biologist, he totally missed the idea of change through natural selection Darwin intended and instead substituted his own social idea of change he called "survival of the fittest," which reflected pure force.

During Spencer's day, Newtonian physics had caught the imagination of virtually every scientist and philosopher. Newton's laws painted a picture of the primary mover and source of change in the universe as a pure unadulterated force. Spencer embraced Newtonian thinking and argued that the most powerful or "fittest" in society should rule because they could exert the most force and therefore create the most change. Suggesting those who lacked power deserved to be dominated by the most powerful.

When *On the Origin of Species* was published 1859, it was a cumbersome text for the layperson to comprehend. Darwin endeavored to explain how natural selection occurs when successful traits are selected by a species so that individuals and the species itself may better "fit" their environment. Darwin wrote how plant stamens and pistils, bee proboscises (tongues), birds' curved beaks, and animals' coat markings were selected to help these plants, insects, birds, and animals better "fit" their environment.[14] Spencer lobbied for a different view of the term: To him, "fit" didn't mean those who best fit their world but instead those with the biggest biceps, fleetest feet, and deepest pockets—those who deserved to rule over the unworthy or "unfit."[15]

Spencer's social philosophy painted a distorted picture of Darwin's biology. But visualizing Spencer's concept of the strong overpowering the weak was more comfortable for the public to grasp than visualizing Darwin's theory of how individuals and species changed by selecting variations that helped them better survive and thrive in their environments. In a 19th-century version of the battle of soundbites, Spencer's social philosophy of change and survival through force overpowered Darwin's biological definition of change and survival through selection.

Spencer's social philosophy led him and his followers to oppose social programs such as public healthcare and public education.[16] They were also adamantly against governmental welfare programs because they would allow the physically and mentally "unfit" to reproduce and populate the next generation. It's easy to see how the phrase became embedded in the fabric of the eugenicists' and white supremacists' goal of breeding a "superior" race.

"Survival of the fittest" also became a slogan for those who justified the use of force and cruel practices in achieving their outcomes.[17] Spencer's social philosophy was particularly welcomed by the elite of the Industrial Revolution. Spencer's misinterpretation of Darwinism set off a chain reaction of misunderstanding. That is exemplified by John D. Rockefeller quote, "The growth of a large business is merely survival of the fittest... This is not an evil tendency in business. It is merely the working out of a law of nature and a law of God."[18]

Even in the 21st century, "Spencerian" views are frequently mistaken for "Darwinism." It is a significant reason that learning and motivational systems worldwide depend on grades, smiley faces, demerits, and money rather than on matching new information with knowledge the individual already possesses. The term "survival of the fittest" has probably done more to cloud the world's understanding of how we learn than any other phrase in history.

Science Accepts Genetic Selection

As you saw earlier, in the early 1950s, biologists and geneticists began to conduct research that allowed them to grasp that adaptation and evolution did indeed occur over hundreds to thousands of generations of a species through the process of selection of genes. They found that the structure and function of DNA were altered by a method that naturally selected gene variations (mutations) that allowed organisms to better fit their environment.

Individuals who came into the world with these favorable variations would pass this genetically learned information to their heirs, who, in turn, would give it to their heirs, until the whole species had selected beneficial survival information into its genetic makeup. Basic instincts such as bonding, nesting, hoarding, imitating, and territorial protection were selected into animal species' genetic composition over millennia.

By the middle of the 20th century, scientists could clearly see how the glacially slow genetic selection process could indeed increase a whole *Species Intelligence/Adaptability Factor*. What they couldn't yet comprehend was how learning that occurred in short periods—seconds to days—could take place through a selective process. Up to this time, virtually all scientists believed that the fast learning and change that took place in organic systems, such as bacteria, enzymes, antibodies, and brains, was directed by an instructive process. The accepted scientific belief reflected the theories proposed by Locke and the behaviorists—that biological systems could learn new information and behaviors without accessing previously encoded information. In fact, whenever scientists were confronted with a problem of how a biological system learned over short periods, they relied on an instructionist's model in every case, probably because scientists had personally spent so much time in instructive institutions themselves. This may have limited their ability to see outside of the box.[19]

Immunology Reveals How the Brain Learns Through Selection

As the 1960s dawned, new discoveries showed that, just like genes, other biological systems, such as bacteria and enzymes, also learned through a selective process. But immunology is the field that gave the most visibility and credibility to this concept that was eventually applied to all learning.

FAST FORWARD *or* DEEP DIVE

If you need less detail, you can skip ahead or skim this section on how immunology gave us insights into how all biological systems learn through selection and go to the next section, "Science Accepts That the Brain Learns Through Selection."

Humans are susceptible to millions of disease molecules called antigens. Up until the 1960s, there was almost unanimous agreement among immunologists that the body could produce millions of different antibodies to fight such a huge number of disease antigens only if the antigen played an instructive role. Scientists believed that the immune process worked somewhat like a teacher instructing a student. The antigen would call out to the antibody, "Hey! Here I am. Let's see if you can learn how to 'fit' your form to mine through my instruction." (In most cases, the antibody doesn't really attack disease molecules, as promoted in the popular press. Instead, its structure fits into the antigen, rendering it harmless and allowing the body to rid itself of the invading virus or bacteria.)[20]

This instructionist model painted a picture of your immune system producing unlimited antibodies to fight all diseases. But eventually, immunologists came to the realization that there were some diseases for which your body could not produce antibodies. No matter how many invading antigens the body encountered, no antibodies would be produced to fit the disease molecules. (This is why HIV is such a devastating virus. Your body can't naturally produce an antibody for it.) When scientists recognized that your body couldn't make antibodies for some diseases, the learning-by-instruction model of immunology broke down. Now immunologists had to look elsewhere for a theory to explain antibody production.[21]

The work of Nobel Prize winners Gerald Edelman and Niels Kaj Jerne helped uncover a two-stage learning process in the immune system: one slow stage at the level of your genes that create antibodies, which took eons to accomplish, and a fast one, which took seconds to days to accomplish. This second stage improved upon the previously selected genetic information that already existed in antibodies.[22]

Jerne and Edelman's discovery was profound: there was no instructive process in the shaping of antibodies; instead, information that had been previously selected from favorable genetic variations was coded into them. Therefore, when the immune system encountered a disease antigen, it would draw on this pool of pre-existing genetic knowledge to select and build upon an antibody with a shape that would render the antigen harmless.

They found that this information on how to build antibodies that fit invaders was already encoded in the genes of the immune system. This process is why your immune system came into the world with pre-encoded knowledge that helps you fight certain diseases, such as chickenpox, mumps, and measles. Jerne, in his now historic paper, "Antibodies and Learning: Selection versus Instruction," wrote:

> *It follows that an animal cannot be stimulated to make specific antibodies unless it has already made antibodies of this specificity before the antigen arrives. It can thus be concluded that antibody formation is a selective process and that instructive theories of antibody formation are wrong."*[23]

Jerne, Edelman, and other dedicated scientists also discovered that the immune system has a second level of rapid learning that increased the *Individual Adaptability/Intelligence Factor* of an antibody after it encountered the disease molecule. They found that after an antibody encountered an antigen, the antibody's shape didn't remain fixed; instead, the immune system could quickly learn how to improve upon the antibody's initial shape, so it better fit a disease antigen. In immunology, the term avidity refers to how well an antibody fits an antigen. It was found that as long as antibodies already contained information that let them recognize invading disease molecules after the first set of encounters with the disease, successive generations of antibodies quickly adapted and develop higher avidity. This is the learning process in action. This quick adaptability is the basis of why we get immunology or vaccine shots. Very low doses of pathogens that cause diseases such as diphtheria, tetanus, and whooping cough are introduced into your body. These low doses will not overwhelm your immune system, but instead stimulate the preselected knowledge in your immune system to release antibodies for these diseases. Once stimulated by the virus shot, your immune system begins to learn to improve the avidity of antibodies by selecting more exact shapes to better fit the disease molecule.

> **The most important principle of selectionism can't be overemphasized; all new learning is dependent upon information that has already been laid down by your genes and previous environmental encounters.**

Later, if you're exposed to a life-threatening disease such as diphtheria, your immune system has already selected refined shapes that can immediately nullify the disease without severe consequences to you. This second level of rapid learning is an amplification of the genetic information that was first encoded in the antibodies. Improved avidity through exposure to antigens is an example of the "enhancement through use" concept that Lamarck introduced fifty years before Darwin published *On the Origin of Species*.

Science Accepts That the Brain Learns through Selection

After Jerne and Edelman integrated their initial findings on antibody learning, they both came to the same conclusion: The two-stage selective learning process in the immune system also occurred in the brain. In the 1980s and 1990s, Edelman, who founded the respected Neurosciences Institute, wrote four highly influential books, and introduced his theory of neural group selection to a wider audience. These works on how we learn through a two-stage selection process revealed a huge piece of the "how we naturally learn" puzzle. This research has now become a pillar of the neuroscientific position of how the brain learns through selection.

These works are quite complex, and Edelman had such a distinct way of writing that it took me over a year to comprehend them. I spent so much time with his works that I felt I knew Edelman on a personal level. I felt so connected to him, and appreciative of the enlightenment he provided on how we learn, that when I heard he died in 2014, I found myself crying.

Today—more than 130 years after Darwin and James proposed that the mind learns through selection, not instruction—the most prominent neuroscientists and biologists of our age embrace selective learning theories. In *Selectionism and the Brain*, neuroscientists Olaf Sporns and Giulio Tononi point out, "Selectionism offers the most promising avenue toward a comprehensive, explanatory, and predictive theory of the brain."[24] Steven Pinker, professor of brain and cognitive science at MIT, says, "The brain… cannot be understood without knowing the forces that gave rise to that organization… particularly natural selection."[25]

Neurophysiologist William Calvin, in his book *The Cerebral Code*, maintains that the brain is so dependent on the process of selection that it should be called the "Darwin Machine."[26] Noted neuroscientists Michael Gazzaniga, Richard Ivry, and George Mangun wrote in *Cognitive Neuroscience: The Biology of the Mind*: "Natural selection is the only known process in which functional organization emerges in a natural world."[27] Gazzaniga and his co-authors also lament that most practitioners of cognitive science don't yet fully appreciate the insights offered by the selective learning model of the mind.[28]

Applying Selective Learning Theories

Although research into how to apply selective theories of learning are in their early stages of development, they have three common sets of principles:

- All new learning has its basis in the information that has previously encoded into your brain structures
- Previously learned information can be enhanced and refined most effectually through real-world environmental interactions
- Variations in our genes and in our experiences build unique

neurological structures, which in turn ensure each of us will select and attach meaning to different elements from our environment. In other words, all 7.5 billion of us on the face of the earth will attach meaning to and learn in different ways

The most important principle of selectionism can't be overemphasized; all new learning is dependent upon information that has already been laid down by your genes and previous environmental encounters. As Olaf Sporns says: "A selective system has a repertoire of pre-existing variable functional components, amplifying those that perform best or match to the environment."[29]

Selection is Based on What You Find Personal Meaningful

Selectionists emphasize you can't learn anything new without this new information in some way resonating with the knowledge you already possess. They point out that new neurological growth, which represents new learning, cannot exist in thin air but must attach itself to existing physical structures, which represent previously learned information. What this means is that all new learning is dependent upon internal referral—that is, on what you already know and hold meaningful.

As pointed out, this view is very disturbing to the instructionist/behaviorist model of learning, which has maintained that force, in the form of reward and punishment, is the primary motivator in producing behavioral change and long-term memory. Unfortunately, most motivational and behavioral programs have been so unsuccessful in helping us get what we want in life because they are dependent on these reward-punishment models. By not understanding the biological process of selection based on personal meaning, we have sucked joy, fulfilment, and efficiency out of our learning and behavioral-change environments. (More on the failure of reward and punishment systems in Chapter 18.)

New Learning Cannot Exist in Thin Air: Explained

To accelerate your speed of learning, you have to understand that all new information must, in some way, be physically associated with existing neural groups that represent past learning—or else it can't be effectively selected into your brain structures. I'm simplifying greatly what was discussed earlier in the book, but for our purposes, we can say that new learning arises from the addition and strengthening of connections between neurons called synapses. Each neuron has an average of 5,000 synaptic receptors on numerous dendritic branches that radiate from it, like spokes on a wheel. A dendrite's primary job is to act as a structure to hold the synaptic receptor buds that represent long-term memory.

Like a tree with splitting branches, a neuron can have up to eight successive branching dendrites. The first, second, and third branches are called lower order; fourth and up are called higher order.[30] In an oversimplification that works for our purposes, we can say that the lower-order dendritic branching holds information coded for by your genes, and the higher-order branching is prompted by environmental interaction. The important thing to understand is that both first- and second-order dendrites hold information on their branches that represent knowledge that has already been acquired. For new information to become part of long-term memory, new synaptic buds must attach themselves to these existing dendritic branches that represent past learning laid down by your genes and previous environmental encounters. This is why we say new learning cannot exist in thin air but must attach itself to your neurological branches that hold past

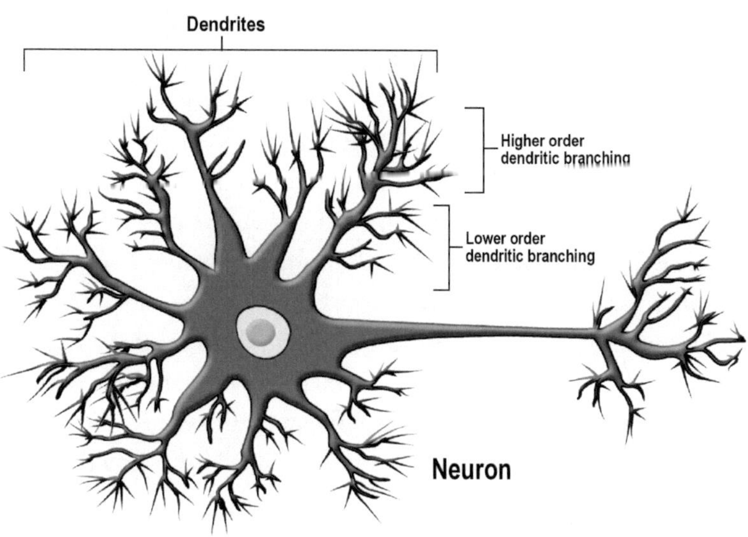

Your lower dendritic branches are in place at birth. They hold information encoded by your genes. Your higher order branches are primarily formed by your environmental interactions after birth.

learning.

The reason that bats can select new information that deals with ultrasonic sounds and cats can select new information about ultraviolet light is that bats and cats have dendritic branches that house these kinds of sounds and sights. Humans will never be able to select ultrasonic or ultraviolet memories because we have no neural structures for this type of information to attach to. In his book *Synaptic Self*, neuroscientist Joseph Ledoux notes:

> *New connections formed by activity are not created as entirely new entities but rather are added to intrinsically determined preexisting connections. Added connections are, therefore, more like new buds on a branch rather than new branches.*[31]

Your Brain's Two-stage Learning Process: How It Works

Like your immune system, your brain has a two-stage selective learning process. The first stage, at the level of genes, took eons to acquire; the second, at the level of the neural tissue, takes only seconds to days.

Each of us comes into the world with genes that are packed full of relevant survival information that has been previously selected by our ancestors' successful interactions with their world. These genes build our lower limbic/stem system, which houses instinctual learning such as bonding, imitating, and mating, as well as basic emotional concepts such as anger, joy, and fear. Your upper, more-evolved neocortex comes into the world prewired by your genes with rough networks that deal with visual, smell, touch, taste, sound, language, math, motor, musical, spatial, and social concepts (see Chapter 9).

To improve on this previously selected information, the second stage of learning takes place as your environmental interactions help you to add more and more synapses and dendritic branches to these existing structures. Your personal experiences then allow refinement of the neural networks that have been prewired by your genes. This refinement of genetic information through experience is very similar to the improved avidity ("fit") that Jerne and Edelman discovered in the immune system.

As an infant, you entered the world with a set of rough neural networks preset by your genes, which allowed you to immediately recognize a cornucopia of auditory, visual, taste, smell, and touch stimuli. Environmental encounters then helped you select new information that resonated with the existing networks and, over time, through experience, refine this information, thus increasing your *Individual Adaptability/Intelligence Factor* and making it easier for you to fit your world.

For example, when you exited the womb, your genes had already set up visual networks that helped you focus on your mother's face. After multiple environmental interactions with other people, your initial visual "mother's face network" was dramatically adapted by adding synapses to existing dendritic branches; eventually, that network would allow you, as an adult, to recognize thousands of faces. Harvard neuroscientist Jerome Kagan says: "No act, idea, image, or word is learned in isolation or ever becomes completely isolated. Every mind consists of nests of interconnected elements that are continually being reorganized with use.[32] Echoing the sophists, Jerne sums up the process of selecting new information: "Learning consists of being reminded of what is already in the brain."[33]

Because instructionist methods approach the brain as if it were an empty vessel to be filled by authorities' directives, information that you've previously acquired and deemed personally meaningful is seen as having little relevance to the learning process. This view perpetuates the idea that if someone can just force you to memorize enough facts—even if they aren't meaningful to you—you'll remember them. Of course, as we all have experienced, this isn't true at all. We'll go much deeper into the value of personal meaning and why it is the Holy Grail of learning motivation and behavioral change in the next chapters.

Cracking the Learning Code

Science has discovered your brain doesn't learn through an instructive process, but a selective one. Therefore, new information must in some way resonate with information you've previously acquired, or it can't be remembered. Most learning organizations in the world are ignorant of this fact, which makes the whole process of creating memory formation more difficult than it needs to be.

1 **PART I– WHAT'S NOT WORKING AND WHY:**
Human Biology at Odds with Traditional Learning

2 PART II– THE SCIENCE OF MEANING:
You Only Learn What is Meaningful to You

3 **PART III– ACTIVATING YOUR UNIQUE LEARNING CODE:**
The Big Picture – Meaning, Concepts, Details

4 **PART IV– KEYS TO THE LEARNING PROCESS:**
The Cycle of Transformation

12

Meaning: The Holy Grail of Learning

When I went to school, they asked me what I wanted to be when I grew up. I wrote down happy. They told me I didn't understand the assignment, and I told them they didn't understand life.

—Attributed to John Lennon, musician and artist[1]

One of the tenets of the Learning Code is there is no learning without meaning. In this chapter, you'll discover the science behind what you intuitively know: you remember what is meaningful to you and forget almost everything else.

Personal meaning isn't just the criteria by which information is selected into your long-term memory bank and the fastest way to increase your *Individual Adaptability/Intelligence Factor*, it also creates the neurochemistry that motivates you and helps you feel joy. From the viewpoint of biology, genetics, and neuroscience, personal meaning is the Holy Grail of learning, motivation, and behavioral change, which makes it a primary key to unlocking your Learning Code. Looking at learning through the lens of science brings another dividend. We're able to answer the question, where does meaning come from?

The Powerful Effect of Meaning on Learning

In this chapter, you'll get an overview of the powerful effect personal meaning has on motivation and learning. In later chapters, we'll go a little deeper into the science, and you'll discover that our traditional authoritarian corporate and scholastic systems, which rely primarily on reward and punishment to impose learning, sadly and unwittingly create a neurological climate that not only can inhibit motivation and long-term memory formation but may even breed out your ability to create personal meaning and construct the

foundations for the psychological condition called codependency.

What I hope to convey is how wasteful it is to try to deliver information to a learner that isn't meaningful to them. What you'll discover is to get what you want in life, isn't all about grit or persistence, but connecting to what is most meaningful. You'll also discover that your brain has a *Meaning Network*, a group of neurological structures that must be stimulated before new information can be selected into its long-term memory banks. If these structures aren't stimulated by what you're learning, not only do you remember very little, but your levels of motivation, satisfaction, and joy plummet.

After you finish reading these important chapters, I'd like you to ask yourself the following four questions:

- How could we have developed an educational system where no one ever asked me what was most meaningful to me as a unique human being in the world?
- If the educational systems had *Wrapped* the knowledge they wanted me to learn in what was most meaningful to me, would it have changed the way I'm living my life now?
- By not activating my *Meaning Network* during my twelve to eighteen years of study, did the educational system damage my ability to uncover what was most meaningful to me in my life? Is this why I often feel unfulfilled?
- What would happen if our learning systems created people who knew exactly what was deeply meaningful to them? So that each one of us could follow that meaning with passion?

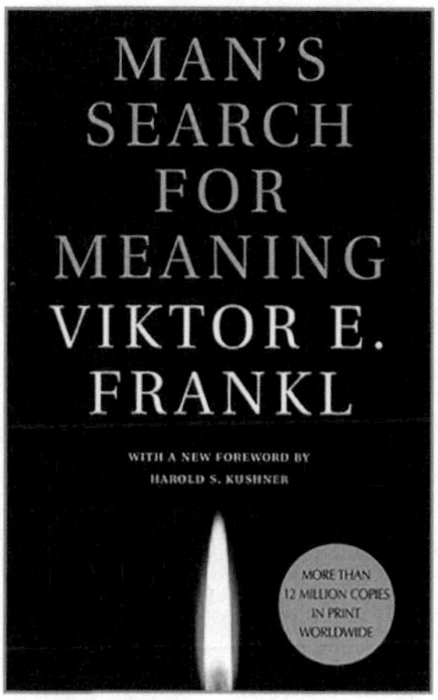

Frankl Shifts Our Focus to Meaning

The individual given credit for focusing the minds of modern researchers on the value of personal meaning is one of the 20th century's most influential psychiatrists. Dr. Viktor E. Frankl's profound book, *Man's Search for Meaning*, described his experiences in a Nazi concentration camp and shows not only that human behaviors are driven by the search for meaning, but in some situations our very survival depends on it. Frankl discovered that of the prisoners who were not put to death in the gas chambers, it was those who found meaning, even in their deplorable conditions, who had the greatest chances of survival. Frankl view was in stark contrast to behaviorist psychological which was dominant at the time. Frankl insightfully maintained that a person's main drive "is not to gain pleasure or avoid pain, but rather to see meaning in his life."[2]

> **In the end, you remember information not because you memorized it or took a test but because of how meaningful it is to you.**

Frankl's experiences allowed experts in diverse fields to begin to recognize the primary motivational force in human existence, and the basis of memory formation is not whether you get a carrot or stick but what you find intrinsically meaningful.[3]

To give you a quick overview of some of the research into meaning just look at the following:

- Philosopher Susan Langer found a pervasive human need is "to invent meanings and invest meaning into one's world" and "to search for and find significance everywhere, to transform experience constantly to uncover new meanings in everything."[4, 5]
- Neil Postman and Charles Weingartner, in their book *Teaching as a Subversive Activity*, also assert that all learners are essentially "meaning makers."[6]
- The physicist David Bohm looked at the topic through the lens of quantum physics and concluded that meaning is the essence of life and the primary factor by which all energy and matter are organized.[7]
- Margaret Wheatley, who uses research into complexity and chaos theories to develop and analyze effective management techniques, writes in *Leadership and the New Science*: "Meaning… serves as a point of reference. As long as we keep the purpose in focus in both our organizational and private lives, [...], we emerge with a discernible pattern or shape to our lives."[8]
- Oliver Sacks, noted author and professor of neurology at Albert Einstein College of Medicine, also describes the "persistent notion

that one may have an intelligence… based on pure logic, without anything so messy as "meaning" being involved," as impossible.[9]

- Accelerated learning pioneers Renate and Geoffrey Caines looked long and hard at the value of meaning in education and maintained, "If we want students to use their brains more fully, we have to teach for meaningfulness."[10]

There's also an extensive body of research that concludes that reading is not a matter of decoding words but a process of extracting meaning. The experts echo a central tenet of Jean Piaget's work—that is, human beings are continuously engaged in making meaning from stimuli presented to them in their environments.[11] In the end, without meaning, your brain has no focal point around which to organize its structure or its functions.

Your Brain Is a Meaning Selector

We already know the animals that made it through the sieve of evolution are those that possess a brain structure adapted to select the information which supports surviving and thriving. Your brain is bombarded with billions upon billions of bits of auditory, visual, tactile, taste, and smell information every second. As you now know, your brain couldn't possibly select all this miscellaneous information, or your head would be as big as a blimp.

One way to look at your brain is as a personal-meaning selector that scans the billions of bits of incoming information bombarding it each second and asks, "Which of these bits will help me survive and thrive, and which won't?" Your brain then creates the neurological changes that allow you to select the important stuff and discard the rest. Peter Senge, director of the Center for Organizational Learning at MIT Sloan School of Management, who has spent most of his adult life researching and implementing learning systems at corporations such as AT&T and Proctor & Gamble, notes, "Learning is about enhancing the capacity to produce results that matter to you."[12]

Some brain-based educators use the acronym WIIFM (what's in it for me?) to describe the criteria the brain uses to select information. As you'll see, one group of researchers who tried to sort out how students remember what they'd read found that a passage's significance to the reader was 3,000 percent more significant than how readable it was![13] In the end, you remember information not because you memorized it or took a test, but because of how meaningful it is to you.

Of course, those who design learning systems are continually looking for ways to trick your brain into remembering—using things such as peg systems, acronyms, and mental imaging in the hope these strategies will help you learn information faster and more effectively. But you don't have to rely on gimmicks to remember information. Neuroscience has shown the easiest and fastest way to create new learning is for any new information to resonate with previously encoded information.

This is because your brain's main job is to continually search the environment for new information that meshes with these potent pre-existing neural networks, which are coding for personal meaning. These networks then act like powerful magnetic hooks, attracting and snagging new information that resonates with them. As these networks of cells incorporate more information, they become larger and larger, making them more potent in attracting more of the kind of information that harmonizes with them. Made out of solid stuff, fats, and proteins (like soft cheese, as I described in Chapter 1), these robust networks have mass and attraction and are why it's always easier to remember information about subjects and activities that are already significant for you.

The more meaning you place on something (whether it's enjoying food, tracking baseball statistics, solving mathematical equations, sewing, fly fishing, acquiring apartment complexes, redecorating your house, or playing the stock market), the easier it is for you to select new knowledge that resonates with what you already consider important. Because of genetic variation, each of us will place a different value on stimuli from our environment, making each person proficient in an area most of our fellow humans are not. As memory researcher Ulric Neisser says, "Every person is a prodigy to his neighbors, remembering so much that other people do not know."[14]

Meaning Grows Your Brain Tissues

As I described in Chapter 2, the creation of new synaptic buds and dendritic growth is driven by an electrochemical genetic process. Any time you encounter information with personal meaning, large numbers of neurons fire simultaneously, which sets off the neurochemical process that blows magnesium plugs out of your NMDA receptors; this, in turn, prompts your genes to express the proteins that build the synaptic structures that hold new learning.

A metaphor for the acquisition of new learning in the brain is how a tree grows. A tree can grow new branches and leaves only from what already exists—the trunk and existing branches. The trunk of a tree divides into two or three major branches, which in turn split into secondary smaller branches, which bifurcate again and again to hundreds of smaller and smaller leaf-bearing twigs. The same is true in your brain.

Your brain's branches are called *dendrites,* and they extend from neurons and hold synaptic buds, which equate to the leaves on the tree. The dendrites and their synapses are the neural structures that hold learning. New learning occurs when new synapses are added to your existing dendritic branches.[15] Dendrites grow in length so that they may hold new synapses, which represent newly learned information. When dendrites reach a certain length, they split so that they may hold more and more synapses.

Also, as discussed, neuroscientists break down dendritic branching into lower and higher orders. *Lower-order branching* consists of the first, second, and third dendritic branches, and the fourth, fifth, sixth, and above branching is called *"higher order."*[16] Neuroscientists point out that most of the lower-order dendrites are in place at birth, whereas higher-order dendrites develop after birth.

Pleasure-inducing Neurochemicals

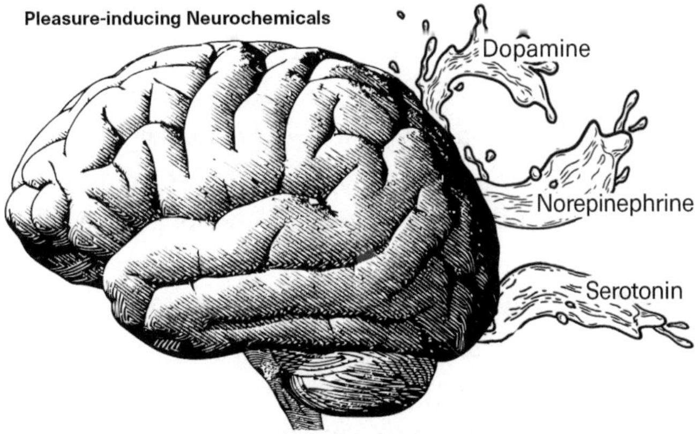

When you pursue what's meaningful to you, you stimulate the structures that prompt the release of three of your brain's powerful pleasure-inducing neurochemicals called monoamines: dopamine, norepinephrine, and serotonin.

In a gross oversimplification, we can say that your lower-order branching is laid down by your genes, and higher-order branching is new learning selected by your environmental interactions. Neuroscientist Arnold Scheibel, former director of the Brain Research Institute at the University of California at Los Angeles, says, "It is not quite that tidy, but is in the right direction."[17] On the level of neural tissue, then, your job as a learner is to add synapses to existing branches, causing higher-order dendritic branching to occur over and over again.

The point is that synapses and dendrites can't stand alone. Neuroscientists don't find synaptic buds and dendrites floating unattached in the brain like so many loose butterflies. (unless there's severe trauma). All new learning—that is, new synaptic buds and new dendritic branching—must attach itself to existing neural tissue. Therefore, we can say that the only way new information finds a place in your brain is by attaching itself to structures that hold meaningful information learned in the past.

What we find significant as a child or adult is directed by these previously encoded neural networks. The brain tissue that represents this meaningful information is continually searching our world to attract and hook information that matches it. As the Cambridge psychologist Fredric Bartlett wrote, about memory, "It's an imaginative reconstruction or construction, built out of the relation of our attitude towards a whole mass of organized past reactions or experience."[18]

When new information doesn't match your pre-existing value networks, the latest data doesn't have much of a chance of being selected into your brain structures. As you saw in Chapter 2, the brains of Lettvin's frogs could select only information that matched pre-existing networks that held value, such as bugs, leaves, and other frogs, and would select out gum wrappers and pieces of pizza, which held no meaning. Nobel Laureate Gerald Edelman points out, "The driving forces of animal behavior are thus evolutionarily selected value patterns that help the brain, and the body maintains the conditions necessary to continue living."[19]

The Neurochemical Pathway to Joy and Fulfillment

Meaning not only sets up the neurochemical climate for you to learn faster but also stimulates chemical pathways that allow you to feel joy and fulfillment.[20] It's critical for all animals to know when they're proceeding in a favorable direction, so neurochemical circuits give positive feedback when you're engaged in meaningful behaviors such as eating, sex, a favorite hobby, or a wonderful job. When you pursue what's important to you, you stimulate the structures that prompt the release of three of your brain's powerful pleasure-inducing neurochemicals called monoamines: dopamine, norepinephrine, and serotonin. The primary reason for seeking out meaningful things and repeating meaningful behaviors is that when you experience them, you stimulate these three potent neurochemical systems.

According to noted Claremont Graduate University psychologist Mihaly Csikszentmihalyi, optimized learning and fulfilment occurs in a state he calls "flow." This uninterrupted psychological state is marked by a sense of timelessness and self-pleasure. In a flow state, you become one with the activity lost in what is fulfilling and meaningful.[21] The evidence suggests that the chemical basis of Csikszentmihalyi's flow state is the stimulation of the norepinephrine, serotonin, and dopamine systems.

One reason that school and work can often feel so boring, unfulfilling and stressful is that your involvement with it doesn't access the vital networks that hold personal meaning, and therefore doesn't fire your norepinephrine, serotonin, and dopamine systems. Hans Selye, who helped discover the physiological basis of stress, often noted that the most stressful condition a person can face is "not having something to believe in."[22]

Your internal motivation also has its genesis in what you find meaningful. The word motivate comes from the Latin *motivus*, meaning "to move." When we're internally motivated, the neurological structures and chemicals that support learning, creativity, and fulfillment are automatically stimulated so that we move in meaningful directions. The Caines write, "Research confirms the search for meaning is at the heart of intrinsic motivation and that much of the energy and drive to pursue goals and engage in essential tasks comes from the search for meaning."[23]

You're able to accomplish so many great things when you're driven by internal motivation because following what's personally meaningful sets up a neurological climate that is joyful and helps you easily select large amounts of new information, which, in turn, helps propel you toward personal success. This is why we encourage those who want to pack their lives with fulfilment to "follow the meaning... not the money."

When your endeavors have personal meaning, new learning can easily attach itself to existing structures, and norepinephrine, serotonin, and dopamine, then motivate you to pursue your goals in the face of adversity and hardship. It is not grit that drives us but meaning. If we look back at individuals who've made a difference in the world, such as Beethoven, Darwin, Edison, Einstein, Churchill, Buckminster Fuller, Helen Keller, Martin Luther King, Carl Sagan, and Stephen Hawking, we can see that their great discoveries and achievements were carried out in the face of huge personal challenges and initial failures primarily because their work was so meaningful to them.

As you'll see in Chapter 18, learning systems that rely on extrinsic motivators, such as reward and punishment, are limited in their ability to increase our *Individual Adaptability/Intelligence Factors* because external motivators fail to stimulate the brain structures that expedite knowledge acquisition and feelings of contentment.

Following Meaning All the Way to the Top

Pro-football coach Jimmy Johnson had one of the most successful NFL coaching careers with the Dallas Cowboys and Miami Dolphins. At the end of the 1990s, Johnson decided to retire, but it was short-lived. Exactly one day later, the owner of the Miami Dolphins convinced Johnson to go back to coaching for another year, even though his heart and neural networks weren't in it. The result of following a path that lacked meaning for him was disastrous. He got into a conflict with his quarterback, Dan Marino—one of the greatest of all time—and managed to alienate half of Miami with his coaching style. He finally exited the NFL in 2000 after suffering the worst playoffs defeat in NFL history. Johnson now does TV announcing. We must ask how different his life might have been if he'd followed what was meaningful to him in 1998 and left the NFL on top!

The Failure of Traditional Education to Support Meaningful Learning

The scientific evidence all points to the greater the meaning, the faster you'll learn, so why doesn't traditional education support this process? Think about your own experience, did anyone ever ask you what was most meaningful in your life? For most of us, the answer is no!

With meaningful self-knowledge, we can more easily take the knowledge from education, teachers, and other sources and use it to self-direct our own learning and motivation. However, by not allowing us to access the deeper dimensions of what is meaningful to us as children, traditional education traps us into chasing extrinsic rewards. The materialistic "golden balls" of clothes, cars, houses, and more "likes" on Facebook… to get brief hits of the neurochemicals of joy. But sadly, this extrinsically motivated materialistic treadmill doesn't lead to fulfilled and joyful lives,

We are all continually striving to stimulate the large pre-existing networks that hold what is personally meaningful, which in turn prompts the production of neurochemicals that allow us to have focus, fulfilment, and pleasure. From the minute you exit the womb until the day you take your last breath, your very existence is driven by making meaning out of the matter and energy your world presents to you.

Cracking the Learning Code

The bottom line is: if you want to learn how to live a life of peace, joy, harmony and success, you need to place yourself in situations that support what's deeply meaningful to you.

13

Where Does Personal Meaning Come From?

What you choose also chooses you.

—Kamand Kojouri, Iranian-born poet and writer[1]

In researching the power of meaning on learning and motivation, two questions arise. The first: "Where does personal meaning come from?" That is, by what criteria do you decide which of the billions of bits of information your eyes, ears, skin, nose, and tongue are bombarded with every second will be selected into your long-term memory and which won't? The second question: "Why is there such divergence in what individuals find meaningful?" Why does one person devote her life to say, fly-fishing, while another would never consider putting his toe in a stream, let alone touching a cold, slippery fish?

Over the past centuries, the debate about the source of personal meaning has been confined mainly to the realm of philosophy, but recent research and discoveries in biology, genetics, and neuroscience means we can now look at the question from a scientific perspective. Through this lens, there can be only two sources of personal meaning and meaning variation: our genes and our environment. These elements have a dramatic effect on what you find meaningful because they actually build your neurological structures from which meaning springs.

How Your Genes Affect What You Find Meaningful

At conception, your genes carry information that, for millions of years, has ensured our species survival. As discussed in Chapter 5, these genes have been pre-encoded to build lower limbic/stem brain structures, which house your instinctual drives such as bonding, mating, nesting, hoarding, imitating, and marking your territory. These are the structures that hold the values that have become meaningful to you and prompt you to move in directions that help you find the right mate, make a comfortable home, collect too much stuff, imitate your peers, and protect your property.

As we'll delve into more deeply later, the environments we inhabit also build neural networks in our upper neocortex, which then become another source of personal meaning. This is why Wisconsinites come to love bratwurst, beer, and the Green Bay Packers; why the Japanese treasure sushi, sake, and sumo wrestling; and why Iraqis value Islam, Italians favor Catholicism, and Cambodians are drawn to Buddhism. Unlike the hard-wired limbic/stem structures, your neocortex is extremely flexible, which makes it susceptible to being shaped by your environments. The longer you're exposed to a particular environment, the more your neocortical networks alter their shape to fit it, and the larger and more powerful these networks become. This shaping and expansion of neocortical networks is the process by which you come to value things. In the end, it can be said, your genes and your environmental encounters have built a brain where every single one of your neural networks stores information that you find meaningful.

As we explored earlier in the book, genetic variation is produced in three major ways: during meiosis, at conception with the formation of the zygote, and through randomly trying different combinations (mutations) of the DNA base pairs (see also Chapter 8, "Jumping Genes"). It is the pressure of these three sources of variation that ensure the chances of two humans having the same genetic makeup is 10^{1200}. To get the idea of how huge a number this is, it has been estimated that there are only 1082 atoms in the known universe![2] This high level of genetic variation ensures none of us will ever find the exact same things meaningful. Therefore, one person with a musically dominated intelligence may favor pitch, another tone, and another, lyrics. The brain stem, limbic, and neocortical structures are so

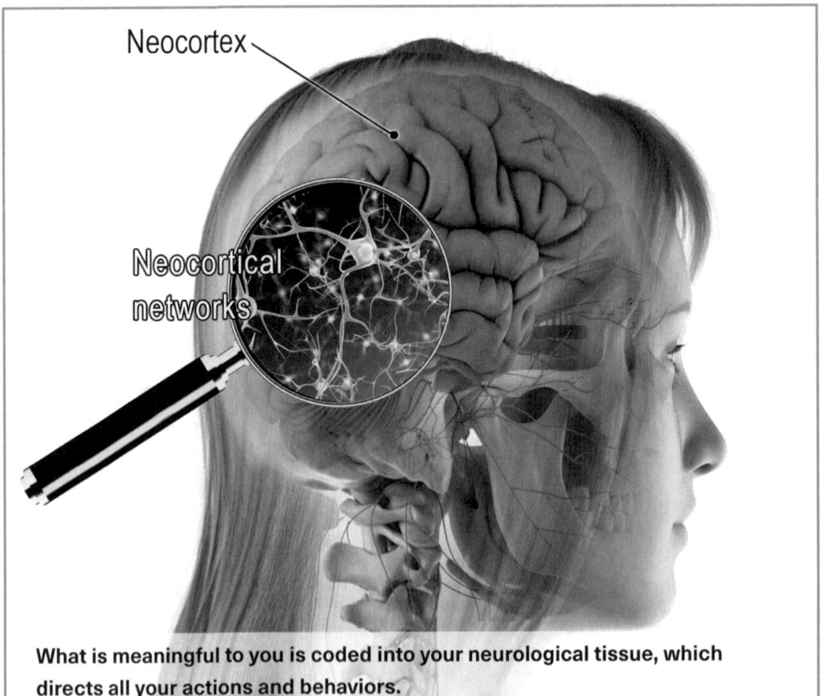

What is meaningful to you is coded into your neurological tissue, which directs all your actions and behaviors.

varied that two people can't possess the exact same neural connections that would allow them to find the exact same things meaningful.

As you roll down the hill of life, the unique neural networks laid down by your genes are made even more unique by your environmental encounters. When this environmental-caused brain variation is laid down over the variation due to genetics, such great brain diversity results that similar brain regions in normal, healthy people vary by as much as 300 percent in size. This is why brain scans show that even identical twins have very different brains. Slight differences in twins' gene expression and slight-to-great differences in their environmental stimuli, such as where each one sits at the dinner table or what side of the bed they sleep on, shape their brains differently.[3]

Accelerate Learning with Hot Networks and Wrapping

When pre-existing networks that code what you find meaningful are turned on and firing, they become what we at The Institute call "hot." As you've seen, large numbers of neurons firing simultaneously, in turn, cause the chemical cascade that produces the synaptic growth that holds long-term memory. Nothing creates simultaneous firing in large groups of your brain cells more efficiently than presenting something to your brain that it already finds meaningful. When your value networks are hot, it's easier for new information to bond to these stimulated structures.

The concept that new learning easily attaches itself to networks that hold personal meaning provides a wonderful opportunity for anyone who wants to create environments for accelerated learning. A process called *Wrapping* purposefully matches new learning with what an individual already holds significant. The job of the educator becomes to pre-fire, or make hot, the student's existing information networks and then introduce or *Wrap* new knowledge in the context of those values.

Metaphor is a powerful form of learning because it takes advantage of pre-stimulated *Hot Networks*. No matter what you want to learn or what you want others to learn—whether it's science, math, reading, management techniques, computer programs, or moral ideas—learning is accelerated when new information is *Wrapped* in meaning. We used this concept earlier. It is easier to understand and remember how synapses grow on a dendrite because you can connect this information to the concepts of trees, buds, and branches that already exist in your brain. So consider that if you want to teach math to a child, and you know that she finds firefighters meaningful, you can most effectively teach her how to count by *Wrapping* the task in the context of being a firefighter. You can show her a picture—or better, a model—of a building on fire and ask her which floor the fire is on. Then you can ask her the number of rungs on the ladder she must climb to reach it.

In the same way, if you want to teach classical physics to a student who loves football, you can show him how Newton's second law of motion (constant acceleration) and Newton's three laws of conservation of momentum apply to more efficient blocking, tackling, and passing. If you want to teach geography to a child fascinated by natural disasters—and many children share this fascination—you can use fires, floods, volcanoes, and earthquakes. For example, you can connect Chicago, and its place on the map, to Mrs. O'Leary's barn and the great fire of 1871.

Wrapping is a disarmingly simple concept. However, it doesn't lend itself to the mechanistic, impersonal, factory model of learning. For *Wrapping* to be effective, both the learner and the teacher must know what the learner values in life. Dialogue and relationship are instrumental in the process, and *Wrapping* works most efficiently when there's a bond of trust and understanding between the learner and the teacher. Those who have applied the *Wrapping* technique find that the teacher's role changes, from trying to cram information into the passive learner, to a mentor helping the student become a fulfilled human being. In my experience, this is what most educators desperately want the learning process to be, anyhow.

Wrapping in the Classroom: Building Greater Cooperation and Empathy

At The Institute, we recommend that each teacher or trainer finds out the three most meaningful things for each of their students and notes that information in their attendance book next to the learner's name. As a teacher, you can ask students one on one or break the students into discussion groups of three to five. The key isn't to try to influence what's meaningful to them, which could be something you don't find meaningful, like video games, superheroes, or a particular sport. What's important is that you gain knowledge about what each child loves—no matter what it is—so you can *Wrap* the new knowledge you want to teach that child in what's already meaningful to him or her.

Creating Immediate Understanding

At one school The Institute was supporting, a third-grade teacher was having great difficulty teaching a learning module on bridges. She found that the boys were highly engaged in building elaborate models, but the girls couldn't care less. One little girl, who we'll call Peggy, was very disruptive and was interrupting the class's ability to pay attention. When I asked the teacher what was most meaningful to Peggy, the teacher said, "I have no idea." I then explained the science of meaning and *Wrapping* to her. She understood immediately.

When I came back to the class a week later, the teacher happily said things had changed. When she took the time to discuss with Peggy what was meaningful to her, she found that Peggy really wanted to become a nurse. The teacher then reframed how she presented bridges to Peggy. She explained that bridges are just like bones and ligaments, and if Peggy wanted to be a good nurse, the more she understood about bridges, the better she could help people heal and the better nurse she could become. The result was that Peggy immediately started paying attention and became engrossed in the bridge module, and class disruptions decreased.

We've found that when teachers take the time to have meaningful discussions with the whole class, first in small groups and then in the classroom as a whole, an interesting dynamic emerges: students begin to see each other for who they really are as human beings. Peer pressure lessens and they begin to focus less on the clothes and shoes they wear, what homes they live in, how rich their families are, and more who they are on a deeper level. Really, isn't this what we're looking for from our fellow humans?

We've even found that these meaning discussions help kids better support one another because they are now aware of what's important to each other. By chance in one school we worked with, the teacher decided to take the list of the three things that were the most meaningful to each student and posted them on the wall. The result? Everyone in the class was constantly reminded of what was of the greatest meaning to every other child. The teacher found cooperation, empathy, and compassion between students rose dramatically. They stopped judging each other on the materialistic and egoic level and started seeing each other in a more profound context.

Another added benefit with shared meaning between students was that behavioral problems lessen greatly. From ego demands, students often act out to be seen and recognized. When each child knows what is most meaningful to their fellow students at a deep level, we find they are no longer driven to act out as much from egoic needs.

Wrapping Works in Corporations

We have found that this process of *Wrapping* works just as well with adults in corporations. The typical way to increase staff productivity is by reward and/or punishment. Yet, when we have teams go through the above process of finding the three things that are most meaningful to each member, not only does cooperation and empathy increase but so does innovation. Why? Because when we are operating in environments where we are more profoundly understood for who we are, safety levels go up, and our stress levels go down. As we will see in future chapters, the research is clear, the lower our stress, the higher our ability to innovate.

This kind of approach takes some training, as at first most people will just say they want to make more money or have more downtime. But when we peel the layers back and go deeper asking team members, "Why do you want more money?" and "Why do you want more vacation time?" we find the real driving force behind what a team member first said they wanted.

The more we know what's meaningful to one another, the deeper the connections we are able to make.

For example, when someone says they want to make more money because they want to buy a bigger, better house, the next question is, "Why do you want a bigger, better house?" It may take some time, but with patience and persistence, if the environment is safe enough, the person may divulge that they want to be seen see as more influential or they want to feel safer. This deeper level of self-knowledge may be a revelation to that person as well as the team.

We call this deeper level of understanding "amygdalaizing." As you'll learn, the amygdala is involved in attaching emotions to meaningful information. To really grasp what's meaningful to someone, you must go deep enough in your explorations to effectively stimulate the amygdala. The result is the learner experiences an emotional reaction. When an individual is willing to follow through with this process, egos become less dominant in teams, while cooperation increases.

For those of us that have slaved away in unfeeling corporate cultures, this may sound a little Pollyannaish. But we have found that when teams see the science behind this approach, they are much more apt to integrate ways of helping team members gain more deep meaning in their lives, which can have a profound effect on the whole corporate culture.

Reading and Wrapping

Reading research also validates the *Wrapping* technique. One of the keys to effective learning is the ability to read quickly and recall what you've read. Research indicates that failures in teaching kids and adults how to read rest not in the reading technique—that is, phonics or word recognition—but in the failure of educators to produce reading materials in which the learner finds meaning.

The unfortunate thing about most reading programs is that, although complex sentence structure, grammar, and phonics are emphasized, little attention is paid to matching what the student finds meaningful in the reading. Just because a storyline is significant to the author, the company that markets it, the school that buys it, and the teacher who assigns it, doesn't mean the book will have any relevance to the student!

Remember that the brain is a selective organ. It chooses the information that it values and filters out information that it doesn't. When students are forced to read passages that hold no meaning for them, they not only select out the meaningless information but because of the associative nature of the brain, that student may also select out the capacity to read efficiently. Who wants to be forced to read passages that have no meaning to them? The opposite occurs when we are exposed to reading meaningful material. Now we have the opportunity to naturally select not only the information in the material but also the process of efficient reading to gain more meaning.

The Best Books in the World

One of my sons went to a school that had a rigid reading program, and only certain books could be used in the classroom and at home to teach reading. By the time he finished third grade, he had said he "hated" reading and was tested at the lowest level of first-grade reading proficiency. At the end of third grade, we removed him from that school, and, that summer, we encouraged him to read only books that he found meaningful, which turned out to be the *Star Wars* series. In three short summer months, he jumped three grade levels. When I asked him to explain why his reading improved so quickly, his answer was, "I didn't like to read, but when I found the best books in the world, I sure did!"

Now in his twenties, this kid who didn't like to read or write at a young age made it his personal mission to read fifty-two classics by masters such as Mark Twain, Hemingway, and Steinbeck in just one year. And he's now writing his third book and second screenplay. Many researchers believe that books that children find meaningful, such as the *Harry Potter* and *Goosebumps* series, as well as fads like Pokémon, have done more to help children learn to read than all the remedial programs put together!

Another couple of personal examples of people who struggled mightily with reading as kids until they found personal meaning in books are my good friend and business partner John MacGregor and myself. John says that as a kid, "I associated so much agony with reading that I felt pain even if I just picked a book up!" John loved football, and at fifteen, he came across the life story of the Native American football player Jim Plunkett. This was the first book John ever read. After he found a book that was meaningful to him, reading became easy and turned into a passion of his. Not only has John used his reading skills to become one of the country's top financial educators and to publish *Wired To Be Rich*, but he also used his ability to read for meaning to make enough money to retire at thirty-seven,

Your Lymbic System "Amygdalaizes"

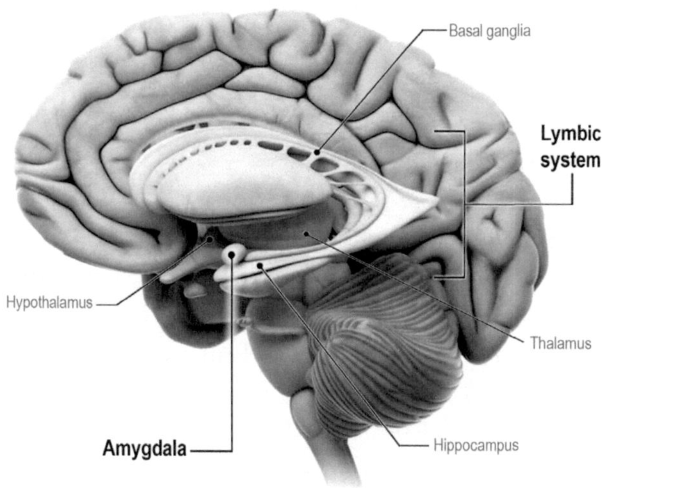

Without emotion there is no meaning. The amygdala is involved in attaching emotions to meaningful information. We call this "amygdalaizing."

if he had chosen to.

Many research projects show how the practice of *Wrapping* accelerates the skill of reading and aids in the ability to remember. One group of researchers was trying to sort through the factors that caused third and fourth graders to remember passages they had read. They found that how meaningful the passage was to the student was an astounding 3,000 percent more important than how readable it was.[4] In another study, subjects were given a list of words and asked to remember them on one of three bases: the visual appearance, the sound, or the meaning of the words. After two presentations, the subjects remembered 300 to 400 percent more of the words based on the meaning.[5] As one researcher puts it, "Reading is not a matter of decoding words but of extracting meaning from text."[6]

The late New Zealand educator Sylvia Ashton-Warner became famous for developing a simple yet brilliant method of teaching reading and writing to the Maori populations.[7] In her system, which was based on what was personally meaningful, children would first learn to spell and read words they personally selected. Soon, a student would have a list of enough personal words he or she could read and write sentences about things that were important in his or her life, such as "I caught uncle pissing behind a tree. He chased me when I laughed at him," and "My dad gave my mum a black eye."

Though the revealing nature of the children's writings shocked New Zealand's educational department, there was no doubt this system of learning was far superior to the previously used "Dick and Jane" models, which relied on less meaningful concepts.[8]

In Brazil, the renowned revolutionary educator Paulo Freire developed an accelerated system of learning for illiterate peasants based on the meaning of political action. He found the people he was trying to teach felt wronged and oppressed by the government, and they found political material meaningful because it would give them a voice in the direction of their government.[9] Freire then created a tremendously successful reading and writing program that had its foundation in connecting the people's desire to have the political power to the material to be learned. Reading researcher J.C. Harste discovered that when children were given props that held meaning for them to help them to write a story, they used more sophisticated grammar and content than those students directed to focus on skills or phonics.[10]

Using personally meaningful reading materials also helps those who have been designated as special educational needs (SEN). A study by the University of Washington and reported in the *American Journal of Neuroradiology* showed having boys with dyslexia read meaningful words helped specific areas of the brain more efficiently decode written language.[11]

For me, struggling with dyslexia, reading was like getting a root canal by a demented dentist. CliffsNotes were my best friends. Then, when I was sixteen, my class was assigned a book called *Hot Rod*. I loved fast cars and devoured it. It was the first complete book I ever read. I still remember how proud I was of myself. That same year, because I discovered the joy of reading and wanted to do more of it, I drove 140 miles twice a week for six weeks to take a speed-reading course. I've been reading for personal meaning ever since. The book you're reading now is a result of that first experience of reading for personal meaning. So, if your kid, or you, doesn't like reading now, have faith—with meaning, there's hope.

Cracking the Learning Code

Your genes and your environment create the neurological structures from which meaning springs. These meaning structures possess mass (neurons and synapses). Like a magnet this mass naturally attracts information from your world that resonates with it. This is why it is so much easier for you to learn and remember knowledge that is deeply meaningful to you.

14
Your Meaning Network: No Emotions, No Learning, No Behavioral Change

There is no separation of mind and emotions; emotions, thinking, and learning are all linked.

—Eric Jensen, former teacher and educational leader[1]

If meaning is, in fact, the Holy Grail of learning, ask yourself, "How is it that something becomes meaningful to me?" And "What is going on in my biology that drives me to search my world for meaning?" Understanding the biology of how meaning is created is critical to developing strategies and systems that effectively encourage learning and behavior change in ourselves and others. Here we'll go deeper into the neurological structures and look at how, without emotions, feedback from your body, and the activation of the brain chemicals that also drive addiction, you can't create meaning in your life.

Before you can efficiently select knowledge into your long-term memory, a complex group of structures and neurochemical systems, that we call your *Meaning Network*, must be activated.[2] In these next chapters, we'll explain how this process takes place. Once you understand even the high points of how your *Meaning Network* functions, you'll be able to more elegantly create systems for yourself and others that make the knowledge acquisition process much more profound, joyful, and efficient. Therefore, what follows in this chapter (and the next ones) is the linear view of how your *Meaning Network* tags and prioritizes information as valuable so you can remember it and, more importantly, act upon it.

What Is the Meaning Network?

Although your brain can't function linearly (the brain is a complex system with billions upon billions of feed-forward and feedback loops, creating endless waves of cross talk among millions of diverse neural groups),[3] the easiest way to understand how your *Meaning Network* functions is to employ linear analysis while recognizing the limitations inherent in this approach.

Your *Meaning Network* uses emotions to tag the millions of bits of information your brain receives each second and selects for attention and focus only on those bits that generate the highest emotional value. Your brain doesn't act alone in allocating emotional value and meaning to incoming information; a recent discovery is that your brain must also experience feedback from your body, like a change in your blood pressure or heart rate. If your brain can't process feedback from your body when you learn something new, then there's no firm biological foundation for clear thinking, memory formation, or behavioral change. After the information is marked with body feedback and emotions, your brain then prioritizes this tagged information into working memory so that the most important knowledge can be selected for inclusion in your long-term memory.[4]

A major element of the *Meaning Network* is the dopamine circuits, which are similar to the circuits that drive addiction. These rivet your attention to meaningful stimuli, forcing you to attach importance to them. Neuroscientific research confirms that if any part of your *Meaning Network* isn't fully activated, you'll be unable to effectively select information into your long-term memory banks, change your behaviors or increase your *Individual Adaptability/Intelligence Factor* in an ever-changing world.[5]

Education's Disconnect with Emotions

Understanding the power of emotions in your learning process is crucial because when new information doesn't stimulate an emotional response, remembering it is difficult, if not impossible. It's also vital to understand the power of emotions in the learning and memory process because our traditional scholastic and corporate education systems virtually ignore them. As you'll see, this omission makes the whole educational endeavor much more tedious and inefficient than it needs to be.

Our existing learning systems have been operating as if emotional competency were of little consequence in our quest for personal success in life. But research shows that in every aspect of life, from job advancement to romantic interests, success isn't directly related to grades, SAT scores, class position, or IQ; instead, it depends on how well your frontal lobes mediate the raw emotions from your limbic/stem system so that you build meaning, enhance self-awareness, feel empathy, and cooperate with others.[6] If we ever hope to fix our corporate and educational problems, we must understand how emotions support your *Individual and Species Adaptability/Intelligence Factors*.

Before we go on, it is important to recognize that emotions aren't only stimulated by your environment like a charging bear, a car wreck, or taking a test. They are also your body's reaction to your own thoughts. You can raise your blood pressure and alter the acid make up of your digestive track simply by what you think about. This is why we say you can change your emotions in the same way.[7]

However, history hasn't been kind to emotions. At best, they've been seen as elements that make us irrational and disorganized; at worst, they've been cast as sinful indulgences that we must purge ourselves of, lest we are driven to perform immoral acts. The concept that emotions are evils that need to be cleansed through absolution and confession has its foundations in the third century BC. with Plato. The idea was refined for the church by St. Augustine 600 years later.[8]

In the 17th century, the founding father of modern philosophy, René Descartes, who gave us the sound bite, "I think; therefore, I am,"[9] published *Discourse on Method*, a work so powerful it has influenced virtually every prominent philosopher since. In *Discourse*, Descartes outlines the case that our senses and emotions disturb the process of creating rational, organized thought. Even in the 20th century, the powerful behaviorist school of psychology considered the gray world of emotions of little consequence in the learning process or its efforts to control human behavior through reward and punishment. But now, neuroscience informs us our emotions aren't frivolous luxuries or interlopers in the process of rational thought. Instead, emotions are the primary organizing factors upon which consciousness, reason, and memory are built.

Today, neuroscientific research leads us to the conclusion that, without emotions, there can be no effective long-term memory formation. Emotional value is what makes information stand out so you can focus

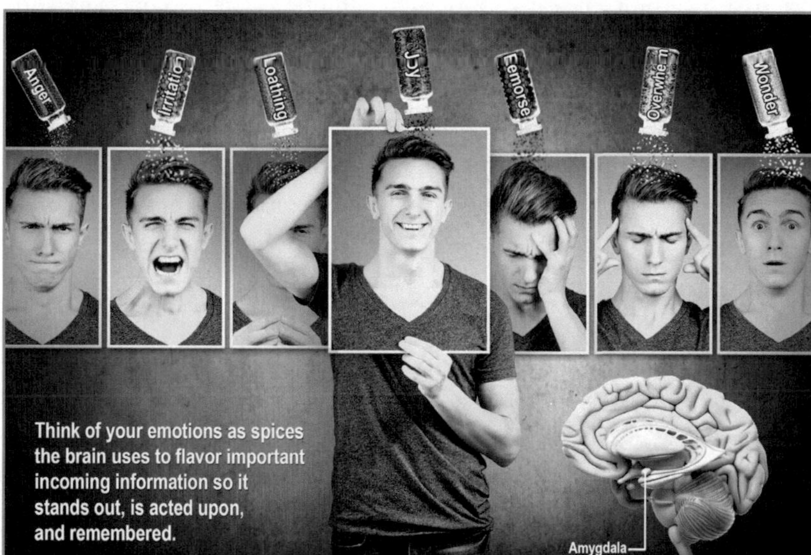

Think of your emotions as spices the brain uses to flavor important incoming information so it stands out, is acted upon, and remembered.

on it, organize it, and remember it.[10] Study after study confirms the more emotion a passage, a picture, or an event evokes, the easier it is to remember. A pioneering leader in defining emotional circuitry, Joseph LeDoux, notes, "Emotions, in short, amplify memory."[11]

If we didn't place emotional value on incoming information, our world would be a gray, bland, and unmemorable place. One way to look at emotions is like spices, allowing the brain to sprinkle important incoming information with flavor so it can stand out, be acted upon, and remembered. Anger, rage, joy, sadness, grief, curiosity, love, jealousy, greed, surprise, embarrassment, disgust, and pride are not intangible or elusive; they're in-your-face spices that prompt you to move in directions that support surviving and thriving.

By failing to recognize that our emotions are invaluable to the learning process, we've turned our educational systems into sterile, mechanistic places devoid of the very emotional charge needed to produce the kinds of thinking and learning we so desperately want to develop. Research is now confirming that, for a corporate or scholastic educational system to produce the learning results we demand, students must feel passion and emotion about what they're learning.

Without Emotions, You Don't Survive or Thrive

Just like breathing, hearing, and sight, emotions are evolutionary gifts because they aid survival and reproduction. In *Cognition and Emotion*, Johnson-Laird and Oatley point out that each emotion "prompts us in a direction which in the course of evolution has done better than other solutions."[12] Emotions are so valuable to survival that the animal species that lacked emotions no longer exist. They never made it through the sieve of evolution.

As the late philosopher Hans Jonas noted, "The capacity for feeling, which arose in all organisms, is the mother value of all values."[13] Joseph LeDoux and Michael Rogan point out in their article "Emotion: Systems, Cells, Synaptic Plasticity" that emotions "can be thought of in terms of computational processes."[14] Emotions help the brain make calculations concerning the worth of information you perceive in your environment.

Possessing a wide array of emotions allows animals and humans to quickly calculate and assemble life-saving knowledge. Emotions are the markers that help you know what to approach and what to avoid. They help you organize your thoughts, from the mundane (such as "What shall I wear today?" or "What should I eat for breakfast?") to the more grandiose (such as "Should I leave my mate?" or "Should I run for president?").

Evolutionary biologists note that as you move up the evolutionary scale from lower mammals like mice to higher mammals like humans, the emotional circuits become much more refined.[15] Many evolutionary psychologists maintain that it isn't the greater memory capacity of man's neocortex that makes humans the most dominant species on Earth; rather, it's our refined emotional circuits, which give us a sophisticated range of emotional values upon which to create the memories that let us know what to approach and what to avoid in our world.

Value of Emotions Recognized

In the 1970s, renowned evolutionary neuroscientist Dr. Paul MacLean was one of the first modern scientists to grasp that neither thought nor memory could take place without an emotional component. MacLean's work in developing the first comprehensive evolutionary theory of brain development led him to conclude that, in the brain, "something does not exist unless it is imbued with an affective feeling, no matter how slight."[16]

In the 1980s and 1990s, neuroscientist Antonio Damasio clarified MacLean's findings, discovering that patients who had damage to the circuits that evoked emotions also lacked both the ability to think rationally and the ability to form efficient memories. Damasio found that, to a greater or lesser degree, depending on the exact location and severity of the damage, these patients were trapped in a world where all information had the same value. Because these individuals couldn't use their emotions to prioritize knowledge into their short- and long-term memories, they lived directionless lives, devoid of deep personal meaning. Some patients with damage to their emotional circuits do have the capacity to remember stimuli from their world, but without emotional tagging, the information gets logged without rhyme or reason and therefore is of little survival value to them when they need it.[17]

Before Damasio's work, the widely held belief was that the brain had one group of circuits for a reason and another group for emotions, and like oil and water, they didn't mix. Damasio's research conclusively demonstrated that emotion and reason share the same neural networks. He writes, "Contrary to traditional scientific opinion, feelings are just as cognitive as other precepts." Now, thousands of research projects using multiple stimuli—from words to pictures to faces and events—have confirmed that if incoming information doesn't elicit an emotional response, your brain will have no basis for effectively selecting information into long-term memory for future use.

Simply put, information that evokes emotion is information that can be remembered.[18] For example, the *Star Trek* character Mr. Spock, relied upon by the crew of the *Enterprise* for his dispassionate evaluation of situations, is a modern hero to many who devalue the role of emotions in thought. Spock is half Vulcan, a species that suppresses emotions almost entirely and is, therefore, able to think more rationally than humans do. In this depiction, the creators of the *Trek* series couldn't have made a bigger mistake. Without emotions, such a species would have no criteria by which to compute which information is important and which is not.[19]

In the enlightening book *Emotional Intelligence*, Daniel Goleman looks at intelligence through the lens of modern psychology. He found that success in life is tied not to "our smarts"—that is, grades, SAT scores, or position in a graduating class—but to what he calls our emotional intelligence, or EQ, which is our ability to effectively perceive, process, and moderate our emotions.[20]

In *The Growth of the Mind*, Stanley Greenspan notes that research leads to the conclusion that "emotions, not cognitive stimulation, serve as the mind's primary architect." Furthermore, "Emotional exchanges… should become the primary measuring rod of development and intellectual competence."[21] TalentSmart, an EQ consulting business, tested large numbers of people and discovered that 90 percent of the top performers had high emotional intelligence.[22] Neurological research is causing many artificial-intelligence experts to insist that a computer will never rival a human brain, primarily because a computer that produces true emotions could never be built.[23]

Not long ago, I gave a keynote speech for The Deming Institute that noted the value of emotions to a bunch of engineers at Iowa State. I was stunned by how many of these analytical thinkers, a large percentage from India, came up to me afterward and said that they'd spent their whole lives trying to be unemotional in their thought processes. They asked me how, based on what they'd just learned, they could become more emotional in their thinking methods about engineering. They now understood how their emotions could help them be even more competent in their study and jobs.

The Lower and Upper Structures of Emotion

The main emotional structures that help you make memory and meaning are situated in your limbic/stem system and your prefrontal lobe. The lower limbic system produces your raw, primal, instinctual emotions, and your most forward brain structure, the prefrontal lobe, mediates and refines these base animal passions.[24]

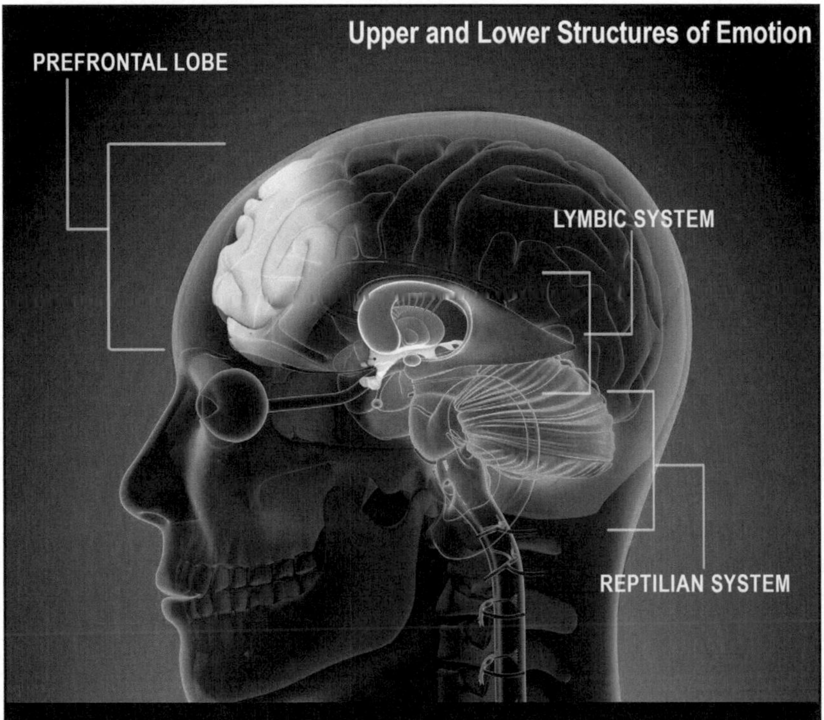

The prefrontal lobe mediates our lower emotional urges, which separates us from the reactive world of our animal cousins.

The whole limbic/stem system has been implicated in creating and housing your rough, instinctual emotions and producing the neurochemicals that give rise to them (especially serotonin, norepinephrine, and dopamine). Most identified with the production of basic raw emotion is the amygdala, an almond-shaped structure at the base of your limbic system. Over the last decade, this ancient structure has been the focus of intense study as the center of emotional reaction, which has its roots in animals' emotional world of "kill or be killed." The amygdala is designed to allow animals to respond with high emotionality without thinking.[25]

If the limbic/stem system were our only source of emotional processing, there would be little to separate us from the reactive world of our animal cousins. The structure responsible for filtering and refining the instinctual urges that surge forth from the limbic/stem system is the prefrontal lobe. The refinement of our raw emotions by this lobe, our newest evolutionary structure, gives humans the greatest range of emotions in the animal kingdom.[26]

The prefrontal lobe's ability to distill your raw emotions enhances your capacity for marking, prioritizing, and logging information into your long-term memory.[27] While many animals do have a prefrontal lobe, it's limited in size, complexity, and sophistication. The small capacity of an animal's prefrontal structure not only keeps an animal locked in a world of emotional reaction but is also a significant factor in limiting memory capabilities.

The idea of amygdalaizing, which I introduced in the previous chapter, can help organizations and individuals remember that if they want the learning to stick, they must embed emotional triggers into their learning and training processes. Without the spice of emotion triggered by the amygdala and mediated by the prefrontal areas, learning becomes a more difficult task.

While your whole prefrontal lobe has been identified in the process of emotional marking and refinement, the ventromedial prefrontal area has been singled out for particular attention. As you read this section, note that I'm overgeneralizing to support a conceptual understanding. When I highlight the ventromedial, I mean that structure and its relation to the other prefrontal areas that support it.

The ventromedial is located behind the middle of your forehead, just about where Eastern spiritual practitioners place the third eye. This frontal and central position is advantageous, for from here, the ventromedial is aware of virtually every activity taking place in your mind or body at any given time.[28] It directly or indirectly connects to every sensual, motor, memory, and chemical circuit available in your brain. As previously mentioned, the ventromedial has especially powerful connections with the amygdala, which it can exert control over by filtering and refining the raw emotions that originate there. In this way, the ventromedial gives humans the emotional refinement and impulse control that animals lack.

The exquisite ventromedial ability to refine, mediate, and emotionally tag data is instrumental in creating the profound capacities that most separate us from other animals. These capacities include insight, abstraction, altruism, anticipation, association, metaphor, judgment, future planning, sequencing, creative thinking, big-picture analysis, and enhanced memory capabilities. This structure allows us to be the most self-aware beings on the planet. The higher your self-awareness, the greater capacity you have to know precisely what's meaningful to you. Emotional nuance allows you to make micro distinctions about incoming stimuli, refining the criteria by which information should or shouldn't be selected for inclusion into your long-term memory banks. Many evolutionary biologists point out that the reason humans are the most dominant species is that we have the most developed ventromedial prefrontal cortex on Earth.

The Power of Empathy

The ventromedial, with other recruited frontal areas, provides another gift that allows you to broaden what you find meaningful in life: empathy, the capacity to feel or experience the emotions, sensations, or thoughts experienced by others.[29] While self-awareness increases your ability to survive, empathy increases your ability to help your fellow humans survive and thrive, thus growing our whole *Species Adaptability/Intelligence Factor*. No discussion of the ventromedial prefrontal cortex would be complete without looking at its role in producing empathy and how it expands your learning and behavioral change opportunities.

Empathetic action can be as simple as supporting a spouse with his or her hobby or as complex as Jonas Salk's development of the polio vaccine to help rid the world of a dreaded disease.[30] When your basis of meaning expands from just self to others, the amount of information that can be marked by emotion and tagged as relevant for inclusion into your neural networks increases dramatically.[31]

Empathy is seen in animals and appears to have been expressed by early humans, but that empathy primarily focuses on the immediate family or tribe. This limited form of empathy is confined to an intimate personal relationship. Modern humans, with their expanded prefrontal areas, are the only mammals with the capacity to help those they don't even know, such as starving children in Africa or people in the war-torn Middle East.[32]

Because empathy is such a survival advantage, you could argue that our greatest asset is our ability to expand our emotional repertoire to include those we don't have personal contact with. Reflecting on empathy and our most-recently evolved brain areas that process it, MacLean asks a spiritual question: "Then why, slowly, progressively, did nature add something to the neocortex that for the first time brings a heart and a sense of compassion into the world? In designing a creature that shows concern for the suffering of other living things, nature seems to have attempted a 180-degree turnabout from what had been a reptile-eat-reptile and dog-eat-dog world."[33] Simply put, empathy expands the amount and kinds of information your brain can select into its structures, thus increasing our *Individual and Species Adaptability/Intelligence Factors.*

Cooperation and social leadership also have their roots in empathy. If you can't cooperate with others, your ability to survive and thrive is limited. Bullies don't rule the world; co-operators do. While bullies like Mussolini and Hitler can rule for short periods, in the end, the co-operators win. This is because to cooperate with others and not judge their feelings limits even the most aggressive individuals' ability to reach their potential.

In early humans, cooperation and social leadership were first seen in endeavors such as tracking and killing prey, preparing food, planting and plowing.[34] Without the ability to use empathetic capacities to cooperate and socialize, we'd still be living in groups of twenty-five to fifty. Over the last 10,000 years, refined empathetic skills have enhanced our ability to cooperate and socialize to the point where we can live in groups of millions upon millions of people (think New York City and Tokyo) and develop cooperative entities such as the United Nations, which encompasses virtually the whole world.[35]

Individuals who experience prefrontal and ventromedial damage or disease are observed to have minimal self-awareness and empathic powers, which, in turn, leads them to disturbingly incomplete lives; they often suffer from failed social relationships, divorce, bankruptcy, employment problems, and the inability to manage their time. Damage to the center of self-awareness, meaning-making, and empathy leaves them emotionally tone-deaf, lacking the ability to organize their thoughts, memories, and actions into what will bring them or others success.[36]

An extreme example of a lack of self-awareness and empathy associated with the malfunction of the ventromedial is seen in the pathological actions of psychopaths, rapists, and child molesters.[37] An underactive prefrontal area has also been implicated in the emotional and irrational behavior of people with schizophrenia.[38]

We Learn from Lobotomy

Science has learned a lot about ventromedial damage from the lobotomy procedure designed by Portuguese neurologist Edgar Moniz in 1936 and used extensively until the advent of psychiatric drugs in the 1950s. The lobotomy made agitated psychiatric patients docile by wiping out their emotional tone. The procedure destroyed the ventromedial either by boring a hole in the forehead and pouring alcohol directly onto the structure or by inserting an icepick instrument through the eye socket and swinging the thing back and forth.[39]

Unfortunately for these patients, destroying their center of emotional tone also destroyed their ability to create personal meaning, empathy, and efficient long-term memory. Neuroscientist Oliver Sacks says many of the lobotomy patients he saw in his career were "psychically dead, murdered by their 'cure.'"[40]

Who Is Minding the Emotional Store?

The ventromedial reaches maturity in early adulthood, making it one of your last brain structures to do so. Because of the incomplete maturation of the ventromedial region, we're much more at the mercy of our limbic/amygdala urges in childhood and adolescence than we are in adulthood.[41]

The first developmental burst of activity in the ventromedial takes place at around age seven. At this time, we see a more pronounced forehead in children, as the skull makes room for the new ventromedial connections. Up until this age, the lack of development in the ventromedial places the child almost totally under the control of his or her raw, instinctual urges. You could say that, at a young age, children don't *have* emotions—they *are* their emotions, which is why reasoning with a young emotional child can be so frustrating.

Without efficient, emotional mediation, a young child's brain can be easily overwhelmed with the chemicals of emotion, thus inhibiting their ability to listen to reason or produce rational thought.[42] In effect, you're trying to have a conversation with a part of the brain that is not only under stress but also not wholly formed. This is why diverting or distracting a child's attention with diversions like candy, or a funny face is much more effective at altering their behavior than intellectualizing with them.

This doesn't mean you don't support your children to self-regulate their emotions. From a neuroscientific viewpoint, you could say that by helping a child or adolescent (the ventromedial doesn't become fully mature in our early twenties) to mediate their emotions and make the right decisions in life, a caring parent or mentor supports the function of the immature ventromedial region.[43] When children and adolescents are left on their own with little adult support in processing their emotions, they can make devastatingly poor decisions. Research shows that children with influential mentors who help them mediate their emotional impulses are much less likely to be involved with gangs, violence, school vandalism, drugs, drive-by shootings, and teen pregnancy.[44]

Although we might well ask, who's minding the emotional store? Our society is becoming one in which neither parents nor other adults have much time or energy to help children develop their emotional intelligence. Meanwhile, the average child spends 33 percent of his or her life getting emotional development from a media babysitter.[45] Because the internet, video games, television, and movies filled with violence, sex, and questionable values end up being primary sources of brain stimulation, the media by default becomes a prime molder of a child's emotional development (see Chapter 29 for an extended discussion of this issue).[46]

We must ask ourselves these questions: by ignoring children's emotional maturation when their circuits are developing their fastest, isn't our society cheating children out of the rich neurological connections they'll need to succeed in life? Is our lack of attention to our children's emotional development inhibiting their ability to make meaning, learn, remember, and feel deep empathy? By ignoring their emotional development, aren't we also prompting them to act impulsively and solve their problems in inappropriate ways, such as by joining gangs or committing drive-by shootings or school massacres?

Joseph LeDoux, who has dedicated his life to researching the value of emotions, points out that our past limited views of brain function as the source of the problem: "… in fact, cognitive science is really a science of only a part of the mind, the part having to do with thinking, reasoning, and intellect. It leaves emotions out. And minds without emotions are not really minds at all. They are souls on ice—cold, lifeless creatures devoid of any desires, fears, sorrows, pains, or pleasures."[47] In looking at the media-rich non-nurturing emotional environments in which our children are raised, David Lykken of the University of Minnesota wonders—a little tongue in cheek—if we haven't produced a society that is "turning out 'little psychopaths.'"[48]

Devaluing emotions in creating learning and memory has limited our educational system's ability to produce anything above mediocre results for our children. To continue to design learning systems that fail to fire and develop our emotional circuits is to continue creating learning systems that will fail to produce the kinds of memory or behaviors that will help our children succeed.

Cracking the Learning Code

Simply stated: without emotions there can be no learning. To create learning without engaging the emotions is a grueling and inefficient process.

15

Without Feedback from Your Body, You Can't Learn!

Thus, it could be said that intelligence is located not only in the brain but in cells that are distributed throughout the body, and that the traditional separation of mental processes, including emotions, from the body, is no longer valid.

—Candace Pert, biophysicist[1]

Once you've grasped that rational thinking, meaning, learning, and memory are indelibly linked to the emotional value you place on the information, the next question is: where the heck do emotions come from? The poet E. E. Cummings had the answer years ago when he wrote, "Feeling is first."[2]

New research into the circuitry of emotions exposes a disruptive phenomenon: Your brain doesn't act alone in allocating emotional value and meaning to incoming information; it must experience feedback from your body. If your brain is unable to process feedback from your body as you're trying to learn something new, there's no firm foundation for clear thinking, memory formation, or behavioral change.[3]

Surprise! Evolutionary Biology Reveals that Your Brain Is Part of Your Body

For more than 2,000 years, the general belief was that the brain and body were separate.[4] One helped us learn; the other helped us move. This view allowed us to create traditional educational systems where almost all our focus has been on sitting at desks listening to lectures and memorizing vast amounts of facts. Neuroscience, genetics, and evolutionary biology are forcing us to re-evaluate our previous position about the body's role in learning. But for six million years, humans did almost 100 percent of their learning while physically moving through time and space. Our ancestors learned while running from predators and after prey. When we finally sat still, it was to rest, eat, and integrate what we'd experienced during the day into our long-term memory.

Over the last 75,000 years, our genes have changed little. We're still genetically wired to learn best when our bodies are involved. Yet not only have we ignored our bodies in the learning process, but because we didn't understand their role, we've abolished things like playground activities and gym to make more time for sitting, listening, and memorizing.

When the body is in motion, the brain is forced to activate more neurological structures and neurochemical systems. And as we know from the Hebbian learning rule (see Chapter 7), the more parts of your brain that are activated during the learning process, the better your memory and the greater your behavioral change. As you will see in more detail later, you don't have to physically move your body to involve it in the learning process. If the incoming information is meaningful and emotional enough, your body will involuntarily become involved, even if you don't physically move at all![5]

I often refer to body feedback using its more biological term "somatic feedback." When I speak of body and somatic feedback (*soma* is Greek for "body"), I'll be signifying the following:

- All changes in the body and organs (viscera)—represented by changes in heart rate, perspiration, respiration, pupil dilation, blood pressure, muscle tone, digestion, and elimination—transmitted back to the brain via the nervous system
- Changes in concentrations of peptides, glucose, oxygen, carbon dioxide, ions, pH levels, and hormones like *Brain-derived Neurotrophic Factors* (which are also produced in the body), ferried back to the brain via the bloodstream

Simply put, if your body isn't providing the correct feedback to your brain, you can't learn, no matter how much you want to.

Your Brain Is Your Body

The importance of your body's feedback in the formation of your emotions and memory is clarified by a view held by evolutionary biologists. During the evolutionary process—from flatworms to fish to reptiles to mammals to humans—the brain-body feedback systems developed more and more sophistication, allowing organisms to more efficiently evaluate and react to information from their environment. While the brain can be considered the most complex element of this system, it's still just an enhancement of the whole system. From an evolutionary and biological perspective, the notion that your mind is separate from your body couldn't be more wrong. It's part of your body, just like your liver, heart,

> **Thus, it could be said that intelligence is located not only in the brain but in cells that are distributed throughout the body, and that the traditional separation of mental processes, including emotions, from the body, is no longer valid.**
>
> *Candace Pert*

and lungs are part of your body. The brain didn't pop up all of a sudden; instead, it's a conglomeration of small adaptations to the body that occurred over billions of years of evolution.[6]

Initially, the primitive nervous system only got feedback from the body when the organism was physically stimulated by its external environment. It was an outside-in approach. But during eons of evolution, the brain developed emotions as a more refined method of evaluating what was going on, not only in the outside environment but also what was going on inside the organism. Over millions of years, the brain advanced such that it could create emotions not only from an actual charging bear; it could also create body feedback by merely thinking about a charging bear. Now the brain had a much more sophisticated system to evaluate its external and internal world and, therefore, could make much more elegant adaption decisions.[7]

Emotions and the neurological structures and chemicals that support them, arose on the somatic ladder because they were effective in helping animals better survive and thrive. You pay attention to the important stuff because you get excited, and your heart rate and blood pressure rise. Body markers and emotions act as limiting factors (see Chapter 5), narrowing your field of attention, thus allowing your brain to more effectively select only information that's meaningful to you.

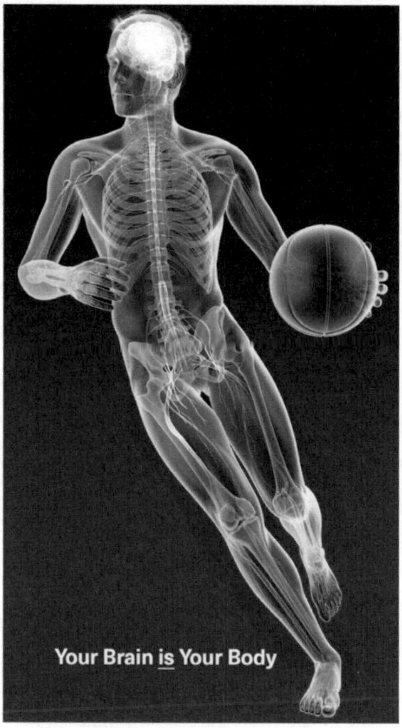

Your brain is not separate from your body. Your body is intricately connected to all your thinking processes.

With the most sophisticated brain-body feedback circuits in the world, humans hold a survival advantage over all other species, not necessarily because we *think* better but because we *feel* better than other species! Even within our own species, research shows that the most successful of us thrive because we're better connected to our own sensations and emotions.

The Latin term for modern man is Homo sapiens, which means "thinking" species. In light of this new research, it might be more fitting to call modern man Homo somata, or "feeling" species.[8]

Brain Science Ignores the Value of Body Feedback

While evolutionary biologists have long grasped that the brain and body aren't separate, up until recently, brain science operated as if feedback from your body was, at best, irrelevant or, at worst, disruptive to thought processes. Neuroscience embraced the position first introduced by Plato, then adopted by the church, and later refined by René Descartes: The mind and body are distinct and separate.[9] This philosophy arose from the belief that because the body was connected to impure animal urges, information from it couldn't be trusted. In contrast, the mind was thought to have direct contact with God, making it the seat of the soul. Therefore, the information generated solely by the brain, without influence from our unchaste bodies, was believed to be pure and sacred.

In his 1637 work *Discourse on Method,* Descartes sums up the mind-body separation view, which ruled philosophy and science for 2,300 years: "The soul by which I am what I am is entirely distinct from my body."[10] He punctuates his position with one of the most often repeated soundbites in philosophy, "Cogito, ergo sum," or "I think; therefore, I am."

At the end of the 19th century, William James, the renowned and beloved pioneer of US psychology who proposed a selective model of learning, was also the first cognitive scientist to suggest that "the body is the theater of our emotions." He proposed that the emotions upon which our thoughts are organized are simply the result of the brain's perception of changes in the body. In his 1890s book *The Principles of Psychology*, he writes:

> *If we fancy some strong emotion and then try to abstract from our consciousness of it all the feelings of its characteristic bodily symptoms, we find we have nothing left behind, no "mind-stuff" out of which the emotions can be constructed.*[11]

James also proposed that emotions feel different from one another only because each has a different somatic tone. Unfortunately, the brain-body separation position of Plato and Descartes was so influential that James's theory on body feedback like his selective theory, while spot on, was attacked and summarily dismissed.

It wasn't until the end of the 20th century that neuroscience began to train its eye on the value of somatic feedback. The two scientists who are given the most credit for popularizing the value of the body in memory and thought are neuroscientist Antonio Damasio and biophysicist Candace Pert.

Damasio, the same fellow who teased out the value of emotions in producing rational thought, also coined the term "somatic-marker hypothesis," which describes the process by which emotions themselves are developed with input from the body. His hypothesis uses modern science to verify what James had theorized over a hundred years earlier: emotion, meaning, memory, and rational thought depend on feedback from the body.[12]

Damasio's work with patients who had experienced brain damage clearly demonstrated that, without the ability to receive body feedback, their minds were unable to form clear emotional distinctions to organize, tag, and log incoming stimuli. In his book *Descartes' Error,* which is now used as a neuroscience primer, Damasio writes that our emotional circuits "...maintain direct and mutual relationships with virtually every bodily organ, thus placing the body directly within the chain of operations that generate the highest reaches of reasoning, decision making, and, by extension, social behavior and creativity."

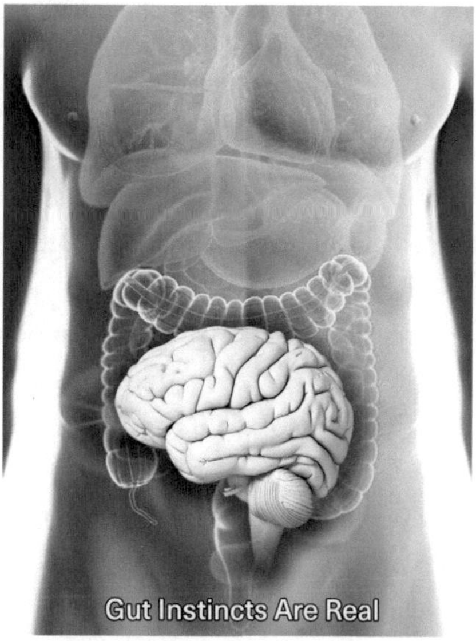

The theory that gut instincts are real relates to the fact that 90 percent of your serotonin and 50 percent of your dopamine are in your digestive tract.

Gut Feelings Are Real

Pharmacologist and neuroscientist Candace Pert's extraordinary career began with her 1972 discovery of the opiate receptor.

Her research lead her to take a chemical route in exploring the body's permanent link to emotions. In her book *Molecules of Emotion*, which draws on her research into neurochemical communicators, she goes a step further than Damasio in saying that not only does the brain create emotions from body feedback, but the body itself creates and stores emotions. She first points out that the neurochemicals that create feelings in the brain are also found throughout the body, including in the kidneys, liver, gut, adrenals, sex organs, and even the immune system. She then states that because memory is stored at the receptor level and because both the brain and body have receptors for these neurochemicals, both the brain and body can store emotions and memory. Her work helped her conclude that "the body is the unconscious mind." Pert writes:

> *Thus, it could be said that intelligence is located not only in the brain but in cells that are distributed throughout the body, and that the traditional separation of mental processes, including emotions, from the body, is no longer valid.*[13]

Pert's theory validates the view of psychosomatic therapists, who maintain that the body must be considered in the process of any psychiatric healing. As an example, your gut produces more than thirty neurotransmitters, such as acetylcholine, dopamine, and serotonin. It is estimated that 90 percent of your serotonin and 50 percent of your dopamine is produced in the gut, which helps explain why a "gut feeling" may be much more than a metaphor.

The work of Damasio, Pert, and other leading neuroscientists sheds light on many puzzling questions in medicine. Surgeons now recognize the reason some patients come out of successful surgery with nonspecific depression is that the somatic markers the body sends to the brain after a major operation are similar to the markers the body produces when we're depressed. As Elmer Green, the Mayo Clinic physician who pioneered the field of biofeedback describes, each "change in the physiological state is accompanied by an appropriate change in the mental-emotional state, conscious or unconscious," and vice-versa.[14]

This body–emotion link may also help explain the curious phenomenon of some transplant patients who report acquiring the tastes, feelings, and even memories of the individuals who donated their organs to them. We now know your gut has approximately 500 million neurons that have dendrites and synapses, just like the neurons in your brain. This is about three times as many as the number of neurons in a rat's brain. And your heart has 40,000 neurons that feel and remember. Some neuroscientists believe that because the transplant recipient inherits the cells and receptor sites of the donor, he or she also inherits the information that has been stored there. As neurosurgeon Richard Bergland writes in *The Fabric of Mind,* "The stuff of thought is not caged in the brain but is scattered all over the body."[15]

Brain Structures That Receive Feedback from the Body

We now understand somatic feedback is so important to the memory-formation process that the brain areas responsible for perceiving feedback from the body are numerous and widespread. The primary structures identified are in the brain stem; the pons and medulla; the limbic system; the hypothalamus, hippocampus, and amygdala, in the neocortex; the parietal lobes; the somatosensory cortex; and the motor cortex.

Recent research has also focused on the dominant effect that frontal-lobe structures have in receiving body feedback and assigning it refined emotional value. The prefrontal areas, especially the ventromedial area discussed in the previous chapter, are ideally suited to analyze somatic feedback—especially from the amygdala and the somatosensory cortexes, located just in front of the parietal lobes, where all past and current body states are continually represented.[16]

Anosognosia patients offer a sad example of the disruption caused when brains can't efficiently receive body feedback. The word *anosognosia* comes from the Greek *nosos,* for "disease," and *gnosis,* for "knowledge." Anosognosia is the inability to perceive disease in oneself.[17] Damage to these patients' parietal lobes, usually caused by strokes, disrupts their somatosensory feedback system, which means they unable to perceive sensations of their own disease. Because these patients aren't receiving correct signals from their bodies, their emotions, thoughts, and resulting actions become irrational. William James believed that the body is the theater of the emotions; unfortunately, anosognosia sufferers have no ticket to the theater.

During their tenures, both President Woodrow Wilson and Supreme

Court Justice William O. Douglas suffered strokes that caused anosognosia and resulted in inappropriate behavior. During the 1919 Paris Peace Conference, Wilson suffered a stroke, and two weeks later, he suffered another, which paralyzed his left side. Despite the paralysis, he claimed nothing was wrong with him. Once Wilson stopped getting correct feedback from his body, his ability to reason and select meaningful information diminished so quickly that his wife and the Cabinet Secretary secretly ran the government without his input until the end of his term.[18]

In 1975, Supreme Court Justice William O. Douglas suffered a right hemisphere stroke, which caused him to deny that his left side had been paralyzed. While in a wheelchair, he asked reporters to go hiking with him and told them that he was in such good health he would try out for the Washington Redskins as a kicker! Douglas's inability to get correct feedback from his body dramatically inhibited his ability to reason and conform to court etiquette. Sadly, one of the great minds of the court was eventually forced to resign by his peers, who couldn't accept his bizarre behavior.[19]

In the last decade, as described earlier, Antonio Damasio did extensive research with patients who experienced damage to the circuits that process feedback from the body. To find evidence of these circuits' importance in the creation of emotion and meaning-making, Damasio developed an

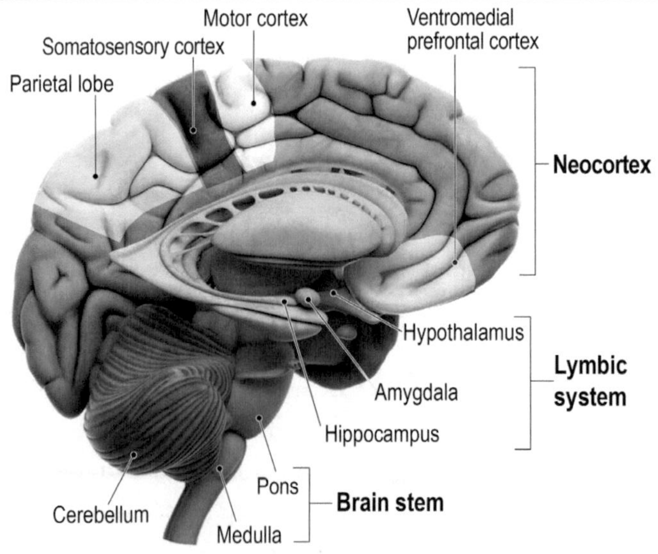

Brain structures that receive feedback from the body are extensive.

experiment in which he hooked subjects up to a polygraph (lie detector). The polygraph works by measuring and graphing changes in galvanic skin response (GSR), or sweat, and attributing emotional value to the changes.[20]

In the experiment, Damasio showed two groups of subjects, one normal and one with body feedback circuit damage, hundreds of mostly bland slides. Every so often, a disturbing picture, such as a homicide or mutilation scene, would appear. The results were unequivocal; when the gruesome images appeared, the subjects without somatic network damage generated abundant galvanic skin/sweat response, while the patients with somatic circuit damage failed to generate any skin response at all. Also, unlike the healthy subjects, these brain-damaged patients who couldn't get clear messages from their bodies were not able to develop emotional responses to the disturbing pictures. The brain-damaged patients knew they should feel something upon seeing these scenes but were unable to generate the appropriate emotional response without clear body feedback.

In another experiment devised by Damasio and his team, two groups—one with body feedback damage and one without—could gamble by selecting one of two decks of cards. In the first deck, the risk/reward ratio was dramatic—"win $100" was written on most cards, with a few cards having massive losses of up to $1,250. In the second deck, the risk/reward ratio was much lower—most cards said, "win $50," with a few cards delivering penalties from $50 to $100. The best and most obvious strategy was to go with the low-risk deck because choosing this deck ensured you couldn't lose. After playing for a while, normal subjects learned—especially after being slapped with one or more $1,250 penalties from the first deck—that the second deck produced the best rewards. But the patients with body feedback circuit damage couldn't learn. They continually chose the riskier deck, which produced huge losses for them.[21]

Damasio found that the patients with somatic circuit damage showed no galvanic skin response before turning over a new card from the riskier deck. The lack of psychological change indicated that these patients were getting no emotional tagging with which to learn and make the correct decision. Without efficient body processing and emotion formation, the brain-damaged patients lost the ability to learn and remember the negative consequences of their previous actions. Their inability to process somatic information kept them trapped in a world of unsuccessful decisions.

Damasio's research has even prompted some people to consider a futuristic learning system in which students hook up to polygraph machines to see whether learning is taking place. The educator could then monitor printouts to see which students were generating galvanic skin response and which were not.

Physical, Emotional, and Body Feedback Helps Unwind PTSD

Why is post-traumatic stress disorder (PTSD) such a difficult condition to heal? Because most experiences that cause PTSD symptoms—like physical and sexual abuse, car wrecks, and war—stem from a powerful combination of the physical and the emotional. These traumatic events burn these horrifying experiences into people's brain structures, bodies, and psyches, which can negatively alter these individuals' behaviors forever.[22] The extended conflict in Iraq and Afghanistan has meant that PTSD has gotten much more attention in the last few years. The damage PTSD causes sufferers and their families is devastating. It's estimated that twenty-two veterans a day commit suicide due to PTSD, and those deaths are just the tip of the iceberg of misery.[23]

From our work with the Warriors Heart healing centers, which offers the first residential retreatment programs for first responders and war veterans, The Learning Code Institute developed a neurobiological approach to help PTSD sufferers manage their symptoms. Most PTSD treatment modalities use talk therapy. We, however, took a reverse-engineering approach. First, we looked at what was happening at the neurological and epigenetic levels to cause PTSD symptoms (epigenetics means a change in gene expression caused by environmental stressors, and is explored in more detail in Chapter 23).[24]

What we found was that during a traumatic event, PTSD sufferers had had potent biological changes that persisted after the traumatic experience ended. In these individuals, there had been significant alterations in the biology of their adrenal glands, immune systems, stress hormones, amygdalae, anterior cingulate, hippocampi, and medial prefrontal lobes, and those biological changes were prompting their symptoms and sorrow.

In attempting to look at how to reverse-engineer PTSD, we had an "aha" moment: because experience had caused the trauma, the experience would be the best way to unwind the trauma. As we came to embrace this theory, it became easier to see why talk therapy alone had its limitations. Remember, the Hebbian Learning Rule we discussed earlier indicates the more neurons that fire, the greater the structural change and thus the greater learning and behavioral change. Therefore, we took the experience-in, experience-out approach. Put simply, new, positive physical experiences in the real world would activate greater amounts of body feedback, brain tissue stimulation, and gene expression than talk therapy alone would. And if done right, this process could create the kinds of biological changes in PTSD sufferers that would support healing and lessen their symptoms.

We then developed a set of tools for the PTSD sufferer that purposefully engaged the affected neurological structures and genes in a positive way, such as heightening awareness of their emotions and feedback from their body before they act, which we called "becoming the observer;" fostering peer mentorships; developing rituals of healing like daily short meditations (one minute to ten minutes depending on their tolerance) getting a grasp of the neurobiology of PTSD (so they understand their feelings are not their fault); and performing purposeful actions in the real world to stimulate the damaged brain areas, based on what I call the *Four-step Cycle of Transformation* (see also Chapter 30).

We also emphasized that to heal effectively, sufferers need to understand that in nature, all transformation happens through a process—that's how seeds become flowers, coal becomes diamonds, and caterpillars become butterflies. And because of genetic variation among individuals, a one-size-fits-all approach won't work. Those with PTSD need to develop a personalized process of self-directed, experiential-based transformation with a therapist that, over time, will alter their biology in the direction of recovery.[25]

To create the safest and most effective environment for these traumatized individuals to heal, we encourage groups that create PTSD programs to hire staff who have suffered from PTSD and who have engaged in processes that greatly healed their symptoms. This is done to significantly lower a new client's stress hormones, which enhance learning and to activate mirror neurons that increase positive imitation behaviors (see also Chapter 28).

Cracking the Learning Code

Unbeknown to traditional learning organizations, if your body is not involved in the learning process there can be no efficient memory formation. The reason it is so much easier to remember your first kiss than it is to remember your last history class, is because that kiss involved your body in the learning process.

16
Working Memory: How It Plays a Trick on Education

There are three side effects of acid: enhanced long-term memory, decreased short-term memory, and I forget the third.

—Timothy Leary, Harvard Professor and LSD Experimenter[1]

Meaningful information tagged with emotional and body markers doesn't just magically appear in your long-term memory banks. It must first pass through your working memory, where a ranking process takes place. This vital part of your *Meaning Network* acts as a staging area where tagged information is processed and evaluated to see whether it meets certain other criteria, so it can be selected into your long-term memory.[2] If the information in your working memory doesn't meet these criteria, it will get jettisoned!

Your working memory, however, can play a trick on your test-based school systems. This is because working memory can hold information for hours to days to weeks, so you can pass a test—without your long-term memory ever being activated.[3] This trick explains why you can get 100 percent of the answers right on a test and weeks later have little or no recollection of the test's content. In this chapter, we'll delve deeper into working memory and how it's a willow the wisp in acquiring long-term learning.

The Misnomer of Working Memory

For many students, grades take on life-or-death significance, so their primary goal is to get passing grades—the higher, the better. Whether the information studied is placed into long-term memory or not becomes a distant and secondary goal. This emphasis on tests creates a mental environment where to survive in the classroom your brain often selects not the information studied for the test, but the strategies used to pass tests.

The result is that you may, for example, have memorized the periodic table as a student and gotten an A on the test but remember very few of those elements weeks later. What does go into your long-term memory? What do you remember? What was most meaningful to your brain? Is it the devices you used to get that passing grade, such as cramming, borrowing notes from friends, visiting your favorite study places, and engaging in homework binges with drug-delivery systems such as coffee or Red Bull? Many students are now using unprescribed attention deficit hyperactivity disorder (ADHD) drugs like Ritalin and Adderall to help them get good grades.

How can we break this cycle of grades on tests being more meaningful than knowledge? Well, once you understand how working memory functions, you'll have the scientific foundation to understand another limitation of our test-centric educational system—and to develop personal and institutional learning programs that will give us the results we demand.

Working Memory's Number One Job: Make Order Out of Chaos

It could be said that your working memory staging areas are where your brain makes order out of chaos. Estimates of how many bits of information your brain processes each second vary, but the numbers are enormous. In his book *Elemental Mind*, Nick Herbert[4] estimates that your brain receives 100 million bits of information per second from each eye, ten million bits of information per second from the skin, and 30,000 bits of information per second from the ears. All this external information is combined with the billions of bits of internal information your mind is processing consciously and unconsciously per second, like "Should I eat something right now? If so, what should I eat? Or should I eat nothing and go work out?"

Herbert estimates that when all that external and internal processing of information is taken into account, your brain is dealing with an incredible 10 trillion bits of information each second. No wonder your brain uses up to 30 percent of your body's energy, and you're a little tired at the end of the day! Working memory is the staging area where all this chaos of reverberating information is converted into order. It provides the workspace where the bits of information that produce the highest degree of meaningful somatic and emotional feedback are sorted through and prioritized for possible inclusion into your long-term memory.[5]

A Change of Name

Up until about forty years ago, most neuroscientists called working memory "short-term memory." In the 1950s, Harvard's George Miller devised psychological tests that showed the mind can hold approximately seven pieces of information in its consciousness for very short periods.[6] Miller is given credit for why US telephone numbers, without area codes, have only seven digits.

In the 1970s, Cambridge's Allen Baddeley reformulated and expanded the notion of short-term memory proposed by Miller, replacing it with the concept of a generalized workspace where conscious and unconscious mental processing was accomplished. Baddeley called this workspace "working memory."[7] Working memory acts like the editor of a book or a movie, deciding which bits of information are relevant and should be retained for later use and which are irrelevant and should be discarded. Today, neuroscientists know the length of time that information can be held in working memory ranges from a few milliseconds to many weeks. Some neuroscientists break down working memory into classifications based on the amount of time information is stored there.[8]

The Three Main Factors of Working Memory

Here we will look at the three important factors that define working memory:

1. It **prioritizes** information based on the intensity of emotional and body feedback.
2. It is **limited** in both the time frame and storage capacity.
3. **Inclusion** into its space depends on information already held by your brain as meaningful.

The following sections of this chapter explore each of these factors in turn and help you understand their limits and advantages.

1. Prioritizes Information: Working Memory Needs Your Emotional and Body Feedback

Your brain selects knowledge into your working memory largely based upon the intensity of the body feedback and the emotions associated with it. As we discussed in the last chapters, the greater the physical change (such as spikes in your blood pressure, skin response, and heart rate) and the higher the emotional charge, the higher the priority that information is given. This is why it is so much easier to remember the bear you see in your backyard than a picture of a bear in a magazine. The bear in your backyard triggers higher somatic and emotional tagging, this is what gives it a higher ranking in your working memory.

Wonder why you so often get bored, and your attention wanders in classrooms or meetings? The answer is if the information has low meaning, it results in low body feedback and low emotional tagging, which, in turn, creates low attention and, therefore, low inclusion into your working memory. And if the information being delivered isn't stimulating your working memory, it won't make it into your long-term memory. In addition, an unstimulated working memory immediately starts to shift its focus to find something that will excite it. This is why you often find yourself daydreaming or sneaking a look at your cellphone in boring classrooms and meetings.

2. Limited in Both Time Frame and Storage Capacity: Working Memory Can Only Hold So Much

Your working memory is only designed to hold information for very short periods—from seconds to days. This is a workspace, like a carpenter's bench—things placed here are not meant to hang around indefinitely. Once your working memory receives information marked with the body (somatic) and emotional tags, it tries to match this data with previously encoded meaningful knowledge. Then if a match is made, the information has a chance of being passed to long-term memory.

What most educators, trainers and bosses don't realize is that when your working memory is full and you add something new, something else must get bumped out to make room for the new information. Unfortunately, that bumped-out information never has a chance of making it into your

WORKING MEMORY CAN ONLY HOLD SO MUCH
When your working memory is full, and you add something new, something else must get bumped out to make room for the new information. Unfortunately, that bumped-out information never has a chance of making it into your long-term memory.

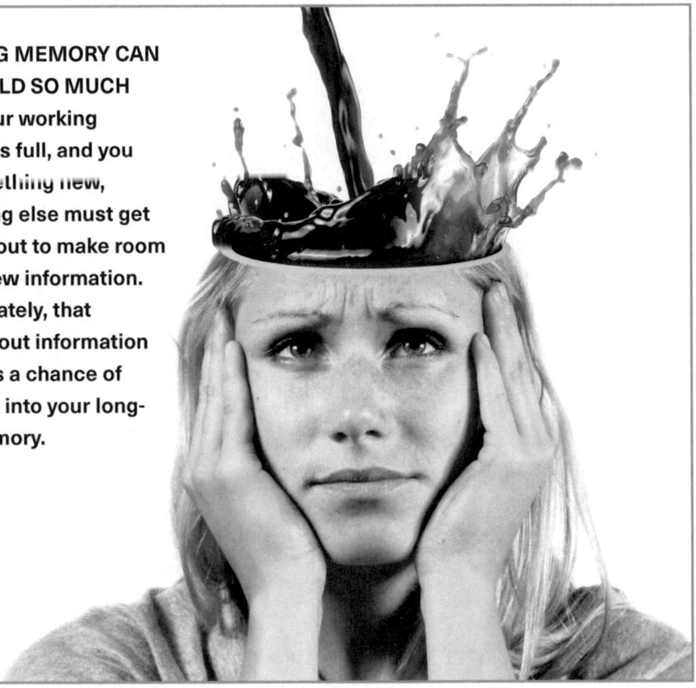

long-term memory banks.

Our traditional educational systems fail to understand this limited workspace concept. This is why they continue to drive students in and out of stressful forty-five-minute get-ready-for-the-test-segments all day long, with little attention to whether that information is firmly implanted into working memory or not before it gets knocked out by new knowledge. Educators then wonder "what's wrong with these students, that they've forgotten so much of what was just presented?"

3. Inclusion Relies on What You Already Know

As you try to grasp the concept of working memory as a limited workspace, one of the most important things to recognize is that working memory isn't a pure product of the here and now. What you select into your working memory is dependent upon the information your brain has previously decided is meaningful to you.[9] Our working memory provides a stable place where new information, marked with emotional and body tagging, can be matched up with meaningful data already held in your long-term memory.

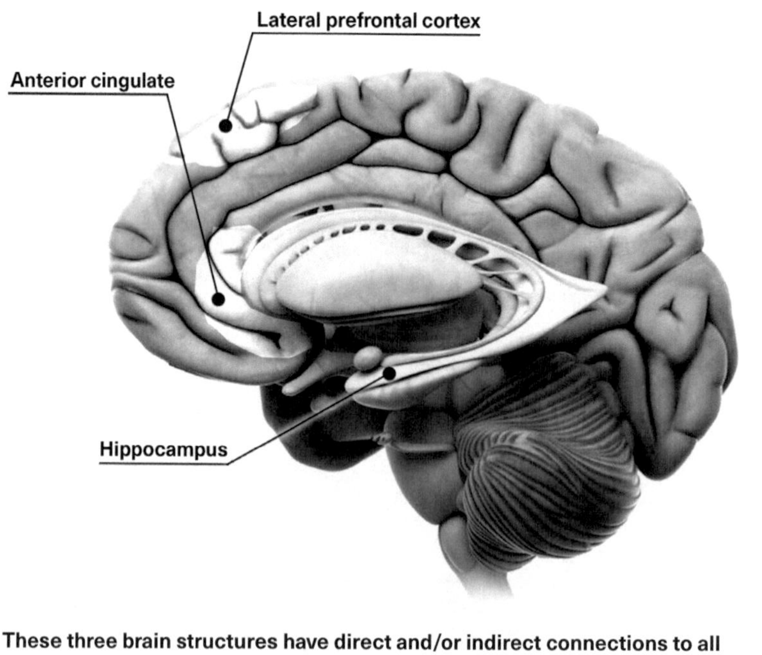

These three brain structures have direct and/or indirect connections to all your physical and emotional brain circuits.

As Damasio says, "The allocation and maintenance of attention and working memory do not happen by a miracle. They are first motivated by preferences inherent in the organism."[10] Neuroscientist Stephen Kosslyn explains that working memory requires "not only some form of temporary storage, but also an interplay between information that is stored temporarily and a larger body of stored knowledge."[11] In other words, if your brain thinks the information will support your surviving and thriving, based on what your genes and previous experiences have encoded as meaningful in your long-term memory, you're more likely to remember that information. If you love model airplanes and walk through a hobby store with a friend who couldn't care less about your hobby, you'll easily select into working your memory lots of information about model airplanes, while your friend will select very little.

Your Structures of Working Memory: A Quick Overview

This section goes a little deeper into the science and focuses on the three main structures of your working memory:

- anterior cingulate
- lateral prefrontal cortex
- hippocampus

All three of these brain structures have direct and/or indirect connections to all your somatic and emotional brain circuits, including the amygdala, ventromedial prefrontal cortex, somatosensory cortex, hypothalamus, and thalamus. Now, as mentioned, your brain works more holistically than linearly, with many actions happening almost at the same time. However, for our purposes, the easiest way to understand how working memory operates is to visualize information as if it enters your working memory on a conveyer belt heading for your long-term memory.

First, information marked by other brain areas with body feedback and emotional tagging is picked up by your **anterior cingulate**, which focuses your attention on this data and does a quick and rough appraisal to see whether it matches previously learned information. The anterior cingulate sits just behind the prefrontal lobe and works like a floodlight broadly surveying your rear associative areas where meaningful long-term memory is held.

If your anterior cingulate finds a general match, a more refined and elegant matching process is then accomplished by your **lateral prefrontal cortex**, located at the side of your head, in front of your temples. Your lateral prefrontal cortex works like a laser beam, providing a more efficient and refined pinpointing of matching information. If your anterior cingulate and lateral prefrontal areas deem the information valuable to your thriving and surviving, they then call on your **hippocampus** in the limbic system, which ensures that this new information gets logged into the correct long-term category networks in your rear associative areas. The hippocampus makes sure if you learn about a new fruit, it gets placed in the category network for fruits, not the category network for cars.

FAST FORWARD *or* DEEP DIVE

If you're seeking only a general understanding of how your working memory helps increase your **Adaptability/Intelligence Factor**, you can skip the next few sections on the structures of working memory and go to the section "Neurotransmitters of Attention and Working Memory: Spark Your Interest." But if you want greater scientific depth, read on.

1. Your Anterior Cingulate: Appraisal and Attention

Located behind the prefrontal cortex, your anterior cingulate accomplishes its job of rough analysis by selectively highlighting some bits of information while keeping others in the background.[12, 13] The anterior cingulate is also known as *the executive in charge of allocating your attention.* Your attention span is lengthened if the anterior cingulate perceives incoming stimuli as meaningful to your surviving and thriving. The more energy your cingulate focuses on incoming information, the better chance the information has of being selected into your lateral prefrontal cortex, where a more-refined analysis can be carried out.

Research shows that if your anterior cingulate is damaged, efficient learning ceases. This is because you can no longer focus your attention on—or make a crude appraisal of—which information in the environment is meaningful to you and which is not.

The anterior cingulate is the first place where you relate outside information to your own personal being. Many neuroscientists believe the anterior cingulate is the source of our sense of "I."[14] Without efficient connections to the anterior cingulate, anosognosia sufferers such as Woodrow Wilson and William Douglas were unable to recognize the conditions of their own disease.[15] Autism also results from the malfunction of the anterior cingulate.[16] Many autistic people are largely oriented to the past and not to what's going on in the present. Piquing their awareness about what's going on in the here and now is often difficult because their attention is stuck on previously acquired information.[17]

2. Your Lateral Prefrontal Cortex: Even Greater Refinement

Your lateral prefrontal areas, at the sides of your prefrontal lobes, have richer and more efficient connections to the long-term memory banks than the anterior cingulate does. These connections allow your lateral prefrontal cortex to carry out a highly refined and detailed matching of incoming information with information you have previously learned. This more sophisticated matching process allows your lateral prefrontal cortex to decide which information should be passed on to your hippocampus for categorization and inclusion into long-term memory, and which should be discarded.

Joseph LeDoux points out that only primates and humans have large lateral prefrontal cortexes.[18] While other animals do have the brain structures that allow them to focus on emotionally and somatically charged information, without a large workspace to connect new information with data they've previously encoded, they cannot make refined distinctions about incoming stimuli; this, in turn, limits their surviving and thriving options. Many evolutionary biologists maintain that humans' superiority over all other animals is directly tied to our lateral prefrontal cortexes.

If your lateral prefrontal cortex and its connections are disrupted, learning and life become chaotic. Patients with damage to the lateral prefrontal cortex tend to repeat mistakes because they lack an efficient workspace to match new information with existing information.[19]

Brain scans show schizophrenic delusions can be traced to underactive neurochemical activity and blood flow in the lateral prefrontal regions. This hypometabolism, as it's called, negatively affects the lateral prefrontal areas. Without sufficient space to make spot-on matches between incoming data and information already held in the rear associative areas, schizophrenics end up connecting new data with information in a chaotic manner, producing garbled communications; this, in turn, results in delusions, hallucinations, and erratic behavior. Drugs that help manage these hallucinations increase blood flow and neurochemical function in the lateral prefrontal cortex.[20]

After a blow to the head, you often can't think straight because the lateral prefrontal area is negatively affected. Sports figures who received numerous concussions, such as the famous Hall of Fame quarterbacks Joe Montana and Troy Aikman, have had to stop playing their sport altogether or risk long-term damage to the lateral prefrontal cortex (as well as other areas).[21] The term "punch-drunk" is used to refer to boxers and other athletes who've received one too many blows to the head, resulting in permanent damage to the lateral prefrontal cortex and working memory.

3. Hippocampus: Your Concept-Matcher

Located at the base of the limbic system, your hippocampus is your concept-matcher. The hippocampus not only helps your lateral prefrontal cortex transfer information from working memory to long-term memory but also ensures that new information gets logged into the correct category networks. It's the hippocampus's job to make sure that if you learn about a new kind of vegetable, it will be stored in the vegetable networks of your long-term memory, not the truck or the fruit networks. This filing of information by the hippocampus in the correct concept network allows for a more compact organization and quick retrieval.

Neuroscientists are aware of the powers of the hippocampus primarily from a patient known as H. M. His case holds one of the most prominent positions in the history of memory research. H. M. suffered from debilitating epileptic seizures and, in the early 1950s, had a temporal lobe lobotomy, which reduced their severity. After the operation, H. M.'s IQ went up seven points to 112, probably because he was having fewer convulsions. While H. M.'s IQ remained intact, his ability to place new information into long-term memory did not. After the operation, H. M. could remember his doctor's name and carry on a conversation while the doctor was in front of him, but if the doctor left the room—even for a short period—and returned, H. M. would have no memory of ever encountering this doctor before and would introduce himself as if he had never met the man.[22]

In later tests, it was established that H. M.'s lateral prefrontal cortex worked normally. He could hold information in his working memory for short periods but was unable to shift this information into his long-term memory. It was suspected that H. M.'s hippocampus had been so severely damaged by the temporal lobotomy that he was unable to consolidate his memories. In the late 1990s, H. M. had magnetic resonance imaging performed, which confirmed that over half of his hippocampus had indeed been removed and that the remaining portions had atrophied.

From H. M.'s amnesia case and the imaging of other brain-damaged patients, neuroscience now recognizes that the hippocampus plays an instrumental role in supporting working memory so that information can be transferred into long-term memory. It's interesting to note that your hippocampus ascribes no value to the information. That job is left to your anterior cingulate and lateral prefrontal cortex, which receives emotionally and somatically charged information from other parts of the brain (such as the ventromedial and amygdala), prioritizes it and then works with the hippocampus to ensure that the relevant information will be selected into the correct long-term memory concept networks.

Toddlerhood Amnesia

One of the reasons most of us have what could be called "toddlerhood amnesia" is the hippocampus is so immature before the ages of three to five that it's unable to support the lateral prefrontal cortex (which is also immature) to complete efficient working-memory function. Without their efficient function, there's no efficient long-term memory formation. This is why our memories of toddlerhood are vague or nonexistent.

Neurotransmitters of Attention and Working Memory Spark Your Interest

Specific neurotransmitters enhance your attention and working memory function. Here we focus on the monoamines: dopamine, norepinephrine, and serotonin. These neurotransmitters are the chemicals that work together to ensure that meaningful information can be tagged as important enough for you to remember. They provide the mental energy to make order out of chaos; they help you attend only to the meaningful bits of information in your world without being overwhelmed by the trillions of bits of unimportant information.

It's the monoamines that provide the neurochemicals that mark information with emotional intensity. Your feelings of joy, fear, love, greed, and anger would cease to exist without the influence of norepinephrine, dopamine, and serotonin. As mentioned earlier, I like to think of these neurotransmitters as the spice rack of your brain. They sprinkle specific bits of incoming information with just the right amount of emotional zest so that your anterior cingulate, lateral prefrontal cortex and hippocampus can effectively perform attention, focus, and working memory functions.

We'll cover the monoamine neurotransmitters and their sources in more detail in other chapters. For now, here are some highlights of how they support you to put the most relevant information into your working memory.

Norepinephrine

When activated, norepinephrine improves the signal-to-noise ratio in your brain by exciting some neurons and inhibiting others. It supercharges the actions of neurons that are firing in response to information that your brain perceives as meaningful, while inhibiting the action of other neurons that are responding to irrelevant information. This inhibition/activation function allows your working memory to focus its resources on the bits of information in the world that have the most emotional and somatic value to you.[23]

Dopamine

Dopamine works hand-in-hand with norepinephrine. After norepinephrine makes the first pass at relevant information, marking it to stand out, dopamine emphasizes this marked data. Using the spice rack analogy, dopamine takes bits of information that norepinephrine has already marked with black pepper and sprinkles on cayenne pepper to really get your working memory's attention.[24]

Serotonin

Serotonin plays an important role in stabilizing cellular communication. Without this stabilizing influence, the billions of bits of information bombarding your brain each second would cause neurons and their connections to fall into a frantic, foaming sea of cross talk.[25]

The calm backdrop of neural firing provided by serotonin is important to your learning because it creates a stable background canvas, allowing information marked as emotionally and somatically meaningful (by norepinephrine and dopamine) to really pop out; thus, there's no confusion about what your working memory should pay attention to and select into its structures.

If the brain structures which produce norepinephrine, dopamine, and serotonin become damaged, your brain's ability to mark and select the most meaningful information into working memory becomes compromised.

Stress Destroys Your Working Memory

Too much stress can make you as stupid as a punch-drunk boxer. When you're under too much stress, the body's signals and emotional feedback from your lower brain stem and amygdala can overwhelm your upper working memory's ventromedial and lateral prefrontal areas. Therefore, these upper structures become inhibited from efficiently processing and prioritizing information. This, in turn, sabotages your ability to pack working memory with the correct information, which prevents you from efficiently making new accurate long-term memories.[26]

Some experts call this stress-induced garbling, caused by high somatic

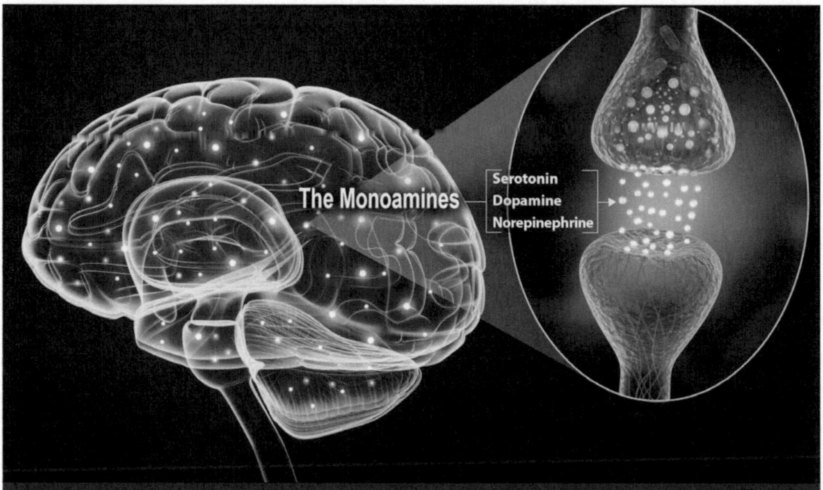

It's the monoamines that provide the neurochemicals that mark information with emotional intensity. Your feelings of joy, fear, love, greed, and anger would cease to exist without the influence of norepinephrine, dopamine, and serotonin.

and emotional feedback "neural static". When you're stressed, neural static causes you to select garbled information into your working memory, which means you then place irrelevant information into your long-term memory banks. The reason for this is that the memory structures of your brain are like the inner workings of a computer—if you put junk in, you get junk out. Neural static also interferes with your ability to access information previously stored. This is why when you feel stressed in an argument, you often can't access the words and facts you know to make your point—very frustrating.[27]

Research shows that stress causes not only neural static but also what we call a *neurological downshift* in your brain's energy. A *neurological downshift* happens when excessive levels of stress prompt the overproduction of your primary stress hormone, cortisol. Excess levels of cortisol shift the distribution of glucose and oxygen away from your prefrontal lobe areas—where the lateral areas of working memory are located, greatly reducing their efficiency. (For more on stress, see Chapters 37–38.)

Downshifting your energy flow also has a profound effect on your attitude. It means that you go from being a rational, caring, and creative human to an individual who relies on your act-without-thinking instincts. This is why you act out of character when under stress. You may yell at your family, say stupid things to your boss, ignore sound advice, make bad plans, or find it difficult to change your inappropriate behaviors. Also, because working memory function is compromised by this *downshift* in energy, you don't learn from your experiences and end up repeating stupid mistakes.

Understanding how neural static and *downshifting* mess up your working memory gives us greater insight into why many of us have test anxiety. Test anxiety is a double whammy, as stress both garbles what you've studied and shifts your brain's energy away from where you've stored the knowledge. People with test anxiety aren't stupid. They don't need more "test skills." They need more "stress skills" or—better yet—a system that recognizes the limits of testing in creating long-term memory formation.

At The Institute, we often use science to help customer service departments become more efficient. Corporate management will almost always blame its representative for bad services, like not providing the correct information or being short-tempered with a customer. After managers understand the neurobiology of stress and its effect on working memory and attitude, they can shift their efforts away from using rewards and punishments to control behaviors to what may be the main culprit—that is, how the stress levels of the customer representatives may be sabotaging all their "best practice" efforts. This is a wonderful "Aha!" moment for these teams, as it gives them a foundation from which to make a real, long-lasting change by reducing stress levels in their cultures.

When we grasp how the working memory element of your *Meaning Network* can be disrupted by stress, it's easy to understand why students and employees who operate under stressful conditions often can't perform up to high standards. The sooner we understand the neurobiology of the *Meaning Network* and elements like working memory, the sooner we can build profound learning and behavioral change systems that create the learning environments we long for.

Cracking the Learning Code

To build learning and behavioral change systems that can profoundly transform people's lives, we must understand the three main factors that define working memory. It prioritizes information based on the intensity of your emotional and body feedback. It is limited in both time frame and storage capacity. Inclusion into its space depends on information already held by your brain as meaningful.

17

Dopamine: Keeping You Addicted to Meaning and Discovery

What is love when it's not for dopamine?

—Dr Saurabh Sharma, psychiatrist and neurologist[1]

In teasing out the brain structures that make up the *Meaning Network*, once it became clear that emotional and body feedback marked meaningful knowledge for inclusion into your working memory, another question arose: why are we compelled to discover new things in life that are meaningful to us? That is, what neurological system is responsible for compelling you to focus on, get pleasure from, and seek out meaningful things in your life, over and over again? For example, why are you so interested in the opposite (or the same) sex? Why do you follow your favorite sports team even when they're losing? Why do you keep channel surfing? When your cell phone pings with a text or Facebook message, why do you feel compelled to check it?

The answer to why you're constantly seeking to discover and connect to things in your life that have personal meaning to you—oversimplified for our purposes—is your dopamine system. Understanding how this neurochemical system prompts you to get more meaning from your world will give you great insight into how you can accelerate the pace at which you learn and change behaviors. As you saw in the previous chapter, dopamine works with norepinephrine and serotonin to mark meaningful information for inclusion into your working memory. The popular press tends to give the impression that dopamine is mainly a pleasure molecule. But the dopamine system does so much more than this. It plays another vital role: compelling you to discover, seek out, and focus on things in your life that are meaningful to you.

I don't use the word "compel" lightly here. It's the dopamine circuits that are responsible for why you feel you must eat that next piece of chocolate, why you'll blow off work to engage in your favorite sport or hobby, and why drug addicts and alcoholics feel forced to take that next hit or next drink. Your impulses to follow what's meaningful to you, good or bad, are driven by the activation of your dopamine system. Dopamine has a powerful effect on your intrinsic motivation. When dopamine activation is high, it's easy to follow what's meaningful. But in many depressives or even bored learners, intrinsic motivation declines dramatically when dopamine activation is low.

When you were in school, and you didn't feel captivated by the teacher's lecture, it was because that lecture lacked meaning to you and therefore failed to efficiently activate your dopamine circuits. Trying to force yourself through willpower to focus on things that have no personal meaning is so difficult because you're swimming against the biological tide. One reason that scholastic and corporate learning systems produce such low levels of efficiency and engagement while producing high levels of boredom and stress, is that the creators aren't aware of the impact that dopamine has on the learners' meaning-making, learning, focus, and motivation. In this chapter and the next, we'll take a deeper dive into how you can use the neurogenetic science to create higher levels of motivation in yourself and others. For now, we'll focus on how dopamine has the power to make you addicted to meaning.

The Evolutionary Roots of Dopamine

It's critical for animals to know when they're proceeding in a favorable direction—and when they're not. Evolution selected genes that produced dopamine circuits that create an inherent drive for animals to seek out, focus on, get pleasure from, and maintain instinctual behaviors that are meaningful, such as eating, bonding, nesting, hoarding, imitating, and having sex. If your biology didn't find meaning in and compel you to engage in these instinctual drives, our species would cease to exist. As we've seen, humans possess more complex and more evolved brain structures than our animal cousins. These

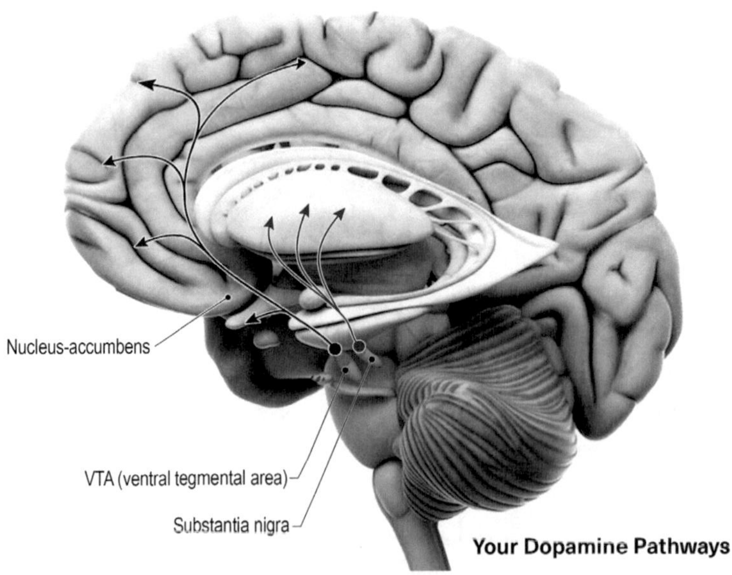

Your Dopamine Pathways

The dopamine system plays a vital role: it compels you to discover, focus on, and seek out things in your life that are meaningful to you.

brain structures allow us to find meaning in a wider and more sophisticated range of endeavors than the instinctual. This is why computers, medicine, math, cooking, writing, composing music, sewing, and golf, for example, become so meaningful to us. Therefore, whether you're engaging in instinctual activities like sex or eating or your favorite hobby or a worthwhile work project, it's the dopamine circuitry that supports the feelings that compel you to keep coming back again and again to the activities that have the most personal meaning to you.

The Discovery That Dopamine Drives Behavior

The first evidence that your brain circuits prompt the discovery and maintenance of meaning was discovered in 1956 by psychologist James Olds at McGill University.[2] Meaningful behaviors in a rat's day consist of seeking food, water, sex, a little exercise, and some self-grooming. Olds, who is considered one of the founders of modern neuroscience, inserted fine electrodes into the midbrain of rats and found if he allowed them to self-stimulate their midbrain by pressing a lever, the rats would press the lever up to 100 times a minute, ignoring all other meaningful activities.[3] If Olds allowed the rats to artificially stimulate this midbrain region at will without interruption, the rats would ignore everything else in their lives, including food, and eventually die from starvation. What Olds discovered was one of the circuits your brain uses to repeatedly seek out and discover meaningful stimuli.

In the decades since Olds' work, our understanding of dopamine's activity has increased dramatically. We now know that Olds' electrodes recorded dopamine activity, which drove seeking of pleasure by direct action upon the ventral tegmental and nucleus-accumbens circuit. The ventral tegmental area rests at the top of your brain stem, and the nucleus-accumbens sits just above and in front of it. Research has shown that this circuit has a profound influence upon the amygdala and strong connections to your ventromedial prefrontal areas, ensuring that powerful positive emotional tagging is associated with dopamine production.

Your Dopamine Anatomy

Three brain structures are singled out as major components of dopamine's effect on learning and behavior are: as just discussed, the ventral tegmental and nucleus-accumbens, and the substantia nigra located in your midbrain which helps direct you to physically move toward meaningful things. A substantial list of genes also influences dopamine processing, like the DRD2, DRD4, and COMT genes.

As we have you've seen, your body is heavily involved in the learning process, and dopamine plays a significant role here—in your organs such as your eyes, kidneys, digestive tract and pancreas; your muscle tone; and the modulation of your blood flow. Therefore, on a micro-level, untangling the role of dopamine in the learning process is somewhat complicated. But don't worry; it may sound counterintuitive, but at The Institute, we've found that one of the simplest ways to gain a quick understanding of dopamine's compelling effect on meaning-seeking and discovery is to look at the five different dopamine receptors which support meaning-seeking, motivation, and grit (your ability to persevere during difficult times).

Five Different Dopamine Receptors

Your neurotransmitters perform multiple functions, and dopamine is no exception. There are five different dopamine receptor sites: D1, D2, D3, D4, and D5. Ongoing research is still untangling the exact function of each of these receptors and how they interact with each other. For our purposes in understanding learning and behavioral change, we can say that when dopamine attaches to these receptors, you'll seek out, focus on, and anticipate the pleasure you'll receive when you engage in personally meaningful thoughts and actions.

Notice, I said, "anticipate pleasure," not just "get." Despite dopamine's reputation in the popular press, research shows this molecule also prompts you to anticipate. It's this expectation of pleasure that forces you to keep engaging in meaningful activities over and over again. This can be a good and bad thing. This is a good thing when you're engaging in a meaningful work project or hobby. The anticipation of pleasure is one of the main sources of your intrinsic motivation and "grit" that keeps you engrossed for long periods, even in the face of great challenges. Without the activation of dopamine, humans would never have had the grit to put a man on the moon or discover the treatment for HIV.

The anticipation of pleasure can also be a bad thing, like when you're trying to stop drinking too much beer or eating too much chocolate. The anticipation of that next drink or next chocolate bar is what keeps your focus on them, and forces you to seek them out even when the willpower part of your brain is yelling at you, "Stop! Don't do it!" This is why we're compelled to repeat bad habits over and over again, even when we clearly see the damage they're causing.

The Dopamine-endorphin Connection

The pleasure you receive from engaging in meaningful acts doesn't come directly from dopamine's activation but from its effect on other neurochemicals like the neuropeptide group called endorphins. Because of the profound pleasurable feelings endorphins produce, they're often called the brain's opiates. Dopamine acts to slow the re-uptake of endorphins, which dramatically prolongs your enjoyable sensations when you engage in meaningful behaviors like sex and eating chocolate.[4] Research has also shown that the release of endorphins, in turn, stimulates even more dopamine production.[5] These two neurochemicals then work in tandem, each supporting the other's activity, dopamine compelling you to seek out meaningful things, and the endorphins giving you pleasure when you engage in them.

Kids Have 40 Percent More Dopamine Receptors

Why are kids running around discovering so much more than adults? Why do they find so many things interesting and meaningful? Neuroscientist Martin Teicher at Harvard's McLean Hospital was one of the first to discover why children have 40 percent more dopamine receptors than adults,[6] which helps them remain in a constant state of discovery. It's why you have to keep your eye on them, or they'll stick their fingers in electrical sockets, and if you try to make them sit still, they squirm like crazy as they try to get free to go discover something more meaningful than sitting on your lap. If a learning system doesn't strive to keep this overabundance of dopamine receptors activated, children will either act out to self-stimulate those extra receptors or get bored and pass out.[7]

This overabundance of dopamine receptors isn't pruned from our brains until we reach our early twenties, and its why adolescents often get into trouble with sex and drugs. Their biology is driving them to seek out many new experiences of self-discovery, good and bad.[8] At The Institute, when we work with educational institutions, one of our primary efforts is to help them understand that if their learning systems aren't allowing students to self-stimulate these dopamine receptors through journeys of self-discovery, not only will learning be impeded, but behavior problems will skyrocket.

Self-discovery Learning vs. Stand-and-deliver Instruction

Yet even with fewer dopamine receptors as you age, the research indicates that, for your whole life, the more self-directed your learning, the greater dopamine activation, which in turn drives more knowledge acquisition.[9] Which means that even as an adult, you learn best through self-discovering what's meaningful to you.

Think of your personal learning. One reason it's so difficult to remember all those lectures from school and corporate training courses is there was so little room to discover what was meaningful to you within them. Most of the information in required stand-and-deliver and online presentations is, in many cases, much more meaningful to the teacher or presenter delivering them than it is to you.

When your brain is limited from self-directed learning, it's hard to jump-start your dopamine circuits, which in turn limits how much knowledge you can pack into your working and long-term memory. Think about it: what do you remember more of? A class you found boring or something you taught yourself because you were fascinated to discover more? We suggest that progressive early-adaptor organizations limit stand-and-deliver

Children have 40 percent more dopamine receptors than adults, which helps them remain in a constant state of discovery.

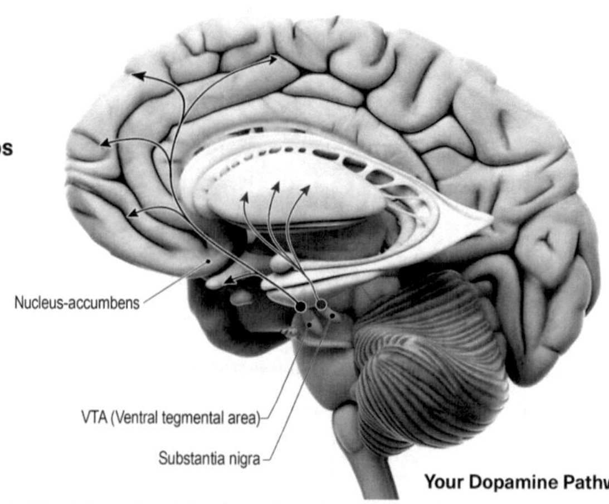

Your Dopamine Pathways

presentations and instead create what I call *learning adventures* to enhance self-directed discovery. Would you rather go to a class, take a course, or embark on a *learning adventure*? Your dopamine circuits find the *learning adventure* option much more charming.

At the scholastic level, when we help create *learning adventures*, we bake into them the knowledge that's going to be on the standardized tests. In these adventures, which we also call quests, we de-emphasize classroom tests while focusing on making the material meaningful to the student so that he or she feels compelled to self-discover more knowledge around the subjects. In other words, the student shifts from being a passive learner to a very active participant in their own journey of knowledge-gathering. In these adventures, the teacher's job transforms from being a deliverer of data to a mentor supporting intrinsic motivation in students, who then find personal meaning in the knowledge to be learned.

At Master's Academy in Canada, standardized test scores are now 400 percent above the average in the excellent category, the material gets covered in 20 percent less time, and behavioral problems have been virtually eliminated, because the teachers and administrators understand how to stimulate the dopamine system through these *learning adventures*. Why would children act out if they're focused on discovering what's meaningful to them?

We also urge corporations to enhance adult learning through the *learning adventure* concept. We counsel them to shift their paradigms and rename their training departments, transformational departments. This change is not meant to be merely semantic. We support organizations to shift their purpose from just training to the enhancement and fulfillment of their members' lives on more than just a financial level. Once they understand The Learning Code science, they can begin to stop focusing on stand-and-deliver and testing and have participants enter adventures of transformation. We suggest they limit stand-and-deliver presentations to around 20 percent of the time. The rest of the time is devoted to self-directed discovery of knowledge to help participants live more fulfilled lives. To accomplish this, we rely heavily on the *Four Stages of the Cycle of Transformation* process you'll learn more about in Chapter 30.

Of course, stand-and-deliver lectures can work well if the information being presented has high meaning to the learner. As the information flows out of a lecturer's mouth, your brain is seeking the bits of knowledge that have the greatest meaning to you. These personally meaningful units of information activate dopamine, endorphin, and other circuits, accelerating the selection of that information into your working and long-term memory banks. But if the presentation has low meaning to you, low dopamine will ensure, and little of that knowledge will be selected or remembered. In a way, with our personal meaning, you're wasting your time.

Addiction Seeking: Usurping Your Meaning Network

You can gain great insight into dopamine's role in your *Meaning Network*, and how it affects your ability to learn, from the world of addiction. When looked at through the lens of science, we are beginning to understand that addiction is a form of learning. And the more we understand dopamine's role in supporting us to become addicted to things like drugs, alcohol, and food, the more we can understand its role in learning and behavioral transformation.

It is vital to understand that drugs and alcohol addiction activates the same brain circuits and chemical systems as other compulsions, including food, sex, shopping, pornography, gambling, and so on. At The Institute, we call those susceptible to addiction *addiction-seekers*, so, if you've learned to seek these addictions, you may learn about your own behaviors in this section.

We've found the lower the meaning in life, the more you desire and seek

Your dopamine circuits can be usurped by drugs and alcohol if your **Meaning Network** is not engaged.

things like sugars and drugs. Why? Because if you lack meaning in your life, the quickest way to stimulate dopamine production and the accompanying hit of endorphins is through addictive things like alcohol, shopping, food, etc. They are quick fixes with dire consequences.

A fascinating study conducted after the Vietnam war by Lee Robins at Washington University School of Medicine found 34 percent of returning veterans had used heroin. Yet after returning to the US, only 1 percent became re-addicted. Why such low numbers? The hypothesis is that once they returned to the more meaningful environment of home, their biology's need to use drugs was dramatically reduced. In other words, meaning replaced the need for heroin.[10,11]

I'm very familiar with the world of addiction. I have a family tree chock-full of crazed addicts. Both my parents and one of my sons died from overdoses (that broke my heart), my sister used to take enough codeine a day to kill a horse (she is in recovery now) and many years ago, before I got sober, I developed several businesses fueled by a fifth of whiskey a day. Also, in the 1990s, I helped develop the country's only holistic treatment center accredited by the Joint Hospital Commission. This successful treatment center was rated in the top 10 percent of all treatment centers and relied on 1,500-year-old traditions from India, such as meditation, yoga, herbs, diet, and bodywork to help people break the bonds of addiction.

Why are drugs, sweets, and alcohol so addictive? To get a huge hit of dopamine, endorphins, and the high they bring, you don't have to engage in meaningful thought or action: just pop a pill, smoke a toke, slam a couple of handfuls of M&Ms, or shoot up. The danger is repeated abuse of addictive foods, and drugs can actually reshape the dopamine circuits so that they prefer to be stimulated by these addictive things over anything else in one's life.

Addicted individuals whose dopamine circuits have been usurped and altered by sugar, alcohol, cocaine, gambling, sex or heroin, find that after repeated use previously meaningful things, such as their jobs, families, sports, and hobbies, don't have as much value to them as their drug of choice. Just like Olds' rats, addicted individuals will give up anything else that has been meaningful to them to focus on and engage in addiction-seeking behaviors, often until their addiction kills them. To solve our addiction problems, we don't need a war on drugs, we need a war to understand **why** we take drugs.

Three Classes of Addiction Seekers

For more insight into the *Meaning Network*'s dopamine-discovery-pleasure-addiction-learning connection, let's look at the three main classes of Addiction Seekers:

1. Those born with low levels of dopamine activity

It's estimated that 50 to 80 percent of alcoholics and drug addicts and many other addiction-seekers possess a variation of a dopamine gene that sounds like it should be a character in the *Star Wars* saga, called DRD2. If you are born with a variation of this gene it prevents dopamine from binding to receptors effectively; so that the rush of pleasure that usually occurs when addicts engage in meaningful acts is muted.[12] The hypothesis is that the people with this gene feel driven to engage in addiction-related activities like overeating that pump up dopamine activation until they feel satisfied. Geneticists David Comings and Kenneth Blum have called the inability to ever feel satisfied "reward deficit syndrome," which they link to the brain's inability to efficiently optimize dopamine activation.[13]

2. Those with negative life and health circumstances

Those whose negative life and health circumstances are such that they come to believe that drugs, food, or risky behaviors like gambling can provide more pleasure and meaning for them than anything else in their lives. Some people have life circumstances so painful that to produce the hits of dopamine that they should be getting from a positive, healthy, and meaningful life, they instead engage in negative addictions. Even if you aren't addicted now, when things weren't going your way, short term addiction may have happened to you at one time in your life. When things go south in life, many of us turn to *addiction-seeking* to pump up our dopamine.

Many addicts also have what's called *dual diagnosis*—that is, their addiction is coupled with an underlying mental health condition such as schizophrenia, bipolar disorder, PTSD, or anxiety disorder.[14] These groups are actually self-medicating, using drugs to get the hits of dopamine that a more meaningful existence should be producing. Some addicts with a dual diagnosis are treated with antidepressants, which can also raise dopamine levels and elevate their moods so that they can find more meaning in life.

3. Those who are recreational users, but it gets out of hand

Those who, through recreational use, alter their dopamine circuits in such a way that they become addicted to their substance of choice. People with normal lives can have their dopamine circuits commandeered by repeated recreational use of drugs, alcohol, sugars, and gambling. Because of the ease and speed with which these things can stimulate the dopamine reward circuits, what starts out as a fun diversion can result in a raging addiction. Because addiction-seeking usurps the *Meaning Network*, an addict seeks more and more of his or her addiction of choice to stimulate more and more dopamine activity.

Why Willpower Doesn't Work and Meaning Does

Family members and friends often have no comprehension of how an addict is compelled to continue their obsessive addiction-seeking behavior.[15] I work with a lot of families of addicts who are severely confused. Why can't the addict just use willpower to stop drinking or taking drugs? They know they can stop easily, so why can't the addict?

Once a family understands from a genetic and biochemical standpoint why their loved one is choosing drugs over their family and friends, they begin to have greater compassion for the hell their loved one is trapped in. This takes a lot of the blame out of the family relationship so that the focus can shift to treatment.

I work with many therapists in the addiction field, and we've found that once they understand the neurobiology of the *Meaning Network* and dopamine's role in it, they start to shift their perspective. They begin to see that just getting the addict off of drugs and into AA or NA, while a good start, may not be enough. They then begin to focus on helping addiction-seekers find things in life other than drugs and drink that can give them meaning and stimulate dopamine circuits.

For example, I supported a twenty-six-year-old alcoholic with his new business start-up. This young man, who is in AA, had been sober for only six months after ten years of wild, heavy drinking. He has some great stories, like one night almost shooting his testicles off while high. The other day he said to me, "J.W., this is the first time in my life that I had something more meaningful to me than alcohol. I can't believe how, since I started this new business venture, my cravings have slowly disappeared. I feel so blessed to understand that I might have had a problem with finding meaning in my life, not just alcohol!"

Now, I'm not saying that finding more meaning in life will be a cure-all for addicts. But we've found that in conjunction with AA and similar programs, if addicts focus on finding greater meaning, their progress is much steadier and more successful because their neurochemistry is more balanced. We hope that the research into the *Meaning Network*/dopamine/addiction connection will support traditional treatment centers to add a "*finding-meaning*" component to their sobriety programs.

Why It's So Hard to Kick an Addiction: Downregulation

One of the reasons why addiction-seekers have such a difficult time kicking the habit is that excessive use of drugs, and even sugar, can eliminate dopamine receptors.[16] When the brain is bombarded for long periods by too much dopamine caused by repeated overindulgence, the result is the dopamine receptors reduce, in a process called *downregulation* (see also Chapter 34). Addicts then need to take more of their drug of choice to stimulate fewer dopamine receptor sites. It can take up to a year or longer of abstinence for the dopamine system of a hardcore addict or alcoholic to normalize or up-regulate itself (add more dopamine receptors). Relapse is such a prevalent problem in addiction treatment because the addict with fewer dopamine receptors sees their drug of choice as the only way out of feelings of depression, anxiety, and meaninglessness.

The Hell of Maintenance

An interesting phenomenon about reward circuits is their capacity to prompt us to repeat drug use even if we eventually get no pleasure from taking them. Addiction specialists point out that, in the state of addiction called "maintenance," addicts' drug use has so damaged receptors in the brain areas that process pleasurable feelings that these individuals get very little or no pleasure from getting a hit of their drug. Yet the brain circuits that compel them to anticipate and seek out their drug of choice are still intact and demanding stimulation. The end result is that these poor addicts are forced to use drugs over and over again without the immediate hit of pleasure that hooked them in the first place. This is the hell of addiction.

Understanding the function of dopamine in meaning-making lets us see that drugs are not an evil unleashed on a gullible society, but something that people who suffer from a genetic predisposition or negative and unhealthy life circumstances have learned to use to stimulate their *Meaning Networks*. New research shows one reason twelve-step groups work so well in helping people get off drugs and alcohol is that the cooperative group behaviors provided by AA and NA stimulate the brain's dopamine circuits. Therefore, a twelve-step group meeting substitutes the boost of dopamine the addict received from his or her drug of choice.[17] If you would like to see a short video The Institute produced to support addiction-seekers and their families understand how our drugs of choice affect our biology, go to YouTube and search "Cracking the Addiction Code."

In later chapters, we'll look at how your *Meaning Network* can also get commandeered by the overuse of extrinsic motivators (reward and punishment) and how this may help build the foundations for millions of us to become codependent—which, from a neurobiological perspective, can be viewed as a form of addiction.

Cracking the Learning Code

Any learning system that does not create heightened dopamine activation in its learners' brains will fail to stimulate a vital element of their *Meaning Network*. The result is less engagement, less focus, less attention and less learning.

18

The Motivational Problem: We Have It All Wrong

Many of the best intentional efforts to foster new learning disciplines flounder because those leading the charge forget the first rule of learning: people learn what they need to learn, not what someone thinks they need to learn.

—Peter Senge, MIT Slone School of Management[1]

Science is finding unsettling evidence that the carrot-and-stick approach to learning, called authoritarianism, which dominates virtually every corporate and educational institution in the world, may be a major inhibitor of the very learning, behavioral change, and innovation these institutions are trying to create. Science continues to force us to recognize the gigantic chasm between what the research reveals and what organizations do.

Of all my research in the last thirty years, the concept that we learn through selection, not instruction (see Chapter 11), has been the most disruptive to me professionally, and the information in this chapter about motivation has been the most disruptive for me personally. Before my investigations, I couldn't conceive of any way to control behaviors, increase creativity, and accelerate learning except through extrinsic motivators. The information presented in this chapter changed how I ran my companies, interacted with employees, and, most importantly, how I related to my children and the rest of my family. It utterly broke my thinking machine. But although this information is profound and has been well-known in the research community for many years, it doesn't seem to have had much of an impact upon how institutions treat people or how we treat one another. My goal with this chapter is to lay out in some depth the damage we do to ourselves and others by not understanding the scientific basis of what really motivates us.

Daniel Pink wrote a wonderful book called *Drive: The Surprising Truth About What Motivates Us*, demonstrating that, except in factory pieceworkers, reward and punishment don't do what we think they do—and often make things worse.[2] Unfortunately, even though his book became a bestseller, it triggered little change in our collective understanding of motivation. I hope the scientific approach outlined here will spark a transformation in how you look at reward-and-punishment-based learning systems. I wholeheartedly believe that if you take the knowledge and apply it to your life, and the lives of those you have an influence upon, a whole new world of opportunities will open up to you.

Extrinsic Motivators Versus Intrinsic Motivators

In their efforts to direct your learning, increase your creativity, and influence your behaviors, traditional authoritarian institutions rely on three major components of extrinsic motivation:[3]

- Control by forces outside of the individual
- Reward and punishment as the primary tools of leverage
- A general disregard for what the individual holds as meaningful

Research now points not to external motivation but to internal motivation as the very best method of creating the neurological change needed for profound learning and behavioral development.[4] In fact, it's fascinating to see that organizations that rely heavily on extrinsic motivation over what's meaningful to the individual tend to decrease, not only the *Individual's Adaptability/Intelligence Factor* but also the *Institutional Adaptability/ Intelligence Factor.*[5] When we look at motivation from this more scientific perspective, we question most of what we thought was "right" about how we motivate ourselves and others. We also begin to understand why most organizational cultures are so stressful and inefficient.

The reason for this is that while external motivation gets you to move in a direction that is meaningful to others, internal motivation compels you to do things that are meaningful to you. This isn't a subtle difference. It is the stimulation of your *Meaning Network*, which we examined in previous chapters, that is responsible for creating internally directed motivation. Conversely, evidence shows that when you're prompted to perform acts with little meaning to you by overly authoritarian institutions, the ability to self-stimulate and enhance those same brain structures and neurotransmitter systems decreases.

External motivators that are heaped upon you by outside sources can also increase your stress hormone levels with dire consequences for brain function, ranging from withering your neural connections and killing neurons, to creating neural tangles, to promoting impulsive behaviors.

Also, external motivators have been linked to boredom, burnout, and the weakening of the bond between mentor and student/boss. The most damaging effect, however, may be external motivators' ability to breed out your very ability to access personal meaning in your own life. And as we will see in the next chapter, research is even showing that this inability to access your intrinsic motivation may be the basis of the widespread psychological condition called codependency.

The "Motivation Problem" Is a Myth

One of the biggest problems in getting students and employees to learn and behave in specific ways is called the "motivation problem."[6] When people fail, we cry, "They're unmotivated," but, in fact, everyone is motivated in some way. The problem is that we aren't always motivated to move in the direction that authority wants us to move in.

Internal motivation springs from personal meaning, whether it's designing a rocket, composing a sonata, writing a computer program, getting a tattoo, goofing off at work, skipping school, or getting pregnant at fourteen. Simply put, the "motivation problem" is a myth. We are all motivated—by our existing neurological structures laid down by our genes and previous experiences—to select new information and behave in ways that match those pre-existing neural networks. The "motivation problem" takes hold when your internal values propel you to select information and move in directions contrary to what an authority deems as appropriate.

Our educational institutions usually see students who join gangs, take drugs, get piercings, skip school, or get pregnant as unmotivated. The truth is these young people are highly motivated. Such rebellious activities all take a substantial amount of effort, energy, and internal motivation. In the same way, corporations see workers who goof off as unmotivated. Yet unproductive employees have to be energized to expend the effort necessary to escape projects, pretend to be working, and hide from the boss. When we say employees and students aren't motivated, we're really saying that they aren't behaving like standardized machines. They are not motivated to do what authority wants them to do in the way authority wants them to do it!

Randall Root, the founder of Root Learning, which helps clients like Mercedes-Benz, Hewlett-Packard, Boeing, and IBM get clear on what's inhibiting employee potential, says:

> *Companies say to us, "We tell people what our mission is, but they don't get it. We must have a communication problem." … Those companies don't have a communication problem. They have a misunderstanding of human nature. If you're a CEO, people tolerate your ideas, but they act on their own ideas.*

Isidor I. Rabi, Nobel Prize winner in physics, says of intrinsic motivation:

> *You do an experiment because your own philosophy makes you want to know the result. It's too hard, and life is too short to spend your time doing something because someone else has said it's important. You must feel the thing yourself.*[7]

Motivational specialist Raymond Wlodkowski notes:

> *I never use the expression "motivate a child." That takes away their choice. All we can do is influence how they motivate themselves.*[8]

The Fleeting Effect of Reward and Punishment on Behavior

We've been duped by *Ancient Learning Theory* into believing that the application of carrots and sticks can make permanent changes in behavior patterns. But modern research from the world of motivational psychology reveals that, in the absence of carrots and sticks, individuals revert to their previous behaviors. In other words, external motivators are, in general, temporary, they do little to alter the neural networks in your brain that are the foundation of permanent behavior change.[9]

Over the years, billions of dollars and countless hours have been spent on learning and behavior programs that use grades, demerits, money, TVs, stereos, vacations, and trinkets to get people to learn and behave in ways that corporations, governments, and schools want. Studies clearly show that, under the influence of rewards and punishments, you can indeed get people to temporarily stop drinking, smoking, and eating excessively; you can even make them start wearing seatbelts, read more, and behave in unique and bizarre ways (think of the contestants on the old TV show *Fear Factor*, eating maggots).[10] But the studies also show that, once the controlling factors have been removed, people revert to their previous behavior patterns. For example, as part of your normal behavior, you would neither eat this page nor suck on your neighbors' socks. But if I put a gun to your head or offered you $10 million to perform these unnatural acts, you would probably do both very quickly. Yet the minute I removed the gun or the money, you would just as quickly and automatically act on your own preferences, which presumably include neither eating books nor sucking socks!

As you've seen in previous chapters, your behaviors are driven by your existing neural networks. Since you don't have networks that prompt you to eat books or suck socks, you have no natural motivation to perform either of these acts. Because extrinsic motivators are inefficient at altering your existing neural networks, they're inefficient at altering your behaviors permanently.

This is shown by prisons which fail to alter criminal behavior patterns (estimates of recidivism in the US reach over 80 percent in nine years)[11] is that the external motivators that prisons rely on (locking people up) do little to change the existing neural networks that caused the inmates to commit crimes in the first place.

In the same way, social programs that pay teens not to get pregnant have also been a well-established failure.[12, 13] Why? Because the external motivation of money does little to stimulate the internal meaning/value networks of a young girl who wants to have a boyfriend, engage in sex, and fulfill her need to bond by having a child. However, successful teen pregnancy programs lead young girls to understand that they can have the lives, relationships, love, and ego satisfaction they want by postponing childbirth until they're physically and psychologically ready. These programs wrap the existing values of young girls in viable alternatives to getting pregnant. Karen Pittman, president of The Forum for Youth Investment, says, "What it seems to boil down to is this: if you have a future, you are

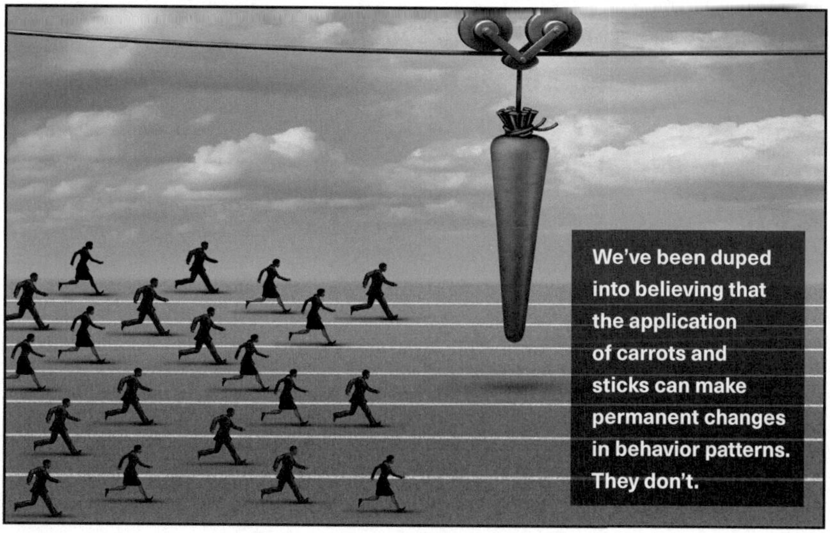

more likely to delay parenthood."[14]

All successful learning and behavioral programs that can produce permanent change alter the existing value networks of their clients. Alcoholics Anonymous, Narcotics Anonymous, and Overeaters Anonymous are successful not because they get their clients to stop drinking, taking drugs, and overeating by the use of threats or rewards. Instead, they create neurological change that leads to intrinsic motivation through a twelve-step approach, which allows addicts to replace their drugs of choice with the joy, friendship, service, and spirituality offered by the programs.

Firms Spend Billions to Fire Up Workers—It's Not Working!

The concept that neural networks must be changed before your behavioral patterns can change goes a long way toward explaining why motivational speakers have such a limited effect on people's performance. A headline in *USA Today* proclaimed, "Firms spend billions to fire up workers—with little luck."[15] The article points out that, contrary to popular belief, exhaustive academic research has shown almost no evidence that motivational expenditures increase worker productivity.

Just think of the all the pump-you-up, rah-rah sales presentations, and seminars that you've attended and how little effect they had on altering your long-term behavior patterns.[16] Motivational researcher Alfie Kohn says of motivational speakers, "At best, the result is a temporary sense of being re-energized, much like the effect of eating a doughnut. When the sugar high wears off, very little of value is left in the system."[17] Motivational authors and speakers can get you pumped up in the short term, but unless their messages can cause change at the level of neural tissue, no behavioral change will result.

> **This coercion (an educational system that relies on reward and punishment) had such a deterring effect on me that, after I had passed the final examination, I found the consideration of any scientific problem distasteful to me for an entire year.**
>
> *Albert Einstein*

Why External Motivators Cause the *Downshift*

To access intrinsic motivation and easily and effectively learn, change behaviors, and be creative, you must engage the highly complex, most flexible structures of your upper and frontal brain areas. But when you're under stress, those vital upper brain areas aren't effectively stimulated. And here lies one of the major reasons learning and creativity are impeded by extrinsic motivators: they cause the kind of stress that shifts neural function down to the lower parts of your brain.[18] And this is not a good thing! We have previously discussed the *downshift*. Here we will look at it from the perspective of extrinsic motivation.

The stress-induced *neurological downshift* is easily created simply by exposing yourself to rewards and punishments. As we've all experienced, when someone is breathing down your neck, prodding you to do something, life becomes stressful and uncomfortable. And nothing affects your brain's blood flow downward as effectively or as quickly as extrinsic motivation.

This *downshift* in your brain's blood flow and energy, when you're under the influence of external motivators, is a major cause of your brain's inability to create long-lasting learning and behavioral change. It also inhibits your ability to be highly innovative. Because this downward shift in brain energy also inhibits emotional mediation, which increases impulsivity, it can also be traced to the high number of behavioral problems we see in so many of our traditional authoritarian institutions.

Through the lens of science, the whole concept that reward and punishment facilitate learning also begins to crumble. You'll probably be astonished to learn that the consensus in the research community is that extrinsic motivators not only impede the learning process but also decrease intelligence. Educator John Nicholas became so disenchanted with the outcome of behavior programs that prompted children to read by giving them rewards like free pizza, he joked that they "produce fat kids who do not like to read."[19]

External Motivators Impair Learning

At Georgetown University, Dr. Bernard Brown showed that external motivators impaired learning, thinking, memory, and problem-solving in over 4,000 students. He also found external motivation slashed intelligence scores by 14 percent in these students.[20] The results of efforts to generate more efficient reading skills through rewards are demoralizing. When external motivators were applied to students, not only did readers have a worse memory for key information, but their ability to recall incidental information dropped to almost nothing. Another study showed that children who were told they would get a toy for work on IQ tests did worse than those who expected nothing. Researchers Ann Boggiano and Deborah Main found that "children who adopted an extrinsic, rather than an intrinsic, orientation toward schoolwork were more likely to do poorly in overall achievement, as indexed by national test scores."[21] Motivational researchers Richard Ryan and Jerome Stiller conclude: "The more we try to measure, control, and pressure learners from without, the more we obstruct the tendencies of students to be actively involved and to participate in their own education."[22]

"Asses Laden with Books"

By relying on reward and punishment and ignoring intrinsic motivation, we've turned our students, not into bright, excited, and expectant learners but into what the 16th-century philosopher Michel de Montaigne's called "asses laden with books."[23] In their study of traditional extrinsically oriented learning organizations, Craig Haney and Philip Zimbardo note "the real tragedies of our school system are [...] not the troublemakers or the dropouts but 'the endless procession of faceless students who go through the system quietly and unquestionably, unobtrusively and unnoticed.'"[24] Hermann Hesse wrote of the struggle between authority and spirit that goes on day after day in learning institutions around the world:

> *The authorities go to infinite pains to nip the few profound or more valuable intellects in the bud. And time and again, the ones who are detested by their teachers and frequently punished, the runaways and those expelled, are the ones who afterward add to society's treasure. But some—and who knows how many?—waste away with quiet obstinacy and finally go under.*[25]

Hesse was right—as we have seen, some of the greatest minds in modern histories such as Darwin, Edison, and Churchill, didn't flourish in authoritarian educational systems. Indeed, the use of extrinsic motivation had devastating effects on what could be considered the greatest mind of the 20th century, Albert Einstein. In his biography of Einstein, J. G. Crowther writes:

> *He considered it a very grave mistake to suppose that "the enjoyment of seeing and searching" can be pushed forward by "coercion and a sense of duty." He commented that even a beast of prey would lose its appetite if forcibly fed under a whip.*[26]

Einstein felt his creativity was damaged by an educational system that relied primarily on reward and punishment to produce it. In an autobiographical essay, he said, "This coercion had such a deterring effect on me that, after I had passed the final examination, I found the consideration of any scientific problem distasteful to me for an entire year."[27]

Brain-based educator Eric Jensen points out:

> *It's not just the case of an unknown number of Mozarts, Louis Armstrongs, or Einsteins being left as undeveloped seedlings; more commonly, it's a case of many potentially productive members of society becoming low-functioning or dependent on others for everyday living. In addition, there have been countless kids that either dropped out or graduated with no sense of competence, direction, or joy of wonder about learning.*[28]

Over thirty years ago, Carl Rogers nailed the essence of learning based upon intrinsic motivation when he spoke about learning. But not the lifeless, sterile, futile, quickly forgotten stuff that has crammed into the mind of the poor helpless individual tied into his seat by ironclad bonds of conformity! I'm talking about LEARNING—the insatiable curiosity that drives the adolescent boy to absorb everything he can see or hear or read about gasoline engines to improve the efficiency and speed of his 'cruiser.' I'm talking about the student who says, "I am discovering drawing in from the outside, and making that which is drawn a real part of me."[29] Jere Brophy and Neelam Kher note that the end result of our heavy reliance on extrinsic motivation over intrinsic motivation creates a climate where there is "little evidence of student motivation to learn in the typical [American] classroom."[30]

Extrinsic Motivation Kills Creativity and Innovation

Sadly, studies show that our reward-and-punishment-driven educational institutions aren't the kinds of environments that support innovative thinking. As Richard Florida points out, "Wherever creativity goes… innovation and economic growth are sure to follow."[31] Unfortunately for a world that demands high levels of innovation to solve its massive problems, research reveals that creativity and extrinsic control are at the opposite ends of the spectrum. You can either produce innovative and resourceful thinking through internal motivation or produce rote, predictable thoughts and behaviors through extrinsic motivation.

There is a disturbing study headed by Dr. George Land, who studied creativity for over thirty years. Land had a unique idea and applied a variation of a standard creativity test used by NASA to a sample of 1,600 children. Land first studied them at five years old and found that a staggering 98 percent of them scored at the "highly creative" level. But after five years in the traditional educational system, this figure fell to 30 percent. And five years later, when these kids were fifteen years old, the percentage fell even further—only 12 percent were "highly creative." In a separate sample of twenty-five-year-olds, only 2 percent were judged "highly creative." What made the difference? The amount of time they spent in the carrot-and-stick-driven traditional educational system. "What we have concluded," Land wrote, "is that non-creative behavior

"Ted, I think we need to re-evaluate your carrot."

is learned." In effect, the system appears to prune out the brain's ability to be "highly creative."[32]

In one survey, 81 percent of knowledge workers recognized that the level of creativity required in their jobs had increased substantially over the last five years. Yet an abundance of studies shows that in everything from painting and writing to puzzle-solving and job performance, creativity suffers in the face of outside rewards and punishments. In thirty-five years of research and numerous studies, Teresa Amabile, now at Harvard University, found that extrinsic motivation represses innovation dramatically. Her analysis of the evidence has prompted her to conclude: "Intrinsic motivation is conducive to creativity, and extrinsic motivation is detrimental."[33] Edward Deci, a pillar of motivational research, found that intrinsic motivation plays the primary role in producing reflective and creative learning,[34] yet in the face of scientific research, our institutions continue to try to drive creativity and innovation through external motivators. After examining the extrinsic motivation research, Renate and Geoffrey Caine note, "After a decade of teaching focused on rewards, punishment, and narrow outcomes, it is easy to concur with educators, researchers, and the general public who lament the fact that our students cannot think."[35]

Alfie Kohn has probably reviewed more research than any other educator in the world on the distressing effects of learning systems that overemphasize extrinsic motivation and underemphasize personal meaning. In his celebrated book *Punished by Rewards*, Kohn paraphrases a famous critical report of how reward and punishment limit the creativity of our youth, noting that if "an unfriendly foreign power had attempted to impose on America a mediocre educational system, it could have devised no better plan than to establish mechanisms for tightly controlling what students do in school."[36]

If we want to cultivate creative, innovative citizens so we can meet the world's dramatic challenges, we need to understand that the very methods our institutions are using to control people are killing the creative and innovative traits we so desperately want them to have.

Why Carrots and Sticks Limit Focus, Attention and Discovery

When compared to personal inspiration, rewards and punishments also limit your focus, attention span, and exploratory behaviors. When prompted, you can and will focus your attention and discovery behaviors on things that help you get the reward or escape the punishment. But your focus is limited, for as soon as the carrots and sticks are withdrawn, your attention reverts back to… you've guessed it—what you find meaningful in your life.

The brain generates about twenty watts of power using glucose and oxygen via the carotid and vertebral arteries. When you are self-motivated, this brainpower is free to discover information in your environment and make all sorts of random connections with existing information in your long-term memory banks; this neural freedom to discover and focus on what's meaningful to you is the source of your creativity and intuition we just discussed. Under external motivators, your brain doesn't want to waste time exploring and trying out new things but wants to know, "Exactly what do I have to do to get my reward and avoid any punishment, so I can get the heck out of here and start doing what's meaningful to me?"

The late psychologist John Condry maintained that external motivators are the "enemies of exploration."[37] There's speculation that overly controlling environments limit exploratory behaviors because they inhibit the expression of the dopamine-producing genes, DRD2 and DRD4, which prompt us to take risks and engage in investigative behaviors.[38]

The Caines note that extrinsic motivation "reduces the desire as well as the capacity of learners to engage in original thought."[39] Research clearly shows that, under external motivation, you become so preoccupied with the outcome instead of the task, that you end up being less attentive to the possibilities that could help you learn and be more creative.[40] Learning directed by reward and punishment doesn't require much, if any, attention to discover the reasons why things happen or what makes things work. As one group of researchers explains it, "Features such as predictability and simplicity are desirable, since the primary focus associated with this orientation is to get through the task expediently to reach the desired goal.'[41] As Jean Piaget maintained almost a century ago, external motivation "renders autonomy of conscience impossible."[42]

Motivational psychologists have known for decades that your concentration lasts a long time when you're self-directed and reduces when others steer you to do what's meaningful to them. In both childhood and adulthood, you can easily maintain your attention for hours to days discovering details you find inherently meaningful. Such meaning driven searching produces sufficient levels of neurochemicals like dopamine, norepinephrine, and endorphins, which allow you to sustain your attention for long periods.

Conversely, when you're forced to perform jobs that have little or no meaning for you, your levels of these discovery- and attention-producing neurotransmitters wane, and you easily become distracted and bored. Mihaly Csikszentmihalyi, the acknowledged expert in understanding optimal experiences, says the most advantageous discovery and learning state is a state of consciousness called flow. Flow is an uninterrupted attentive state where you're lost in the timeless absorbing pleasure of meaningful activity.[43] It isn't possible to maintain this state unless you're sustaining your levels of dopamine, endorphins, and norepinephrine by intrinsically motivating yourself.

When I was fifteen, I had a job in a steel mill, where one day I was assigned to count the metal rods that came off a boxcar. There were thousands and thousands of them. I spent a whole week counting them for the chief of inventory control. It was the most boring and meaningless thing I had ever done in my life. Every day was like getting a root canal! The next week I was fired. Why? When they went to check my work with the invoice, they found I had miscounted the number of rods, not by a little, but by 40 percent. The boss told me I was the biggest idiot they had ever hired!

Extrinsic Motivators Destroy Feelings of Happiness and Fulfillment

Studies clearly show that when you're coerced to perform meaningless jobs or study meaningless things to get a carrot or avoid the stick, you feel a loss of joy in the task you're doing. Would you rather work at a job you hate but where you have the chance to win a TV, or would you prefer to wake up every morning chomping at the bit to go to a job that brings meaning, joy, and a spring to your step?

Nature has set it up so that you have a biological drive to seek out, discover, and engage in meaningful acts that, in turn, allow you to feel biologically fulfilled and happy (see Chapters 12–13). Unfortunately, when extrinsic motivators produce stress, it kills these feelings of happiness and fulfillment. How? When you're under extrinsic motivators for long periods, the increase in your levels of the stress hormone cortisol, in turn, lowers serotonin.[44] As we saw, serotonin is a main neurotransmitter for creating joy and fulfillment. There are fifteen different serotonin receptor sites, and all have unique functions like giving the brain neural stability, but for our purposes in this chapter, it can be said that the higher your stress level, triggered by extrinsic motivators, the lower your neurochemistry of joy and fulfillment will be.

Another way to look at happiness is that it is short-lived if extrinsically driven. Once you get that dress, car, house, or grade you desire, your brain starts looking for the next external thing that can "jack up" your neurochemicals of happiness. It is a never-ending cycle of chasing the next happiness high. Conversely, if your happiness is intrinsically driven, then happiness is long-lived because you have more control over the production of the neurochemicals of joy. This gives you the power to embed happiness into the fabric of your life.

Rewards and Punishments Suck Joy from Your Passions

Hundreds of studies show that reward and punishment can even suck intrinsic motivation and joy out of things you like to do. In "The Undermining and Enhancing of Intrinsic Motivation in Preschool Children," researchers Rosemarie Anderson and her associates write, "Who would have thought that play could be turned into work by rewarding people for doing what they like to do?"[45]

These kinds of studies show that if we reward someone for doing what they like to do, once the reward is no longer offered, the intrinsic motivation and joy from engaging in that activity is greatly decreased. In fact, rewarding someone for what they like to do can lower intrinsic motivation so much that you have to keep rewarding or punishing them over and over again to get them to do it. Educator Joan Roemer writes of extrinsic motivators, "They dilute the pure joy that comes from success itself."[46]

Bill O'Brien, former CEO of Hanover Insurance, laments, "People enter business as bright, well-educated, high-energy people, full of ... desire to make a difference," but "by the time they are thirty, a few are on the 'fast track' and the rest 'put in their time' to do what matters to them on the weekend."[47] In his book *The New Economics for Industry, Government,*

Education, management guru W. Edwards Deming points out the crushing of motivation is an inevitable outcome of most management systems that ignore personal meaning.[48]

Extrinsic Motivators Make Us Feel Powerless and Burned Out

Similar studies with employees show that, even if you're initially interested in your work, your interest will wane if your stress levels are increased by being continuously externally motivated by your boss.[49] This is because extrinsic motivators make you feel powerless. No longer are you getting the neurochemical benefits of directing yourself in meaningful activities. Instead, your stress level is being driven higher by extrinsic motivators.

No one wants to stay in low-meaning, low control, high-stress environments for very long. Because many of us toil at jobs devoid of personal meaning, we often feel like rats trapped in a cage, made to perform tricks for authority figures, as noted by James L. Mursell in *Psychology for Modern Education*:[50]

> *When a person does an intrinsically disliked task because he fears not to do it, it is always necessary to confine him. This is what happens to animals in many psychological experiments… Under such circumstances, the creature is likely to seem more stupid than it really is and to learn less efficiently than it would in its natural habitat when it is positively motivated.*

This helps explain why before the COVID-19 pandemic many companies were reducing the ability of employees to work from home. They felt the need to confine these workers when the work they give them wasn't meaningful.

In the mid-1900s, B. F. Skinner, a pioneer of behaviorism, wanted to show the positive effects of controlled rewards and punishments on a developing mind, so he used his daughter as a test subject. In the effort to minimize input from the outside world and control her interactions, he placed her in a small confined area, he called an "air crib," which became known as the "Skinner Box." About this approach to controlling behaviors, one motivational psychologist labeled today's work environments "giant Skinner Boxes with parking lots."[51, 52]

For a short period, The Learning Code Institute worked with an organization that had just hired a new general manager who was convinced that the higher the reward and punishment, the greater the productivity. As he applied his carrots and sticks to his new charges, their stress levels climbed higher and higher, prompting them to make more and more mistakes and to look much less capable than they were. This company, which had once been proud of its on-time product performance, now started to miss important deadlines.

GM's response? To blame employees, which prompted him to apply for more aggressive performance reviews and utilize ever-greater controls and punishments. GM's ego seemed incapable of importing the new knowledge that his method of management itself was causing the problems. Did anyone in the company ever confront him or explain to him that his management strategy killed the initiative, repressed innovation, and had a destructive effect on product delivery? Of course not. They feared his punishments. Instead, the employees just talked behind his back, made jokes about him, collected their checks, and anticipated the weekends, all while looking for more meaningful work elsewhere. The result? The company's reputation plummeted, and they became well-known as a company you couldn't trust to deliver products on time. Unfortunately, as many of us have experienced, this isn't an isolated case.

Behaviorism Murders Relationships

Another problem with behaviorist learning systems is that they separate the master from his pupils. A teacher is inhibited from looking into the student's eyes and seeing a learner's soul. Instead, all he can see is the glow of the machine that he must force to perform specific acts at specific times to get the student to pass standardized tests. Mechanistic educational systems interfere with the teacher's effort to foster the maturation of a pupil's most precious asset: their connection to their soul and what really holds meaning for them. Behaviorists' educational systems allow little space for a deep master–pupil relationship. Brain-based educator Robert Sylwester laments that, because educators have relied on reward and punishment without regard to the inner process of the student, "Thus we became a profession of behaviorists, whether we liked it or not."[53]

The separation that behaviorism forces between master and students is carried over into the corporate world. Bosses mistakenly believe that they don't really need to have a relationship with their workers to get them to be proficient and innovative. All they have to do is to figure out the right stimulus to get the right response—that is, the right kinds of rewards or punishments that will drive the workers to propel the company to profitability.

Intrinsic Motivation Brings an Added Benefit: Reduced Behavioral Problems

Institutions that rely on what's meaningful to the individual instead of reward and punishment brings an added benefit—a neurological climate that produces a reduction in behavioral problems. Neither kids nor adults want to fool around, cause trouble, or waste time when they believe the information presented will help them more successfully survive and thrive in their personal world. The more an individual's personal *Meaning Network*

is stimulated by the environment they're in, the longer that individual will stay focused on the joy and discovery of learning, which in turn limits their desire to act out. Why act out when you're getting what you want?

And when you're engaged in meaningful activities, your stress levels decrease, and there's a resulting upshift in your neural energy to the upper and frontal parts of your brain. When these structures are engaged, emotional control goes up, and impulsivity goes down. As an example, in high school, there would likely be few male behavior problems in a class titled "How to Impress Girls" and few female behavior problems in a class titled "How to Attract and Keep the Man of Your Dreams." While we may never see classes like this in school, our institutions for children and adults must recognize that unless the information that they want people to learn is personally meaningful to the learners and employees will either act out or want out.

To support companies we work and enable them to keep their employees in *upshifted* states where there is high innovation, low behavioral problems, and high compliance, we have a course called The Neurogenetics of Cultural Transformation. In it, we encourage companies to develop a *Noble Cause*. A *Noble Cause* is a cause that is bigger than any person or the company. We have found that focusing the company's attention on a cause bigger than themselves activates the same Neurological Upshift as personal meaning. Well known fast-growing companies such as Google, Apple, Zappos, and even NASCAR often began with a Noble Cause that is not made public,[54] such as:

- To make information available to everybody in the world. (*Google*)
- To make a contribution to the world by making tools for the mind that advance humankind. (*Apple*)
- Delivering happiness to customers, employees, and vendors. (*Zappos*)
- Everybody's gotta win. (*NASCAR*)

But even if they don't make their Noble Cause obvious to the public, the results sure show up in their bottom line. A strategic partner of ours called the Green Zone Culture Group did an analysis of companies with green cultures, of which one of the elements for most of them was to have a Noble Cause bigger than the company itself. They found that over eleven years, the companies that had Green Zone Cultures had stock price growth of 900 percent vs. 74 percent for those that did not.[55]

Extrinsic Motivators Steal Your Authenticity

Have you ever felt that you can't be who you really are and still impress others or succeed in life? You aren't alone—many of us feel the same way. Have you ever wondered where the feelings that tell you that you can't be your authentic self actually come from?

For most of us, these feelings start in first grade. Think about it: before first grade, what were you doing most of the time? You were running around, stimulating those 3,000 trillion neural connections in your young brain by doing what was most meaningful to you. Life was a joy and marvel to behold. Whether you were banging on pots and pans, eating dirt, chasing your friends, daydreaming, squirting water out of your mouth, gazing at flowers, or playing with dolls or LEGO, you were self-directing all of your discovery and learning.

In general, you were free to fill your "meaning bucket" with what charmed you at that moment. In fact, intrinsic motivation was so important to you that if someone tried to redirect your actions away from what was meaningful to you, your biology could prompt you to throw a giant hissy fit.

Then, everything changed! You went into first grade (or started kindergarten) and were jerked into a whole new world. What did you have to do to survive and thrive? You had to adapt your behaviors to fit the world of the classroom. How did you do that? You had to give up doing what was most meaningful and joyful to you and replace it with what was most meaningful to the teacher and the system. What was meaningful to the teacher and the system? Not eating dirt, not gazing at flowers, not chasing your friends, not spitting water out of your mouth, not daydreaming, not playing with dolls or building forts. Instead, it was sitting quietly at your desk, doing your homework, memorizing things that in general held no meaning for you, taking tests, and getting good grades.

Within the first few weeks of school, you got the message: To survive and thrive in the classroom, what was personally meaningful to you wasn't important at all. Your job was mainly to conform to what was meaningful to others or get punished. Consequently, your stress levels increased as your vision narrowed, while you tried to figure out strategies to get your homework done as quickly as possible while passing tests with the least effort possible, all while anticipating the weekend when you could once again get back to doing things you loved.

As you progressed through the grades, your stress levels grew higher as your motivation become more and estranged from your interests. If you chose the college track, you had to engage in extracurricular activities and join clubs not necessarily because you wanted to but because you needed to get acceptance from the authority that would admit you into college. Little by little, you became less and less authentic as you molded your behaviors to get acceptance from the system.

The end result for many of us? After spending twelve to sixteen years being controlled by extrinsic motivators, we adapted our brain structures so that we lost our true selves. We lost our authenticity—something most of us are still trying to reclaim. Luckily, our brains are highly malleable in adulthood; therefore, we have a tremendous capacity to learn a new way of adapting to our world and to regain the authenticity we lost. As you will see in the next chapter, this altering of our brain structures also helps breed codependency in a huge percentage of our species.

At The Learning Code Institute, we've developed another program called "The Neurogenetics of Authenticity" to help people understand not only the neurobiology of authenticity but also that when you're in your authentic self, you have a more positive effect on other people's behavior because you have more impact on their biology. During the 2016 presidential primary race, many pundits couldn't comprehend how two outliers like Donald Trump and Bernie Sanders could gain such large followings. The answer is authenticity. Most politicians lack it. They say what they need to say to get elected, whether they believe it or not. Then Trump and Bernie showed up, quickly gained passionate followers, and destabilized the whole system simply by seeming more authentic than the rest of the field. You don't have to like someone for your biology to respond to them when you feel they are authentic.

A Very Short Parable

The following very short story is by a favorite author of mine, the renowned and controversial Jesuit priest and psychotherapist Anthony de Mello. He sums up things up rather nicely:

Nasruddin became prime minister to the king. Once while he wandered through the palace, he saw a royal falcon. Now Nasruddin had never seen this kind of a pigeon before. So he got out a pair of scissors and trimmed the claws, the wings, and the beak of the falcon.

"Now you look like a decent bird," he said. "Your keeper had evidently been neglecting you."

The message is this:

"You're different, so there's something wrong with you!"[56]

But as neurogenetics is revealing, there's nothing wrong with you. There's something wrong with a system that employs extended periods of reward and punishment to make you something you're not.

Don't get me wrong—this book isn't proposing educational and institutional anarchy. I don't lay out this grim overview of extrinsic motivation to damn the present system or to imply that we don't need to conform to cultural norms. I lay out the science in some detail so that we may have a conscious awareness of how overly controlling authoritative systems that devalue intrinsic motivation, reduce personal authenticity, and ignore personal meaning ultimately limit the neurological changes necessary to produce deep learning, profound creativity, intense joy, and sustained behavioral change.

Again, it's my hope that as we gain a more scientific understanding of the effects on our biology of our traditional institutions—that rely on what The Institute calls "The Reward and Punishment Approach," —we'll be much more effective at designing learning and behavioral-change systems that increase our *Individual Adaptability/Intelligence Factors* and produce the results we so deeply desire.

Cracking the Learning Code.

Science is revealing the disquieting fact that external motivators (reward and punishment), are a major inhibitor of the neurochemistry that leads to the rapid learning and behavioral change we urgently want to create in people and organizations. Conversely, intrinsic (internal) motivators create a neurological climate that accelerates learning and behavioral change.

19

Do Extrinsic Motivators Cause Codependency?

... as rewards continue to co-opt intrinsic motivation and preclude intrinsic satisfaction, the extrinsic needs—for money, for status, for power—become stronger in themselves. Thus, people develop stronger extrinsic needs as substitutes for more basic, unsatisfied needs... They end up behaving as if they were addicted to extrinsic rewards.

—Edward Deci, Pioneer of Motivational Research, University of Rochester[1]

While extrinsic motivators can reduce your ability to learn, pay attention, change behaviors, and innovate, their most damaging effect may be their ability to alter your neural networks in such a manner that you actually prefer doing what others deem important instead of what you find meaningful yourself. In effect, they can steal your soul.

Because no discussion on learning and behavioral change would be complete without examining how extrinsic motivators can create a psychological condition called codependency, we will spend some time exploring how we learn to become codependent. This is important because estimates by psychologists are there is as many as 90 percent of us with symptoms of codependency.[2] These experts believe that our extrinsically centered society has created a world where generations of people have lost the ability to access intrinsic motivation. We now know that under prolonged exposure to outside control, self-inspiration becomes more and more difficult to generate, and people begin to believe that they can gain fulfillment only by achieving goals formulated by others. Writing of people who are overexposed to external motivators, the Caines report, "In fact, they prefer external forms of motivation and lose sight of internal motivation."[3] It's as if the brain were saying to itself, "If I can't generate my own fulfillment, I'll look outside of myself to get it." This is codependency on steroids.

The Shift

This shift in preference from internal to external motivation can be devastating. It sets up a mental framework in which you may be unable to effectively access meaning in your life. Like pouring water on a flame, instead of rising each morning with a smile on your face and a gleam in your eye, bursting with glorious expectations of fulfilling your dreams, you drag yourself out of bed with the anxious anticipation of being rewarded or punished by others based on how well you live up to their expectations… not your own. When you consider that from first grade until you die, every institution you study or work in keeps you marinated in extrinsic motivators, it's crucial to understand this focus on rewards and punishment, and their capacity to teach you to be codependent.

Laurence Boldt, the author of *How to Find the Work You Love*, says it well:

> *I think part of being an adult is determining your own expectations for your life. For so many of us, the expectations that we have for ourselves were given to us by somebody else. Our parents, our teachers, our guidance counselor, or whoever it was, we live someone else's expectations.*[4]

Marilyn Ferguson, in her groundbreaking book *The Aquarian Conspiracy*, adds:

> *We begin to see the unease and disease of our adult lives as elaborate patterns that emerged from a system that taught us young how to be still, look backward, look to authority, construct certainties. The fear of learning—and transformation—is the inevitable product of such a system.*[5]

Institutions consider the best students and the best workers the ones who can follow the rules and do what the authority deems important. But behind the gold stars, good grades, and grand job titles often reside individuals who feel unfulfilled and in deep pain. You get an A+ for doing what is significant to others and an F for being able to figure out what is meaningful to you.

Your neural networks gain strength, power, and growth based on your usage of them. Like muscles, the neural networks you use become competent and strong, and the ones you ignore become weak and atrophied. When your twenty watts of brainpower is mainly forced to focus on what satisfies others, little neural power is left to maintain and construct the networks that support what matters to you. In the end, when you live in an environment that overuses outside motivators, your neural networks can over adapt to fit others' wishes and under adapt to fit what pleases you. And when that happens, life is not as joyful as it could be. As Sandra Anderson, Emerita Professor at Portland State University points out, "codependency" is a pattern of painful dependence on approval from others, in an attempt to find safety, self-worth, and identity.[6, 7]

Dopamine's Role in Redirecting Your Meaning Focus

The adaptation of your dopamine system is strongly implicated in codependency. As you saw in Chapter 17, you have a genetic predisposition to search for and pursue tasks that prompt dopamine production, which also increases activity levels of endorphins, the brain's opiates. Like all neural structures, the dopamine system adapts itself to fit the most prevalent environmental stimuli.[8] When you live in primarily controlling environments that force you, through reward and punishment, to repeatedly move in directions meaningful to others and not yourself, the dopamine system can apparently adapt itself to prefer external stimulation over internal stimulation.

This isn't an inconsequential shift. This means that your attention, focus, and pleasure come not from stimulating the networks that hold personal meaning for you, but instead from stimulation of the networks governed by outside forces. In the end, the hypothesis is that the extended use of extrinsic motivators can make you addicted—not to doing what you find significant, but to what matters most to others.

Science is now allowing us to take a new look at this shift as a factor in creating the foundations of codependency. In fact, after extended exposure to outside motivators, individuals prefer external forms of motivation and lose their ability to self-direct. Researcher Richard Ryan notes that extrinsic

LEARNED HELPLESSNESS

Even worse than being addicted to external motivators are the extreme cases of people who develop what researcher Martin Seligman calls "learned helplessness." This is when people are dominated by extrinsic motivators to the point where they feel they have no control over their future, so they give up trying to cope with life altogether.

motivators actually "forestall rather than promote self-determination."[9] Eric Jensen says, "Learners who have been on a reward system will become conditioned to prefer it over free choice."[10] You could say that overexposure to external motivation can breed out your ability to access what's personally meaningful to you. Like Pavlov's dogs, you can become conditioned to respond to stimuli that have no substance for you personally.

As discussed earlier from the world of addiction, we've learned that people will go to almost any length to get drugs that will prompt the ventral tegmental to squirt out that satisfying hit of dopamine.[11] Like drugs, external motivators can usurp the dopamine focus, attention, and reward circuits.

This usurping of the dopamine/endorphin circuits by extrinsic motivators have been implicated as the foundation of the love/hate aspect of codependent relationships. The theory is that rewards and punishments can never activate the dopamine/endorphin system as effectively as personal meaning. At a subconscious level, codependent people recognize this, and while they seek dopamine production from the source of their codependency, they also feel unfulfilled by this same source. The result is loving the source for what it gives them and not feeling completely fulfilled by that same source. This is a form of suffering.

Even worse than being addicted to external motivators are the extreme cases of people who develop what researcher Martin Seligman calls "learned helplessness."[12] The research shows that when animals and people are dominated by extrinsic motivators to the point where they feel they have no control over their futures, they give up trying to cope with life altogether. Research shows that individuals who develop learned helplessness don't just give up on the area of their lives that's externally controlled, but they generalize their feelings and relinquish control in all areas of their lives. These individuals stop paying attention and learning new things because they feel they can have little or no effect on their futures.[13]

Tiger Parents: Authoritarian vs. Permissive

In 2011 Amy Chua's provocative memoir, *The Battle Hymn of the Tiger Mother*,[14] stirred up a public debate about which child-raising parent style was the best—permissive or authoritarian? A debate has been quietly going on for a very long time in the world of motivational research. Because while many children of overly authoritative parents often do get better grades, there are tremendous negative consequences. Researchers, such as Qing Zhou, an assistant professor of psychology who heads the Culture and Family Laboratory at UC Berkeley, sums up the research nicely: "Children raised by authoritarian parents are showing maladaptive outcomes, such as depression, anxiety, and poor social skills."[15] Researchers have also found that when compared to children raised by parents who use a more balanced combination of permissiveness and control, children raised by overly authoritarian parents had lower self-esteem, lower self-regulation, were rated as less helpful and less popular by teachers and classmates, were more likely to have experienced bullying as perpetrators and as victims, were less likely to feel socially accepted by their peers, and were less resourceful. And while many authoritarian parents see themselves as champions of morality, numerous studies suggest that they may be getting the total opposite of what they wanted. Their kids are actually less developed when it comes to moral reasoning and are more likely to reject their parents as legitimate authority figures.[16]

As they grow up, things don't get better. For example, the late Diana Baumrind, a respected University of California–Berkeley professor of psychology, found that children of highly authoritative parents had poorer

mental health by their twenties.[17] And a survey of middle-aged Americans found that people who remembered authoritarian childhoods were more likely to report depressive symptoms and inferior psychological adjustment.[18]

While kids raised by authoritarian parents may get better grades and good jobs, the psychological consequences may severely diminish their ability to find deep meaning in their lives. These offspring are greatly conflicted. Intellectually, they understand their parents want the best for them, but get trapped in a world where, because they can't access their own personal *Meaning Networks*, their lives become a stressful quest to make their parents happy. This is the conundrum of the codependent. Living your life for someone else for too long, even if it's your parents, will eventually destabilize your life.

The bottom line is that not only does the overuse of reward and punishment not work the way we thought it did in education and corporations, but it also doesn't work the way we thought it did in our own homes!

A Historical Perspective of Authoritarian Parenting

In the mid-to-late 1800s, the Dr. Spock of Germany was a Dr. Daniel Paul Schreber. In his effort to instruct Germans on the correct way to bring up their children, Schreber wrote several books that embraced a Machiavellian model of reward and punishment. For many German families, these books were the bible they used to guide their child-raising practices. Schreber instructed parents that their primary job was to control the child's will—at any price. He drilled into the German consciousness that acts of cruelty or deception were noble when conducted in the effort to break a child's will and force them to bend to the will of authority.[19]

By the 1930s and '40s, generations of Germans had had their neural networks aggressively adapted to blindly follow authority, without regard to what was personally meaningful or of value to them; thus they became fertile ground for Hitler to induce them to perform the despicable acts carried out during his reign of terror. Alice Miller, in her book *For Your Own Good*,[20] makes the case that Hitler could never have gotten people to perform such inhuman atrocities, which were surely against their base nature, if they hadn't been groomed from childhood to do whatever authority asked them to do. This is an example of nationalism creating a form of codependency.

Teachers, Personal Meaning, and Codependency

As we will see in more detail in later chapters, being trapped in a highly controlling and stressful environment produces high levels of the stress hormone cortisol, while also little capacity to self-stimulate the dopamine/endorphin-focus/pleasure circuits. While overly controlling environments affect learners and open the gateway to codependency, we must also be highly aware that such environments limit teachers' abilities to do their jobs effectively.

One of the learner's strongest allies is someone with a better grasp of the knowledge than the learner—the teacher. But I've been struck by what difficult position teachers are in. In general, they are under the thumb of domineering administrations that operate in fear that students won't attain the correct scores on government-imposed achievement tasks. To ensure that the tests are passed, administrators give teachers little leeway to follow their personal experience and intuition in bonding with and helping students learn. They have little or no influence on what they teach or how they teach it. In general, teachers are impeded from following what's personally meaningful to them, and instead of being focused on enriching the brains and souls of their students, teachers become cogs in the reward-and-punish, Machiavellian education machine. Some teachers can thrive in environments where they exert little control, but most suffer the same results as the students from the learning process: high stress, low joy, and

Instead of focusing on enriching the brains and souls of their students, teachers are forced to become cogs in the reward-and-punish Machiavellian education machine.

reduced efficiency.[21]

Throw in the fact that most teachers are dealing with kids who would rather be anywhere but in a classroom, that they feel underappreciated by parents and the general public, and are underpaid for their efforts, and you have an environment where, if given a choice again, 44 percent of teachers wouldn't become teachers again, and another 20 percent don't know if they would, according to one survey.[22] Research shows that students taught by teachers who felt pressured by authority didn't learn tasks as well as those whose teachers did not.[23] Anyone who has spent time with teachers when they open up and tell the truth about their unfortunate and stressful situations realizes it's amazing that teachers can be as effective as they are!

Most teachers go into the learning profession because they feel they have something of value to give. At a spiritual level, many teachers see themselves as helping a child's soul grow and prosper. The teacher's ability to add value to a student's life is compromised when their primary job is to demand that students sit still, be quiet for six hours a day, listen to information that usually has no meaning to them, and force them to take tests that are given the importance of life or death, imposed by some unseen power. The only way a teacher can get students to comply in this type of environment is to exert control with the threat of rewards and punishments, such as grades, gold stars, demerits, and expulsion. Teachers and students both lose in this type of environment.

Because the teacher controls the power to impose unpleasant things, often they become not the mentor students seek out for honest, caring, strong, and deep relationships but someone to avoid. Regarding the power of teachers to impose rewards and punishments, learning pioneer, Leslie Hart stated, "The power, by its existence, makes every person who has it an active or potential judge and punisher; and for students puts a premium on… 'apple polishing.'" He goes on to say that the use of controls prevents "frank relationships between faculty and students, and encourages teachers to place the blame for learning failure… on the students."[24] Thomas Gordon, educator and author of *Teaching Children Self-Discipline*, once commented, "The more you use power to try to control people, the less real influence you'll have on their lives."[25]

We must recognize that the existing rigidly controlled educational environments can also force teachers' biology to adapt to becoming codependent, just like the students. Teachers end up focusing on pleasing authority rather than doing what their love for children directs them to do. Over the last twenty years, I have had interactions with close to a thousand teachers, and virtually every one of them struggles with this internal conflict.

The Shift from Teachers and Bosses to Mentors

If we want to enhance the quality of learning in our institutions and businesses and break this codependency cycle, we need to take the teacher and boss out of the role of the inquisitor and put him or her into the role of mentor. The best mentors help us follow what's meaningful in our lives and ensure we possess the relevant knowledge to survive and thrive in the world—whether it be multiplication tables, reading, history, science, communications, management skills, or moral values. The wise mentor intuitively recognizes that the best way to impart knowledge is through the pathways of meaning that the student or employee already holds in his biology. (For more on *Wrapping*, see Chapter 13.)

A great mentor doesn't have to be a great orator, as the Greek method of instruction would suggest. In fact, the great masters and mentors of the East used very little discourse to cause transformation. To be a great mentor, one needs only to add to and expand upon what the learner already holds as meaningful. The master has the intuition and ability to impart that information at the correct time in the student's cycle of knowledge acquisition. Think of those who have had the most positive influence on your life, and they'll most probably fit the role of mentor.

Bulgarian scientist Georgi Lozanov was given credit for helping to start the accelerated learning movement. In researching the role of a teacher, he found that the student's opinion of the teacher was critical in the learning process. To effectively impart knowledge, the teacher needed to be worthy of respect and looked up to as a mentor. He found a mentor's competence, the strength of character, persistence, and charisma, increased the student's ability to learn and remember.[26]

Two people whom I consider masters, illustrate the roles of mentors very elegantly. In the foreword to *Embrace Tiger, Return to Mountain,* Alan Watts explains the teaching approach of Al Huang, a Tai Ji master:

> *[The master] begins from the center and not from the fringe. He imparts an understanding of the basic principles of the art before going on to the meticulous details, and he refuses to break down the taiji movements into a one-two-three drill to make the student a robot. The traditional way… is to teach by rote, and to give the impression that long periods of boredom are the most essential part of the training. In that way, a student may go on for years and years without ever getting the feel of what he is doing.*[27]

Best-selling author Gary Zukav, in his book *The Dancing Wu Li Masters,* poetically describes what a master is:

> *A Master teaches essence. When essence is perceived, he teaches what is necessary to expand the perception. The Wu Li Master does not speak of gravity until the student stands in wonder at the flower petal falling to the ground. He does not speak of laws until the student, of his own, says, "How strange! I drop two stones simultaneously, one heavy and one light, and both of them reach the earth at the same moment!" He does not speak of mathematics until the student says, "There must be a way to express this more simply."*

In the end, Zukav says the master doesn't reward and punish the student. He "dances with his student."[28]

In today's educational system, authority's directives, time, and standardized-test pressures don't allow teachers to "dance with their students," but instead, it forces them to become enforcers as they dole out rewards and punishments. This role damages the relationship with the learner. The result is that students can pass tests but lose their authenticity and become codependent on a reward-and-punishment system, trying to squeeze pleasure from it instead of pursuing what holds personal meaning and joy for them.

In corporations, managers also fall back on reward and punishment as the main tool to drive workers. This is because most managers are unaware of the research that demonstrates that if they became more like mentor—helping their workers find meaning in their work and lives—the levels of productivity, learning, and innovation would actually increase much more than with their present carrots-and-sticks approach.[29]

There's a neurochemical reason a mentor has such a profound effect on you. When you work with and learn from someone whom you believe can enhance your personal surviving and thriving chances in the world (your *Individual Adaptability/Intelligence Factor*), the release of the hormone oxytocin increases. Initial research shows that the intensified activation of oxytocin from your pituitary gland, in turn, enriches your ability to learn social skills and exhibit greater levels of empathy, heightens cooperation with the group, and increases trust—all things we want more of in schools and corporations.[30]

Cracking the Learning Code

Extrinsic motivators can not only suck the joy and efficiency out of your learning process, but they can also breed codependency on the very systems, people, and even nations that distribute those rewards and punishments.

20
Why What is Personally Meaningful to You is Devalued in Education

Those who cannot remember the past are condemned to repeat it.

—George Santayana, philosopher, essayist, poet, and novelist[1]

The previous chapters have made it clear that, from a neurobiological perspective, personal meaning is the Holy Grail of learning. You also learned that external motivators in the form of rewards and punishments aren't superior to intrinsic motivation as the most effective way to increase your *Individual Adaptability/Intelligence Factor.*

A question then arises: how is it possible that we've developed educational and corporate institutions that value extrinsic motivation, and operate as if personal meaning were inconsequential to the vital processes of learning, innovation, and behavioral change, rather than their very foundations?

The answer, as you'll see here, is that such systems came from relying on concepts embodied in *Ancient Learning Theory*, put forth by Aristotle, Machiavelli, Descartes, Newton, Locke, and the school of psychology called behaviorism.

History May Not Be Your Thing

This chapter shifts gears from science to history. But even if history isn't your thing, you still might want to give this chapter a scan. Why? Because the way our scholastic and corporate institutions have come to operate as if personal meaning were inconsequential to the learning and behavioral-change processes is having a profound impact on you and your success in life right now.

It's my experience that the more you learn about the foundations of our existing *Reward and Punishment Approach* to learning and behavioral-change systems, the less you'll repeat what doesn't work, and the more you'll start doing what does work. This historical understanding will allow you to have a greater positive influence on yourself, your children, and any organizations with whom you're associated or work for. Understanding the history of how things got the way they are can support us to stop, as W. Edwards Deming said, "recycling ignorance."[2]

Aristotle Started It

The person who could arguably be called one of the most important figures in shifting the emphasis of education away from personal meaning was Plato's famous student Aristotle. All learning systems today that value pouring linguistic facts into a learner's brain in a linear fashion, without regard to the meaning those facts have to the learner, have their genesis in the way the ancient world viewed Aristotle.

The fourth-century philosopher is probably best known for helping foster the stand-and-deliver format of education used at his Academy and for developing the theory of logic used in every school system in the world. This theory demanded that each question could have only one possible right answer, a philosophy that the new sciences of complexity, chaos, and quantum physics have recognized as flawed.

Aristotle was an incredibly powerful thinker who produced 170 works in which he displayed his philosophical, technical, and scientific genius. He surveyed and systematized nearly all branches of knowledge, producing the first ordered accounts of zoology, biology, psychology, physics, and theology. He also addressed every major philosophical problem known during his time. Because of his vast body of work, Aristotle is considered by many experts to be the single greatest influence on Western intellectual development. The great poet of the Middle Ages, Dante Alighieri, called Aristotle "master of those who know."[3] Darwin has been quoted as saying that intellectuals of his time "were mere schoolboys compared to old Aristotle."[4]

The real power of Aristotle stems from the fact that, up until the 1700s, the rate of information acquisition was so slow that intellectuals, philosophers, and theologians believed that the amount of knowledge in the world was static and finite. It didn't increase. Aristotle is an omnipotent force in education because, for almost 2,000 years, his works and the works of a few other "authorities" were believed to be so complete that they contained the sum of all human knowledge. To become an educated person, then, an individual didn't have to go out and experience the world but only needed to memorize the words of the "authorities."

Scholastics See No Need for Personal Meaning

The first extensive effort to set up an organized school system in Europe was started in the 11th century by the Roman Catholic Church. The goal of the Church school system was to ensure that the information in Aristotle's and other authorities' works, which were thought to contain the sum total of all knowledge, was poured into students' heads. This effort came to be known as the *scholastic method of education.*

The scholastic method was spearheaded in the 13th century by the prince of the scholastics, St. Thomas Aquinas, who, through his own writings, ensured that Aristotle and the other authorities would be seen as the virtual source of all knowledge. (Aquinas was another great man who was seen as slow in school, so much so that he earned the nickname "dumb ox.") In the scholastic system, if you wanted to know about bees, you didn't go out and study them; you only read what Aristotle or other authorities had to say about them. Richard McKeon comments, "If you wanted to know what the culture of the 12th century was, you could list, let us say 3,000 quotations that any intellectual would know… you could tabulate it."[5]

In the scholastic school system, personal meaning had no place in the education of students, and the job of a teacher was to simply have the students memorize the information contained in the authorities' works.

"Memorize it; Don't Experience it."

Since it was thought that all the knowledge in the world was contained in the works of the "authorities," you could sit a student down for six hours a day for twelve to eighteen years and have them memorize most of it. Personal meaning and experience in the real world were thus greatly devalued as part of the learning process. What was meaningful to the student or what the student discovered on his own was inconsequential. Why should it matter? Everything that could be discovered already had been. The student's job was simply to shut up, sit still, and memorize it. Then he would be a learned person and a value to the world, but not before.

There's the story of a monk who, while scanning the heavens one night, discovered a new star. When he excitedly told his patriarch what he'd discovered, his teacher told him to look up that star in Aristotle's works. When the student returned and said he could find no reference in Aristotle's teachings, the teacher then told him that the star couldn't possibly exist! This was the power of Aristotle and the authorities. The spread of the scholastic method and its reliance on the authorities was greatly advanced by the invention of Gutenberg's printing press in the 1400s.

The famous European universities of Paris, Bologna, Oxford, and Cambridge all arose out of the scholastic tradition and its reliance on Aristotle and the authorities. Unfortunately, in the 21st century, the majority of all educational and corporate institutions continue to rely on the stand-deliver-memorize systems that worked so well in the past, when all the information in the world was considered static. But as we've seen, today's knowledge in the world is estimated to be doubling at the rate of once every eighteen months, which is a major reason our educational systems have lost much of their effectiveness.

Machiavelli's *Prince*: Spare the Rod, Spoil the Child

The next historical happening that squeezed personal meaning out of the organized school system was the publication of *The Prince* in 1532, by one of the most significant political and military philosophers of the Renaissance, Niccolò Machiavelli.[6] In this renowned work, he laid out the case that it was right and just for authorities to use cunning and unscrupulous methods in their efforts to gain and sustain power over their subjects.

Machiavelli maintained that as long as leaders believed what they were doing was right, they had the moral position, and therefore, the justification for using whatever form of reward or punishment (carrot or stick) was appropriate to accomplish their goals, even if the methods were themselves amoral or horrific. Before the publication of *The Prince,* most works on authority endeavored to glorify humanistic and moral virtues. Pre-Machiavelli, the scholastic school system saw itself as doing the work of God, and a sense of moral benevolence dominated Aristotle's educational philosophy. Post-Machiavelli school systems saw reward and punishment (more often punishment) as the primary method of getting students to

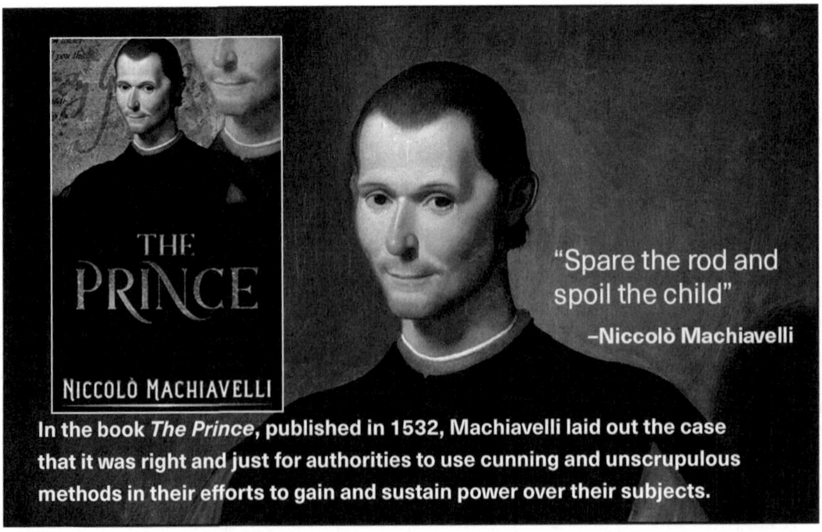

In the book *The Prince*, published in 1532, Machiavelli laid out the case that it was right and just for authorities to use cunning and unscrupulous methods in their efforts to gain and sustain power over their subjects.

carry out the will of the authority. What was meaningful to the student has no place in Machiavellian philosophy. Extrinsic motivators, not personal meaning, was how people were made to learn. The phrase "spare the rod and spoil the child" has its roots in Machiavelli. [7]

From the 1500s until the 1980s, the liberal use of physical punishment on students in secular and religious schools, such as paddling behinds and smacking the backs of hands with rulers, can be directly tied to Machiavelli's efforts. Authority figures in the military school I went to in the 1960s liberally and unapologetically used extended coat hangers and paddles on my young and tender butt, often raising blood blisters, in a Machiavellian endeavor to get me to conform to their will. This philosophy that authority has the right to use force on students still has a strong presence in educational systems today.

Newton's Mechanics: "Might Makes Right"

Moving on 150 years after Machiavelli's *The Prince,* the man who would replace Aristotle as the most dominant scientific thinker in history, Sir Isaac Newton, published what many scholars believe to be the most significant work ever conceived. *Philosophiae Naturalis Principia Mathematica,* published in 1686, created classical physics by putting forth the three laws of motion and the law of universal gravitation.[8] Newton's work not only provided the scientific underpinnings for the industrial and technological age in which we are now living but also laid the foundations upon which today's mechanistic educational and corporate systems were built.

Newton's *Principia* had a profound impact on society. Before this work, God's plan and the guardianship of the angels were seen to have drawn the universe into a smooth working whole. There was magic in creation. To people of the day, the beating of angels' wings kept the planets in motion. Newtonian physics painted a very different picture of the universe: a precise and unchallengeable one, where the world was nothing more than a great, impersonal, complex machine controlled by specific mechanistic laws. No magic here—just mechanics.

Newtonian mechanical laws also introduced the overriding concept that raw, unadulterated force was the primary instrument of causality in the universe. The power of Newton's mechanical force paradigm caused nearly every intellectual, political, social, military, and economic thinker of the 17th through 21st centuries to use it as a model. Herbert Spencer's conceptualization of "survival of the fittest," which led to the "might makes right" style of industrial thinking, comes directly from Newtonian physics.[9] (See Chapter 10 for a discussion of Spencer's misunderstanding of the biological term "fit.") Danah Zohar and Ian Marshall comment in their book *The Quantum Society*, "Mechanistic physics became the touchstone for a whole world view, the central paradigm of the modern world."[10] They also note that, because of Newtonian science, "Today, our perception of social and political reality, our whole perception of 'modernity,' is a mechanistic perception."[11]

The sheer simplicity of Newton's mechanistic laws of force and reaction created a world where society could achieve any goal if you just applied the right amount of force in the right place at the right time. Newton's model allowed society to regard humans as, at worst, cogs in the machine, or, at best, small automatons controlled by a larger force. In the mechanistic force paradigm, human values, emotions, and personal meaning are irrelevant.

Sir Isaac Newton published *Philosophiae Naturalis Principia Mathematica* in 1686. Newton's model allowed society to regard humans as small automatons controlled by a larger force. In this mechanistic paradigm, human values, emotions, and personal meaning are irrelevant.

The Shift from Church to State Education: Students as Machine Parts

What helped ensure that the Newtonian force model would be brought into the educational system was the shift of control of education from the Church to the state in the 17th, 18th, and 19th centuries. During this time, countries began to recognize that the more highly trained and productive the populace, the greater social, economic, and military power the country could possess. With nationalism running rampant, education could no longer be totally the purview of the Church. For countries to become powerful and succeed, learning institutions needed to become instruments of the state. Students then came to be seen as pieces of the nationalistic machine, whose lives took on purpose in terms of the manifestation of national meaning and success, not personal meaning and success.

In the mid-1800s, the Prussians, who had been soundly defeated by Napoleon, developed the first model of mechanistic education.[12] Their hope was that by using a Newtonian machine model of education, they could build a country so powerful that an invader would never again defeat it. The Prussian learning system was built on the factory model. It treated students like machine parts placed in a conveyer-belt like system, shuttled in and out of short classes, where force was applied through the threat of physical punishment. After 10,000 to 20,000 hours of mechanistic treatment, out would pop a finished "unit of production," capable of adding to the national security and enhancing the country's productivity. By the start of the 1900s, Newton's factory force model, as designed by the Prussians, was in place in every school system in the world. Sound familiar? It should. While the world has changed significantly, this mechanistic learning model still dominates today.

Behaviorism Focuses Only on Stimulus and Response; Not Meaning

The next piece of the mosaic that sucked personal meaning out of our learning processes occurred in the early 1900s. Psychologist John B. Watson, who, through his work as an advertising executive, had become fascinated by the ability to manipulate human behavior, placed Pavlov's experiments with animals upon the philosophies inspired by Machiavelli, Descartes Newton, and Locke and developed the school of psychology called behaviorism.[13]

Watson believed that, in its efforts to get humans to perform specific acts, psychology must be totally objective, focusing only upon stimulus and response and totally ignoring the inner world of the individual. Echoing Descartes, he wrote that "sensation, perception, image, desire, purpose, thinking and emotion" had no place in the efforts to get someone to carry out specific tasks. Like Locke, Watson envisioned the brain as an empty black box whose internal workings were totally irrelevant to the scientific explanation of that organism's behavior.

Watson, eventually known as the *father of behaviorism*, also adopted the Newtonian philosophical position that raw force in the form of the correct reward and punishment administered by an authority figure was the best method to control the "assembled organic machine."[14] While Newton had discovered the laws of force that made inanimate objects move, Watson wanted to discover the laws and forces that made humans move! While Newtonian laws could predict the movements of inanimate objects, behaviorists wanted to apply laws that could predict the actions of living organisms.

The behaviorists' desire to control another's life path without regard to what's personally meaningful is clearly reflected in Watson's famous statement:

> *"Give me the baby and my world to bring it up in, and I'll make it crawl and walk; I'll make it climb and use its hands in constructing buildings of stone or wood; I'll make it a thief, a gunman, or a dope fiend. The possibility of shaping in any direction is almost endless."*

To the behaviorists, the external motivators of reward and punishment became the levers by which they could control the human-machine. Behaviorists' reliance on control mimics the hiss of the serpent in the Garden of Eden: "Ye shall be God."[15] Watson's disciple B. F. Skinner sums up the view of behaviorism in two sentences. In 1983, he wrote, "What a fascinating thing! Total control of a living organism!"[16]

Behaviorism Murders Relationships

Another problem with behaviorist learning systems is that they separate the master from his pupils. A teacher is inhibited from looking into the

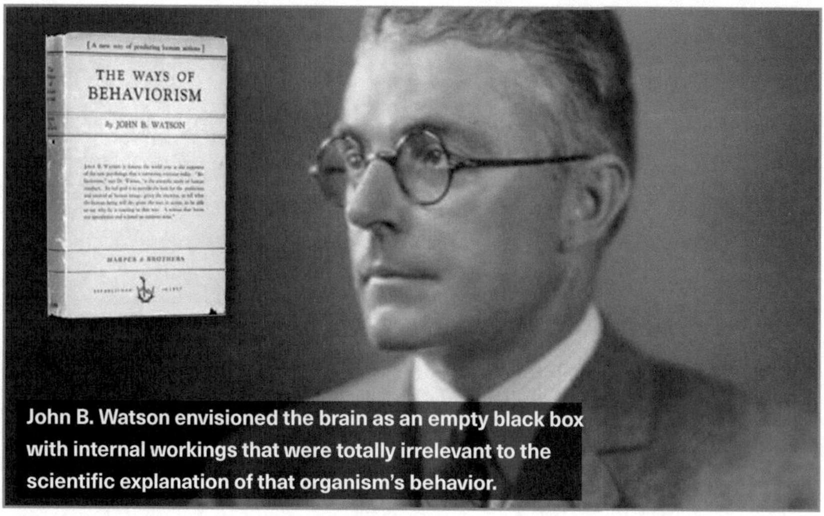

John B. Watson envisioned the brain as an empty black box with internal workings that were totally irrelevant to the scientific explanation of that organism's behavior.

student's eyes and seeing a child's soul. Instead, all he can see is the glow of the machine that he must force to perform specific acts at specific times in order to get the student to pass standardized tests. Mechanistic educational systems interfere with the teacher's effort to foster the maturation of a pupil's most precious asset: their connection to their soul and what really holds meaning for them.

Behaviorists' educational systems allow little space for a deep master–pupil relationship. Brain-based educator Robert Sylwester laments that, because educators have relied on reward and punishment without regard to the inner process of the student, "Thus we became a profession of behaviorists, whether we liked it or not."[17] The separation that behaviorism forces between master and students is carried over into the corporate world. Bosses mistakenly believe that they don't really need to have a relationship with their workers to get them to be proficient and innovative. All they have to do is to figure out the right stimulus to get the right response—that is, the right kinds of rewards or punishments that will drive the workers to propel the company to profitability.

The new sciences are allowing us to look with a skeptical eye upon systems that primarily rely on external motivators as the means of fostering learning, creativity, and behavioral change. Morton Hunt, the author of *The Universe Within,* says, "The wonder of wonders is that so many intelligent and thoughtful people could have believed that behaviorism was an adequate explanation for human behavior."[18]

The Last Assault on Personal Meaning: The Memory Research Industry

The final attack on personal meaning was carried out by memory science itself. What's considered the first modern scientific memory research was performed by Hermann Ebbinghaus from 1879 to 1885. During these years, Ebbinghaus would arduously memorize and rememorize lists of meaningless nonsense syllables, such as *zin, taj, huq, caf, dak,* and *rax.* The result of his painstaking study is now called the *Ebbinghaus forgetting curve.* [19] This curve shows that, even after Ebbinghaus studied and restudied the list of thirteen nonsense syllables more than 150 times, after only thirty days, 80 percent of the list wasn't retrievable from memory at all.

If, after memorizing meaningless information over 150 times, Ebbinghaus forgot so much so quickly, imagine how fast most of us forget information that holds no meaning for us after we memorized it only three to five times in studying for a test! As Peter Russell says in *The Brain Book,* "By stripping away meaning and association Ebbinghaus had effectively stripped away all that was most valuable to memory."[20]Unfortunately, Ebbinghaus established a precedent that has been hard to stamp out: since his first experiments, much of the memory research carried out through the 20th century has been devoid of personal meaning.

In fact, virtually any standardized memory or intelligence test you'll take today totally ignores what you find meaningful in your life. A test taker may have no idea who the Secretary of State is or at what degree of longitude the equator is, thus indicating that his memory and intelligence aren't up to par. Yet that same test taker, who finds baseball and computers meaningful, may be able to repeat every Yankee baseball player's batting average back five years and cite every circuit in a standard computer, indicating that he has a fine memory and a form of intelligence not accounted for by standardized memory/intelligence tests.

Most tests are inefficient at judging memory and intelligence because they have few provisions for accessing what the test taker finds meaningful. By judging intelligence devoid of meaning, we end up labeling many people as less intelligent than they actually are.

Tricks to Remember: Valuable?

Ebbinghaus's research also helped legitimize a whole field of memory tricks called *mnemonics.* Created by the early Greeks to help them remember the long speeches they gave at the Parthenon, mnemonics aid a learner's memory by using tricks such as association, imaging, and acronyms. Today, mnemonics are a great aid to students, especially in a system that values short-term memory retention. For example, if you want to remember the name "Ebbinghaus" for a test, you can visualize a house being swept in and out by the ebbing of the tides. The problem with mnemonics is that unless the memory trick helps stimulate your *Meaning Network,* that new information will be held for only short periods of time in working memory. Consider all the acronyms you used to pass tests—and how many you can't remember today.

Fortunately, things are changing. Today a growing number of memory researchers, psychologists, and psychiatrists are recognizing that memory can't be efficiently studied without including the element of personal meaning. The recognition that meaning is important to learning and behavioral change has come primarily from two sources.

The first was discussed in Chapter 4. In the 1940s, Viktor E. Frankl's research into the survival rate among his fellow prisoners in concentration camps during World War II showed not only that behavior was driven by the search for meaning but that in some cases survival itself depended upon meaning being maintained. Frankl's work opened up psychologists' eyes to the importance of personal meaning in creating learning and sustained behavioral change. He developed a respected revolutionary approach to psychotherapy known as *logotherapy,* the core of which is that humans' primary motivational force is their search for meaning. Frankl's research into meaning's effect on behavior landed a powerful blow to the behaviorist contentions that we learn and change behavior based solely on the amount and kinds of rewards and punishments we receive from an authority figure.[21]

The second source that caused psychologists to rate meaning more highly is recent advances in neuroscience. Through neurosurgery and brain imaging, neuroscientists have noted that when parts of your *Meaning Network* are damaged by trauma or disease, learning, memory, and behavioral change are inhibited. The neuroscientific research has helped many leading-edge memory researchers, psychologists, and psychiatrists recognize that in order to understand memory and how to create sustained behavioral change, we must understand the power of personal meaning and intrinsic motivation in creating the neurological change that's their basis.

What we now know is that while an extrinsically driven learning system, devoid of personal meaning, might have worked well hundreds of years ago when all the information in the world was thought to be limited and could be memorized in a decade of study, that world no longer exists. Now, in an age where information is doubling at a fantastic rate, it's time to change direction and build our scholastic and corporate institutions not on *Ancient Learning Theory* but on sound biology, neuroscience, and genetics.

In the next chapters we will shift from history back to science, first by looking at the biological reasons why, to create profound learning, we must focus on understanding big concepts before we can effectively remember the small details.

Cracking the Learning Code

The historical perspective shows how, by relying on *Ancient Learning Theory*, we've gotten so far off track and caused so much pain and confusion in our efforts to develop efficient learning and behavioral-change systems. It's now time to look back, take what worked and jettison the rest.

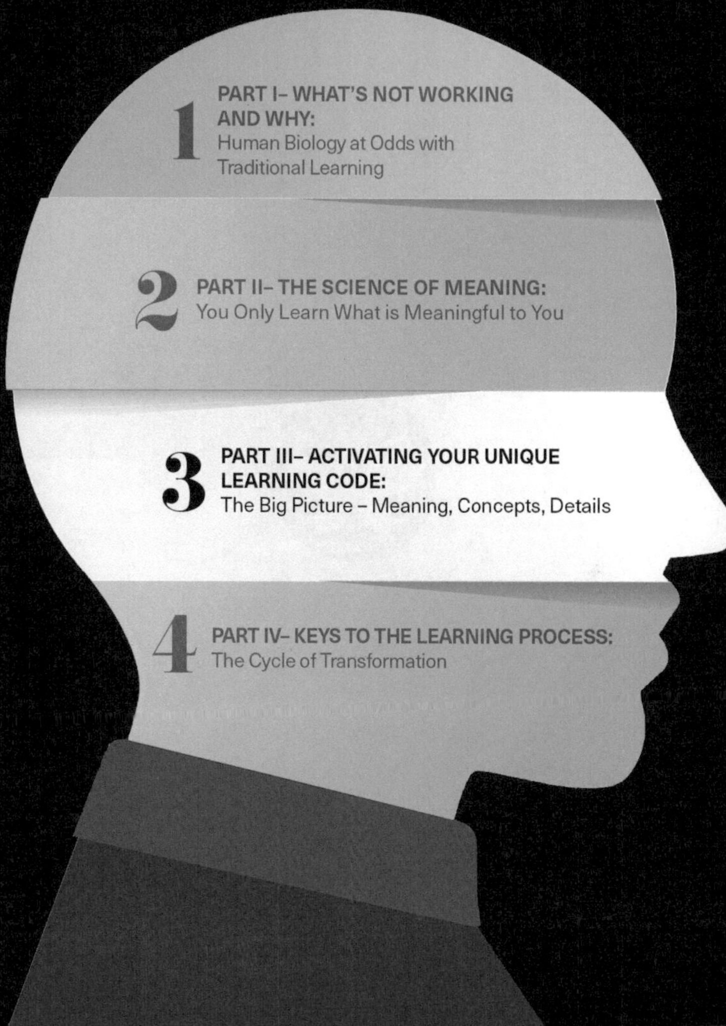
1
PART I– WHAT'S NOT WORKING AND WHY:
Human Biology at Odds with Traditional Learning
2
PART II– THE SCIENCE OF MEANING:
You Only Learn What is Meaningful to You
3
PART III– ACTIVATING YOUR UNIQUE LEARNING CODE:
The Big Picture – Meaning, Concepts, Details
4
PART IV– KEYS TO THE LEARNING PROCESS:
The Cycle of Transformation

21

Why You Must Learn Concepts Before the Details Will Stick

When people think seriously, they think abstractly: they conjure up simplified pictures of reality called concepts, theories, models, paradigms.

—Samuel P. Huntington, American political scientist, adviser, and academic[1]

If the one of the most crucial elements of the Learning Code is to select new information into your brain structures by stimulating your *Meaning Network*, how do you do that? In this chapter, we'll focus on a simple question: How is your *Meaning Network* most easily activated—by details or concepts?

If you've ever had difficulty grasping a specific idea in school, or left a corporate training session wondering how the heck the instructor could have spent so much time on a particular subject when you ended up remembering so very little, it probably wasn't because you're stupid or didn't try. More likely, you experienced the result of typical learning systems that overemphasize details over big-picture concepts, which is in direct conflict with how your brain naturally learns and remembers information.

The Big Picture Helps Accelerate Learning

We know the brain doesn't like dealing with facts and details in isolation. Instead, it prefers to process information in big chunks, also called *gestalts, schemas,* or *concepts.* You might beat yourself up because you're miserable at remembering details, but evidence suggests your brain isn't *designed* to—it is wired for concepts. This is why, at The Institute, we've developed the following model, which we abbreviate *MCD*, to support effective long-term memory formation:

1. **Meaning:** The information needs to be presented in a personalized framework, so the all-important *Meaning Network* of the learner is activated (see Chapters 12–14).
2. **Concepts:** The brain concept network relating to the knowledge to be learned must be stimulated.
3. **Details:** Now, the details can be successfully delivered.

This process is based on the latest research which Leslie Hart summed up when he described the current educational system as making a "giant" and "utterly wrong" assumption "that if a subject is fragmented into little bits, and the student is then presented with the bits… the student will be quite able to assemble the parts and emerge with the whole—even though never given an inkling of the whole!"[2]

Science has led educators and trainers who understand brain function to conclude that trying to learn the details of a subject before knowing the concepts is like trying to do a 5,000-piece jigsaw puzzle without seeing the

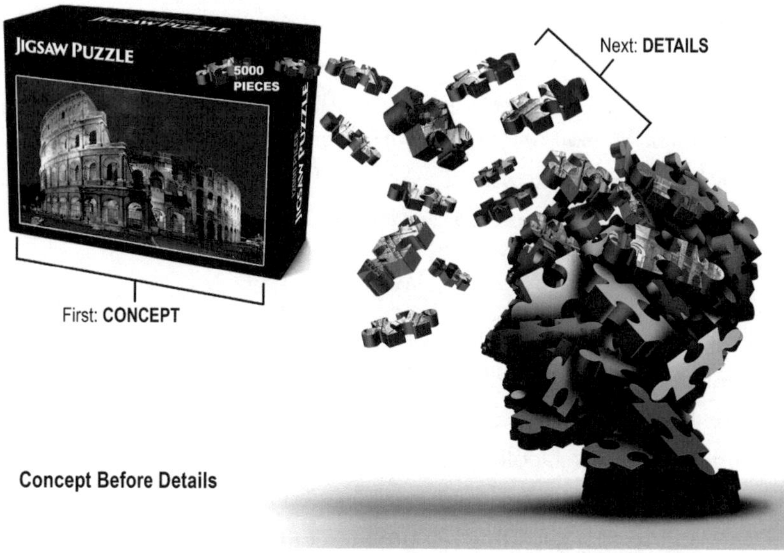

Concept Before Details

Trying to learn the details of a subject before knowing the concepts is like trying to do a 5,000-piece jigsaw puzzle without seeing the image on the box—challenging, if not impossible.

image on the box—challenging, if not impossible. Researcher Richard C. Anderson found connecting new information to previously learned concepts "is a better predictor of comprehension than… an intelligence test score!"[3] From an analysis of the research, it's clear that if our test-centric scholastic and corporate educational systems emphasized the concepts of math, physics, technology, marketing, or management before introducing details, the whole process of education would be much less stressful and much more efficient and joyful.

Try this: take a few seconds to look out the window. Now, if I were to ask what you saw, you'd probably give me a big-picture, concept-based answer. Depending on where you're reading this book, you might say you saw buildings, lawns, or a beach. You wouldn't say you saw a million bricks, or twenty-million individual blades of grass, or seventy billion grains of sand. In the quick scan of your environment, your brain immediately and naturally focused on the big picture, not the details.

To get an even deeper feeling for how big-picture concepts help you encode details, quickly read the following passage:

> *With hocked gems financing him, our hero bravely defied all scornful laughter that tried to prevent his scheme. "Your eyes deceive," he had said. "An egg, not a table, correctly typifies this unexplored planet." Now three sturdy sisters sought proof. Forging along, sometimes through calm vastness, yet more often very turbulent peaks and valleys, days became weeks as many doubters spread fearful rumors about the edge. At last, from nowhere, welcome winged creatures appeared, signifying momentous success.*

If three hours after reading this passage, I tested you on what details you remembered, you'd probably flunk it. Why? Because, to your brain, this passage looks like a bunch of information with no place to put it! But psychologists James Dooling and Roy Lachman found those people who were informed this passage was about Christopher Columbus's voyage to the New World remembered much more than those who weren't.[4] Now, reread the passage. How does knowing the big picture affect your understanding? Now the passage makes sense. Your brain doesn't automatically place details into long-term memory just because you spend hours memorizing them. Concept activation must happen first.

For another example of how having a mental model of something helps you acquire details quickly, look at the image below and take ten seconds to memorize all the letters and symbols. It can't be done. Now, look at image on page 312 for ten seconds. This time you can easily remember all the symbols and letters, but why? Because you have a big-picture model that the details can affix to.[5]

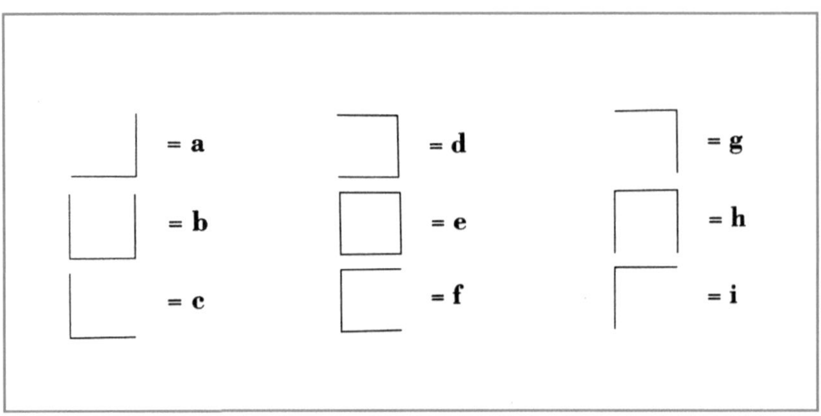

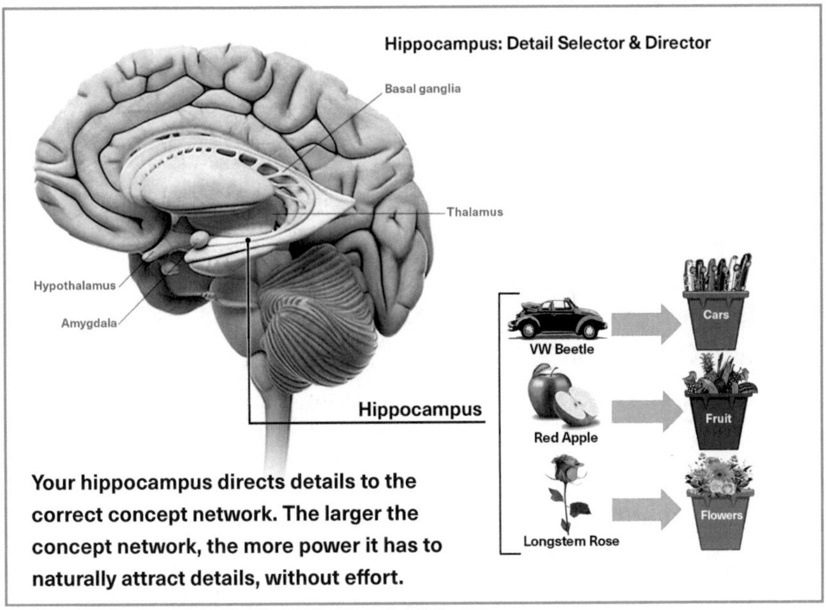

Your hippocampus directs details to the correct concept network. The larger the concept network, the more power it has to naturally attract details, without effort.

The Science of Why Concepts Must Come First and Details Second

Understanding you must first implant concepts into your brain before details can stick is vital to increasing your *Individual Adaptability/ Intelligence Factor*. The science tells us this is true because well-established concept networks, made up of millions of interconnected neurons, provide four components to ensure details can effortlessly be selected into your long-term memory banks:

Physical framework: If you don't have a conceptual understanding of something, your brain lacks the neural material upon which details must physically affix themselves. The large numbers of neurons in concept networks provide a stable landing pad upon which details can easily attach themselves.

Mass: The neurons and connections that make up concept networks in your brain supply mass. Mass attracts mass. Therefore, large concept networks act like magnets, attracting details that resonate with them.

Energy: When your concept networks are activated, they demand large amounts of glucose and oxygen in the brain areas that hold them; this provides the energy necessary to power the neurochemical processes that allow details to be incorporated into your long-term memory.

A Clear Target for The Hippocampus: Neurons firing in unison, as they do when a concept network is activated, supply another advantage: they provide the hippocampus—the part of your brain that supports long-term memory formation—with a clear target to which it can send relevant details. Without neurons firing in unison in specific brain areas, the hippocampus lacks clues as to where to allocate details.

Details Can't Exist in Thin Air, They Need to Attach to Concepts

These components are why we know "learning occurs through selection, not instruction." The small details are selected into your brain tissue by the larger concept networks. As was discussed earlier (see Chapter 9), new learning doesn't exist in thin air. Memory doesn't float freely in the brain, like so many butterflies. Before new information can be selected into your long-term memory, it must attach itself to existing neural connections. Your memory of concepts is made up of a matrix of millions to billions of interconnected neural connections. Your brain is filled with concept networks for things such as faces, fruits, trees, tools, dogs, cars, airplanes, and computers, each having its own unique configuration of millions to billions of neural connections. When you

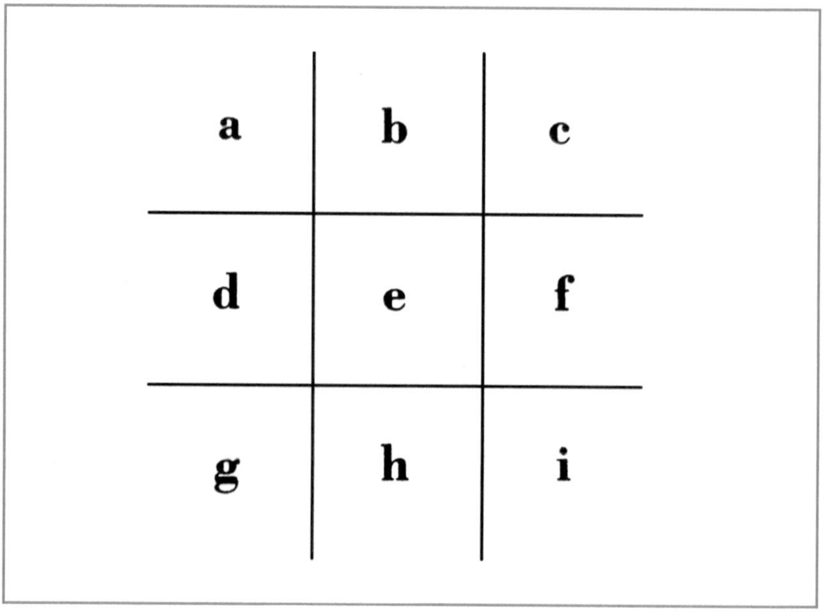

learn a new detail about something, it can't just lodge itself willy-nilly in any old place in your brain. Instead, the hippocampus directs specific details to the correct concept network. This ensures that if you learn a new detail about airplanes, it will get placed in the web of neural connections representing your conceptual understanding of airplanes, not the one dealing with omelets. Therefore, to effectively learn and remember new details about any subject, you must already possess a conceptual framework of that subject, or the hippocampus has no clear target to send this new information.

From a neurological perspective, then, physical memory is less like a container filling up with miscellaneous details and more like a conglomeration of concept networks, each having a unique lattice of hooks.[6] Each hook reaches out to snag details that resonate with it. The more hooks, the more details a concept network can snag. The more robust and established the concept network, the more effective the hippocampus is at allocating new incoming details to that network; conversely, the weaker and less-established a concept network is, the less effective the hippocampus is at helping new details find a place in the brain.

It's like if you went to a remote region of New Guinea and tried to teach someone with no concept of computers about these machines. Their hippocampus would have little success trying to allocate and remember details about them (no matter how much reward and punishment you applied) because that person would have no brain architecture for the new details to fasten to. This is one reason why you forgot so many of the details that you spent thousands of hours memorizing in school. If, as a student, you had a weak understanding of a math concept like algebra, you'd have little neural material for the details of algebra to affix to. As a result, you can't effectively access the details at a later date—like now, as an adult!

Making Completeness Out of Incompleteness

Our preference for thinking in concepts is supported by what many neuroscientists call *completion*—the phenomenon in which only a small number of neurons in a concept network needs to be activated to stimulate the whole network. There's some disagreement about the exact number of neurons that must be stimulated before the whole network is activated; some neuroscientists say only a few, but others agree that at least 5 percent must be stimulated before the whole network is activated.[7]

Completion is a wonderful survival mechanism that allows the brain to make completeness out of incompleteness. For example, if you see the tail of a tiger sticking out of a bush or the back of your mother's head, your brain automatically fires up the whole concept of the tiger or your mother, allowing you to make quick decisions on limited input which affect your ability to thrive and survive—especially if the tiger is hungry or your mother is angry. The completion formula (5 percent stimulation = efficient conceptualization), in effect, forces your brain to prefer thinking in concepts as a survival advantage. This formula is an evolutionary advantage. Our prehistoric cousins who focused on details, like the number of scales on a charging alligator, didn't make it through the sieve of evolution. They got eaten while counting! Therefore, we never got those "details first" genes.

The success of impressionistic painters and political cartoonists to convey complete representations of reality with incomplete illustrations is dependent on this completion formula. Although the completion phenomenon may lead you to jump to wrong conclusions from time to time—like running from a garden hose you mistake for a snake or mistaking a stranger for a friend. More often than not, however, completion quickly supplies you with correct concepts so that you may take corrective action.

Many neuroscientists note that the ability to conceptualize is the neural forerunner of consciousness.[8] Humans, with the most complex neocortices and sophisticated frontal lobes in the animal kingdom, have the greatest capacity to think in gestalts. In the article "Mental Models in Narrative Comprehension," Gordon Bower and Daniel Morrow write, "The ability to construct and manipulate valid models of reality provides humans with our distinctive adaptive advantage; it must be considered one of the crowning achievements of the human intellect."[9]

Experts Say Details Before Concepts Is a Bargain with the Devil

Educators who understand the neurobiology of learning are highly concerned about scholastic and corporate learning systems that continue to follow the "fallacy of the familiar," with a heavy emphasis on detail acquisition over concept formation. In a test-centric society, we've made a bargain with the devil. We've used memorization to force detailed information into working memory to pass tests, but because the underlying concepts to which these details must attach themselves were never effectively put in place, when we try to remember these details later, we find they never became part of our long-term memory, even though we passed the tests that emphasized them so heavily. Think about your primary and secondary education. You were probably asked 10,000 to 50,000 test questions. Of all these questions, how many details do you remember today? Probably very few.

The Caines have done extensive research on conceptualized learning, and they note in *Making Connections:*

> *The more separated information and skills are from prior knowledge and actual experience, the more dependence there needs to be on rote memorization and repetition.*

Emphasizing the storage and recall of unconnected facts is an inefficient use of the brain.[10] They have also found that the focus on details actually shuts down the concept-building functioning of the brain.[11] Gordon Dryden and my good friend Jeannette Vos say in their best-seller *The Learning Revolution*:

> *Many traditional schools still introduce subjects through textbook lectures spread over months. You're taught to read each chapter slowly and deliberately—a week at a time—without ever having the "big overview." That's crazy. It's inefficient.*[12]

One study indicated that ten of the most prominent science textbooks used from kindergarten through twelfth grade are so detail-oriented that they failed to help children learn. George Nelson, director of Project 2061, a math and science reform initiative of the American Association for the Advancement of Science, says textbooks are so focused on unnecessary details that they preclude the learner from understanding big-picture concepts:

Providing bits of information about transmissions, carburetors, fuel injectors, universal joints, and cooling systems doesn't convey a sense of a car as a mode of transportation.

Two independent teams of science curriculum specialists found these textbooks ignore or obscure important ideas by focusing too much on technical terms and trivial details. One evaluator, Jo Ellen Roseman, says, "At their best, the textbooks are a collection of missed opportunities."[13]

Detail Focus Inhibits Your Adaptation to the Real World

Howard Gardner, in his book *The Unschooled Mind,* maintains the inability of our educational systems to embed strong concept networks is the primary reason for the levels of inefficiency we experience.[14] Gardner took a unique approach, looking not only at poor students but also at the supposedly successful ones. He analyzed research in fields ranging from math and physics to literature and history and found that, although good students can regurgitate facts on tests, they cannot truly apply these facts to the real world. He finds that the emphasis on details over concepts inhibits the acquisition of the information educators are trying to get students to grasp in the first place.

Gardner shows that, in study after study, when students are asked to apply the details they just memorized to real-world concepts, they fail to connect the dots. He relates the story of his own daughter, who had been a good physics student in high school but was having trouble with her college physics course. When he prompted her to visit her professor for help, her retort was, "You don't get it, Dad. I've never understood it." Leslie Hart, in his groundbreaking book *Human Brain and Human Learning,* points to a report by the National Assessment of Educational Progress, which showed that students tested strongest in the knowledge of facts and weakest in applying them. Asked how far a car going eight kilometers in five minutes would go in an hour, only 28 percent of thirteen-year-olds and 56 percent of seventeen-year-olds could provide the correct answer. Hart says, "When we consider that a tenth grader had eight to ten years of mathematics, we must marvel at the potency of classroom instruction in preventing learning!"[15]

To get a doctorate degree in any discipline, you must have a deep understanding of the concepts that discipline is built on. But today, it's estimated around 40 to 50 percent, depending on the discipline, never get their degrees despite spending years trying. Some educators point to this high failure rate as evidence that our overemphasis on learning isolated facts at lower levels of education may be developing brains that have a difficult time grasping big-picture thinking. They fear that exposing the young brain to twelve to sixteen years of focused detail acquisition ends up selecting out the brain's ability to naturally and automatically think in the gestalts necessary to acquire a doctorate as an adult.[16]

Years ago, the Carnegie Foundation did a jaw-dropping study, which revealed that the individuals who could get the concepts fastest from a technical manual on how to program a VCR were ten to twelve years old. These youngsters were better than eighteen-year-olds, except for eighteen-year-old dropouts, who were better than high school graduates, who were better than college graduates, who were better than those holding master's degrees! From this study—admittedly isolated—you could conclude that the more educated you are, the more caught up in the details you become, diminishing your ability to grasp overall concepts. Our educational systems, focused on details over concepts, maybe breeding adults who can't see the big-picture concepts even when they're right in front of their eyes.

In companies I've developed, we've discovered that many MBAs (especially the ones fresh out of school) get so caught up in the details of a deal they often totally miss the big picture of the deal. They get analysis paralysis. It's almost as if their detail-focused education cut them off from their own intelligence, common sense, and intuition.

Disease Research Sheds Light on How the Brain Uses Concepts

In the 1930s, British psychologist Frederic Bartlett was one of the first modern scientists to propose the idea that the brain processes information primarily in big-picture concepts, which he called *schemas.*[17] Since then, significant scientific understanding about how our brains process information in concepts first comes from research on individuals called agnosias, synesthetes, and *savants*—people with intriguing neurological disorders that involve concept formation. These disorders have given us wonderful insights into how our brains make and activate concept networks.

Agnosias: People who, through brain damage, have lost complete sections of concept networks. Some agnosias have lost the ability to recognize concepts like tools, vegetables, animals, and even faces. One patient who had "face" agnosia couldn't recognize his own face in the mirror; another could recognize his wife only by the clothes she wore, not by her face.[18,19]

Synesthetes: Some of the most fascinating people in the world—have sensory concept networks that have been mis-wired, allowing them to see sounds and taste colors. Because of this, they see the world in a much different manner than most people. One patient with synesthesia tastes chocolate every time she hears the name, *Richard.* Another sees blue each time he hears the letter *x* and sees brown when he hears the long *a.*[20]

Savants: These are individuals with a form of autism that allows them to amass large amounts of detailed information on one subject. The unique neural processing of savants allows them to overbuild one specific concept network, giving them great intellectual powers in one finite field such as math or music. Unfortunately, their ability to create large numbers of diverse concept networks is severely limited.[21]

Possessing one powerful concept network isn't as good of a survival advantage as having many that freely intercommunicate. Savants become incredibly proficient at (as well as obsessive about) remembering details in specific areas like math, music, and sports statistics. Upon only one hearing, a musical savant may be able to remember and play a sonata perfectly yet may lack the neural capacity to remember how to make a tuna fish sandwich, even after being instructed many times. Raymond Babbitt, the character in the classic movie *Rain Man,* played by Dustin Hoffman, is a good example of a savant who could remember the faces of playing cards very easily, which made him a wizard at poker. But to make it through life, he needed the support of his "shady" brother, played by Tom Cruise. The inability to link clear, distinct concepts dramatically inhibits agnosias, synesthetes, and savants from leading normal lives, but research with these individuals has greatly helped us understand how the brain prefers to learn in concepts before details.

Mass Begets Mass for Generalists and Specialists

A concept in physics says the greater the mass, the greater the pull, and this concept of gravity applies to concept networks: the greater the mass of a concept network, the stronger the pull to the details that resonate with it. This is why the more you've learned in any one discipline, the easier it is to learn more information in that discipline. The idea that mass attracts mass is true whether you're a generalist or a specialist. Our species needs both generalists and specialists to help increase our whole *Species Adaptability/Intelligence Factor.*

We need generalists who can make many diverse connections in many subject areas, as well as those who have rich connections in limited areas. Because generalists have numerous horizontal concept networks that cover many different areas, it's easy for a generalist's brain to snag lots of details about many different subjects. Think project managers, and contestants on game shows like *Jeopardy!* Generalists need to have diverse groups of concept networks in place, which, in turn, attract large numbers of diversified details.

On the other hand, specialists like brain surgeons spend their neural energy building deep, rich vertical concept networks about one specific subject. These networks attract and snag large amounts of details that relate specifically to a single subject. An expert in entomology (the study of insects), for example, may have amassed so many details about the diseases of wasps such that he can write many books on the subject, but this same fellow may have such deprived concept networks about automobiles and plumbing that he may not even be able to change the oil in his car or fix a leaky faucet. Whether you're a generalist or a specialist, learning begets learning.

This view gives an insight into why it's easier for experts in a field to remember detailed information about that field than novices: Experts have

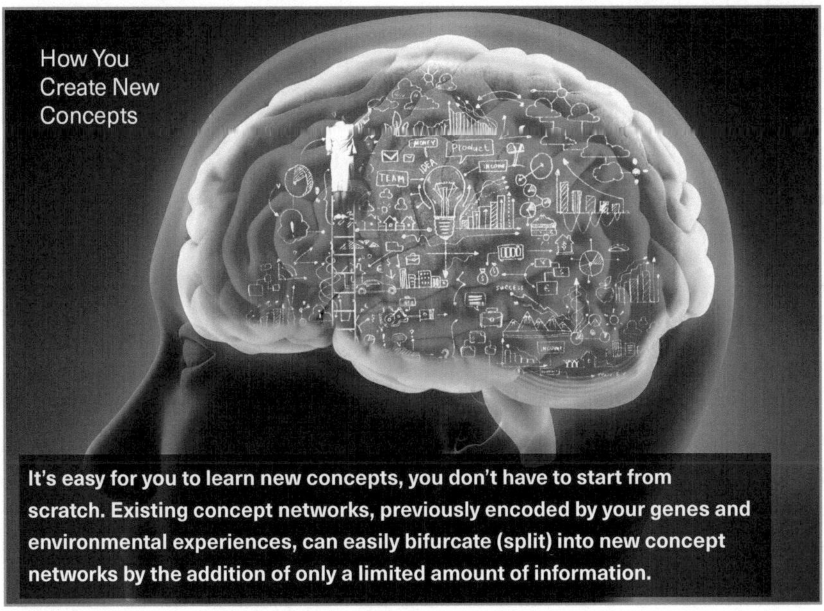

It's easy for you to learn new concepts, you don't have to start from scratch. Existing concept networks, previously encoded by your genes and environmental experiences, can easily bifurcate (split) into new concept networks by the addition of only a limited amount of information.

millions to billions more neural connections in a particular concept network than beginners; an expert has many, many more hooks for new details to attach themselves to.

A famous series of experiments with chess players clearly demonstrated this phenomenon. Both masters and novice chess players were allowed five seconds to view a chessboard displaying twenty-six pieces in a game position. The principal finding was that the chess masters, who had more extensive and established chess concept networks, could remember 400 percent more locations of the chess pieces than novices did.[22] The more brain mass in a particular concept network, the higher the efficiency of that network.

Even if you see yourself as a generalist, you have some concept networks that are larger and more powerful than others. This is what allows you to remember information in one field more quickly than another, whether it's football statistics, Far Eastern history, computer programming, cooking, or farming. Whether or not you see yourself as an expert in a particular discipline, you have dominant concept networks providing an abundance of neural connections that make it easier and faster for you to select details in some subjects rather than others. It is important to understand your largest, densest concept networks help provide the neural material out of which your personal meaning is constructed.

How New Concepts Are Birthed and The Power of Metaphor

It's easy for humans to learn new concepts because we don't have to start from scratch. Existing concept networks, previously encoded by your genes and environmental experiences, can easily bifurcate (split) into new concept networks by the addition of only a limited amount of information. Because your brain is a complex, highly connected system, small changes can have great ramifications.

To understand how small amounts of new information can cause a concept network to subdivide into a new concept network, let's look at how your great-grandfather may have come to understand the concept of television. Because he already had the concept of the radio in his brain before television was invented (or at least widely known), simply adding specific details of a tube that held a moving picture allowed his radio concept network to bifurcate, creating a whole new concept network for television.

The creation of new concept networks by the addition of one or a small number of details is aided by the fact that a single neuron may belong to hundreds if not thousands of concept networks. Neuroscientist William Calvin has hypothesized that neurons in the network dealing with fruit may spend 99 percent of their time helping out non-fruit networks.[23] Neurons are like people who can serve on many committees at one time. A neuron that processes the color "deep red" can be part of numerous networks that represent dresses, sunsets, cars, crayons, shoes, lipstick, lobsters, nail polish, shirts, and blood.

By sharing neurons, dendrites, and receptors, your concept networks can easily separate into new networks with the addition of only a limited amount of learning. This sharing of information saves your brain time, space, and energy. Virtually any new concept can be created by adding a few details to the correct concept network or networks. Years ago, when smartphones first came out, if you had firmly established concept networks that represented television, telephones, and the internet, it was easy for your brain to create a new concept network for videoconferencing. All your brain had to do was add a limited amount of new information that

Forcing your brain to deal with two or more different concept networks at the same time is an ineffective use of neural energy. This is why texting and driving are so deadly.

connected to pre-existing neural concept networks in new ways. The power of the metaphor also comes from our ability to create new concept networks by adding only a few details to existing networks.

Concept networks need fuel to attach details in their structures. As we've discussed, your brain generates about twenty watts of power using glucose and oxygen via the carotid and vertebral arteries. As has also been discussed, your brain needs this energy to create the massive neural firing necessary to make memories by blowing magnesium plugs out of NMDA receptors, commencing the process that supports the foundation of memory.

A large percentage of these twenty watts of fuel goes directly to the concept networks that are being stimulated at any one time. We can see this energy flow in PET and MRI brain scans. The areas stimulated are represented by changes in gradients and colors on a computer screen can indicate when a subject is thinking about a particular idea or concept. (See the *Wrapping* and *Hot Network* processes described in Chapter 13.) When details are delivered in isolation, they don't demand the neural energy necessary to support their incorporation into long-term memory.

Why Multitasking Isn't a Great Idea!

Marcel Just, Director of the Center for Cognitive Brain Imaging at Carnegie Mellon, performed research showing that if you want to use your brain effectively, multitasking may not be such a good idea. His studies revealed that forcing your brain to deal with two or more concept networks at the same time dramatically reduces your brain's efficiency. Science is revealing that you can't really do two things at once! What your brain does instead is quickly and continually switch between the two tasks, giving the illusion you are multitasking.

You'd think that if you were working on a single task that required 7.5 watts of power and then added another task that also required 7.5 watts of power, the two brain areas together would use fifteen watts of the brain's twenty watts of power. But Just found this wasn't so. Using neural imaging, he discovered that when you switch between two unrelated tasks, the brain was forced to use a massive amount of its energy to divide its resources. Therefore, when the brain was trying to switch between two tasks at the same time, each brain area dealing with one of the tasks received **29 to 53 percent less energy** than it would have if it were working on one task alone. "You just can't keep piping new things through," Just says, "and expect the brain to process effectively."[24] This is why texting and driving are so deadly. You don't have enough brain energy left over to focus on potential road dangers.

Try This to See How Task Switching Messes with Your Head!

Try this: open up any book (not this one) and pick a page to read. Then take out pen and paper and place them next to the book. Now start reading a page in the book at the same time you write this sentence three times in cursive on the paper, "I love to eat blueberries and jam!" How did that go? Not so good huh? This is because your brain used up to 50 percent of its energy to continually task switch. Anytime you think you are multitasking, you're really task switching, which dramatically reduces the energy to and the efficiency of both tasks. Now do you see why it is so dangerous to text and drive?

Preheating: Concepts Must Come First!

A study conducted by the National Science Foundation reveals one reason Japanese students continually outscore their US counterparts on math achievement tests is that teachers in Japan spend 217 percent more time teaching math *concepts* than US teachers, who focus more time on math *details*. This Japanese method not only increases test scores but also reduces homework. In the United States, eighth-grade math teachers assign homework for 57 percent of lessons, while in Japan, teachers assign homework for only 36 percent of lessons. You can reduce homework time when classroom learning increases. James Hiebert, the co-author of the study, hopes the findings will help educators "to look at their practice from a fresh perspective.[25]

In addition to math, studies in reading efficiency confirm that the more students focus on the details in a passage, the less understanding they have of the overall concept.[26] In his book *The Learning Brain,* Eric Jensen says:

> *We may have been forcing many to learn in a very unnatural way. We have accidentally created generations of "slow" learners who easily forget what was taught.*[27]

Reports show that 70 percent of students have negative feelings about homework.[28] Is it because homework tends to focus on details, not concepts? When the Raytheon Company asked 1,000 middle scholars if they'd rather do math homework or eat broccoli, 56 percent picked broccoli! And why not? Unless you have a mathematical brain dominance, math homework, in general, is a horrendous experience where you're forced to memorize isolated facts with no strong understanding of the concepts behind them.[29]

At the Institute, we call the idea that you must present concepts at the beginning of the learning process *Preheating*. Kind of like warming up a car on a cold day to make sure you don't damage it. There's now plenty of practical evidence showing the importance of *Preheating*. Years ago, educational consultant Don Brown worked with schools in New Zealand to increase their success on standardized tests. He developed the *advance organizer,* which he says "gives you the picture before you have to look at the bits, and then invites you to bring the bits together to make the picture."[30] Brown says this simple technique has increased scores on national standardized tests by an amazing 40 percent. Research by Marsha Weil and Joseph Murphy prompts instructors to note that pre-exposure to the gestalt of a subject is "especially effective for helping students learn the key concepts or principles of a subject area and the detailed facts and bits of information within these concept areas."[31]

All successful speed-reading courses emphasize that reading the table of contents and skimming chapters to get a general overview of the book before starting to read is a must. Grasping the concept of a book before you've read it is one of the reasons why you can read faster and remember more of what you read when you finish one of these courses. As the words whiz by, you have a conceptualized place in the brain that is pre-stimulated and where details can fit and attach themselves. Emanuel Donchin found pre-exposure to concepts provides a "cognitive waiting room" for detail acquisition.[32] Michael Gazzaniga reports that, in research done at Stanford, pre-exposure led to quicker mental responses.[33] Tony Buzan wrote several books on the power of mind mapping.[34] In this method of taking notes on a subject, the learner literally draws a map showing how all the details connect to create the big picture. Several studies have shown that mind mapping effectively supports detail memory acquisition. I used Tony's mind mapping to develop the outline of this book.

Soviet mathematics Professor Victor Shatalov attracted crowds of teachers to hear him lecture on how he got students to quickly assimilate complex math theories. To get outstanding results, Shatalov's students drew visual maps of what they'd previously learned and then added in the new concepts they'd learned that day. In this manner, the students stayed in touch with what they'd previously learned and the overall big picture.[35]

Learners who don't do well with new information may underperform not because they're slow or "stupid" but simply because the new details haven't been effectively connected with previously learned concepts. As Eric Jensen says, "Unless connections are made to their prior learning, comprehension, and meaning may be dramatically lessened."[36]

Back in the 1990s, I was president of an accelerated learning company whose main course was teaching a semester of college accounting in one short day to Fortune 500 companies like AT&T and IBM. At the time, this course, called the Accounting Game, was recognized as the most effective of its kind in the world. Why could a student come in at 9 a.m. knowing nothing of accounting and leave at 5 p.m. with a competent understanding of complex principles like LIFO/FIFO? Because in the morning, the course started by *Preheating* with the simple concept of a lemonade stand that everyone had concept networks for, and as the day progressed, complex accounting details were skillfully added to this initial concept network of neurons until the class participants were using their knowledge to manage to account for a large soft drink company.

Cracking the Learning Code

To increase learning efficiency, knowledge needs to be presented first in a personalized context so that the all-important *Meaning Network* of the learner is activated (see Chapters 12–14). Second, the concept networks that relate to the details to be learned must be stimulated. First **meaning** then **concepts** then **details = MCD.**

22

The Power of Associative Learning, Metaphor, Trauma, and More

A metaphor is a kind of magical changing room—where, one thing, for a moment, becomes another, and in that moment is seen in a whole new way. As soon as something old is seen in a new way, it stimulates a torrent of new thoughts and associations, almost as if a mental floodgate has been lifted.

—*Mardy Grothe, author of* I Never Metaphor I Didn't Like[1]

What is one of the fastest ways for your brain to learn? By making associations. Without understanding associative learning and how it's created in your brain, the whole process of learning, creativity, and memory formation is a much more unnatural, complicated, and painful process than it needs to be. Here we will examine associative learning by re-connecting to elements we've previously covered, such as The Hebbian Learning Rule, the simultaneous firing of neurons, the power of emotions, and, as we saw in the last chapter, why you must learn concepts before details.

Your brain can't think in isolation. Virtually every mind scientist and philosopher, from Aristotle and John Stuart Mill to modern-day cognitive researchers, has looked at the role association plays in the learning process. In 1879, the same year Hermann Ebbinghaus (see Chapter 20) commenced his six-year research project on memorization and forgetting, Sir Francis Galton, of eugenics infamy, became one of the first modern scientists to systematically study the impact of association on mental processing. What prompted him to commence that research was when he took a short walk down London's Pall Mall. On that stroll, 300 different objects attracted his attention, and when he recorded all the hundreds of associations that each of these simple observations created in his mind, he was overwhelmed.

Since the late 1800s, a great body of research has clarified how the numerous associations that your brain makes every hour of every day are formed. In this chapter, we'll use this research to look at how associations between diverse brain cells allow you to increase your *Individual Adaptability/Intelligence Factor*. Because it is associative learning that supports accelerated long-term memory, new concept construction, creativity, metaphor formation—and because it also forces you to hold onto traumatic memories you'd prefer not to remember, it is vital to understand how it works.

Your Brain Can't Think in Isolation

Your brain—with its 100 billion neurons and 1,500 trillion connections—provides the perfect environment for associations to prosper. Close your eyes and take fifteen seconds to try visualizing *only* the nose on your mother's face. Now take fifteen seconds and try to visualize *only* the stove in your kitchen. You can't do it. Why? Because as soon as you try to imagine your mom's nose, up comes her eyes, cheeks, and chin. And as soon as you try to think of only your stove, up pops the countertops and cupboards that surround it.[2]

There are three primary reasons your brain has difficulty thinking of things in isolation and why it has a tendency to make associations—even when you don't want it to!

- As Donald O. Hebb pointed out over fifty years ago, memory isn't held in a single cell, as the *grandmother* theory of learning once proposed, but instead in widely distributed groups of interconnected neurons.[3]
- One neuron in a network that holds a single memory can be part of hundreds, if not thousands of other memory networks.
- "Neurons that wire together fire together": This phrase, which we'll look at in more detail, means that when a neuron that's part of one memory network is activated, it can automatically fire up other neurons in other memory networks with which it's been previously linked.[4]

This is why firing neurons in the memory network holding your mother's nose will also automatically stimulate neurons in the networks holding the memory of her whole face as well as your feelings about her. In this wet, gooey, highly interconnected, and complex system, where each of your 100 billion neurons is directly or indirectly connected to every other neuron and where firing up one neuron in a memory network automatically fires up neurons in other memory networks, associations become the rule, not the exception.

Your Concepts of the World are Formed through Association

The concept networks we looked at in Chapter 14 are themselves made up of associations. To get an idea of how a single concept network is composed of many different associations in your brain, think of something simple like a lemon. It may fascinate you to know that the conceptual memory you hold of a lemon isn't held in one isolated place; instead, as a survival advantage (so that damage to one brain area won't wipe out complete memories), the lemon is represented by associations between millions upon millions of neurons that are widely distributed throughout your brain.

When you think of a lemon, neurons that hold the round shape are activated in your parietal lobes located just above and behind your temples,

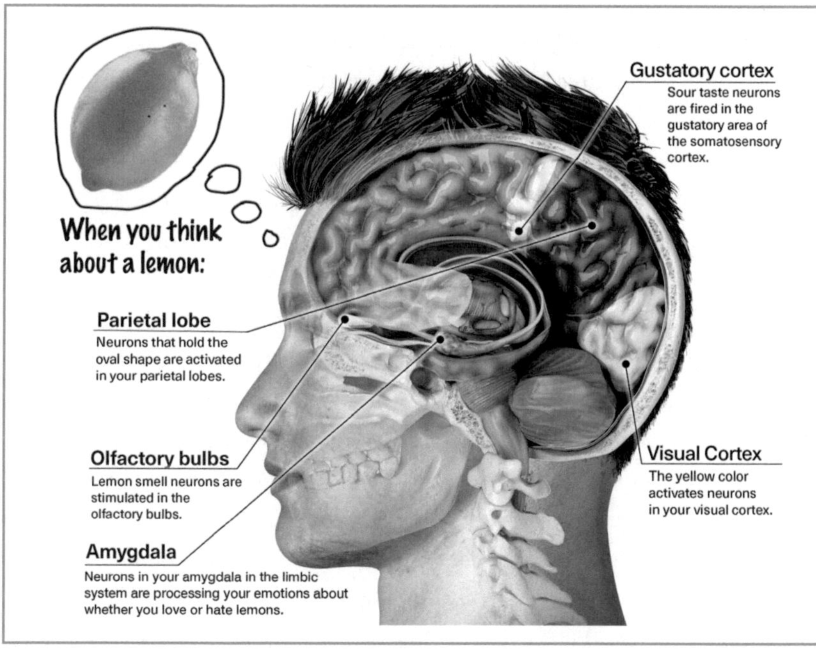

the yellow color activates neurons in your visual cortex at the back of your brain, while at the same time sour taste neurons are fired in the gustatory area of the somatosensory cortex in the upper middle of your cortex. Lemon smell neurons are stimulated in the olfactory bulbs, one on either side of the bottom surface of your brain. While all this sensory association is going on, neurons in your amygdala in the limbic system of your brain and neurons in your prefrontal lobe are processing your emotions about whether you love or hate lemons. As science writers Peter Coveney and Roger Highfield write in *Frontiers of Complexity:* "Every neuronal map, every part of the brain is dynamically… connected with every other, evolving and integrating itself in continuous cross talk."[5]

The key to efficient concept activation is the simultaneous firing of neurons. When you think of a lemon, if all the neurons in the temporal lobe, partial lobe, visual cortex, somatosensory cortex, olfactory bulbs, amygdala, and prefrontal lobe don't fire simultaneously, you'll fail to access the complete conceptual memory of a lemon. In teasing out the best way to create the neurological change that leads to learning, it's essential to understand we not only build memories through the simultaneous firing of diverse neural groups; we also access memories the same way.

Simultaneous Firing Creates Associations

In the 1940s, soon after recognizing that memory is held not in a single cell but in groups of "cell assemblies" Hebb realized that associative learning can be created only when large groups of neural networks are fired at the same time. Modern research has confirmed Hebb's findings, and today the idea that associations are made, learning is created, and memory is cemented when large groups of neurons fire simultaneously in your brain is often called the *Hebbian learning rule.* When neurons fire in unison, memory is enhanced because the possibility is increased than a single neuron will be stimulated at more than one location on its dendrite branches. As you discovered in previous chapters, the memory process commences when a receiving neuron is provoked by other neurons at more than one location. When a neuron is stimulated in this manner, magnesium plugs are blown from NMDA receptors, which in turn lead to the neurochemical–genetic conditions that produce long-term memory.

As discussed, the Hebbian learning rule is considered one of the most important findings in the field of learning neuroscience and, for simplicity's sake, is often reduced to the most repeated phrases in learning neuroscience: *neurons that wire together fire together* or, the converse, *neurons that fire together wire together.*

Grasping this concept helps us understand how new memories can be easily formed by the synchronized firing of concept networks that haven't been previously linked. According to the Hebbian learning rule, when previously independent or weakly connected concept networks are fired at the same time, they'll be linked in the phase sequence. Simultaneous firing is how a child learns to fear bees. When a child is stung, his pain, fear, and bee networks all fire in synchronicity.

The synchronized firing of previously unlinked neurons is the foundation of the conditioned response, which has fascinated behaviorists for decades. Pavlov's dogs learned to salivate at something that had no food value—that was, the ringing of a bell—because he caused the dogs' previously unconnected neural networks for bells and food to fire at the same time.[6] Therefore, once these networks were linked by simultaneous firing, every time Pavlov rang a bell, his dogs would salivate. Without knowing it, Pavlov confirmed the modern neuroscientific adage, *neurons that fire together wire together.*

Why Emotional Learning Beats Memorization: Powerful Association

It's essential to recognize the value your emotions have in promoting associations in your brain. As discussed in earlier chapters, emotions are an instrumental and powerful promoter of long-term memory formation. In general, the higher the emotional value you attach to an action or information, the greater the chance you'll have of forming long-term memory. This is because emotionality enhances the electrochemical activity and gene expression in your brain that leads to efficient learning.

Emotional intensity supports associative learning because the more heightened your love, hate, anger, fear, joy, or jealousy is, the quicker and more powerfully unconnected concept networks that are fired simultaneously will be linked. Emotions act as a supercharger, promoting memory through the strengthening of the bonds between previously unconnected concept networks. The more painful the bee sting was to you as a child, the higher the chance your bee, pain, and fear networks have of being strongly linked. Real-world learning is superior to classroom learning because it generally evokes greater levels of emotion than sitting stationary in a classroom listening to a lecture. Any learning system designed to create long-lasting associations in a learner's mind will be one that promotes evoking emotion over memorization.

Mnemonics Lack of Emotion

Hundreds of books have been written on mnemonic tricks that help us to remember information. A typical method for remembering a person named Bob is you associate him bobbing for apples. Another mnemonic technique is using an acronym, a word formed from the first letters of words you want to remember. HOMES is often used by people who want to remember the names of the Great Lakes: Huron, Ontario, Michigan, Erie, and Superior.[7]

I don't want to dismiss mnemonic techniques as useless. They do help us keep information in working memory long enough to pass tests. But because mnemonic techniques don't elicit profound emotional firing, the linking between networks is very weak, which is why you remember so little of the mnemonic information you once used to pass tests.

Can Simultaneous Firing Be a Bad Thing? Trauma!

Simultaneous firing of neural networks is a survival advantage and generally a good thing, but the linking of previously unconnected concept networks can sometimes cause great difficulty, as seen in trauma victims. Unwanted linking caused by trauma can be debilitating, as with the sexual assault victim who becomes anxious every time they see a shirt the same color as their attacker's. We also see this with the war vet who has PTSD and ducks in terror every time a door slams or a car backfires.[8]

Some of the saddest examples of how simultaneous firing of previously unconnected concept networks can cause psychological distress are the cases of children who are sexually abused by their primary caregivers. A child comes into the world with concept networks for bonding, love, and pleasure pre-encoded by their genes. When the primary caregiver is present, these concept networks fire simultaneously. In this way, bonding, love, and pleasure become linked to the person who shelters and protects the child. This is why a parent needs only enter a room to cause an infant to be flooded with good feelings and break into a grin. If a child is sexually abused by the parent, a whole set of concept networks simultaneously fire that would never usually fire in the presence of a parent.[9]

The theory is that during sexual abuse by a parental predator, the child's networks holding humiliation, sexual pain, anger, and fear are fired at the same time as genetically preset concept networks that hold love, bonding, and pleasure. Because of Hebb's learning rule, *neurons that fire together wire together*, the child matures into an adult with conflicted feelings and ideas around love and sex. An adult who has been sexually abused as a child can have a conflicted time in romantic relationships because their love, bonding, and pleasure networks are simultaneously activated with the humiliation, fear, anger, and sexual pain networks. The high level of emotion individuals experience during trauma mean it's often difficult to completely extinguish the simultaneous firing of networks provoked by these events. But by understanding the underlying mechanisms that link previously unconnected concept networks, therapists are trying out new methods to reduce the intensity and impact of these events.

The psychological school of somatic feedback therapy endeavors to reduce the link between the emotion due to a traumatic event and the event itself. This therapy focuses on reducing the feedback from your body to your brain. As you now know, the amount of emotional intensity your brain allocates to memory is based on the intensity of somatic (body) changes in galvanic skin response (sweat), breath rate, heart rate, and blood pressure. Somatic feedback therapists use forms of meditation, biofeedback, and talk therapy to purposefully lessen the intensity of the feedback from your body to your brain. The idea is that by decreasing the somatic intensity, the emotional feelings linked to the event will also be lessened. The goal is to enable patients to think of the trauma with little or no emotional linking. As Kekuni Minton, a pioneer of this style of psychotherapy, says, "If you lessen the somatic impact, traumatic emotions are unable to get the upper hand, and therefore they stop controlling our lives."[10]

How Simultaneous Firing Helps Produce Creativity and Intuition

The simultaneous firing of previously unconnected concept networks is the wellspring from which your creativity and intuition (the ability to understand something immediately, without the need for conscious

Gutenberg's "Aha" Moment!

Gutenberg's unconscious mind took three existing concepts and associated them in new ways to invent the first printing press.

reasoning) flow. Both arise from new associations between neural networks that are already in place in your brain but haven't yet been connected. While we often consider associative thinking as being triggered by external stimuli, such as bee stings and Pavlov's bell, associative thinking also takes place in the absence of external input. As Gerald Edelman notes, "Unlike the brain areas mediating perceptions, those mediating concepts must be able to operate without immediate input."[11]

Creativity and intuition can arise from random, chance associations between pre-existing neural groups or from more orchestrated efforts. As you will see in future chapters, relaxed states are so valuable to creativity and intuition because they increase the neurotransmitter acetylcholine, which enhances random connections between your brain cells. You can also use your intentions to help you make mental leaps to solve problems. As Edelman comments, "Finally, because concept formation is based on the central triad of perceptual categorization, memory, and learning, it is, by its very nature, intentional." Many books on creativity point to how you can use suggestions to set up the mental framework that allows your brain to make new associations to solve a problem.

Gutenberg Had a Plan!

Let's take a quick look at one of the most important inventions in modern times, the printing press, to see how Johannes Gutenberg used intention to link his previously unconnected concept networks to develop the printing press. Around 1440, Gutenberg was presented with a problem by his neighbors, who wanted to make copies of the Bible on a pilgrimage. He was convinced that if he put his intention to the problem of how to quickly make multiple copies of inexpensive books, he could come up with a creative solution.

Gutenberg's unconscious mind took three existing concepts and associated them in new ways. He had seen playing cards being stamped on paper by inked wooden blocks, had watched grapes being squashed by a press, and had seen coins being minted by pouring metal into molds. After weeks of allowing his brain down-time to sort out the problem, he had a wonderful "aha" moment, and the simultaneous firing of these previously unconnected concept networks led to the world's the first movable type press.

You can feel his excitement in connecting these different concepts in his quote:

It is a press, certainly, but a press from which shall flow in inexhaustible streams. Through it, God will spread His Word. A spring of truth shall flow from it: like a new star, it shall scatter the darkness of ignorance, and cause a light heretofore unknown to shine amongst men.[12]

Metaphor: A Key to Accelerating Learning

As I touched on earlier, the simultaneous firing of previously unconnected concept networks is how your mind creates and understands metaphors. The word comes from the Greek word *metapherien,* which means "to transfer," and indeed, that's what metaphoric teaching does. Metaphors, analogies, similes, and parables are some of the most powerful tools that a holder of knowledge can use to transfer it to someone else. They all accelerate learning by allowing your brain to assemble various concept networks and connect them up under one new banner.[13]

Metaphors and similes have interested philosophers and great thinkers for centuries. Aristotle singled out the ability to create metaphors as the mark of a genius.[14] Great spiritual teachers such as Buddha ("I shall show you how the Dhamma is similar to a raft"), and Jesus ("I am the bread of life") used metaphors and similes in their teachings to speed up understanding among their disciples.[15] Immanuel Kant noted that metaphor makes up the conceptual spectacles through which we view our world.[16] Heinz Pagels wrote in *The Dreams of Reason,* "A good simulation, be it a religious myth or scientific theory, gives us a sense of mastery over our experience."[17] In his book *The Cerebral Code*, neuroscientist William Calvin notes that metaphoric teaching accelerates the association of information between diverse neural groups. "Metaphors and analogical reasoning are the central means by which we project structure across levels."[18]

For example, if you're trying to teach the Bohr Model of atomic structure to someone who already understands the concept of the solar system, all you have to do to is use metaphoric teaching to explain the nucleus of the atom is like the sun at the center, with the electrons like the planets orbiting it. This metaphor simultaneously fires up pre-existing networks that include concepts for mass, center, orbit, and attraction. The learner's brain then can naturally and effortlessly attach the new details of atoms to these existing networks previously held for the solar system.

Association can forever alter the view we hold of a concept. I had a friend who believed that alcoholism was a freewill choice. When he found out his son was an alcoholic, he admonished him for not having the fortitude or willpower to stop drinking. Their relationship became estranged until I had my friend watch a video that made the analogy between alcoholism and disease, pointing out that, like diabetes, alcoholism isn't something you choose.[19] (Search YouTube to see "Cracking the Addiction Code" an update we did to this disease model). This man then found that the "cure" for his son could come from the treatment proposed by Alcoholics Anonymous. Once the father was able to shed his prejudiced thought association, he was able to get his son the help he needed. The piggyback effect of metaphoric teaching is so powerful that Howard Gardner once considered making metaphoric capacity one of his multiple intelligences (see Chapter 10).[20]

Teaching with an emphasis on metaphor has shown very positive results. A school program in Lawrence, Massachusetts, implemented metaphoric thinking and teaching programs. One year after the program started, first graders showed a 363 percent increase in knowledge of letters and sounds, a 286 percent increase in oral comprehension, and a 1,038 percent increase in word reading.[21] Renate and Geoffrey Caine note, "Metaphors are intrinsic to the construction of new knowledge and are at the heart of the acquisition of felt meaning."[22] They go on to say,

> *Metaphors bring with them pre-established sets of relationships as well as positive emotional experiences and rich sensory memories in which new knowledge can be embedded. They, therefore, become means to coherently engage all the systems of the brain/mind.*

Waldorf Schools are probably the only educational system that makes metaphoric teaching an integral part of the curriculum. Metaphors enliven the teaching of the alphabet by attaching a parable to each letter, which helps connect each symbol to the next, forming a profoundly rich basis of symbolic language in the mind of the young first grader. They bring the alphabet to life by telling a story where *M* is for the mountains the *P* (for *prince*) has to climb. There's anecdotal evidence that this form of teaching translates to a more lavish writing style in the middle grades, high school, and college. Metaphoric teaching is also a significant element of the *Wrapping* and *Hot Network* processes discussed in Chapter 13. At The

Institute, as a take-off from *Star Wars,* we often joke to new instructors, "May the metaphors be with you."

Brain-Based View of Association

Educator Pat Nolan, director of integrated studies at Freyberg High, laments the lack of association between subjects in schools. The old method, he says, is too separate from the real world, limiting the brain's opportunity to make rich associations.[23] Whole-brain learning pioneer Tony Buzan points out in *The Mind Map Book* that the linear way we take notes may even be a culprit in limiting our brains' ability to see the big picture and make associations:

> *By its very nature, the linear presentation of standard notes prevents the brain from making associations, thus counteracting creativity and memory. Also, especially when faced with list-style notes, the brain constantly has the sense that it has "come to an end' or 'finished.' This false sense of completion acts almost like a mental narcotic, slowing and stifling our thought process.*[24]

Leslie Hart points out that evolution has built a brain that prefers association. "The human brain," he says, "is not organized or designed for linear, one-path thought… The brain operates by simultaneously going down many paths."[25]

Connecting the curriculum so that associations between subjects can easily be made need only take reframing how we present those subjects. For instance, algebra, which for many learners becomes an extended effort in the painful memorization of sterile, isolated equations, could be more effectively taught if it were associated with its fertile and colorful history, purpose, and even geography. For example, we could make the teaching of algebra richer by asking and answering questions like these:

- Why was algebra created?
- What was Descartes' role in helping create algebra?
- What math concept did Descartes dream of while sleeping in a furnace? What war was going on at the time?
- How did Sir Isaac Newton use algebra as the foundation for his theories of force and movement?

- What role does algebra have in helping build the cell phone in your pocket?
- How can you use algebra this week to enrich your life?

Not only do fertile associations simultaneously fire concept networks and make the subject come alive, but they also activate large numbers of concept networks, creating more neural hooks for details to physically attach themselves to (see Chapter 14 for details).

Real-World Environments Support Associative Learning

Moreover, one of the reasons the real world is a more powerful promoter of your learning than sitting in a classroom listening to a lecture is that experience exposes your brain to millions of bits of auditory, visual, tactile, smell, and taste information every second. This sensory bombardment produces the massive simultaneous firing of neurons, which naturally supports associative learning.

Evidence suggests if learning can't take place in real environments, then learning should be orchestrated so rich associations between diverse subjects can be made. Master's Academy in Canada, which embeds the science of The Learning Code into its curriculum, relies heavily on bringing as much real-world experience into the classroom as possible to help children accelerate understanding. They point to this as one reason their standardized test scores in the excellent category are over 400 percent higher than the province average.

Many brain-based educators fear our conveyer-belt model of education, which emphasizes detail acquisition by isolating each subject from the next, breeds brains that are inhibited from fully developing their capacity to quickly and easily make neural associations. By building learning systems based solely on getting people to pass standardized tests, we may be limiting the brain's capacity to develop efficient associative-simultaneous neural firing capacity. Thus, we end up building brains that are less adaptive because they have limited creativity, intuition, and memory powers. In the following chapters, you will continue to see how your brain highly values associations in creating and cementing memory.

Cracking the Learning Code

Your brain can't think in isolation. One of the fastest ways for your brain to learn and remember new information is by making associations. By not understanding the science of associative learning and how its produced, the whole process of creating memory formation becomes much more tedious, exhausting, and inefficient than it need be.

23 Why Real-World Experience Beats Classroom Learning Every Time

You don't learn to walk by following rules. You learn by doing, and by falling over.

—Richard Branson, businessman and entrepreneur[1]

One of the quickest and easiest ways to access your Learning Code and create profound memory formation and behavioral change is through physical experiences in real-world environments. Ever since life first appeared on Earth, evolution has used physical experience in real-world environments, not words, as the primary basis of change in genes and neural tissue—and it still does. This necessitates a discussion on how new research is forcing scientifically focused trainers, therapists, and educators to rethink learning and behavioral-change systems built primarily on linguistic presentation and interaction.

For ages, humans have intuitively known that experience is the best teacher. The Confucian scholar, Xunzi (340–245 BC), was reported to have said, "What I hear, I forget. What I see, I remember. What I do, I understand."[2] From Lewis Carroll's *Alice's Adventures in Wonderland* (1865), "'Why,' said the Dodo, 'the best way to explain it is to do it.'"[3] In the early 1900s, Albert Einstein is reported to have said, "Learning is an experience, everything else is just information."[4]

We now have enough biological, genetic, and neuroscientific information to understand why experiential learning is a much more powerful promoter of your long-term memory and behavioral change than classroom, training-room, or online learning. But you probably don't need science to tell you this. All you have to do is consider your own history with formal linguistic learning systems devoid of real-world experience. Consider that to get from kindergarten through twelfth grade, you had to pass through thirteen years of approximately 13,000 forty-five-minute lecture segments, put in at least another 2,100 hours of memorization, and study for about 30,000 test questions, of which today you remember very little.

Words Seldom Give Us the Results We Desire

We've probably all experienced trying to get people to change their beliefs or behaviors through words alone. We endlessly lecture our offspring to turn off lights, close doors, wipe feet, brush teeth, and clean up their rooms—with limited success. Spouses lecture each other about similar topics with little satisfaction. Because I'm a disaster in the kitchen, my wife still can't talk me into cooking a tasty meal or putting the plates into the dishwasher the way she wants. On the floors of Congress, Democrats try to talk, berate and cajole Republicans into changing their minds and vice versa, and we all know how well that works! Conservative and liberal TV talk shows spew words over the air to get people to change sides, but what does the research indicate? These shows only reinforce the existing biases of their listeners! And in the workplace, bosses regularly lecture their employees about how to do their jobs and treat their co-workers but end up frustrated because of the limited impact of their words.

In corporate training and classrooms, we try to pour words into peoples' heads like water into jugs. The only difference is that most of the words leak out, never to be remembered. Your brain is a complex biological organ, not a jug. Agents of brain-based change are beginning to recognize the dramatic lack of neurological efficiency of educational and behavioral change systems based solely on a linguistic approach. Merlin Wittrock, a professor of education and psychology at the University of California, notes, "The brain does not usually learn in the sense of accepting or recording information from teachers. The brain is not a passive consumer of information."[5] To the brain, a lecture, blackboard, or book is not reality. Author of *The Brain Book*, Peter Russell, says, "Without direct experience, education can become dry, meaningless, and boring."[6] Author and education researcher Robert Sylwester says it best when he points out that to the brain, explaining "a smell is not as good as actually smelling it."[7]

Linguistic information alone fails to help learners connect the information to real-world experience. When we rely solely on linguistic information, we fail to activate the appropriate brain centers that allow facts to be easily remembered and efficiently used.

Educational pioneer John Dewey was one of the first to recognize that our educational systems would fail if they tried to create learning by locking students away from real-world physical experiences. In 1902, he wrote:

> *From the standpoint of the child, the great waste in the school comes from his inability to utilize the experiences he gets outside the school in any complete and free way within the school itself; … That is the isolation of the school—its isolation from life.*[8]

Consider that by the time students are taught parallel lines in geometry class, they've encountered thousands upon thousands of parallel lines in fences, table legs, door jambs, and picture frames. Yet in classrooms, our educational dogma dictates that a teacher draw parallel lines on the board and supply a written definition as if the students have had no real exposure to them. The relationships between real-world experience and school learning are sadly seldom made.

In the end, we get learners to regurgitate the facts, handing out As, Bs, and Cs that indicate they understand the information. But because linguistic information alone fails to help learners connect the information to the context of real-world experience, we fail to activate the appropriate brain centers that ensure the facts can be remembered and efficiently used.

Experience Increases Your Adaptability/Intelligence Factor

I'll never forget how forlorn I felt after passing a college economics class taught by a visiting University of Chicago economics guru, because I had no clue how any of the information in the class or on the test applied to the start-up lawn care business my football buddies and I ran. Today's students can succeed in math courses because they can memorize equations, but these same students have difficulty making change at the checkout counters of their local stores. The experiential context makes a huge difference in accelerating the learning process, yet most educational systems fail to provide it.

Evolution has designed real-world experience, not words, as the chief architect of your brain. While your genes direct synapses, axons, and dendrites only to their approximate locations in your brain, it's your experience in the real world that shapes and customizes these neural circuits so that you can increase your *Individual Adaptability/Intelligence Factor* (see Chapter 5). As Nobel Laureate Francis Crick writes:

> *It suffices to say that genes… appear to lay down the broad structure of the nervous system, but that experience is needed to tune up and refine the many details of its structure.*[9]

Renowned neuroscientist Robert Ornstein writes, "Our environment and actions shape our brain's internal connections."[10] Pediatric neurobiologist Harry Chugani of Wayne State University declares that experiences at the right time are so powerful that "they can completely change the way a person turns out."[11] Joseph Chilton Pearce estimates the stand-and-deliver, lecture, and test format is so unnatural to the brain that the average learner remembers only 3 percent of a forty-five-minutes class.[12] Leslie Hart writes, "Brain-compatible learning takes place in a free setting. The core of natural learning is the desire to better understand how the world operates."[13] He goes on to say, "Today's students are starved for exposure to reality," and concludes, "Schools have failed to bring reality inside their walls." To the brain, the chalkboard is not reality. Novelist Grant Allen, to vent his

"Brain-compatible learning takes place in a free setting. The core of natural learning is the desire to better understand how the world operates."
—Robert Ornstein, renowned neuroscientist

frustration over how far education was from the real world, had a character state, "No schooling was allowed to interfere with my education."[14]

Six Reasons Experiential Learning Beats Classroom Learning

There are six main reasons your physical experience in rich, natural environments increases the opportunity for long-term memory formation and behavioral change more than classroom instruction:

1. **The need for environmental complexity:** From complexity science, chaos theory, and evolutionary biology, we now know that your brain needs large amounts of environmental complexity to bring about order. One way to look at complexity is as the fuel your brain uses to reach higher states of order. Without this fuel, the whole learning process is compromised. What this means is that the traditional classroom way of feeding the brain small amounts of linguistic information in confined classrooms that have limited environmental complexity is flawed as the primary method upon which to base learning systems.
2. **Immature language centers:** From the viewpoint of human evolution, your brain centers that support linguistic processing are very immature. Consider the foundations of the biological structures allowing you to learn through experience are over 65 million years old, while your brain centers for sophisticated language learning are estimated to be only about 50,000 years old.[15] Research indicates that because these language centers have been present for only 0.008 percent of the time humans have been on Earth, they lack the maturity and robustness to allow us to learn solely through language.[16]
3. **Massive neural firing:** Physical experience fires vast amounts of neurons and their connections in your brain—about 95 percent. Conversely, the verbal presentation alone can fire only about 5 to 20 percent of your neurons. As you've seen, the more neurons firing in your brain, the greater the learning, per the Hebbian learning rule.
4. ***Meaning Network* activation:** Physical experience can dramatically enhance your brain's capacity to activate your *Meaning Network* (Chapters 12 to 14) and increase working memory formation. It does this by stimulating the all-important monoamine neurotransmitter system (dopamine, serotonin, and norepinephrine) by up to 500

percent more compared to sitting in a classroom and listening to a lecture.

5. **Stages of *The Cycle of Transformation*:** Memory is not a single event but a process that can be broken down into four distinct stages we call *The Cycle of Transformation* (which we'll explore in more depth in Chapters 30 to 36). Experiential learning requires: 1) Information, 2) Action, 3) Feedback, and 4) Incubation.[17] Each stage has its own unique neurochemical and neurostructural signature that must be activated to produce efficient learning. To the detriment of the learner, classroom learning focuses on only one stage of the learning process: the information acquisition stage.
6. **Spatial maps:** As you move through time and space, complex real-world experience builds large three-dimensional maps that provide hardy neurological structures upon which new information can easily attach itself. To its detriment, classroom learning can't help the learner build these substantial spatial maps.

The following chapters will delve deeper into why experiential learning beats word learning every time

Cracking the Learning Code

Linguistic learning works when the knowledge delivered is highly meaningful to you. But from a neurobiological standpoint, compared to real-world experience, linguistic learning pales in its ability to activate the brain structures, neurochemistry, and gene expression that help you easily cement learning and transform your behaviors.

24

The Environment: All-Powerful in Your Learning Process

There was a child went forth every day,
And the first object he looked upon and received with wonder, pity, love, or dread, that object he became,
And that object became part of him for the day, or a certain part of the day, or for many years, or stretching cycles of years.

—Walt Whitman, poet[1]

The environments in which you live, work, and play are all-powerful in dictating what you'll learn, what you'll remember, and how you'll act. As discussed (Chapter 11) your brain learns through selection, not instruction, so as far as your biology is concerned, the environment is the only wellspring it can select from to build and shape the neurological structures and chemical systems that determine what you'll remember and how you'll behave. Research also reveals the phenomenon that when your environment changes, so do your biology and brain structures, and therefore, your behaviors. Just as wildebeests on the African savanna think, feel, and act differently from wildebeests in a zoo, so we think, feel, and act differently in different environments.

A vital element of The Learning Code is that when you change an organism's environment, its biology changes, by selecting in new information so its thoughts, feelings, and behaviors can better "fit" that new environment. Without a deep understanding of how the environment impacts your biology and how it can increase or decrease your *Individual Adaptability/Intelligence Factor*, scholastic and corporate educational systems will continue to create learning environments that produce less-than-desired results.

The Environment Approach vs. The Reward and Punishment Approach

At the Institute, we have an oft-repeated saying: "The environment is all-powerful in dictating learning and behavioral change." This keeps our attention on what we call the *Environment Approach* to creating profound learning and behavioral change in cultures.

Before we go deeper into this chapter, in transforming behaviors and cultures, it's important to distinguish between the *Reward and Punishment Approach* and the *Environment Approach.* Why is this important? Because most scholastic and corporate institutions focus heavily on the carrot-and-stick approach, as we saw in Chapter 18. To get people to better fit the desires of the organization, bosses, teachers, and trainers apply pressure directly on workers and students from the outside to force them to change how they're thinking inside their own heads. As you will see, the *Environmental Approach* takes a more holistic approach, looking not at just the pressure authority puts on employees and students but at the impact the whole culture, and all its elements, has on the biology of those within the system.

The *Reward and Punishment Approach* forces organizations to narrow their focus, wrongly concentrating mainly on developing force-based strategies and methods that educators and bosses can use on individuals, so they follow directions, work harder, study more, and motivate themselves better. As you saw, it's more of a Newtonian physics "force makes things happen" methodology. This reflects an organizational belief that if someone doesn't conform to the school or work environment, something is wrong with that person's thinking. To solve this problem, authority uses its strength to put more pressure on that person from the outside to change the way he or she thinks inside—not an easy or pleasant task, as we've all experienced. This tendency forces scholastic and corporate organizations to blame the people—that is, teachers, administrators, students, bosses, and employees—instead of the organization's reward and punishment-based system. This is one of the greatest obstacles to creating profound personal and organizational learning.

The typical force-driven *Reward and Punishment Approach* ends up doing something else: it pits management/administration against employees/students. If you've ever been in school or worked in an organization, you've seen this conflict at work. Management/administration tends to focus on its people for the cause of its failures and doesn't look at the present culture as the real cause of its dysfunction.

The Institute has worked with management teams that tried for years to get their workers and students to change and learn faster with the *Reward and Punishment Approach* only to see little or no increase in learning or behavioral change. Yet, when management recognizes the impact the multifaceted organizational environment is having on people's biology and starts to focus on changing its culture by taking the *Environment Approach*, people change much more naturally and with much less effort and blame. Understanding this *Environment Approach* gives management a broader perspective, which allows them to see more clearly what needs to happen in their culture beyond reward and punishment in order to create transformation.

Going with the Biological Flow

It could be said the most critical stretch of base pairs (A, C, T, and G) in any animal or human's DNA is the stretch that directs neural connections to adapt to their environment. We're not born Christian, Jewish, Buddhist, or Muslim—or Mexican, Polish, Saudi, or American. Instead, we're born as universal blobs of protoplasm, citizens of the world with no religious or national affiliation. What we *are* born with is the drive to fit our world. If that world has stimuli that favor one religion or nationality, then it tends to be what and who we become.

Your genetic plan has built a brain preset to adapt itself to any location in the world. For instance, immediately out of the womb, you can babble every sound found in every language in the world, including the tongue clicks found in some African languages. But your brain's drive to fit your environment forces your neurological networks to select the most favorable language circuits so that you end up speaking fluently only the language or languages that will help you best survive and thrive in the environments presented to you.

This drive to fit your world, neuroscientist Edelman observed, effectively "carves out" your brain structures.[2] This carving process occurs not just when you are a child but also when you are an adult. For example, if you worked at buttoned-down IBM for twenty years and you suddenly found yourself working in the very unbuttoned Twitter culture, within days to weeks, your biology would naturally select information from this new environment that would naturally lead you to learn, think, and act very differently than how you did at IBM. The same thing would happen if you magically woke up tomorrow, not in your western home, but instead in the Syrian Desert surrounded Bedouin tribesmen. You would naturally and immediately start changing your behaviors in order to better fit in this new environment. This is your biology doing what your biology naturally does: automatically changing itself to better adapt to the environment presented to it.

One reason cults can be so dangerous is they bombard the brain with limited, specific stimuli forcing it to select only this "cult" driven knowledge. Often, if people associate with a cult long enough, even those who have no intention of joining can find that their brains have adapted to that environment in a way that makes them members.

Why Thirty-Day Treatment Centers Have Such High Relapse Rates

Another example of the power of the environment in action is found in thirty-day residential treatment programs for drug and alcohol abuse. These programs work very well for substance-dependent people while they're in the positive healing culture, and relapse is estimated to be only around 2 to 5 percent. Yet after the addicts reenter their old cultures, relapse rises to an estimated 90 percent or more.[3] Why this huge difference? Because different environments create different behaviors. The treatment center environment keeps people's brains on track. After people with addictions go home, their environment—family, friends, and work surroundings—reactivate their old addiction networks, and their biology quickly drags them back to their old behaviors, no matter how hard they use willpower to try not to relapse.

You Are Born Incomplete, Your Environment Completes You

Remember that 6,000+ genes are involved in building and directing your brain functions. This limited number of genes can't possibly code for the flood of complexity the world will throw at you during your lifetime. Nature recognized that the only way to deal with the vast complexity of an ever-changing world was to build a brain that had three fundamental elements pre-encoded by our genes. First, your brain had to have an insatiable drive to adapt itself to its environment. Second, it needed its major structures to be incomplete at birth, and third, those structures needed to flexible throughout your life so that rapid adaptation could easily take place.[4]

It is vital to understand that the quality of the input to your brain from your environment has a dramatic impact on the quality of the output. If you live in an abusive, uncaring, or war-torn environment, your biological structures are continually altering their shapes to better fit those situations, and therefore give rise to you becoming more abusive, uncaring, or prone to war. Think Syria.

If you live in a nurturing, caring, and peaceful culture, your biological structures will alter their shapes to adapt better, and you'll have tendencies to become nurturing, caring, and peaceful. Think of children raised in monasteries, like the Dalai Lama. Understanding that the type and quality of the input to the brain determines the type and quality of its output gives us a new perspective on traditional education.

The brain's drive to carve out its structures to fit the environment provides reasons for hanging out in an environment with people who have what you desire most in your life. If you want to make a lot of money, socialize with those who are rich, and if you want to be more spiritual, start visiting churches and synagogues and hanging with meditators. This dramatic biological drive to fit our environment gives us a fresh viewpoint on how, when we spend twelve to sixteen years in academic environments, our brains may be a better fit to succeed in those classroom environments than the complex real world.

Neural Flexibility: Increasing Your Individual Adaptability/ Intelligence Factor

As touched on, your brain's natural drive to fit its world is made easier because your major brain structures (especially in the neocortex) remain highly flexible throughout your life. This allows them to be shaped over and over again to fit changing environments. While your genes lay down the general neurological plan, providing rough emotional, motor, spatial, tactile, auditory, visual, taste, and smell networks, the fit of these networks is improved through your interactions with the present environment.[5] This setup allows your brain to quickly select new wiring which can rapidly and effectively adapt to different situations presented to you. You don't come into the world knowing how to drive a car or write a term paper, but the rough networks are in place for new information to be quickly selected from your environment.

As neuroscientist Antonio Damasio points out:

> *As far as evolutionarily modern brain sectors are concerned, the genome [makeup of all genes] helps set a general rather than a precise arrangement of systems and circuits. And how does the precise arrangement come about? It comes under the influence of environmental circumstances.*[6]

You rebirth yourself daily. Whenever environmental feedback informs us that our actions aren't supporting our survival (remember the chickens pecking at the poisonous pellets in Chapter 6), our neurological structures have the capacity to alter their shape to better fit our world. We see this flexible and adaptable nature of the human brain every time a smoker quits, a bachelor turns into a dedicated husband, or a child burns himself on a stove and forever avoids touching hot appliances.

The drive of your flexible neurological structures to fit ever-changing environments is continually reshaping who you are. If this were not so, you wouldn't be able to change your mind, and instead, you'd be trapped in the world of who you were. A large, complex, and flexible neocortex allows us to continually conform to the world through hundreds if not thousands of daily and weekly micro adaptations. This dramatic flexibility means that even if you feel your family, educational and past environment experiences didn't support you to succeed in life the way you wanted it to, you have a tremendous asset on your side: your brain's profound capacity to quickly create new networks that better fit the world you live in right now. As Gabriel García Márquez says, "Human beings are not born once and for all on the day their mothers give birth to them, but that life obliges them over and over again to give birth to themselves."[7]

Location is Everything to Your Brain Cells

In the real estate business, an adage states that the three most important things are "location, location, location." The same could be said for the brain business. Your location in the world dictates what form your brain will take and, therefore, what behaviors you'll be most apt to engage in. The more positive your environment, the more favorable your learning, actions, and behaviors, and the more negative your environment, the more negative your learning, actions, and behaviors.

You can think of neurons as tiny switches that do nothing unless stimulated by environmental feedback. If your eyes know only darkness, the occipital lobe won't connect for clear vision. If your ears experience only silence, the auditory cortex can't wire to hear sounds correctly. The environment is the fountainhead that your biology drinks from to maintain life. Your cells and neural connections that strike a match with the environment thrive; those that do not fall into disuse, atrophy and die.[8]

The process of your cells altering their structures and functions to better fit the environment commences early in your developmental cycle. Four days after your dad's sperm enters your mom's egg, a clump of thirty-two cells begins the remarkable process called *differentiation.* At this stage of development, the embryo is called a *blastocyst.*[9] In the first example of the importance of location to cell function, the majority of the cells that are located on the outside layer of the blastocyst will become the placenta. Only three to five cells positioned deep inside the blastocyst are in the correct location to allow them to eventually give birth to every one of the trillions of cells that make up your body and brain. What is the difference between the cells on the outside layer of the blastocyst, which are later discarded with the placenta, and the cells deep inside, which give birth to every one of your cells? Nothing… It's only their location that determines their function.

Twenty-three days after conception, the construction of your brain begins to take form, and we see the "location is everything" concept once again. At this time, on the back of the embryo, which is only 3mm long (about the width of a nickel), the neural tube is formed.[10] This neural tube is vital to development because from this tube, mother cells give birth to billions of daughter neurons, also called *stem cells,* which will eventually become your brain.

This process of birthing neurons is called *neurogenesis,* and it's one of the most remarkable feats of nature. For four months, neurons pour forth from the neural tube like sparks out of a Roman candle. The speed of neurogenesis is mind-boggling: cells explode from the tube at the rate of 500,000 per minute. Though some neurons are produced later in fetal life and after birth, it's during this short eighteen-week period that the basic building blocks of the most powerful computer in the world are formed and put in place.

After leaving the neural tube, neurons get to their destination in your brain in a process called migration. During this process, new neurons migrate outward. Like following spokes on a wheel, neurons move along tracks of support cells called radial glia. Just-birthed neurons hop on these tracks and follow various molecular clues to their final locations in the brain.

Location Dictates Your Neurons' Functions

Through intricate staining techniques performed in the late 1990s, Dr. Constance L. Cepko and Dr. Christopher Walsh of Harvard Medical School followed just-birthed neurons as they poured forth from the neural tube to their final locations.

They confirmed otherwise-identical neurons took on different functions simply because of their location in the brain. Cepko and Walsh found that out of the neural tube, all neurons were exactly the same, but once they were positioned in a specific neurological location, such as the olfactory lobe, the audio cortex, or the motor strip, their shapes immediately changed as they took on the characteristics and functions that allowed them to serve the region of the brain in which they had settled.[11]

Adding to Cepko and Walsh's findings, studies conducted at MIT and reported by Dr. Mriganka Sur demonstrated that, once neurons are in place, it isn't merely their location but also the stimuli from the external environment that determine what function these neurons will perform. Sur and his colleagues rewired the brains of newborn ferrets so that signals from the eye would stimulate neurons in the auditory cortex. Amazingly, Sur found the ferrets could see with their auditory cortex![12] This study helped demonstrate that the environmental stimuli washing over neurons are what ultimately determines their functions. The genetic program helps direct cells to a particular location in the brain, and the site itself provides input. But as Sur found, it's mainly the input from the external environment that flips genetic switches and tells a neuron what shape to take and what functions to perform.

Superpowers for the Blind and Deaf

Magnetic resonance imaging (MRI) of visually impaired people's brains confirms the work of Sur and his colleagues. Neurons in the visual cortex of people born blind never receive visual information from the environment. But these neurons don't just sit there and do nothing. Without visual stimulation, billions of neurons in the occipital lobe (visual cortex) organize themselves to take on structures and functions that mimic those of neurons in the auditory cortex, giving blind people considerably more brainpower with which to make sound distinctions. Much like the blind comic book hero, Daredevil, blind people could be considered to have "super hearing."[13]

Imaging reveals a similar phenomenon that occurs in people born deaf: hundreds of millions of neurons in the auditory cortex shape themselves to support visual functions.[14] Controversial research even suggests that if stem cell neurons are transplanted into a brain damaged by a stroke or a disease such as Parkinson's, these as-yet uncommitted cells can develop the structure and function of that brain area solely based on cues they receive from the internal and external environments.[15] As Jon Kaas, professor of psychology at Vanderbilt University, says, "The cortex… is just waiting for signals from the environment and will wire itself up according to the input it gets."[16]

The Environment Shapes Your Synapses from Conception to Death

The concept that the environment is everything to your neural tissue is not limited to neurons. It also extends to your *synapses*—the connections between your brain cells. Although the structure and function of neurons are very flexible in your early years, they become less flexible as you mature; genetic switches have flipped that force your neurons to commit to a particular shape and purpose. The synaptic connections between neurons, however, remain incredibly fluid throughout your life.

The wonderful thing about your synapses and the dendrite branches

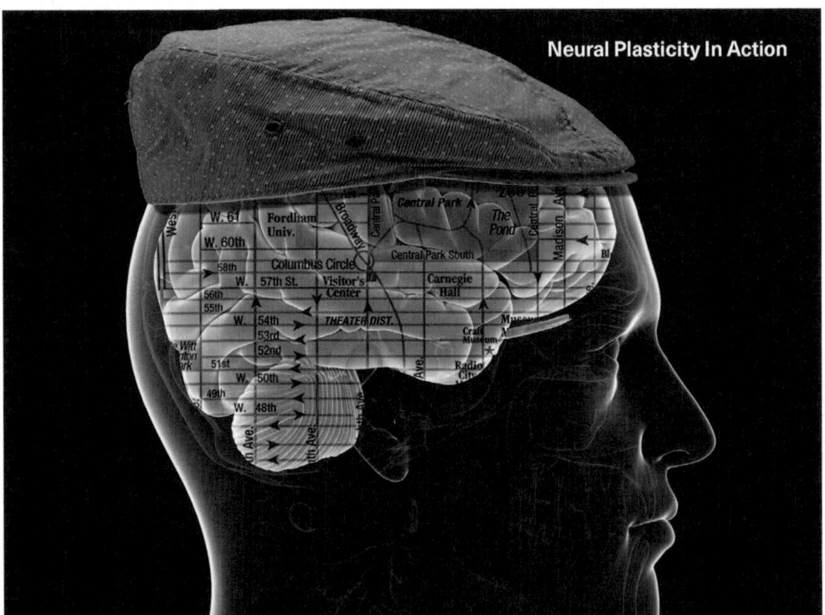

A perfect example of the brain learning from the environment is seen in the brains of foreign-born cab drivers. Regardless of where they grew up, their synapses quickly change to navigate complex new locations like Chicago or New York.

holding them is they remain highly plastic from the day you're born until the day you die. To a neuroscientist, *plastic* refers to your brain's ability to rewire its synaptic and dendritic configurations as it responds to different environmental situations over and over again.[17] Although once neurons have encountered their first bursts of environmental stimuli, the structure and function of neurons are difficult to alter, their synaptic connections can remain highly plastic for life.

It's this highly flexible change—or learning—at the level of the synapse that allows us to successfully adapt to living in New York City, even if we were born on the plains of the Serengeti or the Arctic tundra. A perfect example of neural plasticity is in the brains of foreign-born cab drivers. No matter where they grew up—say a village in India or China—their synapses quickly adapted to efficiently navigate a dramatically different location like Chicago or Los Angeles.

Creating a State of Wonder: One Million Synapses a Second

While in the womb, neurogenesis (the birthing of neurons) is a very short process, completed in four to five months, synaptogenesis (the creation of new connections between neurons) lasts for two years and eight months. Starting at seven weeks after conception and completing at two years after birth, synaptogenesis lays down trillions of synapses at a phenomenal rate of 1,000,000 neural million connections every second! This process is even more spectacular than neurogenesis, ensuring that each of a child's 100 billion neurons will, at their peak, have thick branching of dendrites with an average of 15,000 synapses[18]—about 50 percent more than adult neurons. Neurobiologist Lise Eliot, in her book *What's Going on in There?* remarks:

> *During this period, brain development is physically analogous to the growth of a new forest, when saplings branch upward and outward in their competition for sunlight. Like a rapidly thickening forest canopy, [a child's] cerebral cortex triples in thickness during his first year of life as a result of this enormous dendritic growth.*[19]

At two years of age, a child will possess more than 3,000 trillion synapses in his or her brain. Since I first heard that synapses surge forth in the child's brain at the rate of almost two million a second, every time I see a baby or toddler, I'm spellbound by the wild activity occurring just under their small skulls.

And if you've ever marveled at how kids seem to live their lives in wonder, it's because their brains have 500,000,000,000,000 more connections than adults. In the next chapter, you'll learn how as you become an adult, you lost 50 percent of your neural connections at the rate of twenty billion a day! For our discussion now, you just need to know this biological difference in the number of connections between the child and the adult brain is spectacular. It ensures that the world a child sees is very different from the world you see.

Kids have a biological drive to do the best they can every waking second to stimulate this phenomenal abundance of neural connections. Each time kids stimulate these new synapses, they get a hit of joy in neurochemical

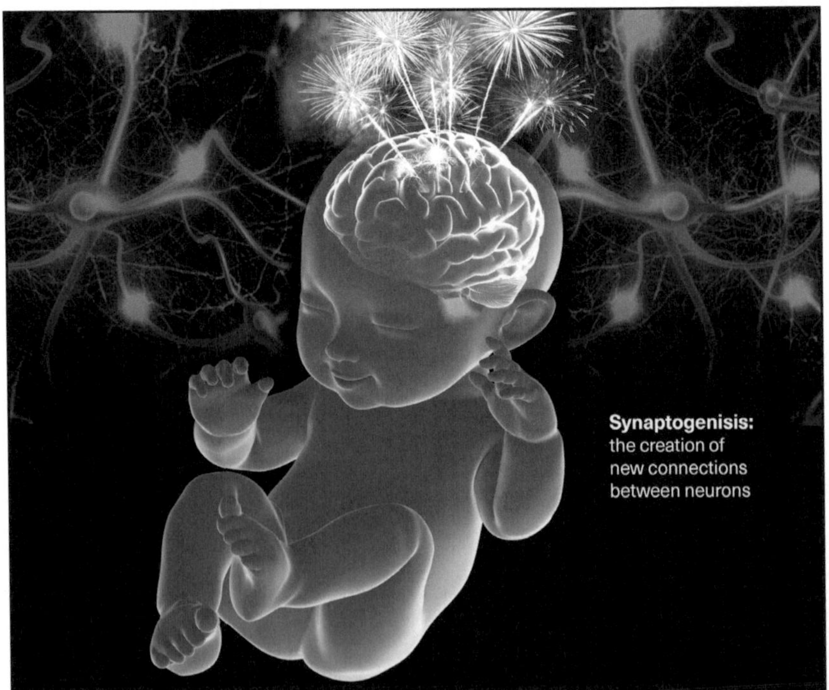

Starting at seven weeks after conception and completing at two years after birth, synaptogenesis lays down trillions of synapses at a phenomenal rate of 1.8 million neural connections every second!

form. Kids don't cross the street; they dance across the street. They don't just look at rocks, trees, and flowers; this bounty of connections drives their focus to be so intense, they appear to become one with the objects of their attention. Spend a little time watching a child, and you'll clearly see they live in a world of astonishment and discovery with almost every new object or event they encounter. They only really get upset when something like hunger, lack of sleep, or not getting what they want gets in the way of stimulating more synapses.

As adults, many of us spend our whole lives trying to get back this state of wonder (which is why drugs are a problem for so many people). Although you may not be able to match the intense wonder of childhood with its trillions of extra connections waiting to be wired up, you can come close. How? It's simple—by learning.

Research confirms that every time you learn something new, you're firing up new neural connections—not as many as when a child learns something new, but as adults, we're still adding to and adjusting our brains' wiring in a way that gives us wonderful hits of chemical pleasure. So, if you want to feel a sense of wonder about life, never stop learning!

The Richer the Environment, the Higher Your *Adaptability/Intelligence Factor*

Over the last fifty years, we've amassed a substantial body of evidence that conclusively demonstrates that new learning occurs when environmental pressures produce massive neuron firing and increase the *Adaptability/Intelligence Factor* in animals and humans.[20] Here's some more evidence about how a rich environment affects the brain:

- In the 1980s, William Greenough at the University of Illinois found animals raised with playmates, toys, and other stimuli had 25 percent more synapses and were better learners than animals deprived of stimuli.[21]
- Using imaging, researchers at Baylor College of Medicine found children who didn't get much physical play or were rarely touched developed brains 20 to 30 percent smaller than normal for their age.[22]
- Scientists at the University of California at Los Angeles found, during autopsy studies, that brains of mentally active individuals had up to 40 percent more synapses than brains of high school dropouts.[23]

- Studies show that laying down more synapses is a buffer against Alzheimer's disease, delaying its onset. Dr. Yaakov Stern, a neuropsychologist at Columbia University, found that people who had low education levels coupled with non-stimulating occupations had three times the risk of being affected by Alzheimer's of those who had stimulating occupations and more education.[24]

Having Appropriate Environmental Signals Corrects Dyslexia

Studies now show that the correct environmental stimuli can even rewire the brains of those with dyslexia so that they can read more efficiently. Typically, it takes neurons in the auditory cortex 0.015 seconds to respond to a signal, calm down, and get ready for the next signal. In some dyslexic children, this process can take five to ten times longer. Thus, short sounds like B and D get garbled in the dyslexic person's brain, and comprehension becomes inconsistent. These people then develop an inability to associate sounds with letters. Scientists take advantage of the plasticity of the human brain by having dyslexic individuals work for twenty sessions, for twenty minutes a day, with a computer that draws out short consonants like a record played too slowly. Bombarding the brain with a form of auditory stimulation different from their usual environment allowed the dyslexic brains to rewire themselves to process sounds more quickly, dramatically

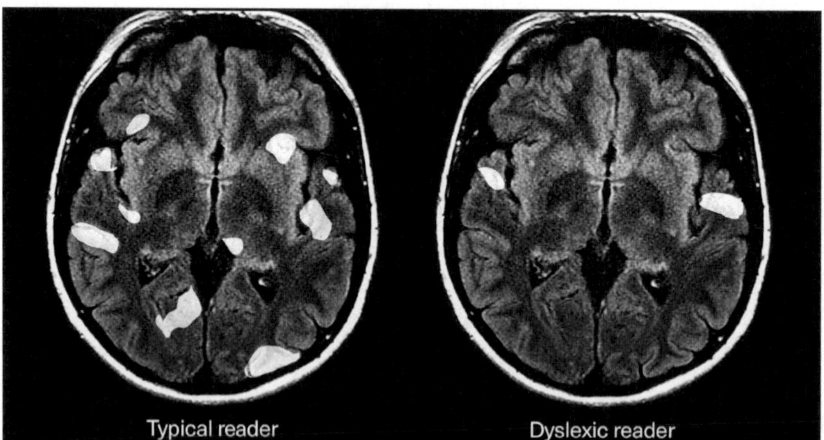

Brain activity (white regions) of a non-dyslexic person (left) and a dyslexic person (right) while reading

improving reading ability. It was found that after only four weeks, dyslexics' reading abilities could jump a full year.[25]

In 2004, Guinevere Eden, using brain imaging at the Center for Learning at Georgetown University, was one of the first to discover that after the specialized reading intervention, areas associated with efficient reading, which had previously shown up dark on scans because of inactivity, started lighting up, indicating new brain activity.[26] (The figure below shows the difference between a regular brain reading and a dyslexic brain reading.) These scientific reading programs alter environmental stimuli to the brain, redirecting energy flow, and thus creating activation in specific brain areas. This suggests that people with dyslexia are not doomed to struggle with reading their whole lives.

Changing Stimuli Reverses Brain Damage and Slows Old Age

To rewire the brain to correct disease- and trauma-induced mental defects, scientists are using all sorts of methods to bombard the brain with specific environmental stimuli. One of these methods was successfully implemented by Edward Taub, a neuroscientist at the University of Alabama at Birmingham. He worked with people who had stroke damage to the part of the brain that controlled hand and arm movements.[27] In a finding that surprised the neuroscientific community, Taub discovered that if he restricted the good arm of the patients for six hours a day, forcing them to use only the hand associated with the damaged area of the brain, the bad hand's dexterity returned, on average, to 75 percent of normal. One patient regained hand dexterity twenty-five years after a stroke! Brain imaging showed that just changing the environmental stimulus to the brain led the part of the brain that controlled the hand affected by the stroke to more than double in size. Taub's discovery offers a perfect example of Neural Darwinism in action.

Adding synaptic connections and dendritic branching also appears to have advantages other than healing and increasing our *Individual Intelligence/ Adaptability Factors*—increased longevity. A study was conducted of 678 nuns, members of the School Sisters of Notre Dame, who were born in 1916 or earlier and who all agreed to have their brains autopsied after death.[28]

David A. Snowden, who headed this ongoing study, found nuns who'd attended higher education and were more mentally active throughout their lives lived an average of four years longer than the less mentally active nuns. Not only did the less mentally active group have smaller brains at death, which he attributes to the lower numbers of synapses that were laid down during life, but they also had higher death rates at every age between twenty and ninety-six.

The Brain Can Only Fit the Environment Presented to It

The scientific evidence is conclusive: the environment physically shapes the brain. If the well is full of junk, all the brain can select is garbage; if the well is full of jewels, the brain can become dazzling. As Nobel Laureate Edelman so eloquently points out, as far as neural tissue is concerned, "Every perception is an act of creation."[29] We end up becoming who we are based on the world we live in, which is why peaceful environments tend to create peaceful people, and violent places tend to create violent people. If the brain is isolated in a segment of society (such as a ghetto, a cult, or an abusive family) and is seldom exposed to what would be considered healthy social norms, the brain will unconsciously do everything it can to alter its shape to fit those unhealthy norms.

There's substantial evidence that seems to confirm negative environments shape minds to behave in negative ways. Consider that daughters of teenage mothers have a 360 percent greater chance of also becoming teenage mothers themselves than daughters of non-teenage mothers do. We also know that 40 percent of daughters born to mothers on welfare will also become welfare moms.[30] In addition, children raised in violent homes or neighborhoods have a disheartening 50 percent greater chance of becoming a statistic in the criminal justice system than children raised in safer environments.[31]

When we firmly grasp that the environment is molding our brains and behaviors, it's easy to see why even with the most aggressive sentencing system in the Western world, the recidivism rate of prisoners in America is over 70 percent by some estimates.[32] Spending years in prison with criminals molds brains that perform criminal behaviors. If we want to change criminals' minds and behaviors, we should sentence them not to spend time in prison with the bad guys but to spend time in churches with the good guys!

If one country wants another country to adapt to its political philosophies, the best tack isn't to prohibit information exchange and communication, because embargoes only limit neurological change. The best way to create new directions in a society is to expose that country's collective mind to new environmental stimuli. It could be said if we want other countries like Iran and Cuba to be more like us, don't embargo them, "send 'em Starbucks and McDonalds."

Your Environment Affects Your Neurotransmitters

Research is also giving us insights into exactly how negative environments shape minds and drive us to behave in negative ways. It happens by affecting our neurotransmitter systems. One function of the neurotransmitter serotonin is to help impede impulsive behavior. Research has shown that individuals with low levels of serotonin have increased propensity toward violence and suicide.[33]

Markus J. P. Kruesi, chief of child and adolescent psychology at the University of Illinois Medical School, found that when children are raised in negative, stressful, and violent environments, their serotonin levels actually decrease. He states:

> *What we are beginning to conclude is that the bad environments that more and more children are being exposed to are indeed creating an epidemic of violence... Environmental events are really causing molecular changes in the brain that make people more impulsive. It is frightening to think that we may be doing some dreadful things to our children.*[34]

One function of the neurotransmitter norepinephrine is to prompt your brain and body to take immediate action in threatening environments. It's often called the brain's adrenaline. Excess levels of norepinephrine help your brain react quickly to dangers by having your senses fixate on the threat and starting the physical stress reaction that leads to increased heart rate, blood pressure, and breath rate—all necessary to deliver power to your muscles so you can take immediate action. Research shows that if the brain is exposed to threatening environments for excessively long periods, levels of norepinephrine can become permanently elevated, which in turn causes people to become more and more aggressive, often leading them to commit violent acts.[35] With norepinephrine levels perpetually set on high, these people's lives are full of hair-trigger reactions.

We see permanently high levels of norepinephrine in PTSD patients who've been exposed to extremely stressful environments. The children raised in the Branch Davidian cult that was attacked by federal agents in 1993 offer a vivid example. After the siege, these children's norepinephrine levels were so high that the kids had resting heart rates of 100 to 170 beats per minute—the average child's heart rate is eighty-four.[36] The higher norepinephrine and lower serotonin levels resulting from threatening environments create a chemical climate in which people act without thinking. Dr. Bruce Perry, who examined the Branch Davidian children and has seen the phenomenon before with inner-city Chicago children, points out that in such environments, "It is adaptive to be impulsive… If you wait, very frequently, you will be victimized. So, it's highly adaptive to be hypervigilant, to be overly reactive and impulsive, to actually act before you're acted upon."[37]

The Environment Affects Blood Flow

As you'll see in more detail in the chapters that deal with stress, negative environments can cause you to "downshift" your brain's blood flow and energy from your higher, most-evolved brain centers where compassion, empathy, and future planning reside, down to your lower brain centers in the limbic/stem system, which house the more primitive instincts of hoarding, deceiving, imitating, and identifying "your" territory. It's also been shown that negative, stressful environments often make us unable to recognize the future consequences of our present actions. The result is we end up acting more like our reactive animal cousins and less like humans.

One reason that sociopaths can perform such antisocial behaviors is that they lack access to their most-evolved upper brain areas. So, in effect, negative, stressful environments can downshift us from acting empathetically to acting sociopathically. If you've ever been cut off in traffic and felt road rage swelling up so much you wanted to humiliate or hurt the source of your anger, you've experienced this *neurological downshift*.

Negative environments are so damaging to the brain that they can also inhibit growth and prevent a brain from ever reaching its potential. A recent study of abused and neglected children revealed that certain brain regions were 30 percent smaller than those of children from healthy homes.[38] The other side of this is healthy environments help shape brain structures that produce positive outcomes. Janellen Huttenlocher, of the University of Chicago, found that two-year-olds with more talkative mothers had a 295-word vocabulary advantage over children with less loquacious moms.[39]

Spatial/3D intelligence includes the ability to find your way through your neighborhood or town, as well as being able to take pieces presented at random and put them together in a reasonable way. Such abilities underlie skills in math, engineering, architecture, and games like chess. Research shows that three-and four-year-olds who took piano lessons for six months increased their spatial abilities 36 percent above their peers who only played with computers, took singing lessons, or did nothing special.[40] In a similar finding, research on the brains of cab drivers show that their constant searching of the environment to find specific locations caused the area that deals with spatial clues to growing to meet the environmental demands.[41]

Craig Ramey at the University of Alabama developed an intervention program that worked with children of low-IQ parents. These children were divided into two groups, and one group was exposed to a dramatically richer environment than the other was. The enriched children ended up with IQ scores that were 20 percent higher than those of their impoverished peers. Another study showed that children who were adopted out of poor homes into more affluent homes had IQs up to 20 percent higher than the IQs of their biological parents. It is believed this occurred simply because the children's environments were more positive.[42]

Finally, it is also true that in the 21st century, no discussion of the environment's impact on the brain could be complete without addressing media. Today, research indicates that the average adult spends amazing eleven hours a day in front of some form of media,[43] the average child seven-and-a-half hours a day.[44] We're just now coming to grips with what happens to the brain when up to half of its environmental stimuli come from the media, which is often full of violent and sexual images. In Chapter 39, we'll look at the effects that exposure to 250,000 acts of media violence and 1,500,000 images of media sex before eighteen years have on the average human brain.

Cracking the Learning Code

The science is clear: your environment influences your neuro-chemistry and "carves out" the structures of your brain so that you best fit that environment. The longer you're in a given environment, the more impact it has on all your brain structures and behaviors.

25
Your Super Windows: How and When You Learn the Fastest

Expecting all children the same age to learn from the same materials is like expecting all children the same age to wear the same size clothing.

—Madeline Hunter, education innovator[1]

In the previous chapter, you learned how the environment is all-powerful in the learning and behavioral-change process. The type and quality of the input to your brain determine the type and quality of its output. In this chapter, you'll discover the special *super-windows of learning opportunity* in your lifetime, also called critical periods, when the speed with which you can select information into your brain structures from your environment dramatically increases. When these windows are open, and you get the correct environmental stimuli at the right times, your ability to increase your *Individual Adaptability/Intelligence Factor* is spectacularly accelerated in specific areas, such as vision, movement, language, and social communication. However, when these windows close, they close forever. Unfortunately, if you haven't received the correct environmental stimuli at the correct time before these super-learning periods shut, your ability to reach your full potential can be severely limited.

Our understanding of super windows was introduced almost a century ago by Jean Piaget, who described how the brain passes through stages of development in a specific order.[2] Today, the worlds of neuroscience and genetics have confirmed and refined Piaget's theory. We now know that a genetic time clock provides for the rapid development of specific brain areas in a precise order. We crawl before we walk and walk before we run. We babble before we speak two-word phrases and speak two-word phrases before we say whole sentences. We get binocular vision at seven months, and the kind of 20/20 vision that allows for easy word distinction and recognition at around six years old. And as Piaget pointed out, we grasp concrete concepts (at ages seven through eleven) before we can understand abstract ones (after age eleven).

Regulatory Genes Open and Close Your Super Windows

To help us quickly adapt to the world we're born into, most of these developmental super-learning periods unfold in childhood, but new neurological evidence shows that there's even a developmental stage later in life that begins to open in your fourth decade as genes lay down the neural material that gives you the capacity for wisdom.

The opposite image shows a general graph of when these *super-learning windows of opportunity* are open and when they begin to wane. There's still much ongoing research into when exactly these critical windows open and close, but we now have enough evidence to confirm that any successful learning or behavioral system must endeavor to present information to the learners' brains when these windows are open their widest.

Special *regulatory genes* drive the opening and closing of super-learning windows of opportunity. During your lifetime, specific regulatory genes are preprogrammed to tell other genes when to turn on and off. When regulatory genes turn on, genes in brain areas that deal with traits such as vision, hearing, language, movement, music, social interaction, and emotions are switched on, allowing for phenomenally rapid learning in these brain areas. When regulatory genes tell certain genes to turn off, the opportunity for accelerated learning ceases.

Throughout the time regulatory genes are performing their magic on your other genes, the activity of a whole host of neurotransmitters that support learning, such as dopamine and glutamate, are prompted to dramatically increase their activity, some by as much as 225 percent.[3] This timed increase in the activity of these learning neurochemicals facilitates the massive neural firing between cells and the blowing out of magnesium plugs from NMDA receptors, which, as you saw in Chapter 7, is instrumental in the process of creating long-term memory.

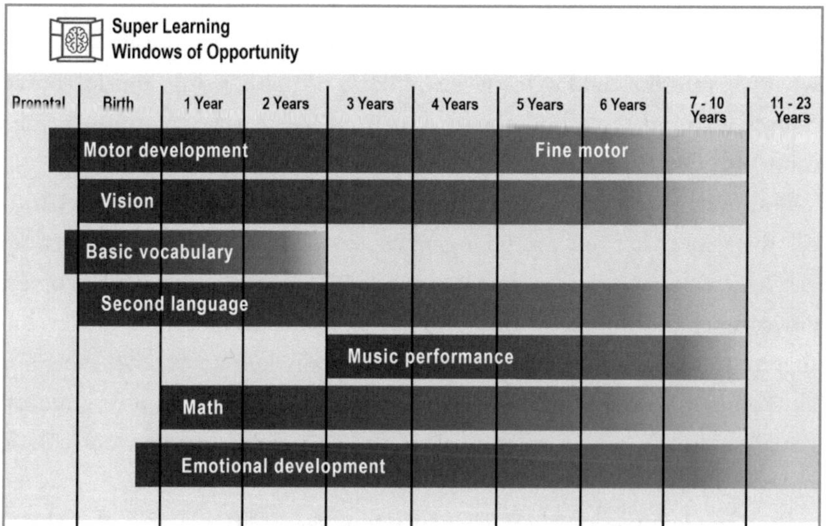

Initial research indicates that most windows of Super Learning are open in your youngest years.

When these genetic windows are open, the increased neural activity allows you to learn at rates and capacities that can never be matched after these windows shut. For example, from eighteen months to six years, your regulatory genes switched on genes in your language center. When switched on, these genes help produce so much neural activity you could select a new word into your long-term memory every two hours. This super-linguistic learning window means that by the time you were three years old, you were fluent in understanding your mother tongue, and by six years old, had amassed an amazing 13,000+word vocabulary as well as the capacity to grasp syntax and all the basic rules of grammar.[4, 5, 6]

The More Natural the Environment the Greater the Learning

The turning on of regulatory genes is a natural, automatic process that prompts your biology to acquire all this profound word capacity by six years old, whether or not you ever enter a classroom. Research has shown the more natural and real-world the environment, the higher word acquisition. This is one reason video games and computer-based learning programs aren't the best way for a child to gain an extensive vocabulary. Instead, the best way for a child to learn a large vocabulary is to live in a family that communicates often and directly with the child using an extensive vocabulary.[7] It's clear that during these super-learning windows, no matter what we want the child to learn—whether it be math, words, music, sports, or even love and empathy—the more natural and real the environment, the more effective the learning.

However, when super-learning periods close and regulatory genes turn off the genes that are producing massive neural activity, not only does accelerated learning cease, but if the brain hasn't yet received the correct environmental stimuli by this time, it loses the chance to ever develop that trait to its fullest capacity.[8] If your brain fails to receive the right environmental stimuli when super windows are open for acquiring motor, visual, auditory, language, musical, social, and emotional networks, these networks will never develop completely.

Bind our legs too long before we turn four years old, and we'll never learn to walk smoothly. Limit visual feedback from the environment before we turn four, and we'll never be able to see clearly. Limit our ability to hear language before the age of three, and we'll never fully develop the ability to speak correctly. The closing of linguistic windows early in life is why it's so hard to learn a second language as a teenager or an adult. Also, consider that if a child with a primary biological musical intelligence is never allowed to play an instrument or refine her singing abilities before he or she is eight years old, they'll never develop the ability to perceive perfect pitch.

And because the majority of super-learning windows are open in your early years, the quality of the stimuli that you receive in childhood has a more profound effect on your future actions and behaviors than at any other time in your life. The reason these windows exist in the first place is so you can quickly and successfully adapt to fit your environment (see Chapter 23). Although your brain remains malleable throughout your life, it will never be as flexible and adaptable as when these super-learning windows are wide open. This is why it can take years of psychological and spiritual work in adulthood to heal the negative networks that were laid down in childhood. Dr. Myron Hofer, a psychiatrist at the New York State Psychiatric Institute, commented on children who were raised in the war-torn Balkans: "… You may become impulsive, aggressive, and you won't want to get close to anyone. You will be beautifully suited to fight a 500-year war in Europe."[9]

> **Neuroscientists now theorize that in humans, the critical super-learning windows for the natural formation of emotional, tactile, and bonding capacities are open at their widest during the first three years of life.**

Sad Lessons from Super-Learning Windows

It's really only been in the last thirty years that science began to get a firm handle on the timeline for the super-learning windows of opportunity. Unfortunately, our scientific awareness has arisen from the experiences of people who were deprived of the correct stimulation at the correct time in their lives.

One of the first documented experiments of the effects of stimulation deprivation on young minds was conducted by the Holy Roman Emperor

Frederick II. Between crusades in the 13th century, Frederick inadvertently conducted one of the first experiments on deprivation of auditory, tactile, and emotional stimulation to children.

Frederick was obsessed with finding what language a child would speak if he or she were raised in an environment without ever hearing adults talk or sing a lullaby. For his experiment, he removed a group of babies from their secure families and handed them over to nurses who kept them isolated. The nurses attended only to their most basic needs—feeding but not holding them, bathing but not hugging them. To keep the babies away from the spoken word, the nurses were instructed to spend as little time as possible with these unfortunate infants. In his experiment, Frederick learned little about language and much about the emotional and tactile needs of children. From the lack of love and physical stimulation, all the babies died. As one 13th-century historian explained, Frederick "labored in vain… for they could not live without the petting." [10]

Throughout history, we've heard echoes of Frederick's experiment. In the 1940s, psychiatrist René Spitz conducted a study on two groups of disadvantaged babies. Spitz compared one group raised in what was then considered a perfectly adequate orphanage with another group raised in prison with their mothers. Except for brief feeding and diaper changes, the orphanage infants were kept isolated in their cribs with the sides draped with sheets to keep out infections. With little to look at, nothing to play with, and minimal human contact, these babies fared almost as badly as Frederick's children did. A large number did not survive to reach two years of age, and those who did were physically stunted, prone to infection, and cognitively and emotionally damaged.[11, 12]

Conversely, the babies raised in prison slept in a nursery and were brought each day to their mothers, who nursed them, played with them, and lavished affection upon them. In comparison to the babies raised in the orphanage, the children of imprisoned mothers developed quite normally. We now know that if infants and toddlers don't receive verbal communication, touching, and love, their abilities to ever speak, bond, or show love in a healthy manner will forever be altered. Spitz's research did a lot to help orphanages eliminate the waiting period for adoption, which at the time was thought necessary to allow a child's "natural" personality and temperament to unfold in isolation.

Give Me Love Right Now!

Over the last three decades, developmental specialists have examined a large number of infants and toddlers who, like Frederick's children, failed to receive positive emotional and tactile feedback early in their lives. In the early 1990s, the political environment in Romania changed, allowing foreigners to adopt Romanian children from orphanages. Because of the overcrowding in these orphanages, infants and toddlers were forced to sit alone in cribs for up to twenty hours a day, sucking empty bottles for comfort. These children developed the habit of rocking back and forth to soothe themselves. Unlike Frederick's and Spitz's test children, whose fate led to death, these children were spared by foreigners who wanted to save them through adoption. It was thought that by smothering these children in love, their adoptive parents could reverse any damage that had been caused by their stark lives before the adoption. Unfortunately, this wasn't so.

Dr. Elinor Ames, a psychologist who studies sensory and emotional deprivation in orphans, notes that years after these infants were rescued, they had problems with emotional processing and developed *social detachment syndrome.* Later in life, when these children were under stress, they didn't know how to reach out for love and support. Instead, they would wrap their arms around themselves and rock back and forth as they did in infancy. Since being rescued, many of these children have come a long way in their emotional healing. However, at this time, it appears that their emotional and social maturation will never be normalized because of what they missed in their infant and toddler years when the bonding window of learning was widest.

To get an idea of what was going on in the brains of children deprived of maternal love and tactile stimulation, Hofer did research at the New York Psychiatric Institute which led to the discovery of "hidden modulators" released during the development of the mother-infant relationship.[13] Hofer found that these chemical modulators, called *trophic factors* (which support the growth, survival, migration, and function of neurons and their connections), are expressed in infants' brains when the infants are loved and touched by their primary caregivers. The research also showed that when baby test animals were deprived of maternal licking and affection during these bonding windows, the babies developed fewer of the synapses that are instrumental in helping direct healthy emotional development.[14] Other studies reveal that children who don't receive appropriate touching and emotional stimulation at the proper time can have brain regions that are a third smaller than those of children raised in loving homes.[15]

Research also shows that not only do missed critical windows of bonding limit synaptic growth, they also cause disturbances in the neurotransmitters regulating emotions. The most marked effect is in the reduction of neurochemicals that stabilize the pathways from the neocortex to the more ancient limbic/stem structures. It's been found that the disruption between the neocortex and the limbic/stem brain regions increases children's aggression and reduces their abilities to appropriately deal with stressful situations. Dr. Bruce Perry, a psychiatrist at Baylor College of Medicine, has done brain scans that show that the brain region responsible for emotional tuning never develops properly in children who don't get their quota of TLC early in life.[16]

Infants and toddlers have such strong preprogramming to receive love during their early windows of emotional maturation that if they don't get enough affection, their stress levels skyrocket. Research shows that if a baby lacks a primary caregiver or if the caregiver is cold, distant, and disengaged, that infant's level of the stress hormone cortisol rises significantly.[17]

Cortisol is a stress hormone that, when you encounter danger, elevates the heart rate and respiration, dilates pupils, decreases blood flow to the organs, and increases blood flow to major muscle groups to allow you to fight or flee a stressful situation. Unfortunately, babies can't fight or flee, and the prolonged presence of a stressful situation produces an overabundance of cortisol, which not only reduces the immune response and interferes with digestion and nutrient absorption but can also cause damage to many brain areas—including the hippocampus, a center of concept formation and working memory; the amygdala, the root of our emotions; and the anterior cingulate, the part of the brain that helps regulate attention. All of these structures are instrumental in normalized emotional function and long-term memory formation. Neuroscientists now theorize that in humans, the critical super-learning windows for the natural formation of emotional, tactile, and bonding capacities are open at their widest during the first three years of life.[18]

You're Born with The Ability to Speak 6,000 Languages

The advancement of science in language acquisition is breathtaking. We've probably learned more about language acquisition in the last twenty years than in the previous 500 years. We now know that the instant an infant makes their entrance into the world from the security of the womb, they can babble all the sounds in all 6,000 human languages.[19] At around the child's first birthday, words start to emerge out of the gibberish. The child starts naming things they're most familiar with, such as *mama, milk, spoon,* and *blankie.* At eighteen months, children's language abilities explode, and most start acquiring new words at the phenomenal rate of one every two hours. For the first time, they also start to use two-word combinations such as "me milk," "all wet," and "no nap."[20]

Once kids can put two words together, it isn't long before they start generating sentences. Around twenty-four to thirty-six months of age, "me

milk" becomes "I want milk," and "no nap" becomes "I don't want a nap."[21] Incredibly, by three years of age, most children are fluent in their mother tongue, and by six years, can understand astonishing 13,000+ words and all the basic rules of grammar.[22] All this super-linguistic learning takes place simply from a selective process, without the need for any instruction whatsoever.

Rapid language acquisition is made possible as timed super-learning windows open and increase synaptic activity in specific brain regions in a back-to-front progression. Around twelve months, babies can start storing a large number of words because of the increased synaptic activity in a part of the brain called *Wernicke's area,* located in the posterior uppermost portion of the temporal lobe. Wernicke's area acts like a dictionary, allowing us not only to store words but also to understand their meanings.[23]

Sentence construction and grammar improves at around twenty-four months of age. Similar synaptic activity increases in part of the brain called *Broca's area*. This language area acts as a grammatical textbook and is located in the frontal lobe just above the fissure that separates it from the temporal cortex.[24] However, if we're not exposed to the correct language syntax before the window for Broca's area closes, we don't fully develop the ability to correctly construct sentences.

Two interesting studies reveal more about how the quality of the environment impacts language-acquisition windows. Betty Hart and Todd Risley, two Kansas City psychologists, did extensive and detailed research with forty families through the first three years of their children's language development. Each month, a member of the research team would go to each family's house and chronicle every aspect of how the parents communicated with their children. At the same time, the verbal advancement of the children was closely monitored. The survey showed that the most powerful influence on children's language development was simply the amount of time the parents spent speaking to their children and the parents' attentiveness to their children's requests for attention.[25] Those children who lived in homes where communication was the greatest had faster-growing vocabularies and higher verbal IQ scores than those whose parents spoke fewer words to them overall.

A disturbing side of Hart and Risley's report was the socioeconomic aspect. In choosing families to study, they found that the number of words spoken to a child dramatically increased as families climbed the ladder of economic and educational advantage. Welfare children heard 600 words per hour addressed to them, as compared to 1,200 words per hour for working-class families and 2,100 words per hour for children of professionals. Hart and Risley also made another disturbing finding: on average, professional parents were heard to praise their children 700 percent more than parents on welfare, and the higher income parents doled out negative feedback only half as much. Another study showed that words heard on TV did not increase a child's vocabulary.[26] To put it plainly, if we want our children to prosper, we must spend time with them, communicate with them, and support them with positive feedback when appropriate, not stick them in front of a video screen.

Missed Windows: "Wolf Children" and Genie Help Us Understand

History has also been an active teacher in helping us gain a deeper understanding of the super-learning windows of opportunity for language acquisition and social development. The first tragic examples can be found in studies of so-called "wolf children." The medical literature contains several documented cases of children who were abandoned early in their lives and raised by nature in the wild.

One well-documented case is of the "wild boy of Aveyron."[27] Just before dawn on January 9th, 1800, a remarkable boy, who appeared to be about eleven years old, emerged from the woods in southern France. As author Roger Shattuck recounted in his fascinating book *The Forbidden Experiment*:

> *No one expected him. No one recognized him. He was human in bodily form and walked erect. Everything else about him suggested an animal. He was naked except for the tatters of a shirt and showed no modesty, no awareness of himself as a human person related in any way to the people who had captured him. He could not speak and made only weird meaningless cries.*

In each documented case of wolf children, one fact stands out: because none of these children were exposed to language at the right age or social convention, none of them ever fully developed language or social skills.[28]

One of the most tragic and studied cases of language and social deprivation occurred in more modern times in Los Angeles in the late 1950s. At twenty months old, Genie was tied to an infant potty seat in a small room by her deranged father. There she lived in complete auditory and social isolation. Genie's blind mother finally escaped with her child, just as Genie was entering her teen years. Genie was kept in the hospital for almost a year after her ordeal ended. Initially, she was incoherent and spat on everybody who came near her. Eventually, she was able to move to a foster home, and her vocabulary increased but only to a kind of pidgin English, and her social abilities never fully developed.[29]

Cracking the Learning Code

You have genetically timed super-windows of learning opportunity when the speed of learning dramatically increases in specific brain areas. When these windows are open, your ability to increase your *Individual Adaptability/Intelligence Factor* is spectacularly increased. However, when these windows close, they close forever.

26

You Learn by the Subtraction of Brain Tissue; It's How You Get Wise

Any man could, if he were so inclined, be the sculptor of his own brain.

—Santiago Ramón y Cajal, Spanish neuroanatomist, Nobel Laureate, and "father of modern neuroscience"[1]

The previous chapters dealt with the process of learning through the addition and strengthening of neural tissue. But to really understand how your brain is shaped by your environment, we need to delve into how your brain learns by the subtraction of neural tissue. Scientists used to think that the refinement of the initial brain plan, laid down by the genes, was accomplished solely by the addition of neural tissue. But painstaking autopsies on the human brain and nervous system have revealed that and learning in your brain also take place by deleting neural tissue. After your *super-learning windows* close, your brain selects out any neural tissue that hasn't been stimulated. As neuroscientist William Greenough notes, "It is the overproduction of synaptic connections followed by their loss that leads to patterns in the brain."[2]

To explain the environmental impact of neural subtraction on developing brains, Yale University neuroscientist Pasko Rakic uses the metaphor of a dressmaker's shears conjuring up the shape from a length of cloth or a sculptor's chisel bringing form to a lump of stone. It's the presence of extra material that expands the range of possibilities, but it's the cutting away of the excess that creates art.[3] Mother Nature deletes neurons and synapses because it's a fast and effective way of shaping your brain to fit your environment. Building new neurons and connections to fit the environment is far more energy-intense—and time-consuming—than just shaping pre-existing neural material. Just like it's easier to prune a bush to fit a particular shape than it is to prompt a bush to grow in a different direction, the brain cuts away old connections more easily than it can grow new ones.

When a particular super-learning window closes, billions upon billions of connections that haven't yet been activated by environmental stimulation will be wholly pruned out. By the time you reach twenty-four years old, neural pruning will have wiped out an amazing 500,000,000,000,000 synapses in your brain. This happens when a group of *regulatory genes* triggers the production of neurochemicals that forever wipe out any of the connections not previously wired up. As you discovered, synaptogenesis—starting soon after conception and extending through toddlerhood—lies down neural connections in your brain at the rate of 1,800,000 per second. This huge number of synapses acts as learning potential, just sitting there waiting to be connected by environmental stimuli. When a particular super-learning window closes, billions upon billions of these connections that haven't yet been activated by environmental stimulation will be completely pruned out.

Pruning Increases Brain's Efficiency, but Limits Potential

Pruning is a good news/bad news scenario. The good news is that by pruning away unused connections, your brain becomes much more powerful and efficient. The bad news is that pruning limits the ultimate potential of brain function.

We see the effect of pruning on efficiency in many areas of our world. More than fifty years ago, economist Joseph Schumpeter emphasized that "creative destruction" was needed to maintain a productive society.[4,5] In farming, we cut back a clogged field so that crops can grow more efficiently. Similarly, the brain gains power and efficiency by pruning out unstimulated neural tissue. During super-learning windows of opportunity, an abundance of neural connections can cause overlap and make information transfer imprecise. Pruning improves the signal-to-noise ratio in your brain, wiping out inefficient connections, which are the cause of neural "noise."

Information noise in your brain can be compared to calling a friend on the phone but also being randomly connected to thousands of other

Neural pruning is the natural process by which your neurons and synapses are efficiently cut away to shape your brain to fit your environment.

people. Your ability to give information to your friend and their ability to reply is greatly diminished by the thousands of other people trying to put their two cents into your conversation. This oversupply of connections is what makes information transfer very inefficient in our younger years. Cell biologist Lise Eliot says:

> *The elimination of stray synapses and the strengthening of survivors is what makes our mental processes more streamlined and coherent as we mature; the party lines sort themselves out into clear, private, efficient channels for information transfer.*[6]

The elimination of stray synapses also allows your brain's energy to strengthen the survivors. Adults have an easier time sorting out facts and taking corrective action than children because adults don't have to deal with as much neurological noise. Children need the support of their elders to make healthy decisions because, in youth, their brains are often dealing with too much noise to make the most-refined distinctions.

There are two main phases of neural-tissue subtraction: one in the womb at the level of neurons and the other at the level of neural connections that occurs after birth when super-learning windows close. The first phase of pruning is swift and brutal. During the first four and a half months of your fetal life, neurogenesis has spread 200 billion healthy neurons throughout your brain. But just after all these neurons are laid down, these healthy cells begin to be wiped out at the rate of 7.4 billion a day. Only one half of their original number will remain by the time you're born. This process of cell slaughter is called *apoptosis,* which in Greek means falling away, like the petals of a flower. The neurons that "fall away" are those that haven't been stimulated to wire up with other neurons. The result of this cellular sacrifice is the first rough map of your brain plan. The 100 billion cells that remain at birth are ready to fine-tune themselves and their connections to fit the environmental stimulation that you'll encounter when you exit the womb.[7]

The secondary phase of subtraction takes place at the level of synapses (neural connections) as various learning windows close. As we've seen, in the earliest years of childhood, your neurons average 15,000 connections per neuron, but by your twenty-fourth year, the shutting of super-learning windows has produced so much synaptic pruning that 40 to 66 percent of those connections are entirely wiped out. The rate of this synaptic slaughter is astounding—at specific periods, you'll lose unstimulated synapses at the astonishing rate of twenty billion a day! Recent research reveals the exciting fact that the protein BDNF (*Brain-derived Neurotrophic Factor*), which helps create brain tissue, also helps control the massacre of unstimulated connections.

Synapses demand stimulation, and lack of stimulation spells their demise. The more electrical and chemical stimulation your synapses receive from your environment during a super-learning window of opportunity, the greater their chance of survival. Environmental stimulation actually triggers the molecular change that stabilizes synapses, cementing their place in the brain and making them resistant to the pruning process.[8] Pruning separates the chaff from the wheat. The connections that are stimulated by your environmental interactions will have the opportunity for permanence, and those not stimulated will wither away and die.

Although pruning gives your brain power and efficiency, wiping out a quadrillion synapse limits its potential. The *field of all possibilities* presented to us in infancy and childhood is greatly reduced by pruning. Though we still possess dramatic learning potential in adulthood, your brain will never

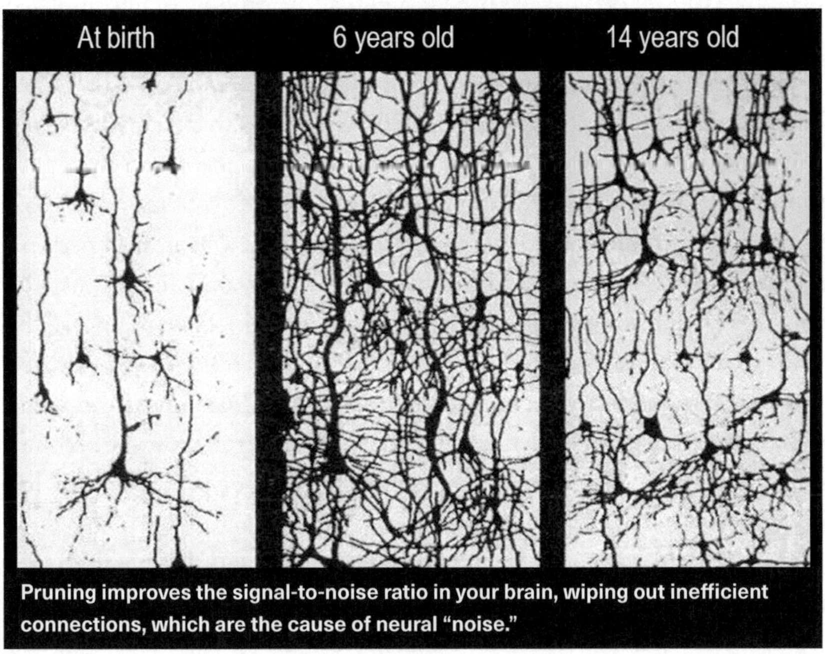

Pruning improves the signal-to-noise ratio in your brain, wiping out inefficient connections, which are the cause of neural "noise."

be as susceptible to selecting information from your environment as it was before the pruning of these connections took place as image on previous page shows:

Brain Disorders and Pruning

Many neuroscientists believe one reason that schizophrenia rarely strikes in childhood is that children have this considerable abundance of neural connections that mask the disorder. After pruning in early adulthood, the disorder can rear its ugly head. [9, 10]

On the other hand, some research suggests that autistic symptoms may be caused by too little pruning. Something goes amiss during the process, and they end up with many more neural connections in specific brain areas than is standard. This overabundance of connections helps explain why people with autism are often oversensitive to certain visual and auditory stimuli and why they often have enhanced memory and skills in specific areas like math and music. This is the biological foundation of why some of these children are known as *savants* (see also Chapter 9).

Something Else Is Shaping Your Brain: Neural Darwinism

Over the last four decades, it's been found that the pruning of unstimulated synapses isn't the only limitation on future learning potential in adulthood. Something else is going on: *Neural Darwinism,* a term coined by Nobel Laureate Gerald Edelman in the 1980s (see also Chapter 8).[11] Neural Darwinism occurs when neurons select the structures and functions supporting the signals that most stimulate them. As you saw in the last chapter, even if the genetic program initially sets up a neuron to perform a specific function like sound processing, if that neuron is activated by completely different stimuli (such as visual stimuli) enough times, the neuron can modify its structure and function so that it becomes hardwired to process this other type of stimulation. Neural Darwinism can cause problems because once a neuron is wired up to other neurons, it becomes difficult to get it to unwire the literally thousands of connections it has made with other neurons.

We've learned about how Neural Darwinism works mostly from the world of vision. Congenital cataracts, though rare (about two in 10,000 births), are the most common cause of childhood blindness. Although light still enters the retina, images are blurry because the signals cannot be focused.[12] During the last century, scientists were very excited about a system they developed that could transplant lenses into the eyes of children who had been born with cataracts. It was thought that this operation should be performed on older children because their bodies would be mature enough to withstand the surgical trauma.

Because these children's brains were judged normal, everyone expected the new lenses would help them learn to see immediately. But the neuroscientific community was shocked to find that, while the new lenses properly focused light onto the retina, they didn't restore normal vision. Instead, not only did the children perceive the light signals as overwhelming and painful, but not one child ever learned how to see. For these sad children, a form of blurry blindness was replaced by bright blindness. Instead of giving them hope for a new life, these new lenses caused all the children to become depressed, and some even committed suicide.[13]

The answer to why the new lenses failed came in the 1960s. In what has been called one of the most important discoveries in developmental neuroscience, David Hubel and Torsten Wiesel unraveled how the visual cortex wires up in childhood.[14] They reached their conclusions after conducting hundreds of the experiments with groups of kittens. In their experiments, Hubel and Torsten "blinded" these animals and then compared the blinded subjects' neural tissue with normal kittens' neural tissue. (These guys were great scientists but obviously not members of PETA!)

One of their first experiments was quite simple: they sewed kittens' eyes shut for periods of their infancy. When these kittens' sutures were removed, it was found that, even though their eyes were quite normal, because the eyes had been shut during critical neurological periods for vision, the kittens could never learn how to see normally. After the genetically timed window for vision closed, there was nothing that the researchers could do to reopen those windows.

Hubel and Wiesel's research helped the medical community recognize that the children who'd had cataracts were unable to see after the lenses were implanted because neurons that should have been wired for normal vision had already been committed to performing other brain functions We now know that specific windows open and close quickly in the visual cortex, starting at a very young age. You were born with fuzzy, black-and-white, monocular, 20/600 vision—thirty times worse than an adult's[15]—and it would take six to seven years of visual windows opening and closing before you would receive full binocular, color-spectrum, mature contrast vision, as well as 20/20, focused sight.

Today, to make sure a child born with cataracts has the opportunity to learn how to see normally, these lenses are implanted as soon as possible, usually within the first sixty days of life, before any of these windows for vision start to close.[16]

The main learning from this is that neurons are always looking for stimulation. They'll quickly wire up to where the action is. This is a neurological principle, and it dominates even after a super window closes.

Neural Darwinism Helps Heal Brains

An excellent example of Neural Darwinism in action in adults was implemented by Edward Taub,[17] a neuroscientist at the University of Alabama at Birmingham. Taub worked with stroke patients who had damage to the sensory and motor cortexes of their brains. These parts of the brain control arm and hand sensations and movements. Neuroscientists know that the right side of the brain controls the left side of the body, and the left side of the brain controls the right side.

In a finding that surprised the neuroscientific community, Taub found after a stroke on say the left side of the brain, which prevents a patient's right arm from moving, those neurons stop getting input from that arm. So what do those neurons immediately do as soon as they start to heal? As they get more and more input from the left arm, so they immediately wire up to support its function. This causes great problems when the patient is well enough to start therapy. The vital neurons and their connections that control right arm movement now can't totally pay attention to right-arm stimulation because they're committed to left-arm work.

Taub's profound discovery was he could get previously committed neurons to pay attention if he changed how they got input. He found that when he tied down the good left arm, the brain would stop getting strong signals from it. Then, if at the same time, he aggressively used therapy to move the right arm, those neurons that had wired up to the left arm would do two marvelous things. First, they would start to unwire (be pruned) from the left arm, and second, they would start to have stronger, healthier connections to the right arm. After this rewiring process, patients' right-arm functions returned, on average, to 75 percent of normal. One patient regained hand dexterity twenty-five years after a stroke! This therapy is now called "constraint-induced movement therapy."

The Society for Neuroscience was so impressed they named Taub's work one of the top ten translational neuroscience accomplishments of the 20th century, and his efforts were hailed by the American Stroke Association as "at the forefront of a revolution."[18] Despite the acclaim, I'm still amazed to find that many medical practitioners who deal with brain damage are unaware of Neural Darwinism or Taub's work.

Want to Learn a Foreign Language? Do it Young!

Understanding pruning and Neural Darwinism has also led brain-based educators to demand that foreign language learning starts in the early years of school. The best time to learn is when the linguistic windows of super learning are open to their widest. Apparently, it doesn't make any difference whether the language is the child's first or second. If a child isn't exposed to a certain sound before the window for learning that particular sound closes, the brain won't wire up to recreate the sounds of that language effectively. During the time linguistic windows are open, the brain selects what it hears most often; it loses the ability not only to speak but even to hear those sounds that it hasn't been exposed to before the window closes. Research in second-language learning indicates that the optimal time to learn a new language without difficulty, without an accent, and with proper syntax is before seven years of age.[19]

As know from earlier chapters, as an infant, you came into the world with universal language capabilities. We're citizens of the world who can perceive all the phonemes (sounds) in all 6,000+ languages. But very quickly, if a child hasn't been exposed to the correct sounds by the time linguistic windows close, his or her ability to possess universal language begins to get selected out. By four months, the ability to discriminate foreign vowels sounds begins to wane, and by ten months, the ability to perceive certain foreign consonant sounds begins to get selected out.

It is your age, not effort when it comes to learning a language. This is shown in the studies of immigrants to the US that indicate their command of the English language is a direct function of their age upon arrival—not the amount of formal training, motivation, attitude, or even the total number of years lived in the United States. Those who arrive on American soil before the age of seven can do as well as native speakers. Performance steadily declines as the age of the immigrant increases.[20] While former Secretary of State Henry Kissinger came to America after his seventh year and speaks with a heavy accent, his brother came before age seven and speaks with little or no accent. By the time you reach the end of puberty, the window for rapidly and effortlessly acquiring a foreign language has been closed. If you try to learn the second language after puberty, there's no correlation between your age and the ease of language acquisition or skill—twenty-year-olds do no better than forty-year-olds in learning a second language.

After linguistic windows close, you speak with an accent when you learn a foreign language because new languages are forced to build synaptic connections on branching dendritic structures that hold your native language—the one you've been most exposed to as a young child. After the language windows close, the neural connections that let you speak a new language without an accent have been pruned out because of lack of stimulation. Once windows of language learning close, all-new language circuits must be built upon your existing mother-tongue circuits. This is why foreign-born personalities such as Arnold Schwarzenegger and Jean-Claude Van Damme maintain their accents even after years and years of living in English-speaking countries. Patricia Kuhl of the University of Washington suggests that the reason we find it easier to learn related languages such as Spanish and French is that the existing circuits can share and do double duty.[21] The new research makes neuroscientist Dr. Chugani ask, "What idiot decreed that foreign-language instruction not begin until high school?"[22]

You Have Windows for the Arts and Sports

We're also finding that super-learning windows affect not only critical life skills that deal with vision, language, and emotional function but also sports and the arts.[23]

A wonderful example is a man often mentioned as the greatest athlete of the 1990s: basketball star Michael Jordan. At thirty-one, Jordan decided to quit basketball and fulfill his lifelong dream of becoming a professional baseball player. However, he didn't do nearly as well as he had hoped. His window of opportunity for acquiring the visual and motor skills that would allow him to compete at the highest level in baseball had closed years earlier. Because of pruning and Neural Darwinism, the predominant wiring in Jordan's visual and motor cortexes were filled with synapses that helped him sink basketballs, not hit ninety-mile-an-hour fast-balls. In Michael's third decade, no matter how much environmental stimulation his brain received about baseball, and no matter how hard he practiced, the genetic switches that would have allowed him to acquire efficient baseball skills and allow him to compete professionally had long been turned off.[24] In the end, Jordan had to admit that he couldn't improve his skills and gave up his dream.

Jordan's experience is evidence that even if you're incredibly gifted in a particular sport, the possibilities of playing at the professional level are greatly enhanced if you start before your motor windows close. At age two, Olympic soccer gold medalist Mia Hamn started kicking a ball; Tiger Woods swung his first golf club at ten months; and the great hockey player Wayne Gretzky was practicing slap shots at three. It's very rare to find a professional golfer, baseball player, gymnast, or skater who started their sport after puberty.

The same is true for the art world. Mozart's parents exposed him to music as a toddler and encouraged him to write his first symphony by the time he was eight. World-renowned cellist Yo-Yo Ma started his musical studies at four. Musicians who can develop perfect pitch never develop this trait if they are not exposed to music before the pitch window closes around seven years old. Few solo concert instrumentalists began playing after age twelve.

Set in Our "Ways" and "Winning Strategies"

After super-learning windows close, the brain becomes much less adaptable. During the process of building up and then pruning down connections, our genetic program essentially says, "I've given you an overabundance of neural material so you can easily shape your brain to fit your present environment. Once you've developed a capacity to efficiently adapt to your environment, it's time to cut out the excess connections."

This rigidity allows us to know who we are—Irish, Jewish, or Arabic—

and what actions help us fit our world—whether we go to church, temple, or mosque. The positions and moral stands we make in our lives are based on the neural networks that have previously been laid down. If our neuro networks were too malleable, we'd become wishy-washy and so susceptible to environmental stimuli that it would be difficult to maintain direction in life. We would never have a sense of self.

But neural inflexibility has its drawbacks. After we acquire a set of neural networks, these networks can become difficult to alter, which can make us set in our ways. We can have a hard time changing our smoking, drinking, and eating behaviors because of this lack of malleability. As Daniel Alkon writes in *Memory's Voice,* "The earlier we are exposed to stimuli (both negative and positive), the more links they (neurons) form, the more entrenched they become."[25] This is one reason 80 percent of children raised by obese parents are also obese and why it's tough for them to maintain healthy eating habits as adults.[26] Interestingly, one report notes that only 20 percent of people change their political affiliations after early adulthood, apparently because of this neural rigidity.[27]

Lack of neuroplasticity can cause us to follow directions in life that don't serve us. An old rugby buddy, financial guru Robert Kiyosaki, who is the author of the best-selling series *Rich Dad, Poor Dad,*[28] insightfully points out that rigid neural representations force people to develop what he calls their "winning strategies." These are strategies that people use over and over again to try to make their lives work for them. He says some people continue to use them long after they cease to work; for example, some people consistently use financial schemes that never make them money or repeatedly enter relationships with unsuitable mates. When asked why they habitually use strategies that lead to failure, these people usually point to small successes they achieved early in life when their brains were forming rather than to the massive defeats they encountered later.

We all know people stuck in failed "winning strategies." I had a friend who maintained he was a genius at understanding the "feminine mystique," overlooking the fact he had been married five times and divorced the same woman twice! Friends, spouses, psychologists, and social workers continually try to get these people to see the folly of their ways, often to no avail. When these people were young, and the brain was more adaptable with its excess synapses, these individuals wired up a winning strategy that appeared to work; in later years, with a less flexible brain, they haven't encountered environmental stimuli powerful enough or science-based behavioral change programs efficient enough to alter these old neurological patterns.

Why Different Generations Don't "See Eye to Eye"

Pruning and Neural Darwinism also ensures we'll seldom see eye to eye with our parents and elders. Because different generations grow up in different environments, each generation has selected in and pruned out different neurological shapes, leading each generation to process information somewhat differently. Whether we're looking at parents of the Depression and hippie kids or Baby Boomer parents and Millennials, each generation has difficulty understanding where the other is coming from. The genetic program for parents to love their children and for the children to bond with their parents is indelible. But the different environments that we've been raised to ensure that our neural networks will hold different representations of the world. Pruning and Neural Darwinism may trap us in a world where parents and children love, but can never completely understand, one another.

Pruning and Neural Darwinism can also be blamed for a lot of the social and territorial conflict in the world. We come into life as citizens of the world, only to be quickly shaped by our immediate environment. If we follow the genetic stream backward, we'll find that the Jews and Arabs in the Middle East, Protestants and Catholics in Ireland, and Croats and Serbs in Eastern Europe are all related, cousins at the core. But solely and exclusively, because of different environments, each religion or nationality sees the other as the enemy. The path to peace in the world has everything to do with changing previously laid neural networks and very little to do with making war on our fellow humans.

You Have a Neurological Window for Wisdom

Up until the very recent past, the elders in society have always held an esteemed position. Not for their speed or strength, which had waned, but for the knowledge they had amassed. Because elders have lived longer, seen more, done more, and made more mistakes, the young have traditionally turned to them for sage advice on how to survive and thrive more effectively. Research now indicates that not only do you know more as you age, but the speed and efficiency at which your brain processes knowledge in specific brain areas also increases. In other words, you have a neurological window for wisdom!

In the 1980s and 1990s, research by Dr. Francine Benes at the Harvard Medical School discovered that, from your fortieth year until well into your eighties, there's a dramatic increase in the neural material called *myelin,* which enhances the speed and efficiency at which your neurons communicate with each other.[29] Myelin is a white fatty substance that coats an *axon,* the neural connection that transmits information from neuron to neuron. Like plastic insulation on a telephone line, myelin helps accelerate signals between neurons. It also prevents "cross talk" caused when a signal from one neuron interferes with signals of a nearby one. The more effective the myelination of an axon, the more effectively your brain can communicate with itself. The problems caused by epilepsy and multiple sclerosis can be traced to poor myelination, which slows signaling and creates too much cross talk between neurons, thus disrupting normal brain function.

Your Window for Wisdom

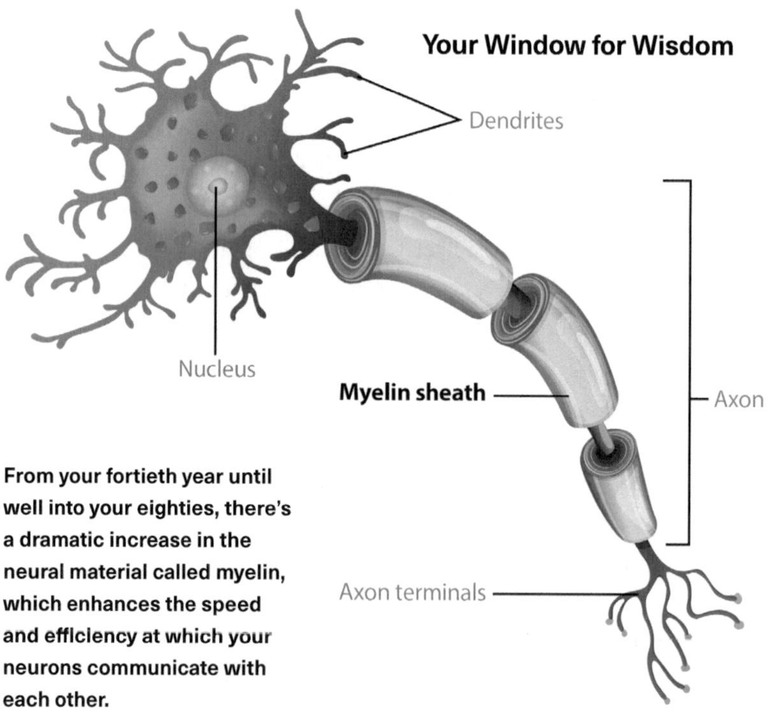

From your fortieth year until well into your eighties, there's a dramatic increase in the neural material called myelin, which enhances the speed and efficiency at which your neurons communicate with each other.

The neuroscientific community had suspected that significant neurological growth occurred in our later years. But it was Benes who painstakingly measured stains around the axons of cadavers of different ages in different brain areas and found that myelination reaches its most efficient level in our sixth decade and stays well above the levels we see in our teens, twenties, and thirties until we're well into our eighties.[30]

As you reach your forties, your genetic program instructs supporting cells, called *glial cells,* to make more myelin available to your axons, thus increasing the velocity and effectiveness of communication between neurons. Einstein's brain, which is sitting in a jar in a research lab, contains many more glial cells than the average brain. This abundance of glial cells, which produced more myelin, is often given as a reason for Einstein's enhanced intelligence.

The increased myelination in our brains as we age enhances speed as well as efficiency of feed-forward and feedback loops between neuron. This is a source of our capacity to be highly creative, intuitive, and conceive out of the box solutions to problems even as we age. Combined with the large amounts of information we've learned, this can be seen as a source of wisdom. This increased internal activity caused by myelination also coincides with a natural decrease in physical activity. This allows an older brain to use the energy that had been directed to the motor cortex and redirects it toward the neural structures that support more in-depth thinking.

Another area of Benes's research focused on the increased myelination between the networks that connect the prefrontal lobe to the limbic system. Your innate emotional responses are housed in your lower limbic structures, and the mediation of these instinctual emotional responses is controlled by the prefrontal lobe. Better mediation and control of your emotional responses are also seen as an important component of wisdom. Benes commented:

> *Perhaps the 'wisdom' sometimes required with the advancing age in humans is the expression not only of accumulated experience but also innately timed programs that give rise to a tighter link between the neocortex in the limbic structures.*

When I shared Benes's research on emotions with a friend, he exclaimed, "So that is why you don't see seventy-year-olds driving around getting road rage and flipping each other the bird!"

Increased myelination in older age may be the reason that many accomplished people do their best work in later years. Thomas Edison, Victor Hugo, Margaret Mead, Claude Monet, George Bernard Shaw, Colonel Sanders, Pablo Picasso, Arthur Rubenstein, Albert Schweitzer, Louise Bourgeois, Pablo Casals, Richard Wagner, Franz Joseph Hayden, and many others succeeded in their fields in their sixties, seventies, eighties, and nineties. It may also be why the dot-com phase of Wall Street in the late 1990s saw many young people in their teens, twenties, and thirties strike upon great ideas that had some merit, but lacked the wisdom to fully develop them. During that time, hundreds of failing dotcoms had new management imposed upon them by the venture firms that had provided their financing. In virtually every case, the financiers replaced the initial young management team with older, wiser management. The financiers recognized that great ideas were not enough—they also needed the wisdom of elders to have a chance of developing a successful business.

Your Aging Brain: Much Better Than the Popular Press Tells You

Recent research clearly demonstrates that as long as you stay physically healthy as you age, many of your brain areas will work much better than they did in your younger years. The popular press paints a depressing, one-sided picture of your aging brain. But studies clearly show that measures of cognition, such as moral decision-making and reading social situations, actually improve in your later years, starting around age forty. One of the longest and largest studies of its kind called the "Seattle Longitudinal Study" tracked 6,000 people for more than forty years; the researchers discovered that as individuals aged, they did better on tests of vocabulary, spatial orientation, and inductive reasoning than they did when they were in their twenties.[31]

The brain's systems that allow you to unwind the intricacies of complex problems and find concrete answers also enhance as you age, as does your ability to connect the dots and see patterns in your world. Also, when confronted with challenges, to find solutions, younger individuals rely on the back of the brain, where raw information is stored. Older adults rely on the frontal parts of their brains. These frontal areas not only have rich connections to previously learned knowledge, but they are much more efficient at making order out of chaos than the rear brain areas. This may be why elders have a much broader perspectives on life than the young.

Age also allows us to *not* "sweat the small stuff," as the part of the brain that drives base emotions, the amygdala, becomes less and less reactive to minor negative stimuli. This gives us the capacity to look on the bright side more. Studies show as we age, we're inclined to shift our attention and remember positive events much more than negative ones; this tendency continues until into our eighties. Brain imaging demonstrates that as we age, the parts of the brain that specialize in daydreaming get more active. Why is that good? Because these are the same brain areas that support creative thinking.

Does Your Memory Really Worsen as You Age?

Memory function in our senior years gets a bad rap, but forgetting is less about having a bad memory and more about having too much stored information. Consider that the average seventy-year-old has probably put their car keys down 46,000 times. The brain stores knowledge in specific networks. When older people can't find their keys, what's often happening is that the last time they put their keys down gets tangled up, in the category network for keys, with the tens of thousands of other times they've put them down. The most recent episode gets obscured by the mass of previous episodes.

Another reason seniors often can't remember something deals with novelty. The brain most effectively remembers novel occurrences. Unfortunately, where you parked your car in the mall at age fifty is not as new and novel experience as when you parked your car when you were eighteen. Although the brain does lose a small number of neurons as you reach your later years, unless there is disease, the major reason elders can struggle to retrieve information is too much data combined with too little novelty.

In the mid-1800s, the Japanese artist Hokusai had an insightful, intuitive grasp of what the research today is demonstrating when he wrote about his work:

> *It is at the age of seventy-three that I have somewhat begun to understand the structure of true nature, of animals and grasses, and trees and birds, and fishes and insects; consequently at eighty years of age I shall have made still more progress; at ninety I hope to have penetrated into the mystery of things; at 100 years of age I should have reached decidedly a marvelous degree, and when I shall be 110, all that I do, every point and every line, shall be instinct with life—and I ask all those who shall live as long as I do to see if I have not kept my word.*[32]

So, take heart! As you age, things don't get worse—they get better! You do indeed have neurological windows for wisdom. The research is clearly demonstrating that any organization that hopes to maximize its impact in the world will need to have a firm understanding of the value of the aging brain.

A Final "Aha" about Super Windows

A few years ago, I had a fascinating "aha" moment about super-learning windows at a Nigerian Economic Summit on Education. I was there supporting the government by offering input on how they could more effectively support education in developing nations by taking a more scientific approach, which they were excited to do. This "aha" moment came to me in a session where I spoke about the windows of learning, offering knowledge very similar to the information in this chapter. During the end discussion, I was watching the adults struggle with the question of how to "fix" education when a girl who looked about thirteen asked me if she could ask a question. I said sure, and her question was so crisp, clear, and wise that it shocked me. I forget exactly what she said, for what I focused on—for the first time ever in all my thirty years of research—was the disparity between the adult brains that were designing learning systems and the child brains these systems were being designed for!

It hit me like a ton of bricks: how could adults ever design learning systems using brains that have up to 500,000,000,000,000 fewer connections than the brains of the children they're designing them for… without the children's input? In business, we're always striving for customer input to make products better. This really doesn't happen in education. And this is exactly why we have so many dissatisfied and unhappy customers (students) in education.

Think about it: most of the educational leaders who are designing our learning systems went to school in the 1970s, 80s, or 90s. They had their brain tissue sculpted in a world where information was doubling at the relatively slow rate of once every fifteen to fifty years. What this means is that all their successes in life, all their innovations, and even what they believe is "right" in education are being processed and created by brains that learned how to learn during a time that no longer exists.

Kids' brains live in a different world today. They switch back and forth between Twitter, Facebook, Pinterest, TV, iPods, texting, etc., etc., all day long. This dramatic rise in the amount and velocity of information has affected kids' and young adults' attention spans. Attention spans have decreased. It's estimated that, on the web, if the young don't find something meaningful to them right away, their attention span is about eight seconds. That's about the same attention span as a bumblebee! There's no question: Children's brain plans are very different from the brain plans that are designing the learning systems for them.

This "aha" moment helps create a philosophy at The Learning Code Institute, where we will never develop a learning system without input from the very brains we're designing the systems for.

Cracking the Learning Code

Your brain not only learns through the addition of brain tissue, but also by the subtraction of brain tissue, in the process called *pruning*. Trying to develop efficient, joyful learning systems without understand the pruning process makes the whole undertaking of creating profound memory much more inefficient and painful than it needs to be.

27

Brain Danger! The Wrong Information at the Wrong Time (And Why Brain Games Don't Work)

The window of opportunity opens and closes as fast as a cameras shutter.

—Warren Criswell, painter, sculptor, and animator[1]

This chapter is for parents and educators, as it mainly focuses on children's brains. These often receive too much of the wrong kind of stimulation, at the wrong time, from programs that endeavor to "hothouse" children as we would plants. As you'll see, forcing brain structures to grow and mature before they are ready is not a healthy idea.

This chapter is also designed for two other groups: first, if you were exposed to very early academic programs before first grade, or feel that your focus in childhood was more on grades than life; second, if you want to know if brain games help your personal learning, please skip ahead to the section titled, "Digital Controversy Not Limited to Kids: "Brain Games" Don't Work for Adults."

Your Brain: Not to Be Filled Like a Jug

As we have discussed, some parents and educators believe that a child is like a huge container. To ensure the child's success, they think it's their job to fill that container with as much information as possible, as quickly as possible. This misconception is damaging the brilliance of millions of our youth. While many brain-based educators are afraid that the information about *super-learning windows of opportunity* won't get out to the general public fast enough, other experts are afraid the information will be misused by parents, guardians, educators, and coaches in a misguided effort to create kids with super abilities.

The research has already plunged many parents and educators into a frantic race to expose children's brains to as much stimulation as possible before these super windows close. But negative results from programs that teach preschool children academic skills are forcing us to question the value of presenting too much neural stimulation to the brain at the wrong time.[2] Researchers are asking an important question about super windows of learning and parents who want to jump-start their children's learning—and our traditional approach to education: could too much information too soon damage the developing brain as much as too little information too late?

In one poll, 87 percent of parents said the more stimulation a child receives, the better off they will be.[3] Matthew Melmed, former director of Zero to Three, a research and training organization for early childhood development in Washington, D.C., says, "Many parents have the concept that a baby is something you fill up with information, and that is not good." He goes on to say:

> *We are concerned that many parents are going to take this new information about brain research and rush to do more things with their babies, more activities, forgetting that it is not the activities that are important. The most important thing is connecting with a baby and creating an emotional bond.*[4]

To get a feel for the frenzy to force very young children to learn quickly, consider the proliferation of books, DVDs and online programs with titles like *Baby Einstein, Baby da Vinci, How to Give Your Baby Encyclopedic Knowledge, How to Teach Your Child to Be Physically Superb*, *Brainy Baby* and a host of others.[5, 6] As you'll see in this chapter, research demonstrates that not only do many of these aggressive learning programs not work, but they also have the potential to be harmful to the developing brain.

Too Much Too Soon Disrupts Regulatory Genes and Energy Flow

Researchers are just beginning to understand that presenting certain kinds of information to your brain before it's genetically and neurologically ready, in the effort to raise IQ, may disrupt its natural development.[7] There appear to be two sources of this disruption. First, as noted by Jean Piaget, children pass through stages of development in a specific order.[8] We crawl before we walk, we babble before we speak, and we grasp concrete concepts before abstract ones. If we try to force a child to walk, talk, read, or understand abstract concepts too soon, we interfere with the natural order directed by the unfolding or our genetic program. Disruption of this process can cause *regulatory genes* to turn on and off at the wrong times, negatively affecting neural development.

Second, as we've often discussed, the brain generates twenty watts of power using glucose and oxygen delivered by the vertebral and carotid arteries. If too much energy is diverted from a particular part of the brain when a related super-learning window is open its widest, that brain area may not have the neural energy it needs to develop normally. As Dr. Bruce L. Miller, director of the Memory and Aging Center at the University of

California, says, "In the brain, we have lots of systems of modules that work independently. When you turn on one module, you often turn off the other."[9]

While the research in this area is limited, I believe there's enough evidence for parents and educators to be very wary about supplying information to the young brain before the genetically timed windows are open to receive a particular kind of information. Parents who want their child to be a great gymnast or a superb academic may have the child spend hours a day practicing or studying; however, they must consider that putting the child in a situation where the majority of the brain's resources are focused on only one brain area—like motor or linguistic—for extended periods when emotional, social, bonding, or spatial windows are open their widest may have severe learning and psychological implications for that child later in life.

Hothousing Children: It Works Better for Flowers!

We're becoming clearer about the results of presenting too much information to the brain at the wrong time in the effort to raise IQ from what educators now call "hothousing."[10] In a hothouse preschool, the child's natural instinct to crawl around to discover, explore, and play is restricted in favor of teaching early academic skills. A teacher calls all the shots, and the child is expected to sit still and absorb information through memorization and other traditional teaching methods for most of the day. It was thought that hothousing would dramatically increase a child's intelligence. The research does not bear this out. While many of these programs did have some positive effects on graduation rates, as you will see, they also had many negative consequences.

Developed in the 1960s, the Head Start program offers one of the first and most famous examples of how forcing early academics on children to increase their IQ may be a fruitless endeavor.[11] Head Start's goal is to provide disadvantaged children with secure environments, giving the kids a basis for becoming productive teens and adults. While Head Start has wonderful elements that deal with health, nutrition, and parent involvement, which reduce rates of teen pregnancy, school dropouts, and juvenile delinquency, the academic portion of the program, designed to increase IQ, hasn't been so successful.[12, 13] Although most preschoolers in the Head

Start program initially experienced a boost in IQ scores, the improvement didn't last. By grade school, the Head Start children's IQ scores were no higher than those of their peers. In the controversial book *The Bell Curve,* authors Richard Herrnstein and Charles Murray labeled all this effort with no return "the dreaded 'fade-out.'"[14] The evidence prompted Yale psychiatrist Edward Zigler, a founder of Head Start, to caution, "The fact is it's very, very hard to raise anybody's IQ."[15]

One of the most cited failures of early academic intervention for preschoolers occurred in Ypsilanti, Michigan, in the early 1960s. Here, the HighScope Educational Research Foundation carried out an intensive program that "educated" preschoolers for four hours a day, each weekday, for two years. Student/teacher ratios were one teacher for every five students. The founders of the program were convinced that intense hothousing would develop super-smart kids. Unfortunately, as in the Head Start program, IQ scores initially climbed, but then faded out. By third grade, these children's

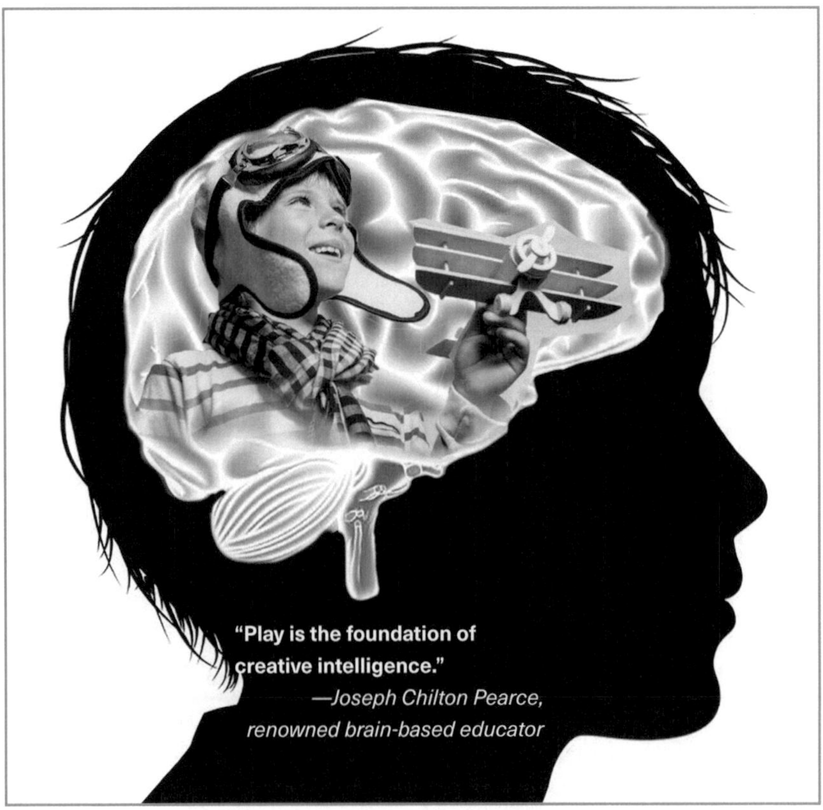

IQ scores were indistinguishable from those of neighborhood children who hadn't been part of the program.[16, 17]

Hothousing Creates Unhappy Kids Who Don't Like to Learn

Educational researcher Kathy Hirsh-Pasek at Temple University did an extensive study on children who were hothoused in preschool. She found that, by kindergarten, these children were no better than their peers in academic skills. In addition to the lack of academic advantage, Hirsh-Pasek found some disturbing side effects. It seems these programs were sucking the joy out of learning for these children. The kids suffered from high-performance anxiety and tended to feel less positive about school than before being hothoused. Worse, hothoused kids with especially hard-driving mothers tended to be less creative than children in non-academic programs.[18] She concludes that it's difficult to justify taking on academics too early on two grounds: first, the benefits are transitory, and second, these kids tend to develop negative feelings about learning.

David Elkind, a professor of child studies at Tufts University, is another critic of academics being imposed too early. In his books *The Hurried Child* and *Miseducation: Preschoolers at Risk,* Elkind warns that teaching academic subjects to children too soon causes fatigue, decreased appetite, lower task effectiveness, and psychosomatic illness. Over the long term, these children can show less interest in learning, less ability to work independently, an increased tendency to worry, and a greater need to compare their intelligence to others.[19] Elkind maintains that exposing preschool children to anything other than self-directed activities can be harmful and dangerous to their psychological development.[20, 21]

Retired Harvard educator Kurt Fischer conducted studies into hothousing that prompted him to conclude, "I feel like it is actually fairly ineffectual."[22] Fischer uses 19th-century philosopher John Stuart Mill to make his point. As a preschooler, Mill was reading in Latin and Greek, but he had a major emotional breakdown in early adulthood, which he attributed to his high-pressure schooling in childhood. In the end, presenting information to the young brain before it's mature enough to receive that information appears to be, at the least, an ineffective way to build resilient brains and, at the worst, psychologically damaging to the child.

Is Play More Important than Academics?

The research is prompting some brain-based educators to ask whether we aren't starting traditional classroom education too early. Many educators maintain that the unfolding genetic plan would be better served if a child spent their first seven years of schooling in free, self-directed, meaningful play instead of being locked to a desk.[23] They point out that play naturally and effectively develops social and emotional skills, and research shows that these kinds of skills are more important than your score on a standardized test when it comes to success later in life. Play also supports the unfolding of gross motor skills before fine motor skills.[24] Some people fear that emphasizing the fine motor skills needed for reading and writing before gross motor skills are firmly in place can have negative consequences on a child's balance and physical development.[25]

"We used to think that play and education were opposite things," Jean Marzollo and Janice Lloyd say in their book *Learning Through Play*:

> *Now, we know better. Educational experts and early childhood specialists have discovered that play is learning, and even more than that, play is one of the most effective kinds of learning.* [26]

Renowned brain-based educator Joseph Chilton Pearce, says simply, "Play is the foundation of creative intelligence ."[27] Pearce also maintains that presenting abstract concepts to the brain before it's ready—as most school systems do—is inconsistent with our biological plan. He comments:

> *Our earliest reading and readiness workbooks… involve formal operations designed by nature to open up once all the other support systems are complete, somewhere around puberty. We can force a child prematurely to employ these abstract intelligences to some extent, but at a price. Demanding that the young child prematurely develop symbolic capacity means programs which nature intended for that period… are severely neglected… Ironically, these early programs are the foundations we need for true abstract symbolic capacity later. … With no neural structure developed for this capacity, a majority of children are defeated before they begin. Guilt and loss of self-image result from their inability, while we continually test them to show how they fail to measure up.* [28]

If you had feelings of inadequacy in school, they might have been driven more by an educational system's inability to embrace the science of learning than by your inability to learn.

If You Read Sooner, Are You Smarter Later? No

Many educators and parents have the misconception that forcing a child to learn to read at a very young age will support them in academics later in life. But the research doesn't support this belief. John Bruer, president emeritus of the James S. McDonnell Foundation, notes, "Even if a child learns to read early, there is no evidence that that accomplishment will transfer into higher grades, intelligence, or success as an adult."[29] Cognitive psychologist Dee Coulter, previously at Naropa University, wonders whether the reason that most of us will need reading glasses before we're fifty is that virtually all of us started to read before our fine motor skills and 20/20-context vision was firmly in place by age six or seven.[30] She believes this early reading effort puts too much stress on the eye, misshaping it.

Genetic Variation Ensures We All Learn at Different Paces

Genetic variation introduces another facet of the problem of delivering too much information to the brain at the wrong time. Genetic variation ensures that not only will all brains be different, but they'll also develop

at different paces. No two genetic plans on the face of the earth unfold at exactly the same rate. For each of us, our super windows of learning open and close at different times. Research has shown that normal brains have up to a three-year difference in developing different skills. Some experts maintain the difference is closer to five years.[31]

If we look at 1,000 tadpoles all born on the same day, we wouldn't expect them all to sprout rear legs and become frogs at the exact same time. But we expect this perfect timing from our children's genetic plans. Our learning systems are set up as if all children developed motor, speech, reading, writing, and abstract thinking capacities in lockstep. We label a three-year-old reader a genius and an eight-year-old non-reader a dunce when all it means is that the specific window that allows efficient reading opened sooner in the three-year-old than it did in the eight-year-old.

And let us repeat: there's no scientific evidence that the early reader will be more successful in life than the late reader. Both accomplished businessmen Charles Schwab and Richard Branson of Virgin Atlantic Airways had very limited reading skills as children but went on to highly successful lives. Brain-based educator Eric Jensen notes that:

No Two Brains Develop at the Same Time

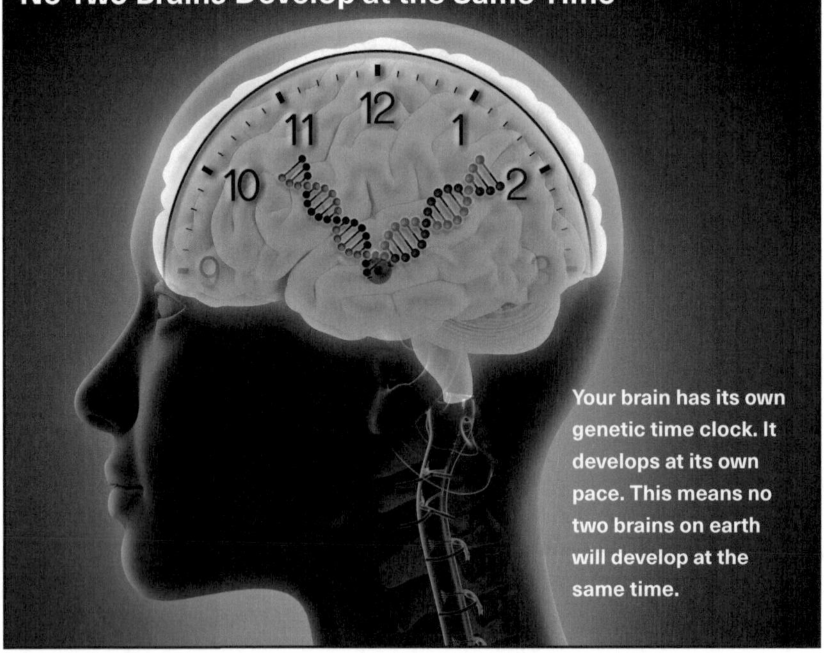

> *All five-year-olds should not be expected to perform at the same level academically, physically, or socially. Statewide curriculums and frameworks which include specific grade-level performance standards are biologically inappropriate. All 13-year-olds may not be neurologically ready for algebra or geometry… Ultimately, some learners simply conclude they're stupid and leave school.*

He says the solution is to "eliminate groupings by age or grade."[32]

The IQ Problem

So are IQ scores wrongly linked to chronological age?

Much of the standardized testing uses specific ages as criteria for determining *normal* on intelligence test scores. For example, an IQ score is determined by dividing one's chronological age by his or her test score and then multiplying by a hundred. Because science tells us that there is a three-year window for normal genetic and brain maturation, intelligence tests that use fixed years to determine normal scores will, therefore, give false readings.[33] We end up collapsing the *field of all possibilities* for enormous numbers of children by labeling them smart or slow based on faulty criteria.

It's been reported that Einstein hardly uttered any words at all until he was three years old.[34] If Einstein had been given a word-based intelligence test before his linguistic skills were developed, he might have done poorly, accepted a negative image of his intelligence, and never gone on to change the way the world looked at physics.

In his book *Brain-based Learning,* Jensen points out that there are many skills, such as reading, where girls develop before boys. If we actually accounted for different brains, we'd suddenly find that up to 75 percent of all boys who are now considered 'developmentally slow' would immediately be reclassified as normal.[35]

The truth is we're just beginning to grasp the fact that each brain has its own genetic time clock and develops at its own pace and that forcing a child to keep up with his peers when he isn't neurologically ready can cause harm.

What If a Very Young Child Wants to Learn Before a Super Window Is Open?

What you do when a very young child (say, ages one to four) wants to learn something before you think a super window is open? We don't have all the answers yet, but it appears that as long as the child is self-directing his or her learning, pursuing that desire to learn is okay.[36] (Learning on electronic media is an exception, as you'll see in more depth in the next chapter.)

Why? Because as children's super windows open up to acquire skills like certain physical movements, math, and reading, their biology drives them to seek stimuli to satisfy those urges. Because of the three-year variation in when windows open up, some kids will want to run, count, or read before others. In most cases, as long as the learning is directed by the young child and not forced on the child by the parents, it's healthy—that is, unless it becomes an obsession that excludes other forms of stimulation the brain needs to develop normally. The problem of providing too much information to the developing brain appears to occur when the parents or educational system force the learning before the windows are open.[37]

Are Digital Learning Programs for Young Children Working?

Because smartphones, iPads, TV, movies and other media have become a major source of environmental stimulation to our brains, in the next chapter we'll go even deeper into the effect of the media on our brains, young and old.

One of the easiest ways for parents to try to accelerate a young child's learning is to buy a DVD or online media learning program. But the research on electronic programs that try to jump-start learning for very young children is disheartening and will greatly disturb the millions of parents who are using them.

In 2007, the *Journal of Pediatrics* published the results of a controversial study done at the University of Washington on programs like *Brainy Baby* and *Baby Einstein*. The researchers reported that infants who watched these electronic learning programs for an hour per day understood an average of six to eight fewer words than infants who did not watch them. "The more videos they watched, the fewer words they knew," says Dimitri Christakis, one of the lead researchers "These babies scored about 10 percent lower on language skills than infants who had not watched these videos."[38] Another study published in the online journal *Archives of Pediatrics & Adolescent Medicine* in 2010 found no evidence that children learned words from infant-focused videos, and, even more alarmingly, they discovered that the earlier the exposure, the lower the scores on measures of general vocabulary knowledge.[39]

Because of these and other similar studies, the American Academy of Pediatrics stated that video screen time provides no educational benefits for children under the age of two years, and they discourage parents from letting their young ones use any media with screens. They also make it clear that there's no such thing as a digital education program for such young children that beats real-world learning.[40]

Digital Media Deprives

The researchers who study digital media's effects on young brains conclude that TV, DVDs, and online programs steal what the young child needs most—that is, caring interaction with others, physical play, and the opportunity to explore real-world environments. The Pediatrics Association estimates that for every hour a child under age two spends in front of a screen, he or she spends about fifty minutes less interacting with a parent and about 10 percent less time in creative play.[41] Digital media robs young children of the activities that their brains need most to develop normally.

Marketing Beats Science!

Unfortunately, the marketing of the baby teaching companies—which generates an estimated $1 billion a year in sales—is so loud that it's drowning out the voice of science. One study shows that only 6 percent of parents are aware of the recommendations made by respected medical organizations and researchers regarding the negative consequences of these programs, which is why you see so many parents allowing their young children to be totally captivated by them.[42]

It must be pointed out that the studies of digital learning programs for young children are just getting started. There's much controversy over the research reported here. For example, Disney, the owner of *Baby Einstein*, sued the University of Washington for how it promoted its 2007 study. Yet because of this and other studies, Disney offered a refund to those who bought their *Baby Einstein* program from 2004 to 2009.[43]

Don't Confuse Addiction with Learning

Just because something is charming to the brain doesn't mean it's good for the brain. If your kids want to use an electronic learning device for long periods and cry when it's taken away, that doesn't mean they're upset because they want to learn more. Something more sinister may be going on. That is, they may be developing a form of addiction that could plague them the rest of their lives.

To hold infants' and toddlers' attention, these digital media learning programs endeavor to keep neurotransmitters like dopamine, norepinephrine, and endorphins at high levels for long periods of time. Parents need to know that the overstimulation of these kinds of neurochemicals can create unintended consequences—such overstimulation is implicated in producing addictive behaviors. Consider what *The Telegraph* reported: "Children as young as four are becoming so addicted to smartphones and iPads that they require psychological treatment." Dr. Richard Graham, who launched the United Kingdom's first technology addiction program, says that young technology addicts experienced the same withdrawal symptoms as alcoholics or heroin addicts. He points out, "They can't cope and become addicted, reacting with tantrums and uncontrollable behavior when they are taken away."[44]

Digital Controversy Not Limited to Kids: "Brain Games" Don't Work for Adults

A 2014 headline in the *Stanford University News* reads, "Scientific evidence does not support the brain game claims, Stanford scholars say." The article goes on to say that "sixty-nine scientists at Stanford University and other institutions issued a statement that the scientific track record does not support the claims that so-called 'brain games' actually help older adults boost their mental powers."[45]

"We object to the claim," wrote the scientists, "that brain games offer consumers a scientifically grounded avenue to reduce or reverse cognitive decline when there is no compelling scientific evidence to date that they do." What the research does show is that if you play one of these brain games repeatedly, you'll get much better at playing that game or games very similar to it. But this increase in a game still doesn't appear to translate into increasing an individual's intelligence or memory in other areas of his or her life. The researchers also point out the time spent on computer games takes away from real-world activities that produce high levels of meaning, purpose, and joy in our lives, such as socializing, gardening, reading, volunteering, and exercising. These are the kinds of activities that have been proven to benefit other cognitive functions and help us feel emotionally balanced and fulfilled.

Again, this kind of scientific opinion doesn't seem to have made it into the minds of the general public. The brain game company Lumosity, for example, has over eighty-five million users,[46] and probably very few of them are aware of the scientific consensus.

Why Aren't Digital Learning Programs Delivering?

The more real the environment, the greater the learning. The main drawback of traditional stand-and-deliver education is that the classroom is it's not real life. The same obstacle hinders electronic learning programs: Digital media isn't real life!

The Hebbian learning rule—which essentially says the more neurons you have firing simultaneously in your brain, the more learning will take place—has become a basis for the 21st-century neuroscientific view of how learning occurs in the brain.[47] So whether you're trying to get a young or old brain to learn, real-world environments will always be most effective because these environments fire more neural tissue than classrooms or digital environments possibly can.

It's also helpful to think about digital media in the context of human evolution to get a more global perspective on why digital media isn't producing profound learning. Humans have been on Earth for about six million years, and for 99.99 percent of that time, we've done all of our learning in the real world—not on computer screens! Our genetic plans and brains have been cultured over the millennia to be activated by stimuli from natural environments. To expect our biology to adapt in a few years to electronic ways of learning like iPads and smartphones—when in the previous six million years, it did all its learning in real environments—appears to be a major obstacle for these digital learning efforts.

Now, this doesn't mean that electronic media doesn't have a place in trying to increase the *Adaptability/Intelligence Factors* of the young and old. But the research does indicate that, as yet, we haven't figured out how to effectively turn on our Learning Code with digital media.

I don't condemn the digital learning pioneers. I applaud them. They're trying to add value to the lives of children and their parents. We're just beginning to understand how to activate our biological Learning Code. In my estimation, every one of the founders of these learning programs strongly believed they were "doing good" with their programs.

It takes a village, and at The Leaning Code Institute, we've learned a great deal from these electronic media innovators. It's my hope that the knowledge in this book will support these entrepreneurs and others in their future efforts to have positive impacts on the lives of children and adults worldwide.

We must also remember that the ultimate task of parents and educators is to use the relevant science not to create super children but to nurture their souls by helping them build meaning and purpose in their lives. Ambitious parents might heed the experience of Yeou-Cheng Ma, the older sister of renowned cellist Yo-Yo Ma. Now a pediatrician, Yeou-Cheng, took up the violin at age two and then went on to suffer a devastating breakdown aged fifteen, just before she gave up her solo career. She has little patience with parents who push their children. "The job of a child is to play," she says. "I traded my childhood for my left hand" (the bow hand).[48] What sadder statement could be made about focusing on the child's achievement at the expense of her soul?

Cracking the Learning Code

Knowledge about super-learning windows of opportunity and the perils of delivering too much information to the brain at the wrong time is invaluable in creating learning programs that provide the correct environmental stimuli in the correct amounts at the correct times. No educational program should ever be developed without a keen awareness of these windows of learning opportunity.

28

Danger! When Media Becomes the Environment

I like to play indoors better 'cause that's where all the electrical outlets are.

—*A tourth-grader's comment in*
Last Child in the Woods, *by Richard Louv*[1]

If our environment is all-powerful in the process of building and shaping the genetic and neurological structures that are the wellspring of our actions and behaviors, what happens if we spend approximately 42 percent of our waking hours sitting in front of some sort of media screen instead of interacting with the real world?[2, 3]

In the last chapter, we looked at why digital learning programs aren't working the way their creators hoped. What follows here is a deeper biological perspective on how TV, computers, movies, iPads, and smartphones are affecting your biology. The research into how the media affects your brain and genes is still in its early stages, with most of it occurring in the last twenty years. And it is continually evolving. What we have done here is to take the relevant research and theories that focus on the link between the media you spend so much time with and depression, loneliness, sleeplessness, addiction, violence, inappropriate sexual urges, attention deficit disorder, numbness to real life, lack of empathy, early puberty and much more. Our goal is for you to gain a new perspective on the real power of the media and its impact on your biology.

The Repercussions of a Media Diet

As a society, we're just beginning to wrestle with what happens when the media becomes the primary source of environmental stimuli to the 100 billion neurons and thousands of trillions of connections in your human brain. It's estimated that the average adult spends an amazing eleven hours per day in front of some form of media.[4] But even more distressing is a report from the Kaiser Family Foundation that found that eight- to eighteen-year-olds devote an incredible seven and a half hours a day to media. That's more than fifty-three hours a week.[5] Even the lives of kids under eight are packed with media—one study shows that 45 percent of kids under eight have their OWN mobile device and 98 percent of under-eights have access to mobile devices at home.[6]

The danger of spending up to a third or more of our waking hours with media comes both from limiting our interactions with the real world and from flooding the mind with huge quantities of sexual, violent, unusual, or worthless information that doesn't support increasing our *Adaptability/ Intelligence Factors*. The Hindu saying *sarvam annam* means "all is food."[7] We must ask ourselves what type of nourishment we're feeding ourselves or a child's brain. Consider before they reach eighteen years of age, their young, once-innocent mind will have witnessed in media over 250,000 acts of violence and over 100,000 sexual scenes.

One report showed that, by the time a child is five, they will have watched more than 6,000 hours of television.[8] That's more time than children spend with their parents! Even seemingly innocent children's cartoons, because of their design, can unwittingly add to the negative nourishment: to keep interest high, most thirty-minute cartoons include at least eight acts of violence.[9] Also, consider that the American Academy of Pediatrics found that the average young viewer is exposed to 14,000 sexual references each year, yet only a handful of these references portray responsible sexual behavior.[10]

Danger: Aggressively Task Switching! Is It Shrinking Your Brain?

But first, no discussion about the media's impact on the brain would be complete without addressing our attempts to multitask, which as we have seen is really the brain task switching frantically. Nature didn't design your

Kids and teens average seven hours a day looking at screens. Could this possibly be good?

brain to get 30 to 40 percent of its waking stimuli from the media or to task switch at the rate you're doing it today. According to the research, the average person now checks their cell phone an astonishing eighty times every thirteen minutes![11] In more bucolic times, the only task switching we did was equivalent to looking at a flock of birds while we plowed the field.

We watch TV and text or surf the web and/or talk on the phone all at the same time, without thinking of the consequences. And our brains are rebelling. Although the research is admittedly in its infancy, initial studies reveal that the consequences of task switching may be much more severe than we'd imagined.[12]

Researchers at the Institute of Psychiatry at the University of London studied 1,100 workers and found people who task switched during intellectual tasks experienced IQ declines of 15 percent. This is similar to what you'd expect if these workers had smoked a few joints of marijuana or stayed up all night! Put another way, task switching lowered these workers' IQ levels to the level of eight-year-olds.[13]

The journal *CyberPsychology & Behavior* reported the more time people spent task switching, the harder they found it to concentrate on a single intellectual task, such as studying a textbook.[14] At the University of California, researcher Russell Poldrack used brain imaging to discover that when you task switch at the same time you're trying to learn something new, the knowledge isn't sent to the hippocampus (which helps you direct new knowledge to be stored) but instead to the striatum, which is terrible at storing facts, ideas, or concepts. This suggests that if you're trying to learn something new while task switching, you're probably wasting a lot of your time.[15]

As we discussed earlier, it takes energy to shift your attention. Remember that your brain generates twenty watts of power and if you try to do two tasks at once, switching between tasks can steal up to 50 percent of your brain's wattage. This means that each task doesn't get half of your neural energy—it gets much, much less. This reduction in neural energy to each area is one reason studies show that, when we multitask, accomplishing a task can take us 50 percent longer on average, and we can make up to 50 percent more errors.

Researchers at the University of Sussex used MRI scans to compare brains of heavy and light task switchers. They found that people who used a higher number of media devices at the same time actually had less gray matter (brain cells and dendrite branches) in the part of the brain known as the *anterior cingulate.*[16]

Why is this bad? Because in addition to other functions the anterior cingulate has influence upon thought and emotional control. It's also one of the few parts of the brain that contain substantial cells called *spindle neurons,* which support the accelerated communication between different areas of your brain. So, if you start wiping out the gray matter in the anterior cingulate, you can end up with a brain that doesn't think, feel, or communicate with itself very well.

How can task switching shrink brain matter? When you frantically task switch, your engaging your anterior cingulate to help you rapidly switch between those tasks. If you're aggressively accessing it hundreds to thousands of times a day, the anterior cingulate doesn't get much time to rest and recuperate. One theory is that this constant switching of your attention causes more stress to that organ than it was designed for, which either kills the neurons and their branches or, in some other way, damages them.

The next section explains how the media directly affects brain functions, actions, and behaviors. As you read it, remember that, for many of us, these effects are being amplified because we're task switching.

The Media Link: Violence, Sex, Loneliness, Depression, Sleeplessness, Addiction, and More

Since the advent of television in the 1950s, thousands of studies have linked the media to everything from a decrease in creativity to an increase in violence and inappropriate sexual behavior.[17] Our children appear to be at the greatest risk. Excess media usage can inhibit social interaction, hinder parental bonding, cause depression, and stimulate the neural chemicals that are the basis of addiction.[18] It may even be a factor in the increasing rates of suicide. Excess media usage is a primary factor in limiting the free physical play that is instrumental in building efficient motor programs, and media use has been directly connected to the obesity epidemic among children.[19] As we will see in more detail later in this chapter, the media has also been shown to influence the hormones responsible for inducing our different sleep stages, during which we dream and which are vital for the consolidating of short-term memories into long-term memories as well as inducing REM sleep, crucial for healthy emotional development.

After decades of research on how the brain selects information from the environment, I'm convinced there's no more significant pollutant for a young child's mind and development than excess media usage. I'm not alone on this point:

- Sixty-five percent of parents believe that television contributes to violence and erodes family values.[20]
- Ninety-five percent of parents think it's vital to control children's exposure to the media.
- Almost 80 percent of Hollywood executives questioned agreed that there's a link between TV violence and real-life violence.[21]
- In a survey of ten-to-sixteen-year-olds, 62 percent said sex on TV had influenced their peers to have sex at too young of an age.[22]

In response to our society's concern over the negative impact of television, the American Medical Association House of Delegates passed this resolution: "The House declares TV violence threatens the health and welfare of young Americans."

In the 1950s, sexual media content was taboo—as the song refrain goes, to see "Mommy kissing Santa Claus underneath the mistletoe" on TV was a big deal. Compare this low level of sexual stimulation to the fourteen scenes of sexual content the average child now gets from the media each day. Here are some other differences that suggest a concerning pattern:

Can Media Overstimulation of the Pituitary Gland Encourage Sex Sooner?

It's believed that when the brain receives too many sexual stimuli at too young of an age, the *regulatory genes* in the pituitary gland prompt the early release of the sexual hormones—luteinizing hormone (LH) and follicle-stimulating hormone (FSH)—causing the onset of puberty and a fixation on sex. Excessive sexual content in the media is like telling the genes in the pituitary, "Hey! Let's get going! There's a lot of sex in this environment, and we need to get ready for it now!" The early release of LH and FSH also prompts pubic hair growth, the development of the gonads (sex organs), and increased production of estrogen and testosterone, which in turn advances the onset of menstruation in girls and first erections in boys.[23]

One of the country's leading researchers on puberty, Dr. Marcia E. Herman-Giddens at the University of North Carolina, says we now know that the media can alter the body's chemical climate: "Looking at something," she says, "can increase the amount of sex hormones in your body."[24] Anyone—young or old—who has ever looked at sexually suggestive pictures or videos knows this to be true.

Many variables other than the media could account for the dramatic increases we've seen in earlier sexual activity now exhibited by our youth. For example, there's speculation that hormones in food and milk may prompt the early onset of puberty and sexual urges.[25] But any good researcher would have to look at the increase of sex in the media as a major potential culprit.

Below are some examples of changes in our youth:

- Fifty years ago, menarche, the beginning of menstruation, happened around twelve to fifteen years of age; now, it's commonly seen in eight-year-old girls.[26]
- Fifty years ago, the age that most boys were able to physically engage in sex was twelve or thirteen; today, a substantial number of ten-year-old boys are engaging in sex.[27]
- Five decades ago, the reported rape of young females by males under the age of ten was zero; today, the number is in the thousands.[28] Today, there are thousands of preteens having babies, and some nine-year-old girls are becoming pregnant.

Light, Puberty and the Pursuit of Screen Time

New research indicates that the artificial light from computers, TV screens, and other electronic devices may help bring on puberty sooner by reducing levels of melatonin.

- In 1950, the average age for the first act of intercourse was after the twenty-first year; today, the average age is fifteen–sixteen years old.[29] One study found that 62 percent of all high school seniors have had intercourse. Another study reports 20 percent of all young teens have had oral sex before they finish ninth grade (many teens say it's "not really sex").

Is the Media Interrupting Super Windows?

The thousands and thousands of sexual images that a child sees before he or she reaches adulthood lead to hormonal changes that may also interfere with the *super windows* of developmental learning (see Chapter 24). Because genes are prompting specific brain areas at specific times, a child can bond with their parents before they bond with a peer, and they're able to learn concrete concepts before abstract ones. When the brain's energy is diverted to the brain areas used to fixate on sex—at the expense of the vital developing brain areas that should be receiving this neural energy—we may be retarding natural development.

The brain has lots of systems that work independently, and as discussed when you turn one on, you often turn others off. We're just beginning to look at how early sexual bombardment through the media interferes with parental bonding, socialization, creativity, and intellectual capacity. Herman-Giddens points out, "It extremely concerns that we have a generation… being exposed to all these factors at once, and we don't know where it's heading." She also points out that because of the amount of sex in the media, "the old boundaries have been blurred between childhood and adulthood."[30]

New research also indicates that the artificial light from computer and television screens may help bring on puberty sooner by reducing levels of melatonin. This hormone not only helps induce sleep but also has been implicated in the onset of puberty. Researchers at Florence University in Italy found that when children watched media, their melatonin decreased by 30 percent. The large decrease caused by media usage is now being looked at as another causal factor in the early onset of puberty.[31]

My Child's Unhappy Experience

As a parent, I have personal experience of how this exposure to sexual images affects a young brain. At around eleven years old, one of my sons and his buddy figured out how to bypass the parental controls installed on our computer and ended up accessing pornographic websites. I knew my son was up to something because this usually easygoing kid had become very irritable. When I finally found out what the problem was and talked to him about it, he indicated that although these sexual images were taking him away from the music and sports he loved, they were so alluring he didn't want to stop seeking them out.

I explained that the images were indeed robbing him of the joy of childhood and potentially redirecting his neural energy away from the specific brain structures—such as his motor cortex and music centers—that he needed to fully develop so he could be the great football player and musician that he wanted to be. This knowledge allowed him to stop seeking these images of his own volition, but there's no doubt in my mind that, at least for a short period, my child's neural development and his view of bonding and love were sadly and negatively altered.

How Does Media Violence Affect the Brain?

In the last 10,000 years, humans haven't had the kind of major genetic change that would lead to a significant shift in how our brains function. Yet the number and kinds of stimuli we perceive in our modern world are far greater and far different than the environment we were built for. No one was designed to see 220,000 acts of violence before age eighteen. Media violence is not benign, as some would want us to believe. It causes neurological changes, and when the brain changes, behaviors change.

The dangers of too much media violence delivered to the developing brain too soon can be best comprehended by grasping the concept of three neurological processes: the emergency response, habituation, and addiction. The next sections offer a quick overview of these brain processes, followed by a more in-depth look at how the media affects each one.

The Emergency Response: The Short Story

Also called the *stress response,* the *emergency response* is an ancient survival instinct that automatically prepares us to fight or flee when we perceive danger. Repeated usage of the emergency response when witnessing large amounts of media violence can cause dramatic chemical changes in the brain, which forces some of us to adopt an aggressive view of the world when we're confronted with stress in our lives.[32]

Habituation: The Short Story

Habituation is the neurological process the brain uses to normalize itself after engaging in numerous emergency responses. This process ensures that the brain will stop producing a stress reaction if it no longer perceives certain stimuli as worthy of reaction. Habituation is why we become numb to media violence and why the media must keep upping the gore to keep our attention.[33]

Addiction: The Short Story

Addiction is defined as a physical and psychological craving that develops into dependency and continues even though it causes physical, mental, and social harm. We now know that the neurological changes brought about by watching violent and sexual media can activate the brain's addiction circuits.[34] This is why so many kids and adults become addicted to video games and porn.

FAST FORWARD *or* DEEP DIVE

If you need a deeper understanding of how media violence affects the brain, please read on. If you believe you have enough knowledge, please skip to the section titled "First-Person Shooter Games: More Real = More Change."

The Emergency Response and Your Brain: The Long Story

The stress response is a complex biological process that dramatically inhibits learning and can cause neurological damage if it goes on too long. We'll look at the stress response in more detail in Chapter 38. In the meantime, here's a general view of how the emergency response triggered by violent media affects your brain and body.

Imagine that you're watching an action movie or playing an action video game. The violence prompts the pituitary gland to squirt the hormone ACTH into the bloodstream. When ACTH reaches the adrenal glands, they release the stress hormone cortisol, which in turn prompts the brain to release a cascade of neurochemicals, including norepinephrine and dopamine. You become hypervigilant as your body tenses, your mouth goes dry, your hands get clammy, your heart rate and blood pressure increase, your pupils dilate, your digestion slows, and your blood flow shifts away from your spleen, liver, kidneys, and other internal organs to your major muscle groups so you'll have the power necessary in your back, arms, and legs to fight or flee the perceived threat.[35] As the action rises and subsides on the screen, your brain and body are continually jerked in and out of this ancient, innate biological reaction designed to help you fight or flee danger. The average child who is exposed to over four hours of television per day sees forty-nine acts of violence.[36] By no

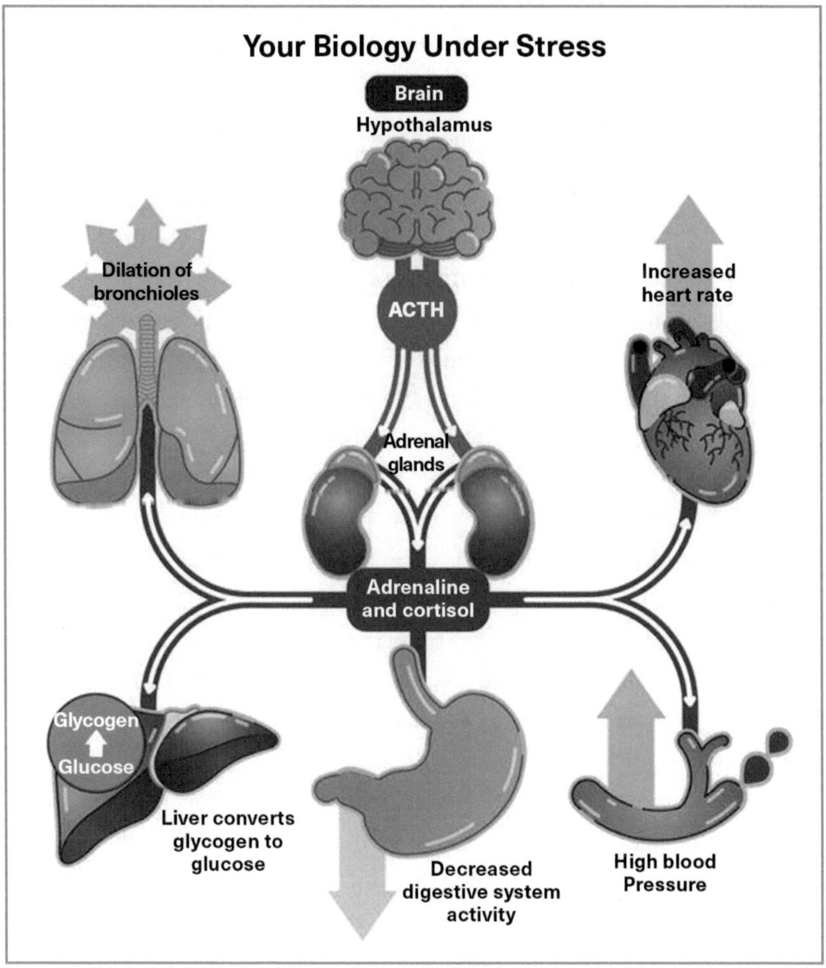

stretch of the scientific mind could this many trips to and from the emergency response be considered a good thing.

Whipping a child's mind and body in and out of the emergency response hundreds and hundreds of times per week can adapt brain chemistry so that the child's cortisol levels can remain permanently high. High cortisol levels create a neurochemical climate that keeps the child on edge and gives him a hair-trigger temper. Excessive media violence and high cortisol levels force the child's brain to see the world, not through peaceful, rose-colored glasses but ominous, dark ones—and react accordingly. Dr. Linda Mayes at Yale University says that if you experience the emergency response "again and again, it changes the structure of the brain."[37]

While both children and adults are susceptible to stress-induced brain changes that become permanent, children are more at risk because of the malleable nature of the immature brain, with its trillions of connections yearning to be stimulated. Megan Gunnar at the University of Minnesota notes, "The kids with the higher cortisol levels score lowest on inhibitory control [and] have problems in attention regulation and self-control."[38]

The Evidence: Media Violence Can Prompt Aggression

Numerous studies suggest that children behave more aggressively after watching violent media. One study from the University of Pennsylvania showed that children who watched violent media were more likely to strike out at playmates, argue, and disobey authority and had less patience than those children who watched nonviolent programs.[39]

In another study conducted by Leonard Eron and his associates at the University of Illinois, children who watched long hours of media violence in elementary school tended to show a higher degree of aggressive behavior when they became teenagers. Observing these young people until they were thirty years old, Dr. Eron found that those who had watched a lot of media when they were young were more likely to be arrested and prosecuted for criminal acts as adults.[40]

Children's shows are no safe haven, either. J. D. Singer and his wife Judith report that a group of two- and three-year-olds "literally fought" after watching violence in the show *Mighty Morphin Power Rangers*.[41, 42]

Habituation: Too Much Violence Can Make You Yawn!

Jerking your brain in and out of its emergency response is bad, but having your brain habituating to violence, eventually perceiving it as normal, is even worse. The scientific description of *habituation* is the loss of sensitivity on the part of a neuron resulting from a prolonged pattern of stimulation. Your brain is designed to react to novel stimuli like violence, but after a new stimulus becomes routine, your brain uses habituation to quell its reaction.[43]

Without habituation, your brain would give equal value to all incoming stimuli. You would be unclear about what is novel and should be paid attention to and what is normal and demands no reaction. Research shows that, even from birth, an infant will show a waning of response to an event that's repeated.[44] Wild birds, squirrels, and deer can be trained to eat out of humans' hands because of habituation. After repeated exposure to humans, these wild animals' brains become habituated to us.

Habituation to media violence has two negative effects. First, it tricks the brain into accepting violence as a normal part of life. Second, once society's collective mind becomes normalized to one level of media violence, media producers and directors must increase the level of bloodshed to keep our attention.[45] In *The Owner's Manual for the Brain*, author Pierce Howard notes that habituation is maintained as long as the level of stimulation remains constant. Habituation can be broken, and novelty and attention rekindled if the quality and quantity of the stimulus are changed.[46] In other words, even if you've been watching—or letting your child watch—violent programming, you can scale back now. After some time away from the media, the brain won't be as habituated to violence.

After our brains have normalized to a level of media violence, producers of movies and video games must increase the level of gore to ensure that we'll stay tuned. This turns into a never-ending escalating spiral of graphic violence, to which our brains continually normalize. From the 1950s to today, each decade has seen an increase in blood and gore in the media. Producers know that showing violence is a simple way to get people to pay attention. Unfortunately, this also causes us to keep habituating to increasingly graphic scenes.

Downregulation is the major process that drives habituation. During this process, neurons literally change their shapes from repeated stimulation. At the synapse, neurons have receptor sites, on which neurochemicals from other neurons dock and transmit their information. If your brain produces too many neurochemical messengers too many times in reaction to a novel stimulus, neurons can reconfigure themselves so that fewer receptor sites are available to these incoming messengers. It's as if the neurons were saying, "Yeah, we already saw thousands of shootings on TV and video games, and it's getting boring, so we're going to close down some receptor sites because the information is old hat."[47] When downregulation occurs, the brain perceives previously novel and abnormal images as routine. The reduction in receptor sites allows the brain to normalize itself to media gore. Writing about habituation, Howard says:

> *Violence on television produces violent behavior in young people. Everyone knows that. Once one has habituated to violence as a way of life, however, everything else is boring.*

One theory of why more and more youth can commit violent acts is that their excessive exposure to violent media has caused these young brains to downregulate so that violence is perceived as normal. The violence they plan and commit in real life, like the Columbine, Virginia Tech, and Sandy Hook massacres, isn't out of the ordinary for them because their brains have stopped receiving the neurochemical messages that would let them know that what they're doing is abnormal.

Media Violence: Opiate of the Masses?

As noted earlier, *addiction* is defined as a physical and psychological craving that develops into a dependency, even though it causes the addicted person physical, psychological, and social harm. The emergency response due to watching violent media activates the dopamine neurochemical system, which has been implicated in addiction. Dopamine is significant because it keeps your attention riveted to events you deem as meaningful, but its release also can numb your pain.

Dopamine blunts your pain by helping to prompt and prolong the effect of endorphins, which are neuropeptides that act as the brain's natural morphine. Nature saw that in dangerous situations, it's a good idea for the brain to release neurochemicals that not only keep us hypervigilant so we can effectively focus on threats but also numb any pain so we can continue to fight or flee if we're hurt by those same stimuli. But in the absence of pain, your brain perceives the numbing effect of endorphins as pleasurable. In the end, the brain likes the buzz that these two neurochemicals provide. The reason drugs are so addictive is that most drugs stimulate both dopamine and endorphins.[48]

The theory is that in the process of witnessing forty-nine acts of media violence in a day, the average child ends up having his or her brain continually exposed to high levels of both dopamine and endorphins. And just like a drug addict who gets hooked on his or her drug of choice from the chemical blast it produces, a child can become addicted to the neurochemical high from watching violent media. Although both children and adults are susceptible to becoming addicted to their own neurochemicals, children may be more at risk. This is because children have 40 percent more dopamine receptor sites than adults.[49] Our understanding of brain chemistry is lending credence to the idea that the media is an "opiate of the masses."[50]

My family witnessed the addictive effects of media on a city youngster, age nine, who had spent an average of four hours a day with TV and video games before he came to visit us in Colorado. At the end of his visit, in which he had spent five days hiking, golfing, horseback riding, and canoeing, his uncle asked him if he wanted to stay longer. "I really want to," the boy said, "but when I get up in the morning, they don't watch TV or play video games, and I feel so uncomfortable without them, I don't think I can stay!" Like a drug addict, this child was experiencing withdrawal. His addiction to the neurochemicals that the media produced was forcing him to choose the media over real-life childhood adventure. How sad!

First-Person Shooter Games: More Real = More Change

Research shows that viewers of television and movies and users of violent video games can exhibit more aggressive tendencies and lower remorse, depending on how much violence they're exposed to.[51] But because video games allow the user to participate in the violent action, many researchers believe that the violent video games called first-shooter games can cause more damage to the brain than violent TV shows and movies. For young, developing brains, these games can be perceived as real life, and, as we've seen, the more real the environment, the more the brain and behaviors change. Dr. Jeanne Brockmyer, a clinical child psychologist and professor emeritus at the University of Toledo notes "The research is getting clearer that over the long term, people with more exposure to violent video games have demonstrated… lower empathy to violence,"[52] This is why a military expert on *60 Minutes* called first-person shooter games "a how-to manual for killing without conscience."[53]

He is right, researchers have found that violent media leaves the prefrontal cortex under-stimulated or depressed. This area of the brain controls compassion and empathy.[54] Therefore, we could say that first-person shooter games can teach children how to kill without stimulating the part of the brain that allows them to see and feel the ramifications of their actions. The two teenagers who performed the cold-blooded shootings at Columbine High School and the gunman at the Sandy Hook School shooting all had access to guns and were fans of first-person shooter games, which gave them clues on how to efficiently carry out their shooting and murdering sprees, apparently without remorse.

TV News Also Has an Effect on Brains

Many parents who don't let their children use violent media routinely let them watch the TV news, believing it's safe. University of Wisconsin psychologist Joanne Cantor surveyed 300 parents of children in kindergarten through sixth grade. She found that graphic news visuals of murders, war, and natural disasters led to nightmares, sleep disturbances, and other lingering problems in their children. In fact, 30 percent of parents said their children reacted with long-term fear to news they had seen in the previous year. Exposure to news violence made these children overestimate the amount of violence they would encounter in their hometown. Cantor concludes that parents shouldn't allow children under eight to watch television news, because neurologically, they're too impressionable and can't separate what's on television from their own reality.[55]

A 1990 study reported in the *Journal of Social Psychology* found that the more violence in a newscast, the more violent the attitudes of children exposed to it. Children who watched the local Northern Ireland newscast developed a more violent attitude than those who watched the BBC national news, which contained a quarter of the references to violence as the

Northern Ireland news. The result was irrespective of whether the children lived in the more-violent Northern Ireland or safer mainland Britain.[56] (Many areas of mainland U.K. receive the local news for Northern Ireland.)

And the effect of the news on our brains isn't limited to kids. So I could keep focused on my research, I stopped watching the news every time I turned on the tube. The news had a powerful and long-lasting effect on my focus. It took my attention away from what was most meaningful to me (you're reading it right now) and got me stuck in an obsessive loop of the US political mess and the world's miseries.

Media Can Steal Relationships, Joy, Social Interactions, and More

Not only can excessive media use have a direct effect on neurochemistry, negatively influencing behaviors around sex and violence, but it can also have an indirect effect on play, bonding, creativity, social interaction, and rates of depression. When a child spends 36 percent their time in front of some form of media, not interacting with the real world, that child's brain is robbed of the opportunity to engage in activities that more appropriately support its healthy biological development.

The human brain has one of the longest maturation processes of any species. It was designed to go through its super-learning windows and development not comatose in front of a media screen, but in the real world, physically interacting with the environment. Excess media usage usurps the time children should be spending developing healthy physiological and psychological networks, which will aid them later in life.

No longer do experts look at the physical play in the real world as frivolous, wasted time. Instead, they see it as instrumental in helping provide our genetic and neurological structures with the stimuli they need for healthy development to unfold. Animal studies have shown that if you interrupt the ability to play, normal development will be retarded.[57]

It's kind of humorous, but researchers take play very seriously. They break play into many categories, including exploratory/sensory-motor play, rational/functional play, constructive play, dramatic/symbolic play, games with rules, and roughhouse play.[58] Extended periods of play in childhood are now recognized as the basis of emotional intelligence, social bonding, physical motor development, cooperation, increased problem-solving, enhanced communication skills, fitting into the societal hierarchy,[59] creativity, coordination, and spatial awareness. As neuroscientist Jaak Panksepp says, "In short, the brain's play networks may help stitch individuals into the social fabric that is the staging ground for their lives."[60] He adds, "Is it any wonder then that play is such fun, perhaps one of the major brain sources of joy?" But excess media usage steals this magnificent source of joy.

Bonding to the Media Instead of to Parents

Statistics indicate that we're developing generations of children who will have spent more time bonding with the media than to their own parents! In their book *Making Connections,* Renate and Geoffrey Caine note that by the time the average child reaches sixteen, they will have spent six hours in front of the media for every one hour spent with their parents. Another survey noted that adolescents spent only forty-seven minutes of quality time with their parents each week, compared to the thirty-one and a half hours spent in front of the media.[61]

Any time that a child spends with the media and not with his or her parents is bonding time lost. Parental bonding stimulates and develops the neurological networks responsible for producing love and empathy. The development of these networks helps us make friends and effectively bond with our life mates and children. The full development of these networks is what makes us most human because these networks are the basis of the compassion we have for each other. But excess media is breaking that bond. In busy households, where both parents work or where there's only one parent, the media often becomes not only the babysitter but also the surrogate parent, from which the child builds his or her worldview. The child ends up bonding not to his or her parents but to the media.

Excess Media Use Limits Creativity and Imagination

Excess media use floods the brain with thousands upon thousands of images a day. When a child forgoes creative play to sit still in front of a screen, this creative playtime lost is time the child could have spent building the neural circuits that will be the foundation of their creative intelligence. Communication researchers J. L. and D. G. Singer at Yale University showed culturally appropriate media to children who were identified as having lively imaginations. After the children viewed programs such as *Mr. Rogers' Neighborhood* and *Sesame Street,* their creative play actually decreased.[62]

Jerome Singer, former professor of psychology and child study at Yale, warns that excessive media use prevents children "from doing what they ought to be doing—learning to create mini-worlds that they can control. Kids who can do this are more cooperative, more likely to become leaders, less likely to be overly aggressive." He says that, conversely, kids who watch more TV and thus engage in less creative play show "negative emotions: not only anger but also more stress or crying."[63]

Dangers of Limiting Physical Action

Marcia Mikulak found that children in nontechnical societies have 25 percent greater sensory awareness than children in the United States.[64] Sitting in front of a screen for 29,000 hours before turning eighteen could be a significant factor in a child's loss of sensory awareness. In the book *Learning Through Play,* Jean Marzollo and Janice Lloyd state that:

> *For a child to understand the abstract concept of "roundness," he must first have many experiences with real round things. He needs time to feel round shapes, to roll balls, to think about the similarities between round objects... It is this combination of action and concreteness that makes play so effective as an educational process.*[65]

Pearce says physical play "is the very force of society and civilization, and a breakdown in the ability to play will reflect in a breakdown of society."[66] Frank Wilson, who is an expert on the evolution of the human brain and a professor of neurology at Stanford, comments on the media learning programs, "We've been sold a bill of goods—especially parents —about how valuable computer-based learning is." He says students struggle to understand certain concepts "because these students have so little real-world experience; they've never siphoned anything, never fixed a car, never worked on a fuel pump, may not even have hooked up a garden hose. For a whole generation of kids, direct experiences in the backyard, in the tool shed, in the fields and woods, has been replaced by indirect learning, through machines. These young people are smart... but now we know that something's missing."[67]

Media Use Steals Personal Meaning

Excessive media usage steals time when you could be doing something meaningful in your life. Doing something meaningful activates the neurological structures that make up the *Meaning Network*. As was discussed, when your *Meaning Network* is stimulated, you feel valuable, fulfilled, and complete. When you spend up to one-third of your life in front of media, you lose time that could have been spent stimulating your *Meaning Network*. Studies have linked excessive media usage to depression. Clifford Stoll, the author of *Silicon Snake Oil,* speaks of excessive internet surfing as a cause of depression: you watch your life dribble out your modem. You sit down for five hours, pop back up for air. Are you a better person? Are you a changed person? The answer's no. You are just a five-hours-older person.[68, 69]

Other studies confirm that levels of depression actually rise after media usage. Research at Carnegie Mellon University found that people who spend more time on the internet each week suffer higher levels of loneliness and depression than those who use the internet less frequently.

Suicide rates in children and teens were up 150 percent between the 1950s and mid-1980s, and increased again by another 56 percent from 2007–2017.[70, 71] Research connecting the media directly to higher levels of suicide is in its initial stages, but there appears to be a correlation between the increased suicide rates and excessive use of media.[72]

Attention Deficit Disorder and Media

A new theory on the cause and/or exacerbation of attention deficit disorder (ADD) and attention deficit hyperactive disorder (ADHD) is being traced to media usage. One group of researchers claims that exposure to media before the first grade limits the brain's ability to focus for long periods in any one area. They point to the fact that to keep our attention, the image on a TV screen changes, on average, once every three to four seconds. Therefore, a child watching a thirty-minute TV program isn't paying attention to anyone's image for more than a few seconds. Researchers say that when you multiply the number of image changes per hour by the 5,000 hours of TV that the average child watches before first grade, you end up with an environment that forces the brain to adapt to extremely rapid stimuli change.

The researchers' hypothesis is that this bombardment of auditory and visual stimuli—at times when these children's genetically timed super-learning windows and development are open their widest—eventually hardwires a preference for rapid stimuli into their brains just like a child born with Attention Deficit Hyperactivity Disorder (ADHD). These researchers fear that the media may create brains that can't process information from the real world and are hard-pressed to follow a slow-moving teacher for six hours a day when the children enter school.[73]

The Media Messes with Sleep

Initial research indicates that artificial light from media screens negatively affects levels of melatonin, the hormone instrumental in regulating the sleep process. Dr. Alessandra Graziottin, director of the Centre for Gynecology and Medical Sexology in Milan, notes that a 30 percent decrease in melatonin levels seen in children who use a lot of media may bring on "sleep disturbances, nightmares, difficulty in getting to sleep and so on."[74] From 2000 to 2004, the spending on sleep aids for youth in the ten to nineteen age group went up 223 percent.[75]

And even once the child does fall asleep, the media they watch is still affecting them. Sleep isn't just downtime for the brain; during sleep, what you've done during the day gets imprinted into your long-term memory (see previous chapters for discussion of the power of relaxed states in the learning process). A child spends much of their sleep time unconsciously processing what they've witnessed during the day—that puts a whole different spin on telling a child to have sweet dreams, especially if they've just seen a violent movie or played a violent video game.

When you slumber, you go through five distinct stages of sleep, which help you consolidate information into your long-term memory banks. Many neuroscientists consider the REM (rapid eye movement) stage when you do most of your dreaming to be the most important. During this stage, with the help of your other sleep stages, not only do you encode much of what you were exposed to during the day into long-term memory, but this stage is vital for emotional learning. You generally go through four to five rounds of REM sleep a night.

When this stage of sleep is filled with media violence, and sex, the neural networks that direct a child's dreams, thoughts, behaviors, and emotional maturation become shaped by these images. The result is children who can become maladapted to the real world and who lack the emotional skills with which to effectively relate to others. Though more research needs to be done in this area, it's believed that if a child is getting 30 to 40 percent of his or her daily stimulation from media filled with sex and violence, then an equal amount of REM sleep is spent processing this sexual and violent data.[76]

Cracking the Learning Code

The environment is all-powerful in the process of building and shaping the genetic and neurological structures that are the wellspring of all of our actions and behaviors. We come into the world incomplete. Do we want the media to complete us?

29

Why Are You A Copycat? It Helps You Survive

Imitation is not just the sincerest form of flattery—it's the sincerest form of learning.

—George Bernard Shaw, British playwright[1]

In the 18th century, the poet Edward Young asked, "Born originals, how comes it to pass that we die copies?"[2] Over 250 years after Young posed this question, we now have the answer: We're copies of each other because we have a genetic drive that compels us to imitate one another. In a way, we're a species of copycats. Stop for a second to check out your appearance in the mirror. Why do you have that particular hairstyle? Why did you pick out that shirt or blouse? Why did you buy those shoes? For that matter, why do you yawn when you see others yawning or feel sad around depressed people?

Whether you like it or not, your genetic predisposition forces you to mimic the behaviors of those around you. Evidence shows that the way you dress, the car you drive, the number of words you know, the slang you use, the traits you prefer in a mate, how you parent, and even your emotional reactions are contagious. They are all influenced by your genetic urge to imitate others. In other words, no, my dear reader, you are not 100 percent in charge of your behaviors and actions; your inherent drive to mimic others plays a significant role.

Your Drive to Imitate: No Imitation, No Survival

Another main component in understanding how you learn and the power of your environment on your biology is grasping your insatiable drive to imitate. Even though most of us are not conscious of it, this urge is as compelling as the instinctual drives of eating, sleeping, nesting, hoarding, and mating. Because our genetic makeup is forcing us all to spontaneously and unconsciously mimic others, imitation must be consciously addressed in the development of any science-based learning system or behavioral-change program. At The Institute, we call all the genetic and neurological elements that drive you to imitate others your *Imitation System*.

Copycatting helps us quickly and easily increase our *Individual and Species Adaptability/Intelligence Factors.* Imitation provides members of a species with a much safer way to learn than the ever-risky proposition of trial and error. Consider your ancestor who, if not able to imitate the successful survival actions of others, learned the hard way that a saber-toothed cat was to be avoided. Having to be attacked by one of these mean critters to learn to avoid them wasn't the most optimal survival strategy for the individual or our species.

Biologist Lee Alan Dugatkin, author of *The Imitation Factor,* says that "the ability to incorporate what others are doing into one's behavioral repertoire" matters more than intelligence.[3] The genes that couldn't help us mimic others' survival behaviors never made it through the sieve of evolution, because the ancestors who carried those ineffective genes became some predator's lunch. Imitation provides such profound survival advantages that every species on Earth today, even lowly bacteria, have powerful gene configurations that predispose them to mimic the behaviors of others.

Why Humans Are Better Imitators Than Animals

You imitate better than any animal in the world. Over 2,300 years ago, Aristotle wrote, "Imitation is natural to man from childhood, one of his advantages over lower animals being this, that he is the most imitative creature in the world and learns at first by imitation."[4] Aristotle was spot-on. When compared to other beings on Earth, humans' capacity to imitate dwarfs all others, allowing humans to learn things that animals could never learn—from how to cook a crêpe to how to fly an airplane.

V. S. Ramachandran, director of the Center for Brain and Cognition and a leader in the neuroscientific understanding of imitation, maintains that the great leap in technology and art that humans have achieved over our animal cousins is directly driven by our enhanced ability to mimic.[5] At The Institute, we have found that five elements support our understanding of

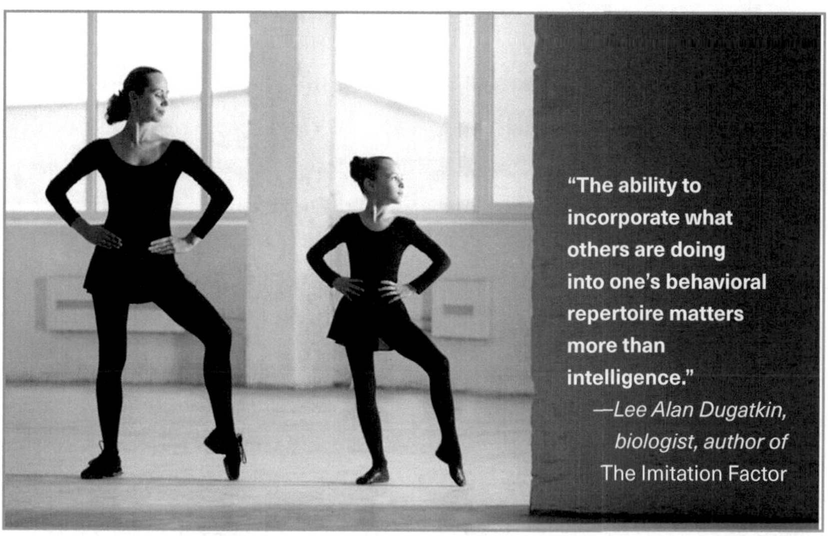

how we imitate others.

We call these elements your ***Imitation System***:

1. **Uncommitted neural tissue:** Because humans are born with more uncommitted neural tissue than other animals, our neural networks are, therefore, more susceptible to being influenced by imitation than any other species. Most animals are born with 80 percent of their neural tissue committed to instinctual drives. Humans, who are born with only about 25 percent of their neural tissue preprogrammed, have an astounding 750 million neurons with trillions upon trillions of uncommitted connections, which all can be wired through imitation.
2. **Long maturation period:** Humans have the longest maturation period outside the womb of any animal, and the consequences of this long finishing process are profound. It allows children an extended period to mimic the thousands of behaviors they witness in their parents, peers, and others.
3. **Long life:** With a few exceptions, such as bowhead whales and some sea turtles, humans have one of the longest life spans in the animal kingdom. This longevity allows you more time to mimic the actions of others. Although in adulthood, your neocortex is somewhat less flexible than in your younger years, it's still so malleable that imitation remains a primary form of learning until your dying day. Just go to a nursing home and see how many elderly women are tinting their hair the same color as their neighbors.
4. **A genetic predisposition to imitate:** While we've yet to identify the exact human genes responsible for creating our drive to imitate, the consensus among geneticists is not only is there a genetic predisposition to imitate but that the drive to imitate is *polygenetic*—that is, many genes work together to build neurons and brain structures that prompt us to mimic.
5. **Mirror Neurons:** Your human brain has a very sophisticated system of special neurons called *Mirror Neurons.* The initial hypothesis is that these neurons automatically fire and prompt us to unconsciously imitate the actions of others.

Before we dive deeper into this chapter, let me point out that the scientific investigation into the biology of imitation is in its infancy. The research is being updated yearly. More than any element in this book, what is revealed one year about the mechanisms of how and why we imitate is challenged the next year. You would think that biological bases of imitation would be one of the easiest elements of learning and behavior change to unravel. Unfortunately, this is not so. Therefore, what we present here is designed to support your general understanding of the value of imitation. We do this by presenting the scientific concepts that we believe will accelerate your grasp of the power of imitation in the learning process the fastest.

Memes: Skinny Jeans, Short Skirts, and Surgical Masks

No discussion of imitation would be complete without addressing the meme. Richard Dawkins became a well-known figure in the field of biological science in 1976 by exploring this concept. A *meme* is best described as a discrete unit of culture[6] that's distributed throughout society by imitation. Examples of memes are rituals, musical trends, fashions, architecture, business practices, and management styles.

In his controversial book *The Selfish Gene,* Dawkins says, "Just as genes propagate themselves in the gene pool by leaping from body to body via sperms or eggs, so memes propagate themselves in the meme pool by leaping from brain to brain via… imitation."[7] The famous saying "When in Rome, do as the Romans do" is all about memes and cultural imitation. The best way for you to quickly survive and thrive in any new culture—be it a country, city, job, school, sports team, or country club—is to look around and do what the "locals" are doing. The faster you imitate the cultural norms of a new environment, the easier it will be to increase your *Individual Adaptability/Intelligence Factor* by quickly fitting into your new surroundings.

A good example of your Mirror Neurons doing their job was during the 2020 Coronavirus pandemic. You might have noticed during the first few weeks of the virus few people were wearing face coverings. Yet as more and more people started to wear them even more people followed. While for many this decision was a conscious one, for others what prompted them to wear a mask, was that either consciously or unconsciously they wanted to "fit" in with the herd.

In most cases, the genetic drive to imitate cultural memes is a survival advantage, but there can be drawbacks. A major reason people get in over their heads financially is that their copycat drive to "keep up with the

Joneses" propels them to buy things they don't need and can't afford. The house and the mortgage that are too big, the too-expensive car, and the fifty-inch flat-screen TV that you "must have" are all examples of the drive to imitate, which can push you toward insolvency (more on the mimicking of the standard of living later).

A quick note: In social media, the term "meme" mainly refers to anything that's shared voluntarily among large numbers of people, including catchphrases, unique images, rumors, texts, and videos (think goofy kittens on YouTube). In this book, when we use the term meme, we are referring to its more biological definition which goes beyond just sharing, here we are talking about unconscious imitation that impacts people's biology so that their behaviors change for extended time periods.

Mirror Neurons Drive Your "Monkey See, Monkey Do" Behaviors

Although geneticists haven't located the specific genes and their relationships that drive us to imitate, neuroscientists have hypothesized about specific neurons that prompt us to mirror the actions of others. In the 1990s, in what many scientists consider one of the most profound neuroscientific discoveries of the last century, Dr. Giacomo Rizzolatti at the University of Parma in Italy discovered neurons that are proposed to propel us to engage in "monkey see, monkey do" behavior.[8] Like many groundbreaking discoveries, this one came quite by chance.

While researching the sensory-motor cortex of monkeys, Rizzolatti implanted electrodes into their brains to measure brain activity. One day, he noticed that when a graduate student who was eating an ice cream cone walked past one of the monkey subjects, feedback from those electrodes spiked in a strange and baffling way. They spiked as if the monkey were not only watching the student eat the ice cream cone but as if the monkey himself were eating the ice cream cone! What Rizzolatti had discovered were specific brain cells that imitate the actions of others. While at first, his findings were ignored, eventually, he became renowned for finding the basis of our herd mentality, what we now call *Mirror Neurons.*

Rizzolatti's pioneering efforts with monkeys gave neuroscientists a place to start looking in the human brain for the source of the imitation of actions, emotions, language, and cultural memes. Today we know that mirroring capacity in the human brain is widespread and includes areas of the frontal lobe, parietal lobe, visual lobe, insular cortex, and *Broca's area*. Ramachandran said:

> *I predict that mirror neurons will do for psychology what DNA did for biology: they will provide a unifying framework and help explain a host of mental abilities that have hitherto remained mysterious and inaccessible to experiments.*[9]

Your Imitation System Helps You Imitate Language

Ever since mirror neuron theory was revealed, neuroscientists theorized that this was the missing link in the evolution of language. At the Brain Mapping Center at the University of California at Los Angeles, Marco Iacoboni used MRI scans to discover that the part of the brain necessary for word production, syntax, and grammar, called Broca's area, has a mirror system. (See Chapter 24 for more on first language acquisition.) The

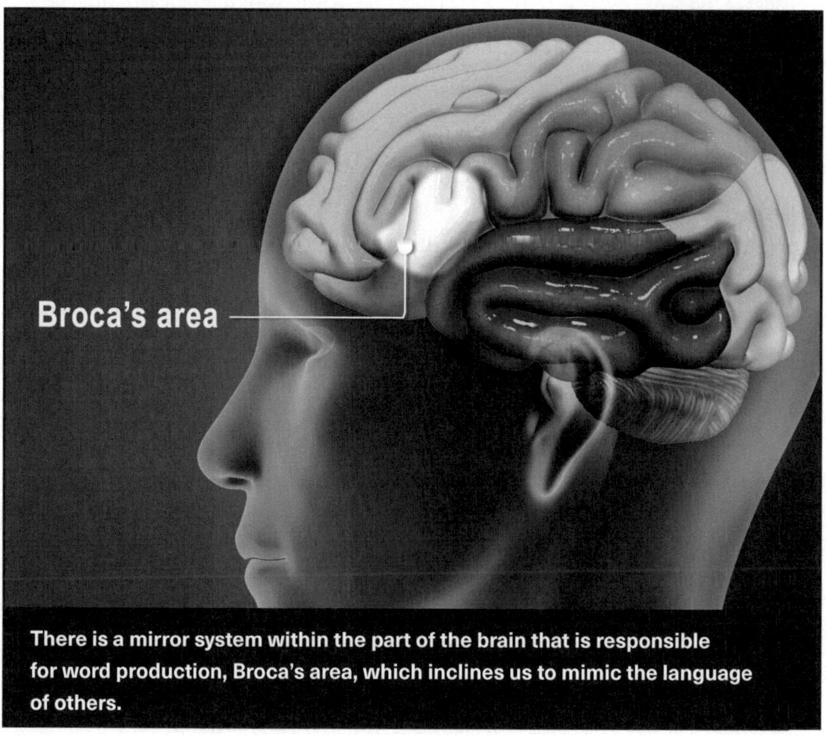

There is a mirror system within the part of the brain that is responsible for word production, Broca's area, which inclines us to mimic the language of others.

discovery of a mirror system in this language area of the frontal lobe indicated that we're programmed to imitate not only the actions of others but also the language of others.[10]

In your first months of life, your genetic program was preset so that you could imitate the sounds you heard in your environment. This allowed you to select into your neural make up the words and syntax of the native language you were most exposed to, thus increasing your surviving and thriving options. Because of the incredible power of imitation, by age six you could speak about 2,600 words and understand an amazing 13,000 to 20,000 words.

The unconscious propensity to imitate language continues in adulthood. One of the reasons it's easier to learn a foreign language living in a foreign land is that, away from home, your *Imitation System* is continually bombarded with language stimuli, which you automatically and unconsciously imitate. Living in a foreign country allows your *Imitation System* to be exposed to more direct and indirect language stimuli, such as background conversations, radio, television, newspapers, and road signs. As with children with more talkative parents (see Chapter 24), the more language stimuli the Mirror Neurons are exposed to, the more bits of language adult Mirror Neurons can imitate.

This unconscious imitation drive is also the reason why, when you move to a new region of the country, you may find yourself mimicking the tone, tempo, and slang of that region's speech. Anyone visiting certain sections Canada for extended periods, where they put the sing-song "eh" at the end of sentences, can attest to the power of mirroring language.

Imitation: Emotions and Empathy

Iacoboni and other scientists from around the world have found that, just as your brain is set up to mimic action and speech, it's also set up to mimic the emotions you perceive in others. The initial theory is that you have Mirror Neurons not only in the emotional centers in your frontal lobes but also in your lower limbic brain areas. It's believed that your Mirror Neurons play a major part in relaying the emotions of others to your emotional centers "as if" they were your own. This means you don't just feel *for* others; you feel *what* others are feeling.

The Mirror Neuron hypothesis of emotions also offers a neuroscientific explanation for empathy and explains why your stomach turns when you see someone eating slugs on a reality TV show. As researcher Mark Thompson at UCLA puts it, our Mirror Neurons are what "allow us to put ourselves in the shoes of another."[11] Empathy is a wonderful gift to our species. When we see others in difficult situations, our Mirror Neurons allow us to feel their pain. The more pain someone is in, the more we feel compelled to come to his or her aid. Because of emotional mirroring, easing others' discomfort eases our own. It is believed by many scientists that Mirror Neurons provide the neurological basis for why humans will reach into their pockets to give billions of dollars of disaster relief to people they don't know and is also the reason that when former US president Bill Clinton said in a campaign speech in response to a question about healthcare, "I feel your pain," he had the mirroring capacity to mean it. And his audience felt it!

Autism and Anosognosia and the Mirror System

Research into the mirror neuron system is providing medical insights into the causes of conditions such as autism and anosognosia. The journal *Cognitive Brain Research* reported on an EEG (electroencephalogram) study conducted at UCLA that indicates why individuals with autism have such a hard time connecting emotionally with others. Lindsay Oberman, who helped administer the study, reports, "The findings provide evidence that individuals with autism have a dysfunctional mirror neuron system, which may contribute to many of their impairments—especially those that involve comprehending and responding appropriately to others' behavior."[12] With their mirror neuron systems compromised, autistic individuals can't understand or empathize with others and are thus socially cut off from the world.

Working with patients with a strange disorder called anosognosia (see Chapter 15), which prevents them from recognizing their own paralysis after a stroke, Ramachandran and others have found some of these unfortunate patients had another bizarre affliction.[13] They not only denied their own paralysis but couldn't see the paralysis in other stroke victims. Brain imaging is revealing that damage to their mirror systems could be a source of these patients' astounding inability to perceive their diseases not only in themselves but also in others!

Attunement: Using Imitation to Refine Your Emotions

In addition to providing the neurological underpinnings for empathy, researchers believe that your *Imitation System* helps you refine your feelings. Although we're all born with a rough genetic program for emotions such as happiness, sadness, and anger, the refinement of your emotional range and tone comes from your ability to imitate the emotions you perceive in those around you. This explains why when you are around depressed people, you feel depressed, and when you are around happy people, you tend to feel happy. Neural biologist Lise Eliot writes, "It appears, then, that we are wired from birth with a way of matching sensory input about another's emotional state to the lower limbic mechanisms that generate emotion in ourselves."[14] What this means is that for children to develop healthy emotional responses, their *Imitation System must* be exposed to caregivers who display healthy emotional responses.

Psychologists have a term for when parents are good at recognizing moods and emotions in their child and then adapting their responses to best support their child's emotional development. They call it *positive attunement.* Studies show that positive attunement is imperative for a child to develop healthy emotional responses to his or her world. Positive attunement occurs when a child falls and hurts her knee, and her dad hugs her and lets her know how sad he feels about her fall. An unhealthy negative attunement occurs if the child falls, and the father yells, "You stupid, clumsy idiot! Watch where you're going."[15]

Emotional Attunement

Brain-based research finds that a child's cognitive abilities, sensory awareness, general alertness, and educability are directly related to the emotional connection between the parent and child.

Researchers believe that when the parent positively reciprocates an emotion, the child's emotional system encodes and imitates how she should respond in similar situations. If a child's parents are usually annoyed at the child or are themselves emotionally detached or disturbed, positive attunement won't be achieved, and healthy emotional circuits in the child will fail to develop, which in turn will inhibit his or her *Individual Adaptability/Intelligence Factor.*

Brain-based educator Pearce finds a child's cognitive abilities, sensory awareness, general alertness, and educability are also directly related to the emotional attunement connection between the parent and child.[16] Studies with orphans and abused and abandoned children show that those not attuned to an emotionally healthy caregiver have retarded social and emotional skills, which limits their abilities to succeed as adults.[17] Bruce Perry at Baylor College of Medicine studied brain scans of neglected children and found that the regions responsible for emotional attunement never developed properly.[18]

A very successful parenting program called "Love and Logic" uses attunement as the basis for purposefully enriching child-adult communications.[19] This program advises that to correct a child's negative behavior, you first need to have empathy for the child's situation. Therefore, if the child loses his homework and gets a zero, you don't yell at him and tell him what a fool he is (even if you want to!); instead, you attune to him and say something like, "It must feel terrible to get a zero. What can you do in the future to make sure you don't feel this way again, and how can I help you do it?" This purposeful focus on attunement helps the young child not only bond more effectively to the parent but also refine his own emotional reactions and capabilities. One of the benefits of early positive attunement is that when a child hits their teenage years and is trying to get out from under their parents' wings, they are much better prepared to handle life's emotional challenges in the absence of parental oversight. My wife and I found this approach to attunement very helpful in raising our boys.

Your Imitation System Lacks Discrimination: Hang with the Winners

Unfortunately, your *Imitation System* isn't very discerning; it can mimic whatever stimulus is available to it—good or bad—as the following examples show:

- Bonnie Klimes-Dougan, Professor of Psychology at the University of Minnesota, has found that toddlers who've been abused by their parents are likely to imitate their abusers by hitting and insulting their crying peers.[20]
- Ulf Dimberg, a Swedish researcher at Uppsala University, used sensitive electronic sensors to show that when people view a smiling or angry face, their own facial muscles show evidence of mimicking the same emotional tone.[21]
- The University of Washington's Geraldine Dawson monitored the brain wave patterns of depressed mothers and their children and found that the children's brain waves mimicked their depressive mothers'. Both mother and child showed reduced brainwave patterns in the left frontal area, which has been identified as a center of joy and other light-hearted emotions.[22]
- Frank Bernieri, a psychologist at Oregon State University, found that, like the flu, romantic partners catch each other's moods—good and bad.[23]
- Studies show the more you hang out with people that overeat, the more you'll eat.[24]

So if you want your *Imitation System* to support your financial, spiritual, and emotional success, it makes sense to "run with the winners," that is hang out with people who are spiritually, financially, physically, and emotionally fit rather than with slackers and drug addicts. It's theorized that one reason twelve-step programs like Alcoholics Anonymous, Narcotics Anonymous, and Overeaters Anonymous work so well is that in these healing environments, new members imitate the positive actions of successful members.

The *Imitation System* might even offer a scientific reason to prompt countries (such as the United States) that want to change the political system of another country (such as Syria or Cuba) to abandon isolationism as a primary method of accomplishing their goals and instead expose the target country to more of their goods, services, and cultural memes.

The understanding of the power of the *Imitation System* might also prompt governments to rethink torture as a means of influencing prisoners of war. If we want to change our enemies' minds, the best "brainwashing" maybe not through torture but through exposure to cultural memes to imitate. A lot of what governments are trying to do with punishment could be greatly enhanced with a better understanding of the neurogenetics of behavioral change. Without it, we continue to do what Edward Deming called "recycling ignorance."[25]

You Started Imitating Right After Conception

To fully grasp the power of imitation, let's look at how your *Imitation System* first developed. Your genetic derive from imitating started very early in your life cycle.

In the Womb and After Birth: Copying Food Preferences and Emotions

Research shows that your copycat behavior starts in the womb. Fetuses can actually sense what type of food the mother is eating and, after they're born, prefer the food ingested by the mother.[26] Other studies show that just hours after birth, newborns imitate the happy, sad, and surprised facial expressions and hand gestures modeled by a researcher.[27] Given how poor an infant's vision is at birth (20/600), it's amazing that babies are capable of imitation at such a tender age. Infants also mimic the crying of other infants, as any nurse in an infant hospital ward will tell you.

Two Months: Obligatory Looking

In the second month of life, you started to overtly exhibit the signs of the genetic quest to learn by imitating others. In a learning window called *obligatory looking,* babies begin to fixate on objects and people for up to thirty minutes. This fixation stage is believed to be our genes' way of ensuring the development of the *Imitation System* in the visual cortex so the child will be ready to mimic the successful survival actions of others as they mature. Many parents say their hearts melted when their child engaged them in obligatory looking. They often describe this prolonged eye contact stage as the first time they really fell in love with their child.[28]

Eight Months: Deferred Imitation

At eight months, you began to develop what's called *deferred imitation.*[29] Starting at this age, children develop a virtual compulsion for imitating other people's actions. We're all familiar with children pretending to cook, talk on the phone, and drive a car like mommy and daddy.

In deferred imitation experiments, a researcher performs an action in front of the young child, and sometime later, the child mimics the action. In a sequence of several actions, a researcher builds a toy out of three metal pieces while the child watches, and after a delay period of days to months, the child is given the props and is observed to see if she'll imitate the previous actions of the researcher. The resilience of deferred imitation is breathtaking and has profound implications. A nine-month-old can spontaneously copy a sequence of actions after twenty-four hours, a thirteen-month-old, a week later, and a fifteen-month-old can imitate a series an astonishing four months after seeing it only once![30]

The vital thing to note here is that this delayed imitation takes place without any training or practice; it's spontaneous. Our *Imitation System* ensures that humans will automatically and unconsciously replicate the actions of others. That a fifteen-month-old child will imitate an action four months after they witnessed it only once goes a long way toward explaining why, decades after we become adults, we're still prone to duplicating some of our parents' behaviors.[31]

This is why parents need to recognize that their actions, good and bad—working, cursing, smoking, exercising, drinking, doing drugs, eating healthy, keeping the house clean—have a profound impact on how their child will behave in the present and the future. Often as children, we vow that we'll

never, ever behave like our parents—yet as adults, we often find ourselves in situations where we act exactly as they did! The powerful influence of delayed imitation is believed to be one of the causes.

Of course, parents and primary caregivers aren't the only influence on a young child. Peers, relatives, teachers, and the media all add to the arsenal of behaviors that a child imitates from his environment on a delayed basis.

Institutions Fail to Understand the Drive to Imitate

As discussed, your *Imitation System* ensures that you're going to replicate the actions, language, behaviors, and emotions of those you're exposed to, whether or not you want to. Unfortunately, our institutions have virtually ignored the tremendous power of imitation that's embedded in your biology. Although educators are well aware that outside of the classroom, students imitate fashion styles, body piercings, tattoos, and even pace and style of walking, our learning institutions proceed as if this most basic of our genetic urges weren't switched on in the classroom.[32] Our corporate and scholastic institutions that want to accelerate learning and transform cultures could make dramatic improvements if they purposefully incorporated imitation into their environments.

Formalized Imitation: Mentorship and Apprenticeship

Purposefully incorporating imitation into learning and behavioral change is not a new idea, though our contemporary educational and corporate systems neglect it. Since life first appeared on Earth 3.5 billion years ago, imitative learning has served all organisms well as a means of quickly and successfully adapting to their environments.

In prehistoric times, all humans had to do to learn how to start a fire with a flint or hunt a deer with a spear was to hang around others who successfully engaged in those behaviors. Their *Imitation System* would eventually do the rest. But over the millennia, as our world grew more and more complex, simple imitation through observation could no longer serve as the primary means of education. To grasp the high degree of intricacy necessary to master complex tasks such as making tools, building roads, crafting jewelry, and running printing presses, more formalized forms of learning through imitation, called *apprenticeship* and *mentorship,* were invented.

The goal of apprenticeship/mentorship was to accelerate the transfer of skills and wisdom from the master to his apprentice. It's safe to say that our modern world wouldn't be what it is without the creation of apprenticeships. Since the invention of the wheel and the plow, apprenticeship was the primary method used to learn everything from technology to healing. In fact, because in 1900, only 6 percent of the US graduated from high school, virtually everyone else learned their life and work skills through apprenticeships and mentorships.[33]

Why does apprenticeship/mentorship play such a small role in the education of children and adults today? The answer lies in the advent of industrialized society. At the end of the 17th and beginning of the 18th century, Newton's Laws of Motion helped to create not only the industrial society but also a machine mentality. During this time, the concept that all production, both mental and physical, could be improved with the application of the machine model was first introduced. The production of goods moved from masters and their apprentices to factories. In factories, workers no longer needed long apprenticeships or mentorships to learn a complete skill. Instead, a worker only required to be quickly taught how to do a straightforward task, like putting a round peg in a round hole, over and over again, day after day, year after year.

In an effort to speed up the transfer of wisdom, the educational system also adopted the machine model. To facilitate your learning, you were put into a belt-like conveyor system, and over the next twelve years passed through 14,000 forty-five-minute segments of instruction that primarily prepared you to pass tests. In a machine-like learning system, the students have been locked away far from the real world and exposure to the real masters who work every day with the very subjects the students are trying to learn.

When the master-apprentice relationship was replaced by the mechanistic model, work and learning became not only less dependent on imitation but also decontextualized. Workers and students became "cogs in the great wheel," merely "units of production" that required no personal relationship to the whole. A worker never has to know what end product he's working on to do the job; he only needs to focus on completing his tasks. It is the same with the student who never has to connect the parts she's studying to the whole… but only needs to cram in enough details into working memory to pass the test.

Harvard's Howard Gardner, for decades a strong proponent of apprenticeships, maintains that the most powerful benefit of mentorship is real-world experience under the watchful eye of a master, which allows the student to practice in the context in which the information or skill will be naturally used. In *The Unschooled Mind,* he writes, "Most crucial for our purposes is the fact that the learning in the apprenticeship is heavily *contextualized.*"[34] He goes on to say, "In contrast to apprenticeships, the school experience is marked by an extreme dissociation from important events or palpable products in the life of the community."[35] Real-world apprenticeships/mentorships not only accelerate the learning process by activating your *Imitation System* but also activate spatial maps and motor centers and produce rich sensory feedback, which helps you quickly place newly learned information into your correct context and concept networks.

Where Have All the Masters Gone?

Our modern society may have evolved to the point where our genetic drive to imitate has started to work against our ability to effectively apprentice/mentor our youth. Let me explain. As discussed, imitation is an instinctual and unconscious urge like eating, nesting, and mating, but it's also indiscriminate. The result is that we're set up to mimic the standards of living and cultural memes we perceive around us. It's our drive to imitate that leads us to obsess and crave that new car, the fifty-inch TV, or the newest smartphone. This biological urge, if left unobserved, can cause us to act as if acquiring these toys were a matter of life or death. To possess them, many of us have abandoned our homes and children. Today, both parents in 63 percent of American households work.

While in many of those homes, both parents must work to keep food on the table, in other homes, it is our out-of-control need to imitate what our neighbors have that propels us to abandon our children. We justify our efforts to work longer and longer hours to "keep up with the Joneses." The concern is our children miss the opportunity to positively activate their *Imitation System* so they can imitate positive parenting skills when they too become parents.

As more and more of our wise patriarchs and matriarchs are sucked into the mouth of the great business machine, the youth of our society suffer because they have less contact with those who possess the most experience and wisdom about how to successfully work and live in the real world. The master elders of our society—aunts, uncles, neighbors, and the like—no longer have much contact with our youth, because, just like the children, they're locked away in sterile concrete buildings doing decontextualized work.[36] One study showed that 70 percent of Americans don't know their next-door neighbors.[37]

Even if our educational systems desired to apprentice and mentor our youth, teachers have no time for it. This is because an educator's day is so consumed with forcing enough verbal and math information into the children's heads to get them to pass standardized tests (so that both teachers and students can escape the wrath of the local school board and state legislature), teachers have no time for it. Unfortunately, just like the children and their parents, the teachers have become cogs in the great impersonal machine, with little time to nurture a child's soul or wisely mentor them in the ways of the world. One study revealed that the average student receives only five minutes of one-on-one time with a teacher per week.

So with all the elders tethered to their corporate bonds and with the teachers unable to mentor/apprentice our youth, who are the "wise ones" that our youth imitate? The answer by default is whomever they spend the most time with. Today, that's the media.

As we've seen, research shows the average child spends approximately forty-five and a half hours a week in direct contact with television, cell phones, video games, or the internet. Compare that to the nine minutes of contact with his or her dad and the five minutes one-on-one with a teacher a week. Because society's wise elders have so little contact with our youth, our youth have little opportunity to imitate them.

Other than media, who do our children have the most opportunity to imitate? The answer is other children. The less time that children are exposed to elders, the greater the role that peer pressure plays on their mirror neuron systems. With the media and other children becoming the primary source of imitation, we end up with a society where tattoos and body piercings can become the norm and wherein some regions of the country a shocking 60 percent of ninth-graders have engaged in oral sex, and a fourth of twelfth graders have taken drugs.[38, 39]

Mentoring and Apprenticeship Make a Difference

To create positive behavioral change, organizations around the country are now looking toward mentoring and apprenticeship programs. One such program was conducted by Milton Shore and Joseph Massimo, who worked with juvenile delinquents after they dropped out of high school in Newton, Massachusetts. Shore and Massimo found that children who had positive mentors they could imitate and model their behaviors and emotions upon had a dramatically greater chance of living productive lives as adults than those without healthy role models. In a follow-up study, 80 percent of the group who did not have positive mentors were in the criminal justice or mental health system. Conversely, 80 percent of the group who had mentors were doing well in terms of work, family, and general wellbeing.[40]

Research shows positive hands-on mentoring inhibits the influence of negative peer and media imitation. Parents who set reasonable limits on everything from the internet and television to music choices to where children go during the day and night raise children who are 61 percent less likely to use drugs.[41] Joseph Califano, President of the National Center on Addiction and Substance Abuse, says, "This shows the parent's power as a pre-emptive tool."[42] Community outreach programs such as the Scouts and the YMCA Adventure Guides Program employ the power of imitation to positively influence our youth.

A few years ago, 1,600 students at Harvard were asked to name the college experience that had the most impact on them. The single most common response was research mentorships.[43] Some universities are beginning to recognize the role of apprenticeships in increasing learning. "We used to pour knowledge into students as if they were vessels ready to be filled," says Shirley Strum Kenny, former president of the State University of New York at Stony Brook.[44] "That's not the way it is now," she says, pointing out that, at Stony Brook, enrollment in classes that have elements of apprenticeship/mentorship has increased 160 percent since 1998.

Imitation: Suzuki and Corporations

Suzuki music learning may be the most well-known method that uses apprenticeship. In the traditional Suzuki method, from birth until thirty-six months of age, a child is allowed to listen to and observe only twenty short songs, which will constitute his curriculum once he begins to work with his instrument. At age two, the child is made part of a music group, but for the first six months is allowed only to observe others playing their instruments. When his desire to play his instrument has reached fever pitch, and his *Imitation System* has been repeatedly stimulated, his master allows him to practice with his own instrument. By the time he's nine to ten years old, the power of imitation enables even the most average Suzuki student to play a Mozart concerto with proficiency. Today, corporate trainers, frustrated with the lack of success of stand, deliver, and test programs, have turned to apprenticeships/mentorships to accelerate the speed of learning in their institutions.[45]

In the 1990s, Marilyn Whalen developed Xerox's Knowledge Interaction and Practice, one of the first technology imitation learning programs. To understand how to better do their jobs, instead of sitting in stuffy classrooms being lectured to or working with a computer learning program, new employees were placed next to someone already doing the job successfully. Here, the new employee could imitate not only how to effectively do the job but also how to successfully navigate Xerox's complex corporate culture. Whalen says people learned "without even seeming to have been paying attention."[46]

Today, advocates of apprenticeships and mentorships maintain that such relationships impart layers of knowledge that could never be transferred through traditional teaching, reading, tests, or computer programs. As Michael Polanyi points out in his book *Personal Knowledge:*

> *The apprentice unconsciously picks up the rules of the art, including those which are not explicitly known to the master himself. These hidden rules can be assimilated only by a person who surrenders himself to that extent uncritically to the imitation of another.*[47]

Although apprenticeship/mentorship looks like a difficult chore to time-starved corporate and scholastic educators, the genetic and neurological data would suggest that any learning system that lacks an apprenticeship/mentorship component will lack efficiency.

At The Institute, our goal is to capitalize on our neurogenetic predisposition to imitate by designing an apprentice/mentorship component into all learning and behavioral-change systems.

How to Stop Keeping Up with the Joneses

Before concluding this chapter, I must point out that although humans have the greatest capacity to imitate any animal on Earth, we also have the greatest capacity to regulate our urges to imitate. Just because we observe our neighbors buying fifty-inch TVs, drinking too much, or eating the wrong foods doesn't mean we have to imitate their actions.

How do you mediate your urge to imitate? How do you stop damaging your financial and spiritual wellbeing by unconsciously mimicking your neighbors and trying to keep up with the Joneses? The first step is to be aware.

Every second of every day, your brain is processing billions upon billions of bits of auditory, visual, tactile, taste, smell, somatic, and emotional information. How many of these billions of bits of data are you consciously aware of? Not very many at all. It's an estimated 99.9 percent of all the information your brain takes in… is processed subconsciously. Most of the time, you're living on autopilot, acting, and imitating without involving your conscious mind at all.

So, how do you start to change things?

Research is revealing that, if you want to change any of your negative thoughts and behaviors, you must first become aware of the moment you start to engage in them. To help people change their behaviors, modern-day cognitive psychology has taken a page from the ancient Buddhist practice called *mindfulness.* When you engage in mindfulness, you focus your awareness on the present moment. You recognize and accept—without judgment—the thoughts, emotions, bodily sensations, and actions you're experiencing in that single moment.

Why is mindfulness vital to transformation? Because the instant you engage your conscious mind by becoming aware of the present moment, your prefrontal lobes, especially the ventromedial and dorsolateral areas, engage. Now not only do you have the capacity to witness your body sensations, emotions, and actions, but you also can regulate them. Now you have a choice. Once you're aware, you now have more resources to choose whether or not to eat that massive piece of chocolate cake, drink that third martini, or buy that expensive new car. With your frontal lobes switched on, you've switched off your autopilot. No longer are you your thoughts and emotions… you are now the *observer* of your thoughts and emotions. Now you can better weigh the consequences of your actions and consciously decide what you want to do next. Now you're awake and can consciously decide if you want to imitate or not.

Cracking the Learning Code

Humans are a species of copycats. As a survival advantage nature has given us a genetically implanted *Imitation System* that compels us to mimic the survival actions of the "herd"... good and bad. Organizations and individuals who want to create profound and rapid transformation must have a deep understanding of the power imitation has on directing our thoughts, actions and behaviors.

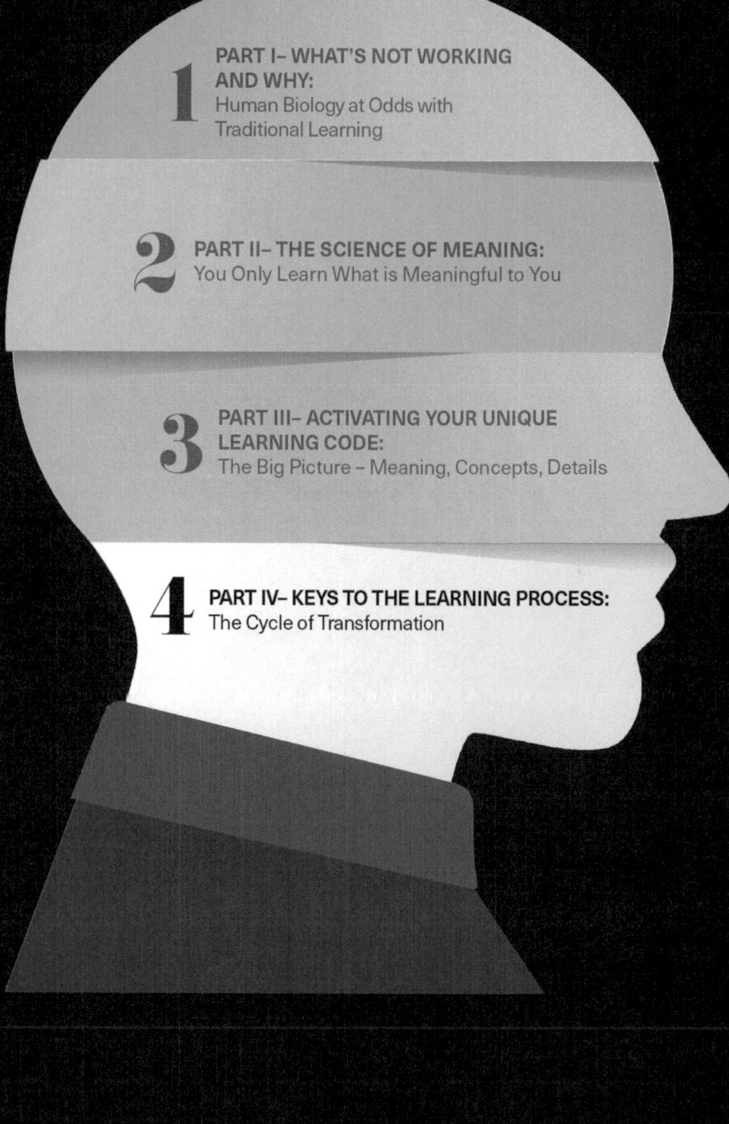
1
PART I– WHAT'S NOT WORKING AND WHY:
Human Biology at Odds with Traditional Learning
2
PART II– THE SCIENCE OF MEANING:
You Only Learn What is Meaningful to You
3
PART III– ACTIVATING YOUR UNIQUE LEARNING CODE:
The Big Picture – Meaning, Concepts, Details
4
PART IV– KEYS TO THE LEARNING PROCESS:
The Cycle of Transformation

30
The Educational System Wants Simplicity, Your Brain Wants Complexity

The illiterate of the 21st century will not be those who cannot read and write, but those who cannot learn, unlearn, and relearn.

—Alvin Toffler, writer and futurist[1]

This chapter dives deeper into how environmental complexity enhances your learning processes and why learning systems that focus only on the part of the brain that processes linguistic information can't deliver the levels of efficiency that our world demands.

Research from complexity science, chaos theory, and evolutionary biology is reinforcing that your brain wasn't designed to learn in sterile, non-stimulating environments. From a scientific perspective, we're beginning to understand that the idea you can lock people away from the real world they need to adapt to, retard their movement, limit complex sensory input, and lecture their genes and neurons into creating the neurochemical changes necessary for profound long-term memory and behavioral change to take place, is a very flawed concept indeed.

It may seem that spoon-feeding your brain small amounts of linguistic information in a linear fashion would accelerate the speed of learning. Because of this belief, we've built learning systems that operate in this way. But the brain is a self-organizing system, and like all highly interconnected, self-organizing systems—such as ecosystems and financial markets—your brain needs complex input from many diverse sources to transform and operate efficiently.

Therefore, if you limit environmental complexity to your brain, you limit its effectiveness. During any one second when the brain is physically moving through complex environments, it's being bombarded by trillions upon trillions of diverse bits of visual, auditory, tactile, smell, and taste stimuli from many different directions. Although the idea may seem counterintuitive, your brain uses this complexity as fuel to build a stable internal order. In a classroom where the brain is confined to one spot and lectured to, environmental stimulation and sensory diversity become limited. Research clearly shows that when environmental stimulation is limited, effective brain processing is dramatically compromised.

Sensory Deprivation Tanks, Solitary Confinement and Decontextualized Teaching

Today, the average child spends six to eight hours in the confinement of the school, another six-and-a-half-hour glued to some form of media, and one to two hours a night doing homework. This leaves very little time for physically moving, and research indicates there are consequences to this lack of real-world complexity.

We know from human experiments with things such as sensory deprivation tanks and solitary confinement that when people are restricted from moving, and deprived of rich sensory stimulation, not only is learning compromised, but they become confused, disoriented, agitated, and depressed. Although a classroom isn't a sensory deprivation tank or solitary confinement, it does limit movement and exposure to real-world complexity. Leslie Hart notes an interesting parallel between how behavioral psychologists research learning in lab animals and the way we teach students:

> *Usually, the first step was to put the animal in some sort of captivity in the best "scientific" tradition of simplifying and controlling the conditions and variables. This, of course, ignored the plain fact that all higher animals have brains developed to deal with complex conditions in a state of freedom, not with strange unnatural boxes in the state of confinement.*[2]

One of our advisors at the Institute, Michael A. Schmidt, Ph.D. (founder of Sovaris Aerospace; a research collaborator with NASA; and course director of George Washington University's Clinical Genomics, Proteomics, and Metabolomics Program), has raised grave concerns about the impact of physical inactivity in both space and academic environments. In a review of his own and other research on limiting physical movement for long periods, he says, "All of these studies have one thing in common: sustained or chronic physical restraint acts as a stressor with significant downstream biological effects."[3] He goes on to conclude:

> *Our great concern about this type of contextual restraint stress in classrooms is that it reduces the brains' production of Brain-Derived Neurotrophic Factor (BDNF), thus limiting the brains' primary activity-dependent growth molecule. A further concern is that the contextual restraint stress of classroom sitting removes the fine and gross motor components that are so critical to developing the complex neural networks typically associated with successful learning.*

Experience Targets 100 Percent of Neural Structures: Providing More Hooks for New Information to Attach

Why can you learn so much more naturally by physically moving through complex real-world environments than sitting still in a classroom listening to a lecture? An oversimplified answer is that experience can fire 100 percent of your neurons, whereas language fires only about 5 percent (**Note:** Please understand this is only a rough estimate and is used only to heighten comparison and make a point.) As we've discussed, the more neural stimulation that takes place in your brain at any one time, the greater the chance you'll have to learn something or change a behavior.

Also as you now know, the Hebbian learning rule shows us the more neurons that are simultaneously fired at any one time, the greater the chance that large numbers of individual neurons will be stimulated at more than one receptor site and therefore the greater chance that learning, memory, and behavioral change will occur. Nothing causes the simultaneous firing of massive numbers of neurons more effectively than physical experience in complex, sensory-rich environments. Also, as you may remember from earlier in the book, new learning doesn't free-float around in your brain, but instead must physically attach itself to pre-existing brain structures that hold previously learned information. This makes your brain less like a container and more like a lattice of synaptic hooks, with each hook attracting and snagging new information that resonates with it. When you physically move through real-world environments, literally trillions upon trillions of neural synaptic hooks automatically become primed and ready to snag new information. The more of your brain structures that are being activated at any one time, the more opportunities that new information has for finding a place to physically attach itself in your long-term memory banks. This is why remembering an accident you were in is so much easier than recalling an accident you read about.

Experience Stimulates 1,028 Connections a Second!

Compared to sitting still in a classroom listening to a lecture, real-world physical experience has so much more of an impact on your learning process because it creates the opportunity to prompt trillions more brain connections into action. To get an idea of how many brain centers, neurons, and neural connections are simultaneously activated at any one second as you move through the real world, let's look at what's going on in your head during a simple stroll in the woods.

During your walk, as your brain takes in the hues and tones of the trees and flowers, the movement of leaves and birds, and as the sun's photons hit your retina directly and indirectly by ricocheting off diverse objects, your brain's visual structures process billions of bits of information each second. As it processes the thuds of your footfalls, the whisper of the wind, the rustling of leaves, and the chirping of insects, your auditory cortex picks up another 30,000 bits of auditory information a second. As your skin takes in minute temperature and humidity changes, your hands brush by branches and your feet shift from walking on stones to the hard earth,[4] your tactile processing centers process another ten million bits of tactile data a second. And as you move through time and space in the woods, millions upon millions of neurons in your motor cortex (which supports physical movement) and millions of neurons in your parietal lobe (which supports 3D orientation) are simultaneously being activated, as are millions of neural connections in your olfactory centers when they encounter the huge numbers of scent molecules given off by vegetation, animals, and the earth itself. And if you stop to put a blade of grass in your mouth or take a sip of Gatorade, millions more taste neurons and connections in your gustatory cortex are activated.

> **It can be stated flatly, that the human brain is not designed or organized for linear one path thought. To try to demand that your brain put aside its mighty resources to go step by step down to one path is to cripple and inhibit it.**

Also, consider that milliseconds after this complex sensory bombardment is processed by your thalamus, this information is transferred to the lower limbic and stem structures, which in turn, via the bloodstream and nervous system, send this information to the body to regulate temperature, heart rate, oxygen consumption, blood pressure, galvanic skin response, pupil dilation, blink rate, salivation, adrenal output, digestion, and the flow of blood to and from muscles and organs. (See Chapter 15 for more on somatic feedback and learning.)

Then consider that all this somatic feedback is sent back to the amygdala and prefrontal lobe circuits, which assign emotional value to the incoming information, allowing you to feel happy, sad, angry, or content about your stroll. Furthermore, consider that while all this neural stimulation is being set in motion by your walk through the woods, millions of more connections in the hippocampus and lateral prefrontal cortex, which support the holding of information in working memory and the placing of new information into the correct concept networks of long-term memory, are also being activated, as are millions of connections in the cerebellum, the part of the brain that helps you maintain your balance. Also, your physical hike through the woods is increasing the activity of the all-important monoamine neurotransmitters—dopamine, serotonin, and norepinephrine—hundreds of percent over sitting still.

So how many brain connections are estimated to be activated during any one second of your simple promenade in the woods? A mind-boggling 10^{28}! And any complex sensory environment outside the classroom can cause this kind of massive neural stimulation.

FAST FORWARD or DEEP DIVE

The rest of this chapter takes a deep dive into why complex experience in the real world beats linguistic instruction, but feel free to skip to the next chapter if you don't need the detail.

Language-based Structures vs. Experience-based Structures

When you compare the numbers of brain connections stimulated by experience versus lecture, it's easy to see why real-world experiences win over hours in the classroom. The primary brain areas activated by language-based learning systems are *Broca's area* in the left frontal lobe, where speech production takes place; *Wernicke's area* in the left temporal lobe, where language comprehension takes place; and the tract of neurons called the *arcuate fasciculus,* which connects these two centers.[5] It's estimated that when these word-based neural structures are activated, they fire approximately 5 percent of your brain's neural connections. Physical experiences cause

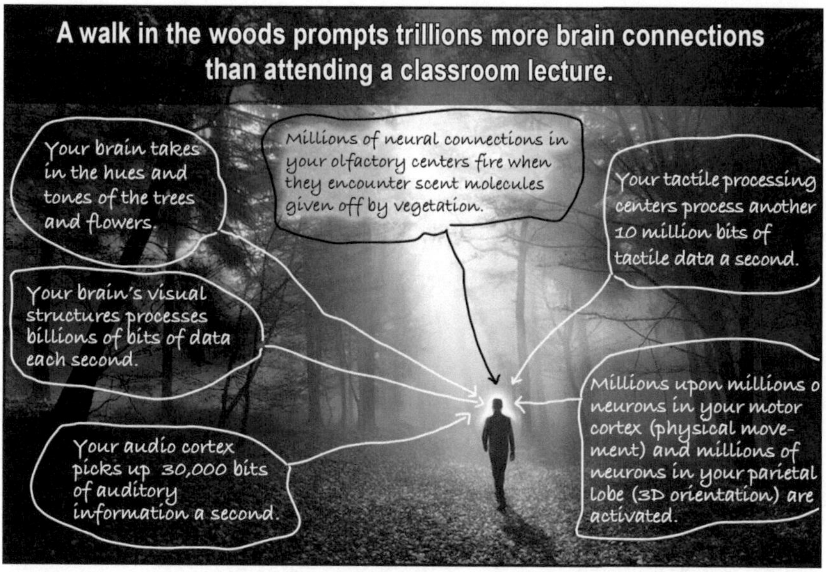

the massive simultaneous neural firing necessary to blow huge numbers of magnesium plugs out of NMDA receptors while causing the activation of enormous numbers of neural synaptic hooks. In the end, you can say that the greater and more widespread the neural firing, the more profound your learning and behavioral change, and that, conversely, the weaker and less widespread the neural firing, the weaker your learning and behavioral change. Of our sensory-deprived traditional educational systems, Leslie Hart says:

> *It can be stated flatly, however, that the human brain is not designed or organized for linear one path thought… To try to demand that this brain put aside its mighty resources to go step by step down to one path is to cripple and inhibit it.*[6]

Evolutionary Biology Reveals: Your Linguistic Centers Are Immature

It's important to appreciate from the evolutionary perspective how immature the brain centers that allow humans to learn linguistically actually are.

Evolutionary adaptations don't become highly efficient overnight; they take eons to refine. It's estimated it took a billion years for the structures that provide taste and smell in reptiles to evolve, that it took 280 million years for the first photosensitive nervous systems in primitive fishes and trilobites to evolve into the efficient visual cortex of mammals, and that it took 200 million years for animals' nervous systems and body structures to evolve so that animals could perform effective upright bipedal locomotion. And ever since the final four layers were added to the neocortex 65 million years ago, all the neurological structures that allow animals to use their bodies and senses to learn how to survive and thrive in the world have been evolving to their present level of perfection.[7]

When compared to other adaptations, the ability to learn through language is relatively new. Many anthropologists suggest the genetic changes that gave us the physical structures—the elongated palate that allows for verbalization of the basic units of speech called *phonemes,* and the neurological structures Broca's and Wernicke's areas that allow for advanced language—first appeared only 50,000 years ago with Cro-Magnon. On the timeline of acquiring and perfecting adaptations, 50,000 years is just the blink of an eye.

From an evolutionary biology perspective, then, the theory we can learn better through experience than words can be reduced to the concept that the neurological and physical structures that allow humans to learn through experience have had 64,950,000 more years to evolve and mature into a high level of efficiency than our language-based structures have had. To think that the adaptations that allow for language production, comprehension, and storage could become efficient thousands of times faster than other evolutionary adaptations such as taste, smell, sight, and bipedal locomotion is contrary to the biological evidence. Hart writes:

> *The implications of this must be seen as staggering for anyone engaged in instruction: the brain we are dealing with was not "designed" by evolutionary needs for logic, manipulation of symbols, or dealing with typed sequential structures or world-systems, which today constitute the main concerns of conventional schooling and much training… They are superimposed, not natural.*[8]

Unless we comprehend the evolutionary implications of our young language centers, we'll continue to be confused about why our linguistic-based educational systems fail to live up to our expectations.

An Experiment: Educate 100 Percent of the Population in a Box

Looking at things in the context of time also allows us to see that teaching solely through linguistic methods is actually very new. As pointed out earlier, up until 1900, only 6 percent of the US population graduated from high school; the rest learned how to increase their *Individual Adaptability/Intelligence* Factors from real-world experience. This makes the experiment to educate 100 percent of the human population in sensory-depriving small rooms only about 120 years old. What this means is that, as the primary method of learning for the masses, solely linguistic learning systems, have been around for only 0.00166 percent of the time humans have been on Earth.

Indeed, our genetic and neurological structures may eventually evolve to the point where linguistic information can be effectively poured into the heads of learners solely through lectures and books and be remembered effectively. Over future millennia, our frontal and temporal lobes may evolve so that we have huge bulges on the left sides of our heads to allow for the excess number of neurons and connections that would permit us to learn exclusively through linguistic means. But to correctly appraise the value of work-based learning systems, we must conclude that this time is not yet here. This evolutionary perspective leads to prudent advice to those designing learning systems: *Go with the evolutionary flow; build learning systems that primarily rely on experience, not words, to create long-term memory.*

I'm not suggesting there isn't a place for linguistic and classroom learning. There is. What we hope this scientific knowledge will impress upon you is that the traditional *lecture, memorize, test system* isn't the most effective or joyful method for creating profound learning and behavioral change. Change takes time, as does integrating this new knowledge. For the time being, my hope is that bosses, corporate trainers, and scholastic educators take this scientific knowledge about the power of experience over language and start to implement it the best ways they can within the boundaries that exist within their organizations.

Quick Look: Learning Systems That Emphasize Experience

Over the last twenty-five years, a movement led by educators and parents frustrated with the limited success and high pain of traditional learning systems has started to go with the evolutionary flow by developing educational programs that de-emphasize traditional linguistic-based instruction in favor of curricula that stress real-world experience in complex, sensory-rich environments.

These programs, which reflect the philosophy put forth by John Dewey in the early 1900s, operate under different monikers—such as *natural, environmental,* and *experiential*—and are springing up around the world.[9, 10] The Association for Experiential Education now has approximately 2,000 members in over thirty-five countries, and these real-world learning programs are experiencing profound success.[11] In a ten-year study in the United States, out of 150 schools in sixteen states which emphasize real-world learning, the State Education and Environment Roundtable found these programs drove student gains in science, language arts, math, and social studies; improved standardized test scores and grade point averages; and developed skills in critical thinking, problem-solving, and decision-making.[12]

In the Environmental Middle School in Portland, Oregon, where teachers employ local mountains, rivers, and woods as part of the curriculum, 96 percent of students meet or exceed state standards for math problem-solving, compared to only 65 percent in traditional schools.[13] The California Department of Education found that in urban California schools that employed experimental education and the use of outdoor classrooms, sixth-grade math and science scores improved by 27 percent.[14] Swedish researchers compared children in two different daycare centers, one surrounded by buildings and the other in a more natural setting bordered by woods and adjacent to a garden. They found children in the "green" daycare center who played outside every day (regardless of the weather) had better motor

New research is prompting more and more schools to implement natural, environmental, and experiential learning programs.

coordination and a higher ability to concentrate.[15]

In New Zealand, at Freyberg High School, the studies of biology, English, history, ecology, geography, and computer science were combined in field trips around the central theme of the preservation issues confronting New Zealand's national parks. The results were very encouraging. When compared to students in traditional classrooms, the experiential students scored 20 to 30 percent higher in geometry and 10 to 15 percent higher in math and science. Most importantly, these students reported more joy at the opportunity to learn in the real world.[16]

In a review by the Organization for Economic Co-operation and Development, Finland outscored thirty-one other countries in literacy and placed in the top five in math, while the United States was in the middle of the pack. Contrary to American efforts to take children out of the real world sooner, Finnish students don't start school until they're seven years old.[17] Commenting on its successful educational programs, which employ real-world learning, Finland's minister of social affairs and health said "The core of learning is not information… being predigested from the outside, but in the interaction between a child and the environment." David Sobel, a researcher who studied the effectiveness of experiential-based learning, found that when it comes to reading, which is considered the gold standard of education reform, environmental learning programs should be considered "one of the knights in shining armor" because students in these programs typically outperform their peers in traditional learning programs.[18]

The Heschong Mahone Group found the more real the environmental stimuli, the better the learning. Its research showed that even something as simple as the added sensory complexity supplied by real sunlight could have an effect. Analyzing test scores for 21,000 students, the group found classrooms with the most sunlight had students progressed 20 percent faster on mathematical tests and 26 percent faster on reading scores than students in classrooms with less sunlight.[19]

Real-world Experience Reduces Behavioral Problems

It also appears that experiential/environmental/natural-learning programs reduce problems.[20, 21] At the Hotchkiss School in Texas, two years before the experiential curriculum was implemented, 560 disciplinary referrals were made, but after the more real-world curriculum was put in place, that number took an unprecedented steep and welcome drop to fifty.[22] Little Falls Community High School in Minnesota reported that ninth-grade students in the environmental program had 54 percent fewer suspensions.

Gerald Lieberman, director of the State Education and Environment Roundtable, sends the simple yet counterintuitive message that, for more effective educational reform to take place, students should be freed from the classroom.[23] As you've seen in this chapter, science is now revealing that these environmental/experiential/natural-learning programs are succeeding because they better fit how the brain naturally learns. What complexity science and chaos theory have brought to light is that the more learning environments replicate the real world, the greater the complexity, and therefore, the more natural the learning and the lower the behavioral problems.

In the next chapters, we'll examine another primary reason real-world experience is such a powerful promoter of learning: it naturally allows your brain to create meaning by passing through the four stages of the learning process, which we call *The Cycle of Transformation*. Here, you'll find if your brain isn't allowed to naturally pass through these stages, your ability to learn and create behavioral change will be severely compromised.

Cracking the Learning Code

If you want to enhance your learning, focus on real-world experience. Physically moving through complex, sensory-rich environments creates more cells firing in your brain than a linguistic-based classroom. Simply put, the more neurons that fire the greater the memory formation.

31
The Cycle of Transformation: Memory is a Process... Not an Event!

There is no end to education. It is not that you read a book, pass an examination, and finish with education. The whole of life, from the moment you are born to the moment you die, is a process of learning.

—Jiddu Krishnamurti, Indian philosopher, speaker and writer[1]

While researching why experience in the real world is a so much more powerful promoter of learning, we found science had made a profound discovery that wasn't generally well understood: that learning and memory don't take place in one fell swoop. Learning is not an event; it is a process. What traditional education and corporate training systems fail to realize is that to create long-lasting learning and behavioral change, your brain must go through a distinct process, marked by the alternating combinations of different neurotransmitter systems in different brain areas at different times. It's the magic of the ebb and flow of these neurochemicals that allows information and experiences to be firmly encoded into your long-term memory banks.

At The Institute, we call this process *The Cycle of Transformation*, and to simplify its implementation, we've distilled it into four stages. Limiting your exposure to any of the four stages of learning transformation hinders the activation of the correct neurotransmitters at the correct times, which in turn limits the efficiency of your memory formation and behavioral change. Understanding the four stages of *The Cycle of Transformation* is a major component in comprehending The Learning Code. It also gives you further powerful clues as to why our traditional scholastic and corporate educational systems are struggling and how to significantly accelerate the speed of learning.

Before we take a closer look at *The Cycle of Transformation*, it's important to understand that in nature, transformation occurs through processes, such as seeds to flowers, carbon to diamonds, and caterpillars to butterflies. So too, your brain transforms through a process. Humans don't change just because they possess information. We all know too many calories make us fat, yet almost 70 percent of Americans are overweight.[2] We all know that changing our behaviors around food will make us healthier and happier, yet most of us can't do it. The information alone about calories and fats has done very little to help the majority of us learn and change our behaviors so that we can lose weight and transform our lives.

Why not? Because, as we say at The Institute, *information alone does not cause transformation*. To transform your existing neurological structures so that you encode new knowledge into your long-term memory banks in a way that increases your *Individual Adaptability/Intelligence Factor*, your brain must go through this four-stage scientific process that literally alters and grows neural networks so you'll learn and behave in more successful ways. This is the kind of learning that allows you to recreate yourself anew. But here's the problem for traditional educational and corporate training systems: science is revealing that any learning system that doesn't incorporate this underlying process principle of memory formation will, from a biological perspective, be an inefficient one.

Origins of The Cycle of Transformation

The concept that efficient learning is a process that occurs in stages arises not only from the fields of neuroscience, genetics, and evolutionary biology but also from the world of artificial intelligence. In researching how to support the brain to transform most easily, we researched the work of renowned mathematicians and physicists John von Neumann and Stanislaw Ulam. In the 1940s, these two scientists helped develop the discipline of cellular automation, or artificial life, based on mathematical and computer modeling.[3] Von Neumann and Ulam discovered that to get

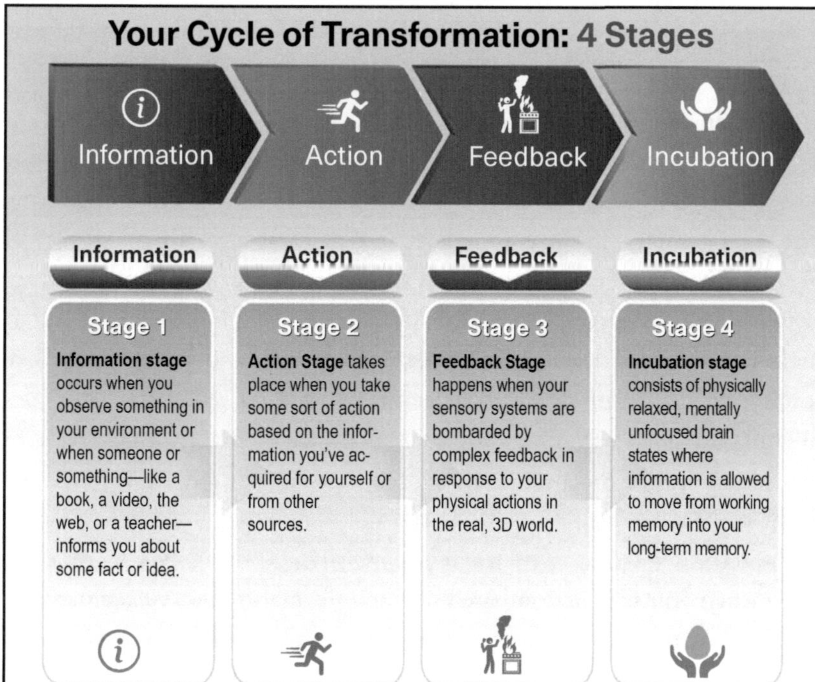

Each stage activates different neurochemical systems, which ensures long-term memory can be encoded.

a deep understanding of how an artificial system could develop, learn, and self-replicate, they needed to break the processes down into distinct stages.[4]

Their pioneering work on artificial intelligence gave me an "aha" moment that, in combination with the biological research, helped me confirm and fully grasp how organic systems like animals and humans also learned and expanded their adaptation repertoire through a process that had distinct stages. Therefore, I knew we needed to define and undertake a full examination of each stage of the learning transformation process before we could build learning systems that activated them!

That learning is a stage-by-stage cycle should come as no surprise: we see change-through-process again and again in nature, from the evolution of life itself to the life cycle of a butterfly. There is no way a caterpillar can transform into a butterfly, carbon into coal, or seed into a flower, without going through distinct stages. And just like those natural processes, learning and behavioral change happen in your brain in more or less the same way.

The four stages of ***The Cycle of Transformation*** learning process are:

1. Information
2. Action
3. Feedback
4. Incubation

The Information Stage occurs when you observe something in your environment or when someone or something—like a book, a video, the web, or a teacher—informs you about some fact or idea. This first stage of *The Cycle of Transformation* is where the typical education system focuses virtually 100 percent of its energy. And herein lies the problem. Without understanding that learning is a four-stage process and not an event like reading an article or going to a lecture, traditional education struggles mightily to create learning; it experiences unacceptably high levels of frustration and inefficiency because it concentrates on only this first stage of the process.

The second stage in *The Cycle of Transformation*, the *Action Stage,* takes place when you take action based on the information you've acquired for yourself or from other sources. Eric Jenson, the author of *Brain-Based Learning,* sums up the research on the action stage simply by saying, "New brain structures (both cells and dendrites) grow fastest with physical activity."[5]

The third stage, the *Feedback Stage,* happens when your sensory systems are bombarded by complex feedback in response to your physical actions in the real, three-dimensional world.[6]

These first three stages—*Information*, *Action*, and *Feedback*—allow information to be efficiently placed into working memory, where it's stored temporarily. But before information can be firmly implanted into your long-term memory banks, you must pass through a fourth stage of the learning process, *The Incubation Stage.* This last stage consists of physically relaxed, mentally unfocused brain states where information is allowed to move from working memory into your long-term memory banks.[7] Passing through a complete cycle of the four stages of learning provides your brain with the correct levels of different neurotransmitters, in the correct structures, at the correct times so that efficient long-term memory and behavioral change can easily take place.

The best way to understand the four stages of *The Cycle of Transformation* learning process is to imagine yourself as a young child, learning not to touch a hot stove. The *Information Stage* consists of observing heat coming from the stove in your kitchen or your mother yelling, "Don't touch that stove. It's hot."

Of course, being a little human who needs to discover the world for yourself, you take the *Information* (stage 1) that you've acquired and then take *Action* (stage 2) by touching the stove. Immediately, you get sensory

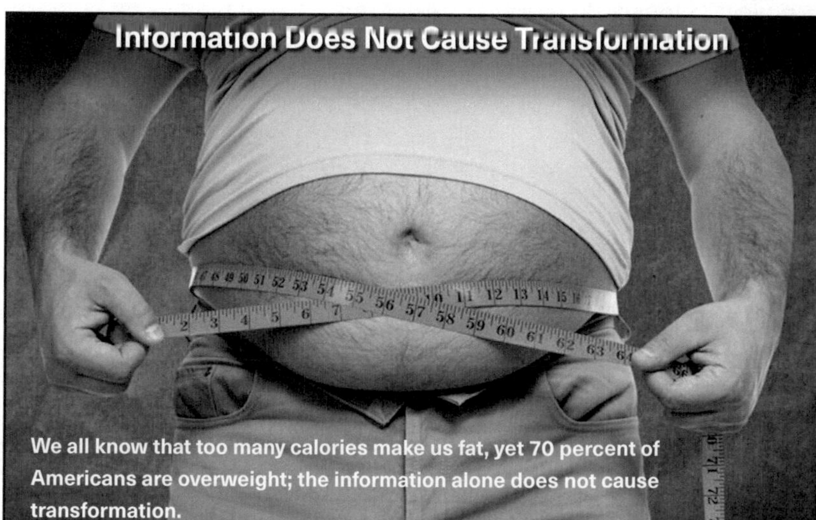

We all know that too many calories make us fat, yet 70 percent of Americans are overweight; the information alone does not cause transformation.

Feedback (stage 3), resulting in a burned finger, and after an *Incubation* period of relaxed downtime (stage 4), you integrate this important information into your long-term memory banks.

Depending on the intensity of the burn and your genetic makeup, temperament, stress levels, and previous experiences with hot things, you could learn everything you need to know about hot stoves in one cycle of learning, or you may need to pass through *The Cycle of Transformation* many times before you learn not to touch them.

Creating Meaning Where Little Existed Before

One of the greatest challenges of any education system is how to make newly introduced information personally meaningful to the learner. As we saw in Chapters 12 to 14 on meaning, the higher the personal meaning, the more effectively and easily your brain can select new knowledge into its structures. This is because when you encounter something in your world that has personal meaning to you, a chemical climate is produced in your brain and body that accelerates the whole learning and behavioral change process.

In the above example, initially, as a child, stoves held little meaning for you—until you reached out to touch a hot one and activated *The Cycle of Transformation*. Once the four stages of the cycle—*Information, Action, Feedback*, and *Incubation*—were complete, you not only had the neurological foundation for making hot stoves meaningful to you, but you could easily associate other hot things with your experience with stoves. As we explored in Chapter 14, once a strong concept network is formed, it can be easily linked with other concept networks without having to go through all four stages of the learning process.

The Ebb and Flow of Neurochemicals During The Cycle of Transformation

Passing through the four stages of *The Cycle of Transformation* provides the wonderful benefit of exposing your brain to the natural ebb and flow of different neurotransmitter groups, which in turn allows relevant information to first be easily stored in working memory and then be transferred to your long-term memory banks.

The chemical states produced in specific brain regions at specific times during the different stages of learning are complex and just beginning to be understood. It's estimated that, in addition to the monoamines and acetylcholine, there are over 100 neurotransmitters and hormones, including glutamate, GABA, endorphins, and cortisol, all interacting with one another in highly complex algorithms in specific brain regions at specific times.[8]

To facilitate your understanding of *The Cycle of Transformation,* here I'm going to paint a simplistic picture of the neurochemistry that directs your different stages by focusing on the monoamine system (dopamine, norepinephrine, and serotonin) and the cholinergic system. In later chapters, we will go much deeper into the biochemistry of these stages and why learning systems that fail to stimulate them to fail to create profound learning and behavioral change.

Stage 1: Information

If the information you encounter in this stage has little meaning to you, there will be very little neurochemical activity and, therefore, little memory formation. For example, let's say you have no desire to learn how to fish. When a friend tries to tell you how to fish, very few of your neurochemicals of learning like dopamine and acetylcholine will be activated; consequently, you'll have very little memory of what that person tried to teach you. But imagine your friend takes you out fishing on a lake on a hot sunny day. Even if you hate touching slimy bait worms and squirming fish, a dramatic amount of learning still occurs. Why? Because your brain wasn't exposed to information alone. It had the opportunity to pass through the *Action* and *Feedback Stages* of the learning process.[9]

Stages 2 and 3: Action and Feedback

The *Action* and *Feedback Stages* turbocharge the learning process because the neurotransmitters of the monoamine system—dopamine, norepinephrine, and serotonin—become highly engaged.[10] As you saw in Chapters 12 to 14, the monoamines are responsible for helping maintain your attention, refine your focus, and mark information with emotional value so that it can be prioritized for inclusion in your working memory.

Any time you take action, you get rich, complex sensory feedback from your environment, and the activity of the monoamine system (dopamine, serotonin, norepinephrine) can increase by as much as 500 percent, substantially supporting your memory formation and behavioral change. This insight into how real-world experience enhances dopamine, norepinephrine, and serotonin activation helps us gain a greater understanding of why traditional academic and corporate educational systems that rely only on a presenter standing in front of the room have such a minor effect on your monoaminergic system and ultimate learning.

Stage 4: Incubation

During *The Incubation Stage* of learning, as you enter a physically relaxed state, dopamine, norepinephrine, and serotonin begin to loosen their control and your cholinergic system, which relies on the neurotransmitter acetylcholine (ACH), begins to gain more control. Now the parasympathetic system, which controls your relaxation response, becomes more engaged. You become calmer and less focused during periods of the day or enter sleep at night. It is in this stage that personally meaningful information gained in the *Action* and *Feedback Stages* of learning which has been placed into working memory can now be sent to your brain's rear associative areas for inclusion into your long-term memory banks. Any interruption of the *Incubation Stage* of learning will compromise long-term memory formation. And how much time does education and corporate America allow for this vital downtime stage of learning? None might be the common answer. And this one major reason we forget so much of the information we struggled so hard to memorize and were later tested on.[11]

Creating New Connections and Brain Cells

Ultimately, engaging in all four stages of the learning process should be a primary goal of any learning or behavioral-change system: creating neurological change and growth. Not only do the four stages of *The Cycle of Transformation* help create the best neurochemical climate for learning, but they also build the physical structures of memory formation and behavioral change—new connections between brain cells (synapses), as well as brand-new neurons.

For over six decades, scientists have known that new synapses can be produced through learning, but until the 1990s, there was a firm belief in the scientific community that no new brain cells could be created after you were born.[12] This consensus was shattered as more and more studies began to reveal that the brain not only creates new synapses but also new neurons when you learn.[13]

In the late 1990s, at the Salk Institute, neurobiologist Fred Gage made a fascinating discovery about new neuron growth and physical experience. When he looked at the brains of mice allowed to run on an exercise wheel, he found that physical action and sensory feedback increased the neurochemical-electrical activity and synaptic growth that leads to long-term memory formation.[14] But that wasn't all—it also increased brain cell numbers.[15] Other neuroscientists later discovered that humans who physically moved in their environment not only had greater brain tissue growth than those of us who were more sedentary, they also had greater learning and memory capacities.[16]

We now know that *The Cycle of Transformation* also activates specific genes that promote learning and behavioral change. As an example, the BDNF gene provides instructions for making a protein found in the brain called a *Brain-derived Neurotrophic Factor*. This protein promotes the survival of your brain cells by playing a role in the growth, development, and maintenance of brain tissue.[17] The advantage of allowing your brain to physically pass through the four stages of the learning process is your biology gets the support it needs to more easily create not only the right neurochemical climate but also new synapses and new neurons, which we now know are the structural foundations of all your learning and behavioral change.

Combining Meaning and The Cycle of Transformation: Learning on Steroids

Let me bring up something I'm often asked. If the knowledge you're trying to learn is highly meaningful to you, do you need the four stages of *The Cycle of Transformation* to effectively learn it? The answer is no.

Why not? Because personal meaning itself has created the neurochemical climate in which learning is optimal.[18] Take the fishing example we used earlier. If fly fishing is meaningful and enjoyable to you and someone tells you about a new fly-fishing technique that will make you a better fisherman, that meaningful knowledge has a good chance of being selected into your long-term memory. This is because meaning has already built the brain infrastructure for the new knowledge to easily stick to.

The Cycle of Transformation comes into play when you're trying to learn something new and haven't yet established the chemicals or structures in your brain that label specific new information as deeply meaningful. In effect, by entering *The Cycle of Transformation*, your brain starts to build the neurochemical and structural foundation for meaning to be enhanced. It is why reading about a pinprick is not as effective at creating meaning and memory as experiencing a pinprick. So, one way to look at *The Cycle of Transformation* is as a way to create the neurobiological foundation for personal meaning where little meaning existed before.

But it's also important to recognize that combining personal meaning and *The Cycle of Transformation* can give your brain double the chemical and structural benefit. This combination may not always be possible, but when it is, implementing the four stages of *The Cycle of Transformation* will bring significant neurobiological benefits.

For example, if I'm researching dopamine-producing neurons and I read about them (stage 1, *Information*), I'll have a good chance of remembering much of what I read because this information is so meaningful to me. But if I take the action of looking at dopamine neurons with an electron microscope and moving them about in the petri dish (stages 2 and 3—*Action* and *Feedback*), I'll get a double bang for my learning buck—the neurochemical benefits of meaning and *The Cycle of Transformation* combined. Know that all the structural and neurochemical benefits the cycle provides will be supercharged if deep, personal meaning already exists for what you're trying to learn.

Now that you have a general understanding of this process, in the next two chapters we'll dive much more deeply into the structures and neurochemical systems that are activated when you enter the *Action* and *Feedback Stages* of learning. What follows in the next chapters is designed to give you a deeper understanding of the structural and neurochemical benefits of *The Cycle of Transformation*.

Cracking the Learning Code

The Institute has developed several businesses that range from healing metabolic diseases like polycystic ovarian syndrome (PCOS) to online programs such as Cracking the Learning Code for Teachers and Homeschool Parents. All use this more natural four-stage process to learning. You can learn more about these online programs where educators and homeschool parents can get a professional development certificate from the prestigious Claremont Graduate University at , www.thelearningcode.com

32

No Feedback No Learning: Making Order Out of Chaos with a "Touch of Spice"

Feedback is the Breakfast of Champions.

—*Ken Blanchard, author of* The One Minute Manager[1]

In this chapter and the next, as you come to understand even more about the *Action* and *Feedback Stages* of *The Cycle of Transformation*, it will become even clearer to you why learning systems based only on information can't live up to our expectations. Although we briefly looked at the monoamines—dopamine, serotonin, and norepinephrine—in the chapters above, here, we'll peel away one more layer to gain a deeper look at how these neurochemicals impact how you learn.

FAST FORWARD or DEEP DIVE

If you feel you have a good general understanding of the interactions of the *Action* and *Feedback Stages* of *The Cycle of Transformation* you can skip to the rest of this and the next chapter and go to Chapter 34 "The *Action* and *Feedback Stages* Build Huge 3D Maps in Your Brain." But if you want to go into the nitty-gritty of the neurobiology of why these two stages are so vital to the learning process, please read on.

Making Order Out of Chaos: Your Spice Rack

Science now allows us to see that our experiment over the last 100 years of trying to just deliver information to the brain to get it to learn and change has been flawed. When we look at the timeline of life on Earth, we can clearly see that, for 3.5 billion years, organisms have learned how to increase their *Individual Adaptability/Intelligence Factors* by taking physical action and getting sensory feedback from their real-world experiences.[2] It's only been within the last 100 years, in our efforts to educate 100 percent of the population, that we've developed educational and corporate training systems that have gone against the evolutionary flow by locking people up in small rooms, delivering only information and dropping the *Action* and *Feedback Stages* of the learning process. Before this, the majority of our species learned through apprenticeships and mentorships in which action and feedback are natural components.

To get a grasp on the power of the *Action* and *Feedback Stages* of learning over information-only based learning, it is important to understand that when you are taking action in, and getting feedback from, real-world interactions, billions upon billions of more bits of sensory information are bombarding your brain and body each second. This is much more stimulation than your brain receives from passively sitting in a classroom.

As we saw in the previous chapter, this bombardment of sensory information activates massive numbers of neurons and neural connections. To deal with this complex neural stimulation, Nature gave your brain a special neurotransmitter system that could impose order on the chaos created by all this massive firing. That system is the monoamine system (dopamine, serotonin, and norepinephrine), which we have been exploring.[3]

The neurons of your monoamine system are located primarily in and around your brain stem and the limbic area. Though they consist of only a tiny number of your total neurons (less than a million), they have a profound impact on the direction, coordination, and intensity of the neural firing among your other 100 billion neurons. This is because of dopamine, serotonin, and norepinephrine act as *neuromodulators*—that is, they modify the way other neurons act. Like the volume on your

Monoamines– Your Brain's Spice Rack

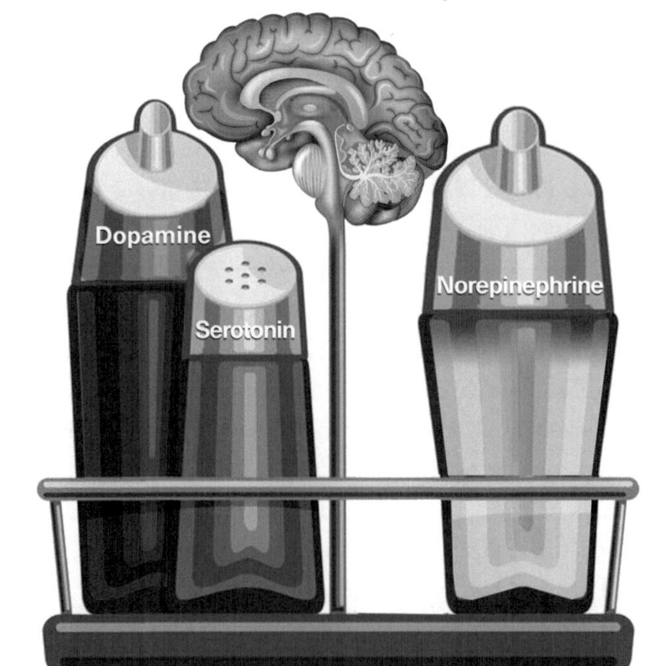

Your monoamines act like a spice rack in your brain, sprinkling specific bits of meaningful incoming information with just the right emotional zest so you remember them.

stereo, the monoamine system superimposes "gain" control, either amplifying or muting the activity of other neurons.[4]

As you take action and get feedback from your world, this modulation function of dopamine, serotonin, and norepinephrine inserts order into the billions of bits of information you're bombarded with each second. James Austin, professor emeritus of neurology at the University of Colorado and author of *Zen and the Brain*, nicely sums up the role of the monoamines, saying they "further emphasize the brain's basic response functions, inserting suitable pauses, occasional italics, and sometimes multiple exclamation points."[5]

In a "rising tide lifts all boats" concept, because the *Action* and *Feedback Stages* of learning naturally increase the monoamines which help your brain make order out of chaos, they also increase these neurochemicals which allow you to attach more emotional value to what you are doing in the real world. As we discussed in the chapters on the *Meaning Network* (see Chapters 12 to 14), it's your monoamine system that also provides the neurochemicals that mark information with emotional intensity so they can be remembered. Dopamine, serotonin, and norepinephrine are the neurochemicals that make you feel alive; your intense feelings of joy, fear, love, greed, and anger would cease to exist without the influence of your monoamines. They act like the spice rack in your brain, sprinkling specific bits of meaningful incoming information with just the right emotional zest, so you remember them.[6]

Action and Feedback: Natural Ritalin and Prozac

Another way to look at the *Action* and *Feedback Stages* is as natural forms of Ritalin and Prozac. The main function of Ritalin and Prozac is to increase your attention and mood by increasing the action of monoamines. Ritalin increases the action of the monoamines norepinephrine and dopamine,[7] thus helping you focus. Prozac's job is to stop the brain from taking up serotonin; this makes more serotonin available to neurons, helping you feel more content.[8]

Like these pharmaceuticals, the *Action* and *Feedback Stages* of the learning process automatically pump up your monoamine activity and thus naturally pump up both your attention span and your moods. The increase in monoamine activity caused by your actions in the sensory-rich world is one reason that, when you're unfocused, down, or blue, something as simple as taking a walk outside can perk you up and make you feel better and more alert. (The research makes you wonder why more psychiatric hospitals don't make exercise and outdoor activities a part of every treatment plan.)

An individual in a coma gives us an extreme example of a compromised monoamine system—the person is unable to move, feel emotions, or make memories because his or her monoamine neurons have been damaged by disease or trauma.[9, 10] Remember that systems dictate behavior; therefore, learning systems that fail to effectively activate the monoamine system can produce environments where people often feel like they're in a coma! Boredom rules, memory flatlines, and depression can flourish.

Taking a Deeper Dive into the Feedback Stage

The *Information* and *Action Stages* of the learning process are rather straightforward and easy to understand. Using our stove analogy again, the information is that the stove is hot, and the action is that you touch it. The *Feedback Stage* has several interesting elements that add complexity and are important to understand in our efforts to create learning systems that produce profound results.

Feedback is any process where the result of your action serves to modify future actions.[11] In 1863, at the genesis of the understanding of the electrical nature of the brain, Ivan M. Sechenov, the "Father of Russian scientific psychology," who had a great influence on Pavlov's famous work, was one of the first scientists to recognize the power of sensory feedback in creating memory. In *Reflexes of the Brain,* he wrote, "The internal cause of any action always lies in external sensory stimulation, because without this thought is inconceivable."[12]

Since Sechenov's time, we've learned a lot about the power of sensory feedback from our actions. Today, there's even a field for the study of feedback. It's called *cybernetics,* which is derived from the Greek for "to steer or navigate."[13] The fundamental point of cybernetics is that systems can have absolutely no control of themselves without feedback from their world. We now know that to learn effectively, we need potent sensory feedback to produce enough dopamine, norepinephrine, and serotonin to create potent memory. We also know that the very expression of the genetic code itself is dependent upon loops of feedback between an individual and their environment.[14]

The More Immediate the Feedback, the Greater the Learning

What many memory researchers consider to be one of the most important discoveries about learning and sensory feedback is that the more immediate the feedback from an individual's action, the more effective the learning; conversely, the longer the lag time between actions and feedback from the environment, the weaker the learning.[15] To use an earlier example, as a child, you quickly learned not to touch a hot stove because you immediately felt a burning sensation the instant you touched one. If you didn't feel the pain until hours after you touched the stove, your brain would have had a difficult time sorting out which of your previous actions caused the pain feedback, which, in turn, would have made learning to avoid hot things a much more difficult task.

MIT systems analysis guru Peter Senge notes that good management and personal responsibility are often hard to implement in large corporations because the lag time means the actions you take in the corporate world and the feedback you get from those actions are usually disconnected. He says, "Herein lies the core learning dilemma that confronts organizations. We learn best from experience, but we never directly experience the consequences of many of our most important decisions." The most effective learning and behavioral-change systems always provide for immediate feedback.

Mihaly Csikszentmihalyi, Distinguished Professor of Psychology and Management at Claremont Graduate University, researched the optimized state of learning, which he calls *flow*. This uninterrupted state is recognized

The more intense the feedback, the greater the learning.

by its timelessn°ess and pleasure.[16] It's the phase of learning where you become one with the task you're engaged in. This is where greatness occurs. Csikszentmihalyi points out that this wonderful state can't take place without immediate sensory feedback.

Feedback from Your Body Creates Learning

Sensory feedback is also effective at creating efficient learning because it prompts your monoamine system (dopamine, serotonin, and norepinephrine) to create immediate somatic changes in your body.[17] As we discussed, somatic changes are alterations in your blood pressure, heart rate, galvanic skin response, pupil dilation, breath rate, distribution of blood flow, and more. The key here is that somatic feedback is a slave to sensory input. You get sensory feedback, and your body has to respond—it has no choice. As you saw in Chapter 15, your brain uses somatic/body changes as the material from which it constructs emotions, meaning, memory, and behavioral change.

The more intense the sensory feedback, the greater your somatic/body changes, and therefore the greater your learning. You remember your first kiss or a car accident you've been in so well because of the strong somatic feedback caused by these events. You forgot so much of what you studied in school because the classroom failed to produce strong feedback from your body that was either immediate or intense. Therefore, your brain lacked enough of the neurochemical materials to construct emotions, meaning, and memory. In the end, your brain flourishes on feedback from your environment and body because it's the fuel that helps you survive and thrive in your world.[18, 19]

Sociopaths and Drugs Help Us Understand the Power of Body Feedback

It's believed that one of the reasons sociopaths fail to behave appropriately in society, even after intense therapy, is that their brains are unable to effectively process immediate sensory feedback from their actions.[20] In response to sensory stimulation, the sociopath's brain gets faulty signals from the body and therefore can't produce suitable emotions upon which to base acceptable actions. Heart rates and other body markers of chronic domestic abusers, rapists, and sociopaths actually drop in response to the violent sensory stimulation caused by their actions.[21] With sensory feedback failing to produce the correct somatic signals, sociopaths fail to learn how to alter their sadistic ways.

The opposite occurs in the drug addict. One reason people become so rapidly addicted to powerful drugs such as cocaine, heroin, and methamphetamines is that taking a hit produces high levels of serotonin, dopamine, and/or norepinephrine, which in turn creates immediate, powerful, pleasurable body feedback.[22] This instant gratification helps cement the learned behaviors, which keep an addict coming back again and again to his or her drug of choice.[23]

In the next chapter, we will take a deeper look at the neurochemicals of the *Action* and *Feedback Stages* of learning, motivation, and behavioral change. In the chapter after that, we will discover how those two stages build huge spatial 3D maps in your brain, providing more brain mass for you to more easily select new knowledge into its structures.

Cracking the Learning Code

Understanding how the neurotransmitters—dopamine, serotonin, and norepinephrine—impact the *Action* and *Feedback Stages* of *The Cycle of Transformation* is vital to comprehending how to create more impactful and transformational learning systems.

33

A Deeper Dive: The Specific Roles Your Monoamines Play in The Cycle of Transformation

When dopamine dips, your unhappy chemicals get your attention. You get that "do something" feeling, and you have to decide what to do. One option is to accept the unhappy chemicals and figure out what's triggering them. Another option is to quickly distract yourself from them with a behavior you expect to trigger more dopamine.

—*Loretta Graziano Breuning,* Meet Your Happy Chemicals: Dopamine, Endorphin, Oxytocin, Serotonin[1]

This chapter goes more deeply into the monoamines' role in *The Cycle of Transformation* by helping us comprehend their relationship to smart drugs, alcohol, LSD, Ecstasy, Parkinson's Disease, ADHD, and how they can create calmer institutions. The insights presented here are designed to give you an even firmer foundation for understanding why personal and organizational learning systems must activate the *Action* and *Feedback Stages* of learning to create profound memory formation.

If you're reading this book to gain only a general understanding of the science, you can just skim or skip the next pages, as the earlier chapters give you the big picture model of the roles dopamine, serotonin and norepinephrine play in the *Action* and *Feedback Stages* of your learning process. If so, you can jump ahead to the next chapter.

Going Deeper for Deeper Understanding

The neurochemistry of *The Cycle of Transformation* would be much simpler if each neurotransmitter of your monoamine system had only one function, but each neurotransmitter can have many. For example, fifteen different serotonin receptors have been identified.[2] Although the general public is aware that serotonin plays a role in mood stabilization, most of us are unaware that its roles also include initiating movement, modulating anxiety, controlling your impulses, and providing a stable platform for cell firing, so meaningful information can stand out. You also have five types of dopamine receptors, each playing different roles in pleasure, meaning-making, the repetition of meaningful behaviors, and other functions involved in the learning and memory process.[3] Neuroscientists have even created a specialized field just to keep track of all the new subtypes of receptors and to assign a function to each of them.[4]

To add to the complexity of teasing out brain function, your monoamines have almost infinite direct and indirect reciprocal moderating interactions with themselves and the 100 or so other neurochemicals in your brain. This complexity makes it difficult to paint a simple picture of the effects of the monoamine system. To facilitate your conceptual understanding of their functions during the *Action* and *Feedback Stages* of the learning process,

I'll paint with a broad brush the roles that dopamine, norepinephrine, and serotonin play in the learning and behavioral-change process.

The Short Story of the Monoamines and The Cycle of Transformation

Here's the short story of the monoamines' role in *The Cycle of Transformation:* Norepinephrine says to your other brain cells, "Perk up and pay attention to this relevant information right now." Dopamine then acts as a spotlight on a dark night, highlighting what norepinephrine has already marked as meaningful and making it stand out even more. While all this is going on, serotonin helps make order in your brain by creating a very stable, calm background of cell firing so that what norepinephrine and dopamine have marked as meaningful and important can be easily identified and selected into your long-term memory banks.

Norepinephrine Shouts, "Pay Attention to This Right Now!"

During the *Action* and *Feedback Stages* of learning, norepinephrine's main job is to super-sensitize your brain so that it knows which of the billions of bits of information bombarding your brain are the most meaningful and worthy of attention. Most of your norepinephrine cells cluster in the *locus coeruleus*—Latin for "a blue place"—named for its azure color—in the brain stem.[5] On activation, the 45,000 to 60,000 norepinephrine cells located here say to other cells in your brain, "Hey, stop paying attention to all that other input and pay attention only to what we mark as emotional, somatic and meaningful."

Norepinephrine does this in two ways: excitation and suppression. It's best known for making your neurons already activated by specific, meaningful stimuli fire even more aggressively. But norepinephrine also has a filtering effect, suppressing the spontaneous firing activity of neurons that don't enhance your brain's capacity to focus on meaningful stimuli. These two functions improve the signal-to-noise ratio in your brain, allowing your working memory to select only the important information that stands out.[6]

Norepinephrine's repression and excitation effect are widespread, affecting virtually all your brain areas, including the thalamus, which processes incoming sensory information, and the parietal lobe, which helps you form spatial-map representations[7] of your world, so you know where you fit in.

Norepinephrine also plays an important role in stimulating your somatic/body changes that are the foundation of emotions and learning.

When you take purposeful action and get feedback from your environment, norepinephrine helps you become more engaged by altering heart rate and blood pressure, dilating (opening up) airways in the lungs and, shifting blood flow from the digestive organs to the muscles. Your hypothalamus, which also plays a role in the mediation of somatic feedback and learning, has a high concentration of norepinephrine receptors.[8] The feedback from these somatic/body signals to your brain is vital in helping you make meaning out of the important information in your world so you can remember it.

Also, norepinephrine supports emotional marking because it excites neurons in the central area of your amygdala (the seat of instinctual emotions) and stimulates receptors in the parts of your frontal lobe that mark information as emotionally relevant. Research shows that physical activity and sensory feedback increases norepinephrine activity by 40 percent in your hippocampus, the part of your brain that works with your lateral prefrontal cortex to maintain information in working memory and place relevant knowledge into the correct long-term memory concept networks.[9] Activation of norepinephrine during the *Action* and *Feedback Stages* is one reason it's so much easier to remember information you

Norepinephrine's main job is to super-sensitize your brain so that it knows what's most meaningful and worthy of attention.

actively encounter in the real world than what you try to learn passively by reading or listening.

Relaxed states and sleep rejuvenate your norepinephrine levels. This is why it's often easier to focus earlier in the day and why, as the day wears on, your focus diminishes, your mind begins to wander, and you become susceptible to unfocused daydreaming. Because caffeine increases the available norepinephrine in the brain, you may reach for coffee or Red Bull in the afternoon to help you stay tuned in and focused during work and study sessions.

What ADHD Can Teach Us about Norepinephrine's Learning Function

Research involving people with attention deficit hyperactivity disorder (ADHD) has enhanced our understanding of the effects of norepinephrine on *The Cycle of Transformation*. Although many people think that those with ADHD don't pay enough attention, in most cases, the opposite is true. These individuals' difficulties arise because they pay too much attention to too many things for too short a period. In many individuals with ADHD, learning difficulties have been linked to the brain's inability to filter out unwanted stimuli due to low levels of norepinephrine. Without enough norepinephrine, their brains flit from stimulus to stimulus.

Amphetamine drugs such as Ritalin and Adderall, which increase norepinephrine activity, can decrease ADHD symptoms in some people (though their prescription is controversial). Many students who don't have ADHD are now illegally taking these "smart drugs" because they raise norepinephrine levels and do indeed enhance the students' focus during study sessions and tests.

The Concern: Smart Drugs

I want to emphasize the concerns about the legal and illegal use of these "smart drugs" not only to point out the dangers but also to deepen your understanding of norepinephrine's role in the learning process. Although we need more research on the long-term effects of these drugs on the growing brain, there are currently four main areas of concern about taking amphetamine drugs like Ritalin and Adderall that raise norepinephrine to enhance learning:

Affecting the adult brain in unknown ways: The first concern is that "smart drugs" may help the young brain focus while damaging the potential of the adult brain. The fact is we really don't know what effects using these drugs as a child will have on your brain as an adult. Research on the effects early use of these smart drugs have on the maturing brain is very limited.

Promoting compliance: Because amphetamines can calm the hyperactive brain, they're often used as a control mechanism in schools. As you saw in Chapter 3, for the 13,000 total hours of K–12 instruction, only a small number of us have brain structures that allow us to sit still in classrooms for these long periods. Most children's genetic makeup prompts them to prefer to run around in the real world and learn. Ritalin and other amphetamines are often wrongly prescribed, especially to kinesthetic- and social-dominant children, to drug their brains into compliance to better fit the school system. (Starting in third grade, I was one of the kinesthetic kids who was prescribed Ritalin. They so disrupted how I felt in the world that I used to hide mine in books in our home library or flush them down the toilet. We sold our old house with all the books in it. Bet the new owners got a surprise when they opened some of them!)

Setting up students for drug addictions: The third concern is that Ritalin and Adderall stimulate the same brain circuits as cocaine and methamphetamine. Initial research indicates artificially stimulating these circuits at a young age may set up these individuals' brains for lives afflicted by various addictions. This is because of the overabundance of synapse connections that are made by these "smart drugs" in children's addiction networks during their formative years.

Rushing maturation: The fourth concern stems from a lack of understanding of the maturation of norepinephrine cells in the locus coeruleus. The cells in this brain stem area don't mature until around puberty, which is the main reason that as a young child, you happily ran about focusing on so many different things for such short periods.[10] The concern here is that children's brains don't mature in lockstep (see Chapter 27). There can be a three-or-more year difference in the maturation of specific brain areas among different children. This is why one child can start reading well at age five and another not until eight or older, yet both kids turn out to be completely normal students and readers. What we call hyperactivity and attention dysfunction in some children may simply be a late maturation of the main source of norepinephrine production in the locus coeruleus. Drugging these types of late bloomers is surely not a viable solution. The maturation of the locus coeruleus in puberty is why most ADHD children can stop taking amphetamines as they get older.

Dopamine: A Spotlight on a Dark Night

Dopamine and its different receptor types have four major jobs related to the *Action* and *Feedback Stages* of learning:

- Highlighting emotional and somatic/body-charged information that norepinephrine has already marked as meaningful so that the information has an even better chance of standing out and being picked up by working memory
- Controlling purposeful motor function
- Allowing you to get pleasure from discovering and learning meaningful information
- Prompting you to repeat meaningful behaviors over and over again

Because Chapter 17 covers the third and fourth functions of dopamine in some depth, here, we'll focus on dopamine's role in emphasizing information that norepinephrine has marked as important. Also, we'll cover dopamine's role in purposeful motor movement during the *Action* and *Feedback Stages* of *The Cycle of Transformation.*

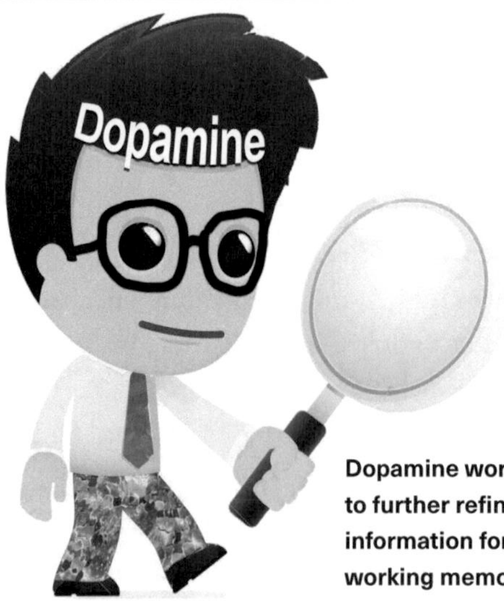

Dopamine works with norepinephrine to further refine the clarity of relevant information for inclusion into working memory.

At the top of your brain stem sits a small cluster of darkly pigmented dopamine cells called the *substantia nigra* (Latin for "black substance"), and just above these neurons sits another cluster of dopamine neurons called the *ventral tegmental area.*[11] From these sources, dopamine's connections fan out to all your brain areas that regulate focus, attention, and working memory as well as provide for the full range of motor movements necessary for purposeful actions.[12]

Dopamine, in its role of maintaining attention and enhancing the tagging of information for inclusion into your working memory, works hand in hand with norepinephrine. Norepinephrine makes the first pass at relevant information, marking it stand out by improving the brain's signal-to-noise ratio. Like a spotlight on a dark night, dopamine then refines this ratio even further, dramatically illuminating relevant information so it can be more easily recognized for inclusion into working memory. Continuing with the spice rack analogy, dopamine takes information that norepinephrine has already marked with black pepper and sprinkles on cayenne pepper so that it really gets your working memory's attention.

During your *Action* and *Feedback Stages* of learning, dopamine also allows you to make smooth, purposeful motor movements. Experiments show that physical activity increases dopamine's production by 20 or more percent over being in repose.[13] This is a key to understanding *The Cycle of Transformation* because when dopamine is inhibited from doing its job in any way, both your purposeful actions and working memory formation are compromised.[14]

Parkinson's Disease and Schizophrenia Give Insight into Dopamine's Function

Looking at Parkinson's disease and the psychiatric disorder schizophrenia can give you a clearer understanding of dopamine's role in creating learning during the *Action* and *Feedback Stages* of the learning process. These disorders give us insight into why learning is compromised if your brain isn't getting the right amounts of dopamine in the correct brain areas at the right times. In a grand oversimplification of what causes these two very complex diseases, we can say that too little dopamine activity in one part of your brain leads to Parkinson's motor dysfunction and dementia, and that too much dopamine activity in the part of your brain that controls focus leads to schizophrenia.

Parkinson's disease attacks and kills dopamine-producing neurons, mainly in the midbrain, where the substantia nigra is located. Without the support of healthy dopamine function, both motor function and mental focus suffer. The shakes, tremors, and rigid actions seen in Parkinson's patients are caused by the lack of dopamine. The decrease can also inhibit these patients' abilities to focus on and learn meaningful information. In advanced cases of Parkinson's, this lack of dopamine activity can lead to the mental confusion of psychosis and what is called *Parkinson's dementia.*[15]

Neuroscientist Oliver Sacks, in his famous book *Awakenings,* described the reaction of Parkinson's patients when they were first treated with the prescription drug L-DOPA, which substantially increases dopamine activity.[16] Many of these patients had been unresponsive, almost frozen since they were young adults. The increase in their dopamine levels caused by the administration of L-DOPA resulted in a dramatic awakening in their capacity to focus their motor movements and mental abilities. The increased levels of dopamine also heightened their interest in the world, elevated sexual attraction, and increased meaning- and sensation-seeking. Just like what L-DOPA does for the damaged brain, the *Action* and *Feedback Stages* of the learning process can increase this valuable learning neurotransmitter for the normal brain.

In another great oversimplification, we can say that schizophrenia is linked to either raised levels of dopamine or an increase in the sensitivity of neurons to dopamine, both of which induce levels of arousal and focus that are way too high.[17] This occurs specifically in the limbic structures of the ventral tegmental area and the nucleus accumbens.

Too much dopamine activity in these areas creates a mental environment where everything seems important to the schizophrenic. The intense dopamine activity forces the schizophrenic's mind to flit from input to input, never being able to maintain attention or focus on a single stimulus for very long.[18] It's as if the spotlight that highlights information in the schizophrenic's brain is out of control, swinging wildly from stimulus to stimulus, shining brightly on a bewildering array of data. When schizophrenics speak, they often start a sentence talking about one subject and finish the sentence talking about another. Focusing on so many stimuli at one time creates chaos in their lives. Many cocaine addicts develop schizophrenia-like symptoms because cocaine increases the dopamine activity in these brain areas to levels far above normal.

In the 1950s, it was discovered that two drugs, clozapine, and chlorpromazine, would help alleviate the symptoms of schizophrenia by reducing dopamine transmission.[19] Daily use of these types of drugs slows dopamine activity in the brain,[20] which then brings the schizophrenic person's focusing abilities back to a normal range, freeing the mind from its bizarre and crippling thoughts.[21] Today, schizophrenics (such as Nobel Laureate John Nash, who was featured in the movie *A Beautiful Mind*) can regain coherent focus in their lives by taking drugs that reduce dopamine activity. Unfortunately, if not monitored, the prolonged use of certain psychotic drugs that reduce dopamine can result in so little being released in the brain areas controlling movement that schizophrenic patients develop Parkinson's-like symptoms.

To summarize, when a normal brain enters the *Action* and *Feedback Stages* of the learning process, dopamine activity is naturally increased in the correct brain areas at the correct times so that important information can be spotlighted, and motor function is purposeful. If the information you encounter is meaningful enough, dopamine will prompt you to anticipate and get pleasure from it (from its effect on the pleasure molecules endorphins) as well as prompt you to repeat the learning process over and over again. Any learning system that isn't supporting your dopamine function is one that will make the whole learning process much more challenging than it needs to be.

Serotonin: Creating a Stable Platform for Learning and Behavioral Change

During the *Action* and *Feedback Stages* of *The Cycle of Transformation*, serotonin plays an important role in stabilizing your cellular communications. Without this steadying influence, your neurons and their connections would fall into a frenetic, foaming sea of disorganized cross talk, compromising both your learning and your behaviors.[22] Yale researcher Thomas Carew calls serotonin "the bandleader who choreographs the output of the brain."[23] As we've seen, every second that you move through sensory-rich environments, billions of bits of sensory information are stimulating billions of your neurons, and neural connections.

Serotonin and its fifteen different receptor types regulate this sensory bombardment by modulating the effects of your other neurotransmitters, which, in turn, helps provide a tranquil background of cell firing. This allows information that has been marked as meaningful by norepinephrine and dopamine to stand out with limited competition from other stimuli, lets your working memory easily identify the most meaningful bits of information, and more effortlessly selects them for inclusion into working memory. Besides, serotonin's modulating effect on neuron firing also has a potent impact on mood, impulsiveness, self-esteem, and your ability to initiate physical action—all-important to the learning process.

Though you have only 500,000 serotonin neurons in an area called the *raphe nuclei,* buried deep in the core of your brain stem,[24] they have a dramatic calming effect on the rest of your 100 billion neurons. This is because a single serotonin neuron can influence as many as 500,000 synapses.[25] To maintain a stable backdrop of cell firing during the *Action* and *Feedback Stages* of learning, serotonin activity in this tiny group of cells accelerates dramatically.

Initial research demonstrates that serotonin levels can rise as much as 500 percent when your brain passes through the *Action* and *Feedback Stages* of *The Cycle of Transformation.*[26] With this intense increase, is it any wonder you learn faster from real-world experience? Two different mechanisms appear to be involved in this increase: motor activity increases the firing

Serotonin plays the important role of stabilizing your cellular communications like a band leader conducting an orchestra.

rates of serotonin neurons, and there's also a corresponding increase in the processing of tryptophan, from which serotonin is produced.

Interestingly, studies have shown that natural sunlight will also raise your serotonin levels.[27] The measure of one unit of the illuminance is called a *lux*.[28] There's around 400 lux in the average classroom, but outside, indirect sunlight produces around 25,000 lux, and direct sunlight produces 100,000 lux. You can see why it's easier to increase serotonin levels outside in the natural world than in a classroom.

Why You Can't Learn Well Under Stress: Lower Serotonin

One of the reasons you become confused and have a difficult time learning and remembering when you're under stress is that stress prompts the release of the hormone cortisol from your adrenal glands, which in turn lowers the amount of serotonin available to your brain. With cortisol inhibiting serotonin, neural stability is lost, and your learning and thinking become jumbled.[29] Forgetting your words when you're giving a speech, being unable to think of what to say next in an argument, and getting so anxious on a test that you can't retrieve what you just studied can all occur because high-stress levels are causing your brain to fill with neural static as serotonin's stabilizing activity is compromised.

LSD, Prozac, Ecstasy, Food, and Alcohol Help Us Understand Serotonin's Function

You know why LSD and psilocybin (magic mushroom) users have wild hallucinations? Because these hallucinogenic drugs are excellent blockers of serotonin receptors. These drugs create delusions by stopping serotonin from doing its job of reducing cross talk between neurons.[30] With no calming constraint on neural firing, wild, disorganized thoughts arise. As you'll see in Chapter 36 when we talk about sleep, an LSD trip is very dreamlike because, in the dreaming sleep stage called REM, serotonin activation is virtually shut down.

Serotonin's most well-known effect is on mood. It's credited with prompting feelings of satiation, which make you feel that all is right in the world and that your life is worth remembering. Serotonin's relationship with depression is well-documented. Prozac and similar drugs are prescribed to approximately thirty million people in the US every year because they're believed to be effective at alleviating depression by increasing serotonin in the brain.[31] Some researchers think that the reason depressed people can commit the unthinkable act of suicide is that their low serotonin levels not only make them despondent but also destabilize cell firing so much that their thinking becomes confused, thus making suicide seem like a rational act.

One reason that the use of the drug Ecstasy (MDMA) is so disturbing is that it's very effective at destroying the fine serotonin axons branches in the raphe nuclei.[32] To self-medicate, Ecstasy addicts repeatedly take the drug to elevate their moods, which results in a dangerous downward spiral of addiction, axon destruction, and greater depression.

Food and drink can become very addictive—and your center of attention and focus—especially if other areas of your life lack meaning. Like Prozac, they initially increase serotonin levels and thus temporarily elevate our moods. The urban myth is we feel content, sleepy, and happy after a huge turkey dinner at Thanksgiving because turkey is full of tryptophan, and serotonin is synthesized from tryptophan. The real reason we feel so relaxed after a Thanksgiving meal is because the carbohydrates, the stuffing, and potatoes, raise serotonin activation substantially.

This gives us insight as to why we crave carbs when we are feeling down. Eat that muffin or piece of cake, and you jack up your serotonin for a bit. Why do we feel so good and relaxed after those first few drinks? The same reason. Because alcohol initially raises serotonin levels. Why do we feel so depressed after drinking too much? You guessed it: because after the first few drinks, alcohol depresses serotonin levels.

More research must be carried out, but early studies indicate meditation raises your mood levels.[33] Why? Because meditation may improve serotonin activity.

Higher Serotonin, Calmer Institutions

Serotonin also affects learning because it acts as one of your brain's major impulse modulators. When The Institute is working with organizations that are dealing with high levels of behavioral problems, we encourage them to develop methods that increase serotonin levels in their population. This is because low serotonin activity has been implicated in increased levels of noncompliance and aggression.[34] When your serotonin activity is low and neural stability is lost, you can feel anxious and on edge and then act out of character. Many researchers believe that the majority of the prison population is plagued by chronically low levels of serotonin, causing them to habitually act out in socially unacceptable ways.[35]

It's interesting to note that the reduced number of serotonin receptors found in Alzheimer's patients is believed to be the major cause of the impulsive and aggressive behaviors that are often seen in later stages of the disease.[36] Evidence also links serotonin with self-esteem.[37] Research shows the higher an individual's serotonin levels, the better they feel about themself. Low serotonin levels caused by high stress have also been blamed for impulsive behaviors in children and adults who live, go to school, or work in high-pressure environments.[38]

Serotonin, Physical Movement, and Learning

In addition to its many other functions, serotonin plays a major role in the *Action* and *Feedback Stages* of learning by helping you initiate physical movement. The influence of serotonin allows you to get off the couch and take the actions that will help you learn more in your life. It's been shown that serotonin neurons fire not only when you're moving but also right before you decide to move. Therefore, if your serotonin levels are low, initiating action turns out to be difficult, and becoming a couch potato becomes very easy.

Physical movement and exercise have been shown to raise all the monoamines and thus help increase impulse control, learning, and mood. This is why you feel so much better about life after a nice walk in the woods. But unfortunately, a despondent person often doesn't have enough serotonin activity in their brain and may not be able to start moving—movement that could help them get rid of their dark moods and take more positive actions in their life.

Cracking the Learning Code

All you really need to know about the *Action* and *Feedback Stages* of *The Cycle of Transformation* is that during these stages of the learning process, the neurotransmitters of your monoamine system—dopamine, norepinephrine, and serotonin—become highly engaged. These neurotransmitters are responsible for helping you maintain your attention, refine your focus, and mark information with emotional value so it can be prioritized for inclusion in your working memory. The monoamines are also vital in the process of allowing you to make meaning where no personal meaning existed before.

34
The Action and Feedback Stages Build Huge 3D Maps in Your Brain

True intuitive expertise is learned from prolonged experience with good feedback on mistakes.

—Daniel Kahneman, psychologist and economist[1]

In this chapter, you'll gain an even wider perspective of why you remember so very little of what you so painfully memorized in school, and why you remember so much more of what you experienced in the real 3D world (see also Chapter 33). The short story is that classroom learning and memorization don't build three-dimensional maps of the real world in your brain. And without spatial maps, new information that's important to your personal surviving and thriving in the real world has a much harder time being selected into your long-term memory banks, because it has less brain material to attach itself to. The spatial maps of your world are held in very large brain structures called parietal lobes located just in front of your visual cortex.

Whether you know it or not, you spend most of your life running around trying to figure out how you can best fit into your 3D world—that is, how you best fit into your family, social, school, and work environments. As you navigate through time and space, information that comes into your brain from sight, sound, touch, smell, and taste must be put together so you can build accurate maps of your world. Without these maps held in your parietal lobes (see image on page 509), your life would be a buzzing cauldron of confusion. In her wisdom, Mother Nature designed brain structures and neurochemical systems that allow you to quickly and efficiently build spatial maps of the world.

You could say that those of us who best understand and can best manipulate our different environments have the greatest surviving and thriving chances—that is, we have high *Individual Adaptability/Intelligence Factors.* These valuable spatial maps are constructed fastest and easiest when we actually enter the second and third stages of *The Cycle of Transformation*, taking physical action in and getting rich sensory feedback from our real, 3D worlds.

The Value of Spatial Maps Uncovered

It wasn't until the 1970s that neuroscientists started to focus on the value of spatial mapping in the learning and behavioral-change process. In their famous scientific tome, *The Hippocampus as a Cognitive Map,* John O'Keefe and Lynn Nadel point out that everything in life takes place in time and space.[2] Their comprehensive labors laid the groundwork for future efforts to locate which brain areas were responsible for spatial mapping. We now know that the hippocampus in the limbic system and parietal lobes in the neocortex work hand in hand to build a spatial representation of your world.[3] As soon as location maps have been built, the frontal lobes allow you to personalize these maps and manipulate them to best fit your purposes.

Building 3D spatial maps in your brain's parietal lobes provide three advantages.[4] These advantages are similar to the ones we saw in Chapter 33 when we discussed why it is so important to present concepts to the brain before the details. Let's review the similar advantages here:

- **Hooks:** New information doesn't float freely in your brain. Instead, before it can be selected into your long-memory, new information must hook itself to pre-existing networks. Spatial maps provide these valuable hooks.
- **Mass:** The neurons and connections that make up spatial maps supply mass, and mass attracts mass. Therefore, large spatial maps of your world act like magnets, attracting large numbers of details that resonate with them.
- **Demand energy:** When spatial maps are being built or activated,

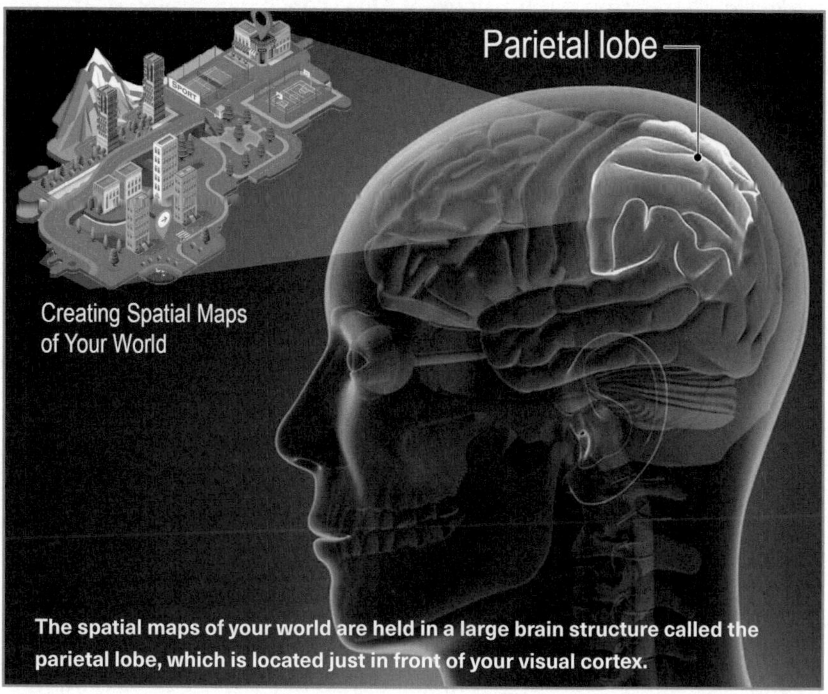

The spatial maps of your world are held in a large brain structure called the parietal lobe, which is located just in front of your visual cortex.

> they demand large amounts of glucose and oxygen, which, in turn, provide the energy necessary to power the processes that allow details to be incorporated into long-term memory.

Education Fails to Build Real-World Spatial Maps

The prior advantages reveal why it's so much easier to remember what you experienced on a trip to Beijing, China than what you read in a book about it. Traditional classroom learning and memorization are decontextualized from your real, 3D world. Without the *Action* and *Feedback Stages* of learning, the typical classroom isn't an effective location to create real-world spatial maps. And with no real-world spatial maps to attach itself to, information that you worked so hard to memorize for a test can quickly drop out. Is it any wonder we spend so much time in classrooms and remember so very little?

Renate and Geoffrey Caine point out that our traditional educational method "actually shuts down map-building functions because dynamic maps are always complex and personal. Maps constitute an internal organization of information and always contain much more than the information packaged in a lesson or textbook."[5] They conclude, "Emphasizing the storage and recall of unconnected facts is an inefficient use of the brain."

This inability of our traditional educational system to create and activate real-world spatial maps is the main reason why so many of us graduate with so little idea of how to successfully survive and thrive in the real world. It's why a student can do well in school memorizing math rules but find it impossible to employ these same rules working at a hardware store where they must operate the cash register. We've had MBAs work for us who got fantastic grades in school and seem almost clueless in the business world. This failure of the educational system to integrate the *Action* and *Feedback Stages* of learning, which builds real-world maps, is one reason that nearly 38 percent of former students are delinquent on their student loans, and at least another 25 percent have already defaulted.[6] Why are these percentages so high? Possibly because these graduates can't transfer the information that was delivered in the classroom to the real world higher-paying jobs, so they can't repay their loans.

The lack of real-world spatial map-building also accounts for corporate training that can succeed at getting employees to pass written tests on new management, customer service, and technology methods when tested in the classroom, but fail miserably in getting these same people to implement these techniques within the company. Also, when we look at the media in the context of building real-world spatial maps, it's easy to see the dangers of having the average child spend approximately six-and-a-half hours a day with TV, cell phones, video games, and other media, not the real world.

Action and Feedback Stages Create "Hot Maps" Supporting Linguistic Learning

As we've seen, the *Action* and *Feedback Stages* of *The Cycle of Transformation* heighten the activation of the monoamines (dopamine, serotonin, and norepinephrine), and when the structures that create spatial maps are also activated, your ability to create the neurological foundations that hold learning and meaning-making is profoundly enhanced. By understanding the "Hot Maps" concept, education and corporations can purposefully use real-world *Action* and *Feedback Stages* and combine them with a linguistic presentation to improve classroom learning.

For example, our emphasis on "Hot Maps," combined with stand-and-deliver presentation, was one of the reasons the Accounting Game Course, where employees at Fortune 500 companies were taught what amounted to the concepts of a semester of college accounting in one day, produced such dynamic learning. We didn't start by verbally presenting accounting principles or handing out worksheets to be read! Instead, we started by activating spatial maps with a game. The game, which we looked at in Chapter 20, when we discussed *Preheating*, started off with participants running a lemonade stand, just as they probably did in the real world as children. The game relied heavily on action, feedback, and spatial mapping. Throughout the day, we gradually added more and more complexity and spatial mapping to the *Action* and *Feedback Stages* of the game, until we created enough hooks and neural mass in participants' brains for the new accounting knowledge that was presented to easily attach itself. By the end of the game, after only six to eight hours, the students understood the accounting principles needed to manage a large soft drink company. This concept of making spatial maps "hot" and then attaching new, relevant information to them can be a very simple and beneficial element to add to any learning program where the goal is to accelerate the speed and depth of learning.

A Missed Opportunity to Create Hot Maps

When one of my sons was in sixth grade, his class took a camping trip to the Great Sand Dunes National Park in southern Colorado to help the kids get to know one another outside of school. There they hiked the huge windswept dunes, which resemble the Sahara, and played in the bubbling river that flows at the bottom of these white mounds. The kids were amazed by this real-world experience, and their actions and sensory feedback from this natural environment created Hot Maps in their brains. If, while at the dunes, they had also been presented with knowledge about wind flow, barometric pressure, soil composition, and river/water erosion—all of which influenced the creation of these wondrous dunes—their retention level of this information would have been dramatic and probably lifelong. But instead, because of core curriculum requirements, it would be years before these kids would be instructed to memorize much of this kind of information. And without the benefit of attaching new linguistic information to their special maps when they were Hot, sadly, so few of those memorized facts would be remembered in adulthood. From a biological perspective, that school trip, while indeed a bonding experience, was also a missed learning opportunity.

The Ratio of Experience to Language: 70 to 30 Percent

At The Institute, we now recommend a ratio of 70 percent experience (action-feedback) learning to 30 percent linguistic presentation. This ratio of experiential to linguistic learning brings another benefit: it can help alleviate boredom. When your monoamine system and parietal spatial 3D maps aren't effectively stimulated, your attention wanders, your mind loses focus, and learning is lessened. By creating education systems that endeavor to keep learners' monoamines activated and spatial maps engaged, we automatically prompt learners to become more naturally engaged and more joyful in their learning.

Corporate trainers and educators may ask, "How will we ever afford to devote so much time to experiential learning?" Some brain-based educators suggest that without the benefit of real-world experience, we forget more than 95 percent of a forty-five-minute class shortly after we hear a lecture.[7] To support a learning system that has a 5 percent efficiency rating could only be called lunacy. Considering how inefficient solely linguistic learning is compared to experiential learning with a language-association component, the question might be "How can we afford *not* to develop learning systems that have a real-world action and feedback components?"

This new scientific information isn't going to transform traditional education overnight. But it's our experience that just allowing educators to understand *The Cycle of Transformation's Action* and *Feedback Stages* supports their ability to be creative in how they use it to enhance their classroom learning environments (see Chapter 30 for an example).

Negative Effects of Too Little Action and Feedback: Dis-ease

The absence of the *Action* and *Feedback Stage* of learning in real-world 3D environments doesn't just hamstring learning. Many researchers believe it's a major cause of the increased stress levels and mental disorders we see in the world today.

Consider that since 1995, bike riding, fishing, swimming, touch football, and other unorganized outdoor activities among our youth have decreased by approximately one-third,[8] and coinciding with this downturn in physical activity in real-world environments is an increase in mental disorders. A survey in the journal *Psychiatric Services* found the rate at which children are prescribed antidepressants more than doubled from 1998 to 2002, and for the first time,

The Human-Environment Research Laboratory at the University of Illinois found that activity in green outdoor spaces reduced stress, fostered creative play, increased learning, and decreased symptoms of ADD.

spending on antipsychotics, benzodiazepines, and antidepressants has surpassed the spending on antibiotics and asthma medicines.[9] Also, between 2000 and 2003, spending on ADHD treatment for preschoolers increased by 369 percent.[10] It's now estimated that eight million children in the United States suffer from mental disorders, of which ADHD is one of the most prevalent.[11] Another study published by the Children's Hospital and Regional Medical Center in Seattle found that for each hour of TV preschoolers watched per day, the likelihood that they would develop concentration problems and other symptoms of ADHD by age seven increased by 10 percent.[12]

New studies are revealing that we may be able to decrease stress levels, alleviate depression, and increase the attention spans in our children simply by getting them outside, moving through three-dimensional time and space. In *The Human Relationship with Nature,* Peter Kahn points to over 100 studies that confirm that one of the main benefits of spending time in nature is stress reduction.[13] The Human-Environment Research Laboratory at the University of Illinois found that activity in green outdoor spaces reduced stress, fostered creative play, and decreased symptoms of ADHD.[14]

Other researchers found an interesting factor; the more natural the setting, the greater the relief. Nancy Wells at New York College of Human Ecology found being close to nature helps boost children's attention spans. She writes, "By bolstering children's attention resources, green spaces may enable children to think more clearly and cope more effectively with life stress."[15] Cornell University environmental psychologists report that "Our study finds that life's stressful events appear not to cause as much psychological distress in children who live in high-nature conditions compared with children who live in low-nature conditions." Outdoor physical play in 3D environments is important to the young mind because it naturally activates the spatial neurological structures and neurotransmitters that allow children not only to learn quickly but also to feel good about themselves.

Richard Louv, the author of *Last Child in the Woods,* comments: "ADHD may be a set of symptoms aggravated by lack of exposure to nature. [...] but the real disorder is less in the child than it is in the imposed, artificial environment. To take nature and natural play away from children may be tantamount to withholding oxygen."[16] Several studies lend credence to Louv's position and suggest that the media is a major source of the problem (see also Chapter 19).

Cracking the Learning Code

The research is clear, if we want to increase learners' span of attention, enhance their focus and increase their learning efficiency we must ask... why would we ever develop learning systems that lack the *Action* and *Feedback Stages* of the learning process?

35

The Incubation Stage of Learning: If You Can't Relax, You Can't Learn

What is required is to make oneself tranquil, rhythmical, and peaceful in order to make... memory distinct.

—Hazrat Inayat Khan, Sufi spiritual master[1]

As discussed in the preceding chapters, memory formation doesn't take place in one fell swoop. Learning is a process, not an event. Today, research reveals that in *The Cycle of Transformation,* once you've encountered data in the *Information*, *Action*, and *Feedback Stages* of learning, something else must happen before this information can be selected into your long-term memory banks.

That something else is a period of relaxed downtime, in which your biology is altered in very different ways than it was in the other stages of *The Cycle of Transformation*. We call this more relaxed brain and body state—where information in working memory is sent to your long-term memory banks and where creativity increases—the *Incubation Stage*. As discussed, information placed into working memory is very fragile. Like a delicate flower bud, precious, meaningful information gained in an activity must be cultivated before it can become a hardy part of your long-term memory networks. The *Incubation Stage* of learning provides for the change in your brain and body states that are required for this cultivation period to take place.[2]

The interruption of the *Incubation Stage* through stress, overwork, or in any other way will lower your *Individual Adaptability/Intelligence Factor* by compromising your brain's ability to select vital information you need to succeed in life. When you rob your brain of essential downtime, the science is revealing it's like putting your nose to the grindstone for too long, you end up grinding it off. You don't get smarter; you get stupider.[3]

Incubation = Creativity + Innovation

The *Incubation Stage* of learning also supplies the neurochemical and brainwave environment where your creativity and innovation can flourish. Our traditional scholastic and corporate factory-model educational systems are often criticized for not producing enough innovators. However, organizations inadvertently disrupt creative processes by acting as if periods of relaxed downtime and sleep are the enemy of productivity. By keeping workers and students trapped in highly focused and attentive brain states for too long, organizations actually sabotage the very learning, behavioral changes, and innovations they're trying to implement. Referring to the power of relaxed downtime in the memory and creative processes, Bobbi DePorter, founder of the world-renowned accelerated learning program

SuperCamp, counterintuitively implores you to "give yourself permission to do 'nothing' until incubation gives way to illumination."[4]

As you'll see in this chapter and the next, by creating learning environments that don't allow people to naturally pass through this fourth stage of *The Cycle of Transformation*, organizations unintentionally produce people who not only have lower *Individual Adaptability/Intelligence Factors* but also lower creativity.

Relaxed States: Not Your Enemy, Your Salvation

In what brain state have the most wonderful technical discoveries and inspirational ideas of our modern world been made? Most people would guess something like "aggressively attentive, highly focused, wide-awake states." But research is showing that response to be very, very wrong.

As I explored the history of hundreds upon hundreds of innovations, I was amazed that I couldn't find any great creations of humankind that were not made in the *Incubation Stage*. I don't want to overwhelm you with examples, but because our modern world so aggressively holds onto the belief that any time the mind is relaxed and "offline" is wasted and unproductive, I feel compelled to list at least a few instances that stand out to quash this misconception. As you'll see, if so many of the world's great thinkers hadn't intuitively recognized how important the incubation stage was to their learning and inspirational processes, we wouldn't have the marvels of our modern age.

- As far back as 400 BC, Socrates described his firm belief that a period of mental incubation was paramount to gaining a better understanding of any concept or difficult problem.[5]
- Around 250 BC, one of the first recorded examples of how an incubation period aided technical innovation comes from Ancient Greece, when Archimedes, who while relaxing in his warm bath, unraveled how to measure the volume of a solid object.[6]
- In 1455 Johannes Gutenberg created the first printing press. From his letters, we know he instinctively understood that he needed to let the information he'd amassed about how to build a printing press time to incubate so that his mind could put all the pieces together. He wrote: "For a month my head has been working; a Minerva, fully armed, must issue from my brain."[7]

- In 1596, René Descartes reputedly had the general principles of science "revealed to him" not in heated discussions with other philosophers but alone, while shut away from the hubbub of the world, relaxing in a "stove heated room."[8]
- In 1665, after years of focused intellectual effort at Cambridge, Sir Isaac Newton was forced by the onset of the Great Plague into a prolonged vacation at his mother's country home in Lincolnshire. One biography states that here, "He was able to submit the contents of his mind to eight months of undisturbed meditation."[9] During this relaxed downtime, not only was Newton able to discover the principles of general gravitation, but he also developed important theories of light and color and invented mathematical calculus.[10]

Here are some other accounts of inventions, discoveries, and creations that came out of the *Incubation Stage* of learning:

- The invention of the sewing machine, which helped propel the world into the industrial society, came to Elias Howe not while he was aggressively putting together machine parts but instead, as he dreamed of natives carrying spears with holes at the end.[11]
- Charles Darwin developed his "great discovery" in restful seclusion at the graceful Down House in Sussex, England. Here, he consolidated the learning he'd acquired aboard his long voyage on the HMS *Beagle* into the theory of evolution by natural selection.[12]
- Fredrick Nietzsche got his most profound ideas not while poring over a desk but on leisurely strolls. He wrote, "Only those thoughts that come by walking have any value."[13]
- After a party at the poet Lord Byron's house, Mary Shelley had a dream that gave her the storyline for the horror classic about Dr. Frankenstein and his monster.[14]
- Robert Louis Stevenson regularly willed himself to dream the storylines of novels like *The Strange Case of Dr. Jekyll and Mr. Hyde*.[15]
- Thomas Edison came up with a large percentage of his 2,000-plus patents as he emerged from regular power naps of five minutes or longer under the stairs of his Menlo Park lab. He recognized that these short periods took his brain offline and dramatically enhanced his learning and creative abilities.[16]

- Henri Poincaré solved the profound problem of Fuchsian functions as he emerged from the relaxed state we often encounter just before we go to sleep or wake in the morning.[17]
- Friedrich August Kekulé attributed his discovery of the ring structure of the benzene molecule to a dream he had of snakes biting their own tails.[18] Kekulé received so many of his good ideas in relaxation that he, like Edison, eventually scheduled the incubation stage of learning into his daily work program.[19]
- Gordon Gould, the inventor of the laser, says that his idea came to him during a relaxed drive in the country.[20]
- Stephen Hawking discovered the event horizon of black holes while climbing into bed, half-asleep.[21]
- Albert Einstein firmly believed quiet, relaxed periods were necessary to stimulate a creative mind.[22] In fact, he hit upon one of the most profound theories of the 20th century, the theory of relativity, while daydreaming on a grassy hill on a warm summer day.[23]

There may be many more unrealized Newtons, Darwins, and Einsteins among us, but because institutions keep the majority of us trapped in active mental states for too long, we may never know how many. One of the most respected neuroscientists of our day, Antonio Damasio, notes, "Few things can be as salutary, once you find an intellectual hurdle, as giving yourself a vacation from the problem."[24] In your own life, you may be inadvertently inhibiting your learning, creativity, and your potential in life by failing to give enough respect to the incubation stage of learning, when your brain goes offline.

New Research on Incubation Confirms the History

Ever since the publication of *The Relaxation Response* by Harvard physician Dr. Herbert Benson in 1975, medical research into the neurochemistry, genetic expression, and brainwave patterns that give rise to the incubation stage of learning has expanded dramatically. Now, prestigious organizations such as the National Institutes of Health, Oxford University's Mindfulness Centre, UCLA's Mindful Awareness Research Center, The Mind-Life Institute (in collaboration with the 14th Dalai Lama), Stanford School of Medicine, and the Max Planck Institute for Human Cognitive and Brain Sciences are making wonderful progress in uncovering the underlying biological mechanisms of relaxed brain and body states. What follows in this chapter and the next ones reflects what these and other researchers have discovered about the biology and value of your *Incubation Stage* of learning.

The Incubation Stage Has Two Phases: Awake and Asleep

Research first conducted in the 1960s by the father of modern sleep research, Nathaniel Kleitman, found that about every ninety minutes during both the waking and sleeping phases of incubation, your brain cycles in and out of different relaxed states.[25] Newer research shows those relaxed states are created by unique neurochemical combinations, which in turn create the brain waves of relaxed states called *alpha, theta, delta,* and *REM*.[26]

Here are the four main benefits of the incubation phases, which:

- Allow information held in your working memory to be passed into your long-term memory banks.[27]
- Promote the protein synthesis that provides for the construction of the physical structures of learning and behavioral change so that new learning has a place to attach.[28]
- Produce the brainwave rhythms that allow for heightened creativity and the consolidation of learning into your long-term memory.[29]
- Replenish the monoamine system (dopamine, serotonin, norepinephrine) so that when you re-enter the *Information*, *Action*, and *Feedback Stages* of learning, you're more focused, attentive, and innovative.[30]

In the following sections and the next chapters, we'll look more closely at how these benefits are created in Phase 1 (awake) and Phase 2 (sleep) Incubation States.

Moving Information into Your Long-term Memory Banks

The scientific definition of *incubation* is maintaining a chemical or biochemical system under specific conditions to promote a particular reaction.[31] The *Incubation Stage* of learning does just that for you. For newly acquired information to gain durability and promote creativity in you, the *Incubation Stage* of learning bathes your brain in a neurochemical environment that's very different from the ones that your brain experiences during the *Information, Action,* and *Feedback Stages* of the process.

In previous chapters, you saw that during the *Action* and *Feedback Stages* of *The Cycle of Transformation,* the brain's major emphasis is focusing its attention on the most meaningful external stimuli to be selected into working memory. In these active, external processing stages of learning, the monoamine system exerts control over your brain's function.

During the *Incubation Stage*, your brain literally changes its chemical mind.[32] In a broad generalization, we can say that as you go into deeper and deeper phases of the *Incubation Stage*, the monoamine system begins to relinquish its control of your brain function; meanwhile, the cholinergic system, of which acetylcholine (ACH) is the primary neurotransmitter, becomes more influential. In the day and night *Incubation Stages* of learning, when the monoamines ebb and acetylcholine begins to flow forward, a neurochemical environment is created where the most meaningful information that has been packed into your working memory can now be effectively selected for inclusion into your long-term memory banks.

Acetylcholine: The Band Leader for Incubation in Your Brain and Body

Many neurotransmitters and hormones, such as GABA, adenosine, orexin, and melatonin, are involved when you enter the *Incubation Stage*, but to

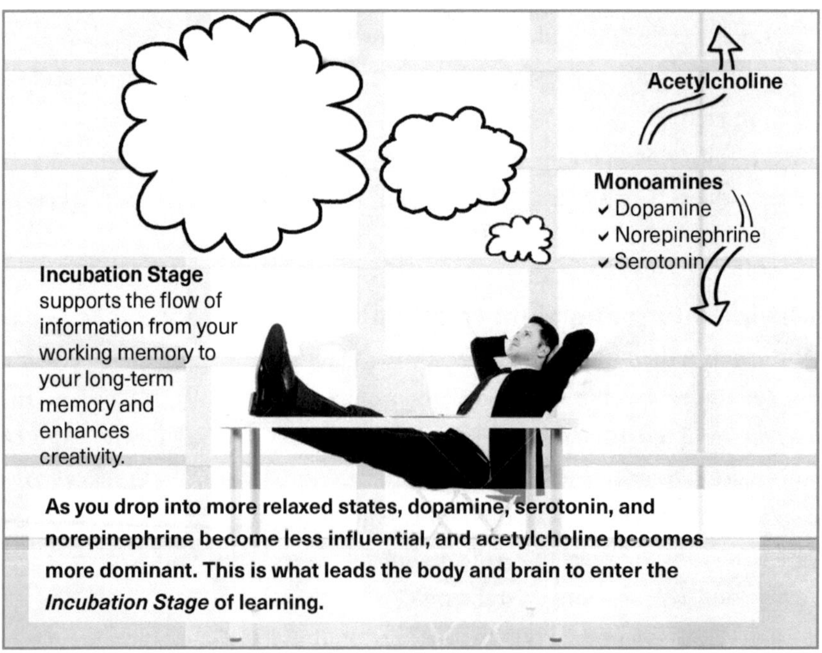

As you drop into more relaxed states, dopamine, serotonin, and norepinephrine become less influential, and acetylcholine becomes more dominant. This is what leads the body and brain to enter the *Incubation Stage* of learning.

simplify, we'll focus on acetylcholine (ACH), which is so influential in your brain and body during this stage it could be called the band leader of relaxed states. Research demonstrates that without its influence, you couldn't relax, learn, or be creative.[33] Nor would the cascade of other neurochemicals that are released when you relax, be released. The influence of acetylcholine is widespread. As many as 10 percent of all your brain synapses use acetylcholine, and it's the primary neurochemical of your parasympathetic (relaxation) nervous system.[34]

As you drop into more relaxed states, dopamine, serotonin, and norepinephrine become less influential, and acetylcholine becomes more dominant.[35] This is what leads the body and brain to enter the *Incubation Stage* of learning. Not only does the *Incubation Stage* support the flow of information from your working memory to your long-term memory and enhance creativity, but it also gives your brain and body the vital time and space they need to rest, digest, and recuperate.[36]

Acetylcholine: Necessary R&R for Your Body

The sympathetic nervous system—the part that serves to increase adrenaline, accelerate heart rate, constrict blood vessels, and raise blood pressure—is controlled by your monoamines.[37] In both the day and night phases of the *Incubation Stage* of learning, when your sympathetic nervous system becomes less engaged, the parasympathetic nervous system, controlled by acetylcholine, becomes more domineering.

As this shift to your parasympathetic nervous system occurs, signals to your body cause profound physiological changes, such as decreased strength of heart contractions, reduced heart rate, the dilation (opening up) of blood vessels, reduced respiration rate, decreased blood flow to muscles, increased blood flow to digestive organs, stimulation of human growth hormone and efficient activation of your immune system. All this occurs so that your body can rebuild, refortify, and revitalize itself.[38] Acetylcholine cells are so crucial to relaxed states they're embedded in every muscle and organ of your body.

Acetylcholine: Necessary R&R for Your Brain

When you enter a more relaxed parasympathetic body states, your brain is also allowed to restore and rebuild. During the *Incubation Stage*, as acetylcholine becomes more and more governing, the monoamine neurons reduce their activity. Therefore, they have time to rest and recuperate.

Studies show that with acetylcholine becoming more influential, your brain up-regulates (increases the number of) dopamine, norepinephrine, and serotonin receptors.[39] The result of more available monoamine receptors is that, when you re-enter the *Information, Action,* and *Feedback Stages,* you're much more focused, attentive, and capable of packing working memory with relevant information. This is why, when you miss sleep or work too long or too hard, you not only forget so much more of the information you want to remember but also become much less creative and innovative. This is also why your brain loves those wonderful naps, profound meditations, and a great night's sleep: your biology is much more efficient afterward.

There are two major clusters of ACH neurons in the brain: one in the brain stem in the lateral tegmental area and another just above the brain stem in a cluster of neurons called the basal ganglia.[40] In *Zen and the Brain,* which many people consider to be one of the most comprehensive texts on the neurochemistry of relaxed states, author James Austin writes of the role of acetylcholine, stating simply, "ACH is a transmitter to remember."[41] The devastating memory loss caused by Alzheimer's disease is directly linked to the massive loss of cells in the ACH system (primarily in the basal ganglia),[42] which also disrupts the brain's ability to effectively drop into critical incubation phases of learning. Drugs that boost acetylcholine levels and the ability to efficiently enter the incubation phases of learning often improve the memory of Alzheimer's patients.[43]

The Main Neurochemical of Relaxed States, Discovered in a Relaxed State!

What we're labeling the primary neurochemical of relaxed states, acetylcholine, was itself discovered in the *Incubation Stage* of learning! Otto Loewi, the German-born physiologist who discovered acetylcholine, won the Nobel Prize in Physiology or Medicine in 1936; he wrote:

> *I awoke, turned on the light, and jotted down a few notes on a tiny slip of paper. Then I fell asleep again. It occurred to me at 6 o'clock in the morning that during the night, I had written down something most important, but I was unable to decipher the scrawl. The next night, at 3*

> *o'clock, the idea returned... I got up immediately, went to the laboratory, and performed a single experiment... according to the nocturnal design.*[44]

This coincidence confirms my belief that there are no coincidences!

Innovative Educators and Trainers Embrace the Incubation Stage

Over the last several decades, a handful of scholastic and corporate educators who understand brain function have insisted their learning systems include periods of reflection and incubation. The Caines stress that "reflection is a critical aspect of all sophisticated higher-order thinking and learning."[45] Joseph Vaughan, in his article "The Potential of Reflective Practice," points out that "perhaps the single most glaring omission... is adequate and explicit attention to the ultimate end of reflective practice."[46] Donald Schön, educator and author of *The Reflective Practitioner* and *Educating the Reflective Practitioner,* advocates that "reflection on action"[47] periods be made part of all learning systems. Educators Karen Fellows and Nancy Zimpher state, "Essentially the act of reflection establishes within the teacher *and* the students simultaneously an open-mindedness and discernment, rational judgment, and creativity—all characteristics of the 'educated' person." While Jensen observes that periods of "incubation for learning" are vital because "learning

"The act of reflection establishes within the teacher and the students simultaneously an open-mindedness and discernment, rational judgment, and creativity—all characteristics of the 'educated' person."

—Karen Fellows and Nancy Zimpher, educators

is best when focused,[48] then diffused, focused, then diffused. Constant, focused learning becomes increasingly wasted over time."[49]

In the next chapter, we'll look at the unique brain waves created by this change in your neurochemistry when you enter the *Incubation Stage* that make you smarter and more innovative. You'll also gain even more in-depth insight into why scholastic and corporate organizations that ignore the *Incubation Stage* of learning end up not building up but dumbing down those within their walls—and what we can do about it.

Cracking the Learning Code

The *Incubation Stage* of learning creates the changes in your brain and body that are required in to encode memory effectively. This is why taking time out to relax and build in downtime is vital to your ability to learn and transform. Unfortunately, most existing learning systems are unaware of this fact.

36

Incubation: Waves of Learning and Innovation

Sometimes we just need to put down our phones, close our eyes and take a few deep breaths. Ideas are often in-flight patterns around our brains, just waiting for clearance to land.

—*Sam Harrison,* Creative Zing! Spark Your Creativity and Powerfully Present Your Ideas[1]

Up until this point, I've been focusing on the ebb and flow of the monoamines (dopamine, serotonin, and norepinephrine) and the neurotransmitter acetylcholine. But there are actually 6,000+ genes[2] and over 100 different neurochemicals and hormones[3] all interacting in different ways at different times to initiate and maintain the various stages of *The Cycle of Transformation*. To complicate these complex processes even more, the method of measuring neurochemicals in the brain is very invasive. To even get close to efficient readings of which neurotransmitters are present at any one time in your brain, it's necessary to perform painful spinal taps, or drill holes into the skull and insert probes into neural tissue.[4]

Neuroscientists find it is safer and easier to measure the different electrical brain pulses that are produced by the changes in your neurochemistry during the *Information, Action, Feedback*, and *Incubation Stages* of learning. These waves can be measured noninvasively with an EEG (electroencephalogram) machine by placing small, sensitive electrodes on your scalp.[5] Over the last fifty years, research using this type of measure has produced a wealth of information on the stages of learning.

Just like a radio, your brain uses a definite band of signals with specific frequencies and amplitudes to transmit information. Your brain definitely has rhythms. In general, the more physically active and externally focused you are in the *Information, Action*, and *Feedback Stages*, the faster, lower, and less synchronized your brain wave states. Conversely, the more physically relaxed and internally focused you are during the *Incubation Stage* of learning, the slower, higher, and more synchronized your brain waves.

Changes Propelled by Your Neurochemistry

In the *Action* and *Feedback Stages* of learning, when you're physically active and your focus is external, the high levels of monoamines help produce very high-frequency, low-amplitude, unsynchronized brain waves called *beta waves;* they range *13–30 hertz* (Hz), or cycles per second.[6] In the *Incubation Stage*, as the monoamines and sympathetic nervous systems begin to wane and acetylcholine and the parasympathetic nervous system become more dominant, your brain waves shift from unsynchronized, high-frequency, low-amplitude beta waves to more synchronized, lower-frequency, and higher-amplitude. In descending order, these waves are called *alpha waves* (8–12Hz), *theta waves* (4–7Hz), and *delta 1* and *2 waves* (0.5–3Hz). These brain wave states are often abbreviated, as SW (slow-wave) states. In general, it can be said the lower the hertz (Hz or cycles per second), the deeper the incubation. The exception is the stage of REM sleep. We will look at this more deeply in future chapters.

What is vital to remember as you read the following chapters is that it is your neurochemistry which is propelling these slower, higher waves of learning. Without chemical changes in your brain, there are no waveform changes. These slower waveform changes take place in Phase 1 and 2 of the *Incubation Stage* of learning. In Phase 1, when you're awake, these slower

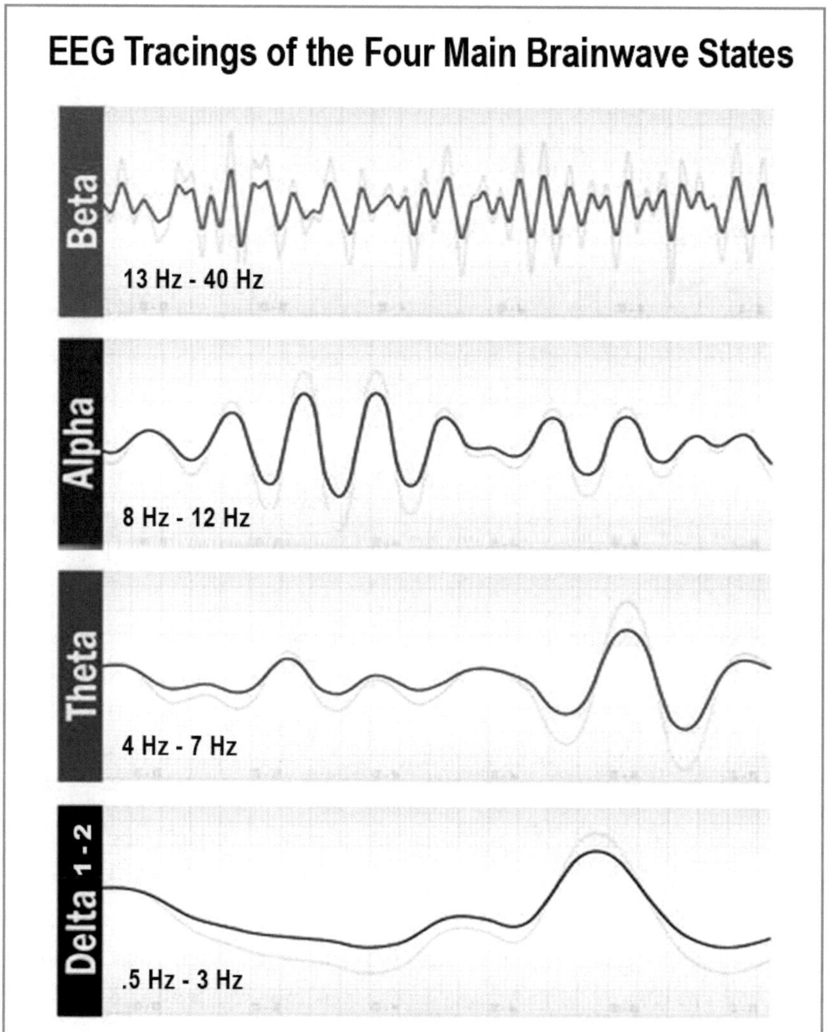

incubation waves show up in smaller, localized brain areas. In Phase 2, when you slumber, these slower waves take over large swathes of your brain matter.

Different Waves for Different Kinds of Learning

Over the last twenty years, research into the incubation phases of learning has exploded. It's been found that although all the waves of incubation help consolidate information into long-term memory, each state may have its own specific kinds of information that it specializes at consolidating.[7] That having been said, for our purposes, because all types of learning—consolidating episodic memories (life experiences), spatial maps, emotional memories, verbal memories, and procedural (movement) memories—are strengthened by all the different waves of incubation,[8] we will only go into specific functions of different waves when it adds to understanding.

Slow Brain Waves Help Move Knowledge into Your Long-Term Memory

Every ninety minutes or so in Phase 1 and Phase 2 of the *Incubation Stage*, your brain naturally cycles in and out of different rhythms. In fact, it's this very shifting between different brain wave cadences that helps your mind integrate information held in working memory into your long-term memory banks. In *Landscapes of the Night,* Christopher Evans draws a parallel between the brain's need for relaxed downtime and the need for a computer to go offline so new programs can be installed and old ones modified.[9]

As discussed, the hippocampus is a major structure of your working memory and also the part of the brain that helps place relevant information into the correct concept networks in your long-term memory banks. Research is revealing that in the *Action* and *Feedback Stages,* when monoamines and beta wave states dominate, meaningful information flows into your hippocampus; during the *Incubation Stage* of learning, when acetylcholine is more prominent, alpha, theta, and delta waves take center stage, and knowledge flows in a different direction, from the hippocampus into your long-term memory banks in rear associative areas of your brain.[10] So, if you don't allow your brain to enter the *Incubation Stage* of learning, information has no way to get out of your hippocampus and into the correct long-term memory networks. This is why we say that keeping your nose to the grindstone can make you more stupid.

Like a closet that you can pack only so much stuff into, your working memory has a limited storage capacity. When your working memory is full, something has to get knocked out for something else to fit in. To give relevant information a chance, it needs to be cemented into your long-term memory banks (before it gets knocked out), so parts of your brain are naturally shifting into slower wave rhythms every ninety to 100 minutes.

Neural Knockout is the term we use at The Institute to label what happens when you try to cram too much information into your working memory without allowing your brain the time and space it needs to drop into slow-wave (alpha, theta, and delta) states. If you get trapped in active, awake beta states for too long, the information you've stored in your working memory can easily get knocked out before it has a chance to be marinated in the chemical stew that makes up slow-wave waking states—that is, before it has a chance to be set up for inclusion in your long-term memory.

The fact that you find your mind wandering or that you have short daydreams throughout the day is simply your mind naturally shifting from external to internal processing as your brain's rhythms shift from beta waveforms to slow-wave forms that allow information in working memory to be passed to long-term memory.

Therefore, when kids are spacing out, it's not as if nothing's going on in there. If students are actively engaged and find the information they're trying to learn personally meaningful, spacing out is simply a necessary part of the learning cycle. As Pierce Howard notes in *The Owner's Manual for the Brain*, "Neural connections formed by learning need time to fix and strengthen themselves without competition from additional novel stimuli."[11] Genetic and neurological variations ensure that one size does not fit all; therefore, each person needs slightly different lengths of awake and asleep slow-wave incubation time to effectively integrate newly learned knowledge.

Note that when you're in waking states, these slow-wave rhythms don't take over your whole brain. If they did, you would immediately drop into sleep. Instead, so that you can stay awake while information is gaining resilience, isolated patterns of alpha, theta, and sometimes even delta waveforms occur in localized brain regions. These waking slow-wave forms coexist in localized brain areas with the more widespread beta waveforms. At this time, researchers speculate that the specific location of the slow-wave forms in your brain during waking stages indicates where long-term memory is starting to form.[12] If your brain is encoding linguistic information, slow waves will appear in your frontal and temporal lobe; if it's encoding taste information, these slower waveforms will be produced in your gustatory cortex; and if you're encoding visual information, they'll appear in your visual cortex.

Slow Waves Turn on Genes of Learning

Research also indicates that slow-wave states support your long-term memory formation by enhancing the availability of the CREB activator molecule[13] (see Chapter 7). The CREB molecule is important for turning on the right genes that produce the proteins and enzymes necessary for building the structures of long-term memory formation: receptor sites and dendritic branching. Because your brain has only a limited amount of CREB molecules available at any one time, too much widespread fast beta activity for too long uses up the available CREB molecules, compromising learning. Slow-wave states appear to help replenish your CREB molecules so that when you re-enter more-active states, learning and long-term memory formation can take place with higher efficiency.

Shock Therapy, the Slow-wave States and Learning

The scientific evidence that the brain needs slow-wave incubation periods to integrate information from working memory into long-term memory had its genesis in the 1940s. Back then, electroconvulsive (shock) therapy, abbreviated as ECT, was seen as a humane alternative to the lobotomy and was used as a last resort for severe psychiatric disorders.[14] During ECT, a strong electric current is applied to a patient's skull, causing violent brain seizures. The hope was that these brain seizures would help mental patients become more functional by kick-starting critical neurotransmitter systems. After large numbers of patients had received ECT, psychiatrists began to note that after ECT was applied, patients would develop a type of short-term amnesia and couldn't remember events that had occurred just before the shock was administered.

More than thirty years after the first ECT trials, James McGaugh at the University of California found that although long-term memory is impervious to almost anything short of brain damage, the information in working memory is extremely delicate.[15] Working with his rats, McGaugh established that before information could be transferred from working memory to long-term memory, a certain amount of time had to pass.

After a rat was given a painful shock when it stepped off a platform onto an electrical grid, McGaugh would immediately disrupt the rat's brain by giving it ECT. After the ECT, the rat would instantly forget about the shock and put its foot back on the grid again. McGaugh also found that if he gave the rat time (an hour or more) to incubate and associate the source of its foot pain with the electrical floor grid, the rat could learn from one quick experience not to step on it. When given the ECT after the *Incubation Stage* had had time to be activated, the rat would easily remember the previous foot shock learning and avoid the grid.

Today, because of McGaugh's brain-wave research, we know that the rats' brains needed time to learn from their experiences and also what their brains did with that time. Parts of their brains were allowed to enter slow-

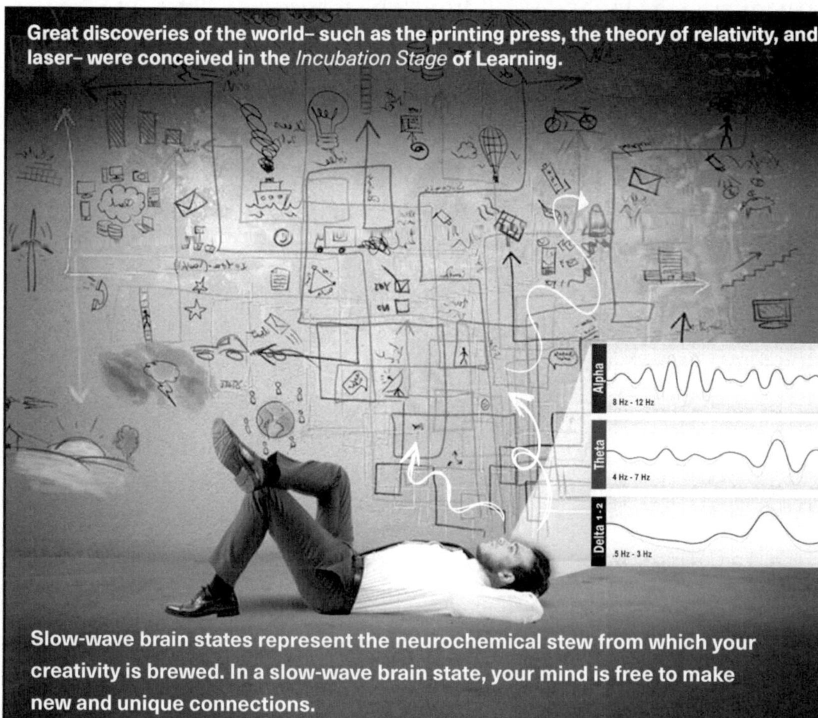

Great discoveries of the world– such as the printing press, the theory of relativity, and laser– were conceived in the *Incubation Stage* of Learning.

Slow-wave brain states represent the neurochemical stew from which your creativity is brewed. In a slow-wave brain state, your mind is free to make new and unique connections.

wave (alpha, theta, and delta) states, which permitted the information gained in experience to become more resilient. Because of the work of McGaugh and others, today patients are given time to naturally drop into waking slow-wave states to integrate any important learned information they gained that day before technicians apply ECT.

Institutions that keep students and employees trapped in activities that produce beta states for too long must understand that they are knocking out meaningful learning in the same way electroconvulsive therapy does. The research is now conclusive: to increase the efficiency of learning and behavioral change, institutions need to embrace slow-wave incubation brain states as part of the natural learning and behavioral-change process.[16]

Slow Waves Create the Foundation for Innovation and Creativity

By definition, *creativity* involves breaking off or going beyond the standard frame of reference by making new perceptions, associations, and patterns. The information you already possess in your brain structures is the fountainhead from which creativity flows. In Chapters 12 to 14, you saw that for creativity to take place, previously wired neural connections must link up in new ways. Slow-wave brain states represent the neurochemical stew from which your creativity is brewed.

When slow-wave brain states are present, your mind is free to make new and unique connections. As Ludwig Wittgenstein once observed, "We gain insights mostly through new connections of things we already know."[17] Previously, we took an in-depth look at how extrinsic motivation kills creativity. Because slow-wave brain wave states are the wellspring of creativity, studies show not letting your brain drop into slow-wave states has an even more devastating effect on your creative process than extrinsic motivation. Some neuroscientists call acetylcholine "neural lightning"[18] because of its ability to increase the speed and intensity at which your neurons communicate. In slow-wave states, with the monoamines' control waning and acetylcholine getting a stronger foothold, neural processing becomes less restrained, and cross talk between neurons begins to take place. The "aha" moments that appear when you emerge from confusion to understanding are a result of the less-restrained neural cross talk that takes place in slow-wave brain states.

The lifting of controls on neural cross -talk in slow-wave brain states is the primary reason the majority of the great discoveries of the modern world, such as the printing press, theory of relativity, and laser, were conceived in the *Incubation Stage* of learning. Researcher Pierce Howard is so convinced that a period of incubation is vital to the innovation process that he urges, "Always take a break between preparing for your creative act and actually trying to execute it."[19] Psychologist Eugene Gendlin calls purposefully dropping into relaxed slow-wave states a "deliberate letting"[20] so that the brain is free to make associations and connections it would never make under the iron grip of focused beta-state concentration.

"Far-Out" View of Slow-Wave Brain States

Some researchers, who might be called "on the fringes" or maybe even "over the edge," note that our creative slow-wave states resonate with many frequencies found in outer space. The implication is that our creativity is somehow enhanced by having our brain rhythms resonate with the heavens.

In 1957, German physicist Dr. W. O. Schumann calculated that the most dominant frequency of Earth's ionosphere was about 7.83Hz, which corresponds to the border between our alpha and theta brain-wave states.[21] Later, it was found that the inner Van Allen radiation belt (a zone of energetic charged particles that held are held in place by the earth's magnetic field) has a frequency of 4Hz, which correlates to the brain's theta rhythms, and that the plasma sheet opposite the sun has a frequency of 0.8Hz, which corresponds to our delta brain waveforms.[22] In 1977, the relationship between brain-wave rhythms and the vibratory frequencies of space became the basis for Itzhak Bentov's well-received book *Stalking the Wild Pendulum: On the Mechanics of Consciousness*.[23] I really don't know if we're connecting to the heavens in slow-wave states, but I do find the correlations, while far out, interesting to ponder.

Trapped in a Beta World: You Learn and Innovate Less

Our modern world is obsessed with the idea that "more and faster" is better. The result is our lives become a frantic blur of activities, which keeps our brains locked into fast-paced and unsynchronized beta brain wave states far longer than nature designed. Because of this, most of us habitually experience *Neural Knockout* and a lack of creativity. When you're burned out, not only do you forget essential things, but you also often lack the creativity to come up with innovative solutions to work and life problems.

Educators and corporates have wholeheartedly embraced this "beta is best" philosophy, to the detriment of efficient learning and insightful creativity. They design learning environments as if relaxed and reflective activities (which promote synchronized slow-wave alpha, theta, and delta brain states) were the enemy. Even though researchers have found that reflective cognitive styles of learning are associated with higher levels of intelligence,[24] surveys indicate that most of us still wrongly assume that faster is not only better but smarter.

Intelligence tests reflect this assumption too. To pack in more activities that produce more beta waveforms, schools have shortened lunch hours and dropped slow-wave-producing activities like phys-ed and music, even though, as Jensen notes, the research illustrates that "in fact, the whole notion of "time on task" is biologically wrong and educationally irrelevant."[25] Some preschool programs, to keep their little students' heads whirling in beta wave states, have even eliminated nap time!

Unfortunately, after you spend twelve to sixteen years in an academic environment where only beta brain states are valued, you end up believing that fast action is the only path to success. Peter Senge, founder and director of the Center for Organizational Learning at MIT's Sloan School of Management and author of *The Fifth Discipline* points out corporate managers are so steeped in fast-action states they see them as the only way to success.[26] Because of this "beta is best" viewpoint, Senge found that "most managers do not reflect carefully on their actions" and therefore learn little from them.

Senge discovered that when given a problem, managers quickly take on a fast-action beta wave strategy. If the first action doesn't work, they immediately switch to another and another and another. They often run through six different strategies without ever examining why the previous plan failed. Senge says, "The 'ready, fire, aim' atmosphere in American corporations have been fully assimilated and internalized by those who live in that atmosphere."

Bill O'Brien, the former CEO of Hanover Insurance, complains that corporate America has banished time for reflection and calls the frenzied pace in corporations the "chain-gang" style of management, and "the speed of the boss sets the speed of the gang." Senge mentions one corporate manager who laments this mentality, saying, "In America, we assume that when a person is sitting quietly, they aren't doing anything important." Yet no one ever won a Nobel Prize or received profound inspiration by running around with a brain that was in a constant state of beta waveforms. As accelerated-learning pioneer Terry Wyler Webb says, beta wave states are useful for amassing information, "but they inhibit access to the deeper levels of the mind... It is in the alpha and theta states that the great feats of super memory, along with heightened powers of concentration and creativity, are achieved."[27]

The only global system of education we're aware of that acknowledges and emphasizes the *Incubation Stage* of learning is the Waldorf school.[28] It was created over 100 years ago by Rudolf Steiner, who was reportedly a very intuitive individual. Although he knew nothing of EEG wave states or today's science, Steiner took advantage of the fact that children spend more time in slow-wave and REM sleep states than adults. He saw that because young children are already in creative fantasy states so much of the time, the best way to get them to learn was to wrap new information in an imaginary world (see Chapter 13 for details on *Wrapping*). In kindergarten through fifth grade, the Waldorf system teaches language, math, and science by wrapping scholastic information in fairy tales, parables, and the like.[29] This style of teaching allows Waldorf children to spend more extended periods in slow-wave states than children in traditional schools, who are forced to learn mainly in beta states.

As well as omitting phys-ed and music classes, some school districts are also looking at eliminating or shortening the summer vacation to keep children in beta states longer. Before they go there, it would behoove them to look at scientific research. At The Institute, we ensure that all learning and behavioral programs have the *Incubation Stage* of learning deeply embedded in them.

Meditation Brings You Slow-wave States and Brain Coherence

Research indicates meditation may be one of the most effective ways for you to immediately change the chemical climate of your brain, allowing it to drop into slow-wave (alpha, theta, and delta) activity.[30] Ever since the publication of Dr. Herbert Benson's *The Relaxation Response* and Harold Bloomfield's *TM: Discovering Inner Energy and Overcoming Stress* (TM stands for *transcendental meditation*) in the 1970s, meditation has become more accepted by the general and medical populations as a credible method to reduce stress and improve overall health.[31]

Over the last five decades, a plethora of studies have shown the benefits of meditative practices on learning. Research at the Stanford University School of Medicine separated subjects into two groups. One group was taught meditative relaxation techniques, while the other group was given a lecture on positive attitudes. Then each group attended a three-hour memory training course. When tested, the group that had practiced meditative reflection scored 25 percent better than the non-meditating group.[32]

In a memory study performed by Allan Abrams, it was found that meditators had better long-term memory retention than non-meditators. Abrams found that people who had practiced meditation for two years had 62 percent better long-term recall on tests that dealt with word lists than non-meditators.[33] Abrams attributes this impressive increase to the more optimized internal environment of meditators. In another study that covered several universities, David Orme-Johnson found that the grade

Research indicates that meditation is one of the most effective ways to immediately change the chemical climate of your brain, allowing it to drop into slow-wave (alpha, theta, and delta) activity.

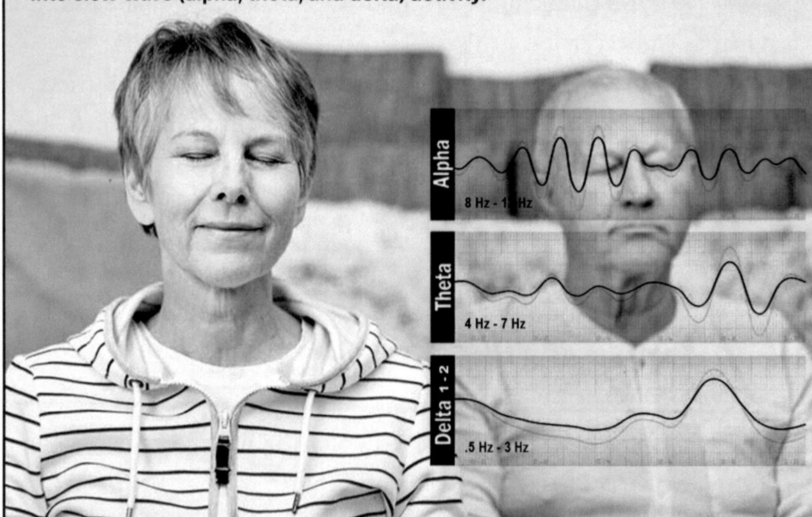

point averages of students who started meditation practice jumped 36 percent over the previous semester's GPAs,[34] presumably from the extended periods the students spent in slow-wave states. Other research has shown an improvement in reaction time and increases in perceptual ability on the part of meditators, which indicates that downtime in meditation is effectively replenishing and possibly improving the monoamine system's efficiency.[35]

EEGs and brain imaging reveal that the positive impact of meditation on memory formation comes directly from its ability to allow practitioners to consciously and immediately shift from fast, unsynchronized beta wave states into synchronized, slow-wave states. At any time, experienced meditators can quickly and effectively access the wellspring of alpha, theta, and delta brain wave states.

As an example, back in the late 1980s, when I worked in New York City in the *Mad Men* media business, I had an acquaintance who was the publisher of *Bride's Magazine.* He had started Transcendental Meditation at about the same time as I had. Because at the time meditation was seen as something weird for businesspeople to do, in the late afternoons, when he felt himself running out of "brain juice," he would sneak off to a quiet hotel lobby to meditate. He'd come back to work thirty minutes later, refreshed and ready to deal with the hectic world of publishing till late into the night.

Another advantage of meditation is the widespread coherence of brain wave patterns. Research shows that not only do meditation, chanting, and similar relaxation techniques increase overall slow-wave brain activity, but the whole brain also begins firing in harmonic wave coherence.[36] First, it was found that when we enter meditative states, populations of neurons on the left and right sides of our brains begin to fire in synchronicity, entwined with one another until large numbers of neurons on both sides are recruited into the same rhythmic patterns. Then it was found that this meditation-induced brain wave coherence isn't limited to the left and right hemispheres but also extends up and down to the frontal cortex and the lower limbic system.[37] Dr. Bernard Glueck at the Institute of Living in Hartford, Connecticut, who did the initial research on this phenomenon, says this slow-wave coherence indicates a profound improvement in neural communication throughout the brain.[38]

It was once thought you needed to meditate for at least twenty minutes to get the benefits. Recent studies have shown that much shorter periods can also be very beneficial.[39] At the Institute, we work with many high-anxiety and hyperactive individuals who find it almost painful to sit still and meditate for long periods. We've had great success by having them start with just three to five minutes of guided meditation morning and night and then slowly increase the time. Many quite good short guided meditations can be found on the web at sites like Insight Timer. These short meditation sessions calm the biology of anxious and hyperactive individuals who find that they can quickly and joyfully extend the time they spend in mediation after only several weeks' practice.

What is Orchestrated Incubation Time?

At The Institute, we call Orchestrated Incubation Time any time while awake you purposefully take yourself away from activities that demand a high level of concentration, which creates beta waves, and shift to engage in activities that produce slow-wave, alpha, theta, and delta activity. Waking incubation states are produced when you do things such as listen to music, take long drives, garden, zone out, do yoga, go on leisurely walks, meditate, or engage in sports and hobbies like knitting and jogging.

To enhance his creativity and learning abilities, Thomas Edison developed a rather bizarre method of giving his brain a break. Sitting in a comfortable chair, he would place a metal ball in each hand and let his arms dangle at the sides over two metal pans. He would then allow himself to drift off until he was in a state close to sleep, where he would be startled into waking from the balls clanking into the pans. Edison reported that his most creative ideas came in these moments.[40] I find that riding a stationary bike allows me to immediately slip into the alpha state. The choice is subjective, depending on your temperament and genetic variation.

How Much Incubation Time Do You Need?

The question I'm often asked is, "How much and how often should I engage in relaxed downtime for optimum slow-wave incubation time?" Unfortunately, there's no definitive answer. Even though the ideal number and length of downtime periods haven't been established, research does indicate the more complex the information to be learned or understood, the greater the duration of downtime may be needed.[41]

For example, while writing this book, I could effectively maintain a focused beta wave state for about three to four hours in the morning. The earlier I rose to write, the longer I kept a focused beta state. At the end of that time, my attention and creativity would become depleted, and I would go lift weights, ride a stationary bike, jog, or do yoga—all activities I can do in a very unfocused state. Back at my office, I would meditate for approximately thirty minutes. Usually, I could then productively write again for anywhere between one and three hours. If something stopped me from taking my Orchestrated Incubation downtime, I found that when I returned to writing, it was usually a very unproductive session.

To counter activities that produce intensive periods of beta-wave action, some researchers say the ratio of beta activity to slow-wave waking states must be at least 70/30; others say a 50/50 ratio. Several brain-based educators say that no adult should spend more than twenty-five minutes in a beta state without a relaxation break. Jensen says ten-year-olds should spend two-and-a-half to five minutes of incubation time for every ten minutes spent in focused attention.[42]

Although experts give can provide guidelines about how to deal with short, intense, focused, attentive beta periods, what about when you've been ensnared in extended beta states for weeks, months, or even years? While the specific amount of time you should spend in extended incubation periods hasn't been determined, the evidence is clear: if you aggressively focus on a project or learning experience for too long, you may impede the very neurological systems that produce the learning, understanding, and innovation. As more research is completed, we'll develop a better comprehension of just how long waking incubation periods should be, but until then, base your decisions on your own intuition and anecdotal evidence.

The impact of the taking extended time off from a beta focused project is reflected in two of the most important medical discoveries in the last 100 years: the polio vaccine and the brain's opiate receptor. Despite his years of effort, Jonas Salk was stuck. Every research effort led to a dead end. He figured he needed downtime and retreated to a 13th-century monastery in Italy where he found the "spirituality… was so inspiring… I intuitively designed the research that I felt would result in a vaccine for polio. I returned to my laboratory in Pittsburgh to validate my concepts and found that they were correct."[43] Neuroscientist Candice Pert had a similar experience. After months of failing to locate the brain's opiate receptor, she took eight weeks off. When she went back to work, she had a eureka moment. Pert wisely suggested, "It's often a good idea when faced with a busted experiment, to put the mind on vacation and wait for the unconscious to come up with the right answer."[44]

My dear friend Robert Kiyosaki built an educational and financial empire and sold over forty million books based on ideas he dreamed up during a year's sabbatical, which he spent reflecting on his previous twenty-five years of business activity.[45] The clarity and vision he displays in the *Rich Dad, Poor Dad* series sprang from the creative downtime period he spent in peaceful Bisbee, Arizona, where for a year he allowed his brain to incubate the information gained in years of business dealings. If he hadn't intuitively known to orchestrate that downtime into his life, not only might Robert be poorer for it, but so too might the millions of people who've increased their wealth by applying the wisdom he gained in slow-wave states.

Because of genetic variation, all of us have different reactions to environmental pressures, and we're all wired to prefer a different kind of activity to produce slow-wave states, so specific guidelines are difficult to impose. What we can say is if you want to enhance learning, understanding, and creativity, it makes sense to give your brain an extended period of incubation if you've experienced a very long period of beta time. Like everything else in life, there must be a balance.

Cracking the Learning Code

It is vital for you to allow your brain to drop into slow-wave states during the day. These slow-wave waking states allow learning to gain some level of memory permanence and durability. But these daytime slow-wave waking states aren't nearly long enough or widespread enough to ensure learning permanence. For this to happen, you must enter the most potent incubation phase... sleep.

37
Your Longest and Most Powerful Incubation Stage: Phase 2, Sleep

I love sleep. My life has the tendency to fall apart when I'm awake, you know?

—Attributed to Ernest Hemingway, American novelist[1]

When you're wrestling with a difficult problem and a friend urges you to "sleep on it," there's a lot of wisdom in those words. While Phase 1 (awake) slow-wave (alpha, theta, and delta) incubation phases allow learning to gain some level of permanence and durability, these waking slow-wave states aren't nearly enough to ensure the information is retained by your long-term memory banks. For your learning to stick, you must enter the most potent incubation phase of all: Phase 2, sleep.

Please note that different researchers have somewhat different criteria to designate the various stages of sleep. Also, studies on sleep are advancing every day, and the new data paints a very complex picture of sleep cycles and what causes them. We have chosen the approach you see here, as we believe it gives the highest clarity and is the easiest to understand, helping you can apply this knowledge to your own life and the lives of those you love.

The Five Stages of Sleep

Just as when you're awake, in sleep, your brain cycles in and out of different brain rhythms. Neuroscientists have identified five stages of sleep: four slow and synchronized sleep stages alpha, theta, delta 1, and delta 2, plus the fast and desynchronized waveform called REM.[2] You fall asleep in stair-step fashion, first descending into the four synchronized wave stages from fastest to slowest and then ascending those same stages until you reach desynchronized REM sleep, which is so unlike the first four stages that it's also called paradoxical sleep. During a typical eight-hour night, you'll cycle through all these stages four to five times.

Phase 2 sleep is a super-incubation period and is much more potent for promoting learning and creativity than our waking incubation Phase 1 because of three main factors:

- Longer slow-wave periods
- More widespread slow-wave periods
- REM (Rapid Eye Movement) stage sleep

When you sleep, you spend longer in slow-wave brain states than when you're awake. In a typical sleep cycle of eight hours, you average six-and-a-half hours in the chemical stew that produces slow, synchronized alpha, theta, and delta waveforms that are widespread in your brain. As you saw in the last chapter, slow-wave states represent the chemical combinations that allow information in working memory to be passed to your long-term memory banks. But although waking slow-wave states do indeed give meaningful information some resilience, in most cases, these phases aren't long enough or widespread enough to ensure that new information will be firmly consolidated into your long-term memory.

The neuroscientific term *memory consolidation* refers to the process whereby a memory becomes resilient and, therefore, resistant to interference from competing and disrupting factors. Without a full night's sleep, it's very easy for the information you learned during the day to degenerate. We've all experienced forgetting things we needed to remember because we failed to get a good night's sleep. New learning is very delicate, and while it can garner some resilience from our shorter localized waking slow-wave states, it needs the longer periods of widespread slow-wave brain states provided by sleep to ensure firm consolidation.

Slow-wave sleep states provide another benefit: they support the widespread distribution of large amounts of information into your long-term memory. While you're awake, only localized sections of the brain drop into slow-wave states, while the rest of the brain maintains active beta waveforms so you can also pay attention to life as it happens. In sleep, there's no such limitation, your brain doesn't "zone out" but instead "zones in."

It's believed that while you're awake, only those specific areas of your long-

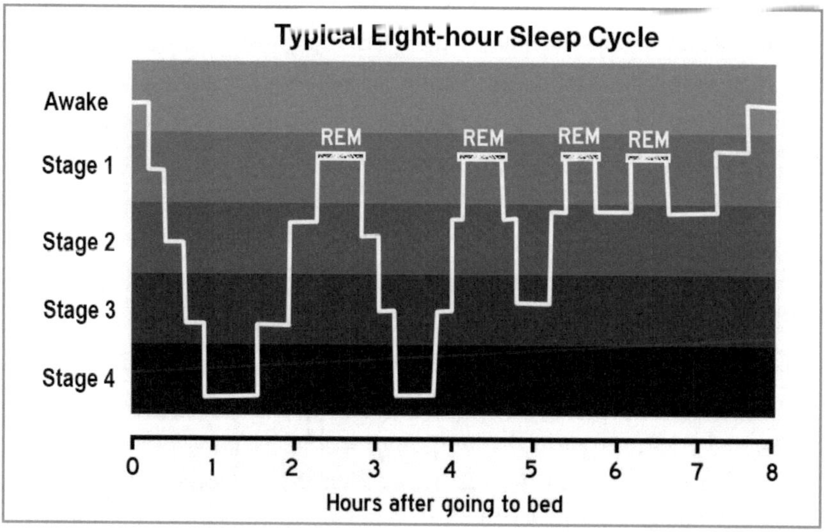

term memory banks where the slow waves occur receive information from working memory. But in slow-wave sleep, alpha, theta, and delta waveforms dominate, and it's believed this widespread activation ensures the equally widespread distribution of information into many different brain areas. This is critical because your most effective memories are housed in multiple areas. Think of an orange. If you remembered only what an orange looks like, it would have less relevancc to you than if you remembered what it tastes, smells, and feels like.

REM: Your Chaotic Waves of Incubation

The phase of incubation called REM is a relatively recent discovery. Until the middle of the 20th century, grand figures in science and psychology like Ivan Pavlov believed that, like the lights at bedtime, our brains shut off during sleep, and we were truly dead to the world.[3] Well, they had it partially right: we were dead to the world, but only the outside world.

In 1953 in a sleep lab in Chicago, Eugene Aserinsky and Nathaniel Kleitman conducted a series of experiments that discovered a unique phenomenon. Under closed eyelids at certain times during the night, the subjects' eyes were observed to move back and forth at incredibly high rates of speed, as if the individuals were engaging in some sort of wild physical activity or watching a hyper-fast tennis match. Yet their bodies remained motionless.[4] What Aserinsky and Kleitman discovered eventually became known as *rapid eye movement* sleep, or REM.

Later, when researchers measured the EEGs of subjects in REM sleep, they found a curious contradiction. While the body was completely immobile, the brain was producing unsynchronized, high-frequency, low-amplitude waves that could be 200 percent or more vigorous than those seen in active beta waking states.[5] In beta states, the brain's unsynchronized wave activity ranges typically from 8–40Hz (cycles per second), while in REM sleep, the frequency increases to 90Hz and beyond. In fact, during specific periods of REM sleep, storms of neural activity that are so intense they resemble mini epileptic seizures sweeping through the brain.[6]

It's been found that to support this enhanced neural activity, blood flow increases by 47 percent in the brain stem and by 41 percent in the cortex.[7] Dr. E. Clark was one of the first neuroscientific pioneers to discover this increased blood flow in REM sleep. During an operation, he noted the brain of a sleeping patient became so engorged with blood that it protruded out of the patient's opened skull casing![8] No, your mind is not "off" in REM sleep; instead, literally, trillions upon trillions of neural connections are being activated to such an intense level that it looks like all hell is breaking loose in there.

REM research over the last two decades has given us a clearer understanding of what's happening at the neurochemical level to cause this frenzy of unsynchronized brain activity. During REM sleep, the spigots in the raphe nuclei and locus coeruleus, which allow the monoamines serotonin and norepinephrine to flow, are virtually shut off, while the spigots in the lateral tegmental and basal ganglia, which provide acetylcholine to your brain, are turned to high. Remember, the monoamines act as modulators controlling the way your neurons fire. With serotonin and norepinephrine virtually shut off and acetylcholine flooding your brain, there's little constraint on the neural-lightning effect of acetylcholine. The result is that your neurons fire wildly.[9]

The effect on neurons when these monoamines are cut off is kind of like the effect on a classroom full of pent-up third graders when the teacher leaves the room: unrestricted cross talk breaks out. In slow-wave (alpha, theta, and delta) waking and sleeping states, the monoamines are reduced but still exert some control on neural firing, which helps support slow, organized, and synchronized waveforms. But in REM sleep, with your serotonin and norepinephrine spigots shut down, chaotic and unsynchronized neural wave activity ensues.

Dreams, Psychosis, LSD, and Dream Amnesia

The unreal quality of your dreams in REM sleep is a direct result of this unrestricted neural cross talk. The Greeks believed a dream spirit invaded the brain during sleep, causing wild hallucinations.[10] In 1641, René Descartes wrote, "I am accustomed to sleep and in my dreams to imagine the same things that lunatics imagine when awake."[11]

Harvard's Allan Hobson, one of the world's leading sleep researchers and author of *The Chemistry of Conscious States*, writes, "What the visual brain does with the acetylcholine signals during REM sleep is hallucinate."[12] In REM sleep, with little support from norepinephrine and serotonin, neural stability is lost, and neurons propelled by acetylcholine are free to make wild neural connections, which they could never make in the presence of higher concentrations of the monoamines. In this chemical dream stage, we can have all sorts of wild associations, with our dream selves doing anything from kissing snakes to surfing clouds. Hobson says of REM sleep, "Dreaming is not a model of a psychosis. It is a psychosis. It's just a healthy one."

Too much acetylcholine activity is implicated in the wild deliriums of many psychotic and schizophrenic patients. These patients have delusional thoughts because their brains' neurochemistry has tipped in favor of acetylcholine. Under the influence of hallucinogenic drugs, like LSD, psilocybin, angel dust, and mescaline, individuals think and act like psychotics because these drugs cause users to experience wild REM-like dreams while they're wide awake. This phenomenon occurs because these drugs hamper the firing of serotonin cells, which in turn destroys neural stability. But at the same time, these drugs increase the firing of norepinephrine cells, which supply some focus to their drug-induced deliriums and keep users fully awake. The end result is that hallucinogenic drug users end up wide-awake and crazy as loons.

It's estimated you have four to thirty dreams per night.[13] So, why do you remember so few of them? As you saw in the preceding chapters, the monoamines are the neurochemicals that allow you to place information into working memory. In REM sleep, with the monoamines' spigots turned down, knowledge can't be effectively put in your working memory. Therefore, most of your dreams go completely unremembered. Yet, upon waking, you often can remember some elements of a dream. Why? This is because you usually awaken directly out of REM sleep. And what causes you to wake is the serotonin and norepinephrine spigots turning back on. Therefore, in the morning, as these monoamines come back online, you have a chance of getting snippets of the dream you were just having placed into your working memory so that some parts of it may be remembered.[14]

REM: Your Hyper Creative Brain State

New connections between neurons are the source of your creativity. Many researchers consider REM to be our most creative brain state because more random neural connections can be made in REM than in any other state. This is why so many of humankind's great discoveries have occurred as individuals emerged from a good night's sleep.

In REM sleep, with the controlling monoamines absent and acetylcholine's neural lightning effect dominating, your neural networks are prompted to make hyper associations, which provide the opportunity for a dramatically higher degree of random connectivity than can be achieved in the other brain wave states where the modulating effects of serotonin and norepinephrine are present. In REM sleep, these undirected hyper associations become the source of profound creativity. As noted in an earlier chapter, Otto Loewi attributes his Nobel Prize-winning discovery of the function of acetylcholine to a dream.[15]

REM: Your Evolutionary Gift

REM sleep can be seen as an evolutionary gift that allows your mind to develop super creative solutions to life's most challenging situations. Research shows that when you're presented with complex circumstances in life, REM sleep can be both longer and more active, as represented by faster, longer periods of eye movements and EEG measurements.[16]

REM stage sleep provides your brain with the opportunity to reprogram itself by making new and unique connections so that you can more effectively adapt to life's ever-changing state of affairs.[17] Because most of the monoamines aren't present in REM sleep, you're unconscious and unaware of these new creative connections being made. But when you awaken and the monoamines are turned back on, these new connections can result in the "aha" and eureka moments where solutions to life's problems seem to appear spontaneously. Fish—not efficient, quick adapters—have minimal creative powers. They sleep to recover physically but most never enter creative REM stage sleep. The theory is that because these creatures don't have a flexible neocortex, little new information is logged into their brains, and therefore there's little reason for REM sleep to manipulate what isn't there.

The Paradox of REM: Brain On, Body Off

As discussed, REM sleep is also called *paradoxical sleep*. This is because while your brain is engaging in wild, chaotic activity, which is much greater than what's seen in your waking beta active states, your body is as limp as a wet rag. In REM, with monoamines turned down and acetylcholine surging forward, the parasympathetic nervous system is entirely free to prompt your body to drop into its most relaxed state. While you may dream of running from a wild, scaly, one-eyed, fire-breathing monster, your feet don't move, because with the monoamines cut off, the brain lacks the necessary neurochemicals to urge your body into action. It's interesting to note that comas are the result of the brain stem's monoamine sources being cut off by trauma or disease. Therefore, in effect, your body is indeed comatose during REM sleep.

Contrary to popular belief, sleepwalking doesn't take place during REM sleep; it can occur only in the alpha, theta, and delta stages. This is because the brain produces enough of the monoamine activity necessary for the body to take physical action only in these stages. Patients with the sleep disorder narcolepsy can literally fall flat on their faces at the onset of an episode. Their sudden loss of muscle tone is brought on from an instantaneous shift from a beta waking state directly into REM sleep. The immediate increase in acetylcholine and the withdrawal of the monoamines cause their muscles to go instantly flaccid.

Something else is off in REM sleep: your temperature control. In the hypothalamus, the cells that control body temperature are slowed so much that temperature regulation is suspended.[18] This is why mountain climbers who have lost their heat source in the winter try to stay awake. If they fall into REM sleep, the body cannot maintain a temperature of 98.6 degrees, and they can quickly freeze to death. Some researchers believe that this periodic suspension of movement and temperature regulation during REM sleep allows for more energy to be utilized by your brain to maintain the wild neural activity that occurs in paradoxical sleep.

Although REM sleep is best known for its rapid eye movement, REM could also stand for rapid *ear* movement. In REM sleep, your vestibular system is activated, and there are corresponding rapid twitches in the auditory upper radius of the middle ear, indicating that hyper-auditory processing is also taking place.[19]

Another paradox of REM sleep is that while the major muscles are flaccid, the metabolism has increased 20 percent.[20] This increase is believed to be due to the dramatic rise in brain activity in comparison to other sleep stages. Not all body parts are flaccid in paradoxical sleep. REM sleep turns on the genes that direct sexual function. This results in erections for males, vaginal engagement for females, and wet dreams for adolescent boys.[21]

Emotional Learning and REM Sleep

Brain imaging now shows that during REM sleep, the seat of emotion, the amygdala, lights up, indicating emotional learning is taking place.[22] It's been found that after an emotional episode, the number of your eye movements in REM sleep increases. Which indicates that emotional learning is taking place. As you saw in the chapters on the *Meaning Network*, your emotions are vital to the process of rational thinking. When emotional learning is inhibited, efficient meaning-making and correct thinking suffer. Although limiting REM sleep has negative consequences on emotional learning in all age groups, it's most devastating for children because they spend so much more time in REM sleep than adults.

How Drugs, Alcohol, and Stress Disrupt REM Sleep and Learning

For decades, psychiatrists have recognized that drug addicts, alcoholics, and people experiencing long-term stress all have poor emotional development and poor judgment. Research into REM sleep is giving medical professionals clues as to why these individuals have such difficulties. Drugs, alcohol, and stress all hinder your capacity to fall into REM sleep because they inhibit acetylcholine and your parasympathetic nervous system from taking control during your sleep cycles. When the brain can't experience REM stage sleep, emotional learning, and thus, emotional development and maturation suffer.

Research into REM sleep supports what twelve-step programs have been saying for years: an individual's emotional development stops the day they start heavily indulging in their drug of choice. They say if you start drinking at fifteen and stop at fifty, you're middle-aged but with the emotional maturity of a teenager.[23] Like characters in the TV show *Arrested Development,* individuals experiencing extended periods of REM deprivation from drugs, alcohol, and stress can end up emotionally retarded.

Amphetamines such as cocaine, crack, speed, and, yes, caffeinated drinks such as coffee and Coke, all increase the activity of norepinephrine, which is a major neurotransmitter of activity. A problem arises when you overindulge in them because acetylcholine can't get an effective foothold, and the REM stage of sleep becomes compromised. Science now knows that the all-nighters you pulled in school, fueled by stimulants like Monster Energy,

Mountain Dew, and espresso, interrupted your REM sleep stages and thus did little to improve the long-term retention of the very information you were trying to pack into your long term memory.[24] And this is another reason you remember so little of all those tests you studied for, no matter what your grade was.

Too much alcohol also interrupts REM sleep. Trying to fall into a good night of REM sleep cycles when your belly is full of alcohol is like trying to drive a car with one foot on the brake and the other on the accelerator. This is because alcohol is both a depressant and a stimulant. Initially, alcohol can help you fall asleep as it amplifies the effect of the brain's main inhibitory neurotransmitter, GABA (gamma-aminobutyric acid). But alcohol is also a stimulant, increasing the levels of available norepinephrine, which in turn inhibits your ability to fall into acetylcholine-dominant REM sleep.[25]

When you drink, the liver also breaks down alcohol into toxic byproducts. One of these is formaldehyde, which gives new meaning to the phrase being "pickled." If you overdrink, your brain becomes so bombarded by these toxins that they prevent the cholinergic (acetylcholine) system from effectively engaging. Thus, when you're drunk, although you can enter slow-wave sleep states, you fail to enter REM sleep.[26] Alcoholic delusions, like pink elephants and snakes climbing out of keyholes, are the brain's effort during the day to make up the REM stages of sleep lost at night, thus making the sorry alcoholic's waking hours a real nightmare!

Alcohol also inhibits your ability to remember because as excess imbibing interrupts REM sleep, that in turn stops NMDA receptor sites from functioning efficiently. In Chapter 7, you saw that for efficient memory formation to take place, a neuron must first be stimulated at multiple locations, which in turn blows magnesium plugs out of NMDA receptors. This sets off the chemical chain reaction that provides the structural bases of learning and long-term memory formation. If you've experienced an alcohol-induced faulty memory after a night on the town, it's because the excess alcohol and toxicity create a chemical soup where NMDA receptor sites stop working. If you experience a blackout… you can't go back and retrieve the memory because it was never effectively formed. Scary stuff!

Just like drugs and alcohol, stress unbalances the ratios of the monoamines and acetylcholine, inhibiting your ability to fall into deep REM sleep cycles. When you have significant stressors in your life, like natural disasters, war, relationship difficulties, or job dissatisfaction, the top of the adrenal glands produces a potent stress hormone called cortisol. One of cortisol's main jobs is to increase the level of norepinephrine in your brain so that you can more effectively take physical action and focus on the problems at hand. When your worry levels become too high, too much cortisol is released, prompting excess norepinephrine expression, which activates your sympathetic nervous system; this, in turn, inhibits your brain's ability to drop into REM sleep. Although your cortisol levels do naturally increase throughout the night (increased levels of cortisol are partially responsible for helping you wake in the morning), too much of this stress hormone destroys your ability to fall into REM sleep, thus compromising your ability to remember and be creative when awake. (For more on stress and learning see Chapter 38.)

Procedural memories refer to learning skills such as how to ride a bike, swim, type, or play a musical instrument. Research shows that if learners are allowed their normal four to five cycles of sleep after learning a new skill, they're actually more proficient at that task the next day than right after the experience.[27] Studies show that missing REM has devastatingly harmful effects on your ability to encode procedural memories. Many athletes who play games like football, tennis, and golf and who burn the candle at both ends or overindulge in alcohol or drugs fail to improve at their game. This occurs because their hectic lifestyles and substance abuse limit the amount of REM sleep necessary to effectively enhance memory consolidation and therefore limit their performance.

Researchers have also found that REM sleep may be the most vital of all incubation stages for the replenishment of your monoamine system.[28] At this time, it's understood that in REM sleep, when the monoamines are turned towards off, they are refortifying themselves so that you can focus your attention more powerfully and effectively when you emerge from sleep.

Validating this hypothesis, research shows that during REM sleep, when the norepinephrine cells cease firing, there's a direct increase in the number of norepinephrine receptor sites in the brain (upregulation).[29] When REM sleep is interrupted by drugs, alcohol, or stress, even if you've entered slow-wave sleep stages and the bedside clock says you've slept six to eight hours, you can awaken feeling totally exhausted and lacking the ability to focus on life's issues because your monoamine system hasn't been refreshed.

Your Hippocampus and REM Sleep

No discussion of the value of REM sleep in memory consolidation would be complete without looking at the function of your hippocampus. The hippocampus is well understood as a major player in the process of transferring working memory into the correct concept networks of long-term memory. A fascinating phenomenon occurs in the hippocampus during REM sleep. Although the rest of the brain responds to the dominance of the acetylcholine system with wild, desynchronized waveforms, the hippocampus goes its own slow, synchronized, theta-wave way.[30]

As you saw in Chapter 7, new learning can't survive in thin air but must physically attach itself to existing neural concept networks before it can be integrated into your long-term memory structures. It's believed that, during REM sleep, while the rest of the brain's neurons are wildly firing, the hippocampus's theta-wave action serves to enhance its capacity to connect new information gained in activity and held in working memory with the appropriate concept networks in the neocortex. Robert Miller has spent decades studying the hippocampus, and notes that theta rhythms act to strengthen concept loops between the hippocampus and the neocortex's rear associative areas, where long-term memory is held.[31]

Studies show another exciting event going on in the hippocampus during REM sleep. The same regions of the hippocampus that are activated during

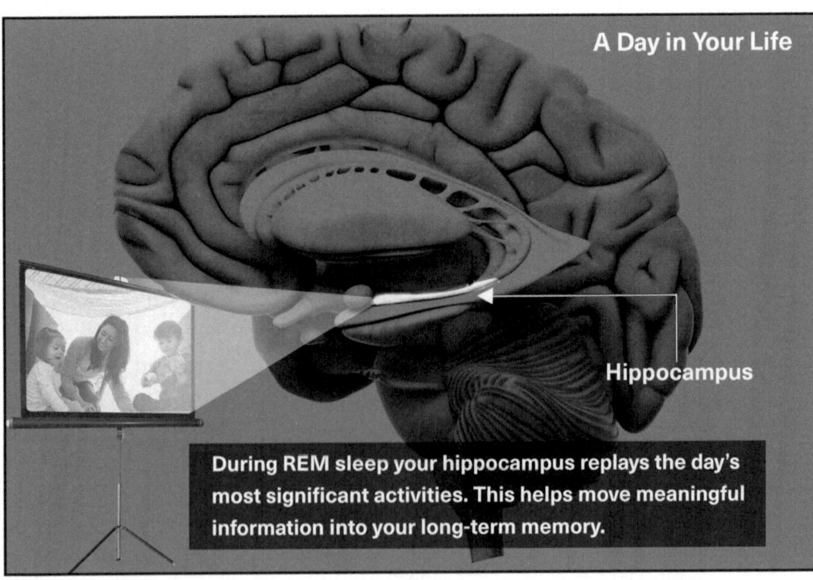

waking learning Phase 1 states are also activated in Phase 2. This suggests the hippocampus is literally replaying the important events of the day. For reasons not yet understood, the hippocampus replays the day's most significant activities a little more slowly than real-life experiences, but the sequence of cell firing is just the same. This exact cell re-firing is believed to be another element that supports the solidification and consolidation of meaningful information into your long-term memory banks.[32] When you limit your REM sleep stages, you curtail the length of hippocampal replay and thus limit your ability to encode new knowledge.

Why Children Need Longer REM Stages Than Adults

The amount of time we spend in REM in our early years is directly tied to how much more we learn as children when compared to adults. At twenty-six weeks, the fetus is in REM sleep twenty-four hours a day. An infant spends sixteen hours a day in sleep, half of which is REM.[33] And toddlers and children spend 30 percent of their total sleep time in REM. In comparison, adults spend only 20 percent of their sleep time in REM, and those over fifty spend only 15 percent of their sleep time in REM.[34] Premature babies compensate for the lack of sleep in the womb by spending more time in REM sleep than babies are born at term.[35]

Your Genetic Plan and REM Sleep

Many researchers take the view that when you were a fetus and a baby, you spent so much time in REM sleep because it activated genetically predetermined programs that held ancestral learning.[36] Your genes are the receptacles of instinctual, inherited learning (see Chapter 4). It's believed that REM facilitates the transmission of learning—such as sucking, imitating, hoarding, protecting territory, fighting, and fleeing—from your genes into your neurological circuits.

As an adult, you need less REM sleep than children not only because you've learned the basic language, motor, and social skills to get you through life but also because your instinctual memory has already been transferred from your genes into your neural tissues.

The Media and Sleep

A few words about the media and sleep here. As we've discussed, the average child spends six-and-a-half hours a day with some form of the media—which exposes them to over thirty-six acts of violence and eighteen acts of sex every day.[37] Since the job of the hippocampus in REM sleep cycles is to replay knowledge gained during the day, what images do you think are being played in the brain of that sweetly sleeping child? Yep, the same disturbing images they perceived on TV, the web, or a violent video game. For many children, the media has raised their anxiety levels so high and made their dreams so disturbing that the prescription of pharmaceutical drugs to help children sleep is becoming normal. From 2000 to 2004, the spending on sleep aids to children has increased by over 85 percent.[38]

Children aren't the only ones that are affected by the media when they sleep. I often tell clients that if you want to sleep like a baby, don't give your brain violent input (this includes the news) before you hit the hay; instead, give it pleasant, relaxing, soothing input before you go to sleep.

Also, whether you're a child or adult, using a tablet, smartphone, TV, or other electronic devices right before bed disrupts your sleep by disrupting your body's internal clock (circadian rhythm). This is caused by the short-wavelength artificial blue light that's emitted by these devices. More than any other wavelength, blue light suppresses the sleep-inducing hormone melatonin. So, if you want to fall asleep easily, its best to stop using electronic media an hour or more before you go to bed. As I type these words at 11 p.m., I have my blue blocker glasses on, so I can fall fast asleep as soon as I quit working on my computer.

The School Day and Sleep

New research is uncovering disturbing evidence that suggests the early start times at junior high and high schools are inhibiting the slow-wave, and REM cycles teens need to ensure efficient academic and emotional encoding.[39] The biological clock of teenagers isn't like that of adults. As children reach puberty, they naturally begin to want to stay up later and sleep longer. This shift is caused by a change in the way the pineal gland releases the hormone melatonin, which is instrumental in producing the conditions for the onset and maintenance of sleep. When you're asleep, melatonin blood levels are ten times higher than when you're awake. At puberty, the pineal gland automatically begins to shift its release of melatonin to a later time. This change results in adolescents naturally wanting to stay up later.

Adolescents can't force themselves to sleep even if they want to. This is because the chemical climate needed to induce sleep, which is partially prompted by melatonin, isn't activated until later in the night. Years ago, Dr. Richard Evans of Johns Hopkins University found the threat of an early start to the school day does little to get teens to go to bed early and they went to bed at the same time whether they had to be at school at 7:30 or 9:30 a.m.[40] Before both of my sons turned twelve, they used to naturally go to sleep before 9 p.m., and a short bedtime story put them right to sleep. After they turned twelve, I would have to tie them to the bedpost to get them to bed by 9 a.m. Some researchers speculate that this shift in sleep time helps teenagers foster individuality by placing them in a time zone which is separate from younger children and adults.

Mary Carskadon, a professor at Brown University and head of the Sleep and Chronobiology Research Laboratory, says of our early school start time, "These kids are in shift-work conditions. They're being asked to perform at times when their biology is out of sync with expectations."[41] Stephen Sheldon, a pediatric sleep therapist at Northwestern University Feinberg School of Medicine, has done research showing that too little sleep by students causes classroom underachievement, daytime hyperactivity, short attention spans, crankiness, and mood swings.[42] Pity the poor teacher who's forced to educate a classroom full of kids who haven't had their full cycles of short-wave and REM sleep. The obvious solution is to use this research and have schools work with students' sleep rhythms, not against them, by starting schools an hour or two later, as some schools are starting to do.

Sleep: Two Hours Makes a Difference!

Dr. Robert Stickgold, professor of psychiatry at Harvard Medical School, found that we have a sleep window of opportunity for consolidating memory. His research found that learners who cut short their sleep for as little as two hours at night were unable to remember as well as those who had gotten a full night's sleep.[43] Stickgold discovered that students who were trained to do a visual task and were then tested three to twelve hours afterward showed no improvement in learning, nor did those who were allowed sleep only six hours a night. Only those students who'd had more complete cycles of slow-wave and REM sleep, sleeping more than six hours a night, showed learning improvement. The research illustrates that late to bed and early to rise is not a wise learning strategy. "In any kind of learning," Stickgold says, "it's a great idea to sleep on it."

Stickgold has made another interesting discovery. If you miss your REM sleep stages directly after a learning experience, you can't go back and open the window of opportunity for memory consolidation. He found that if he allowed subjects who were initially deprived of their normal slow-wave and REM stages to then sleep normally for two nights, their "catch-up" sleep didn't help improve their learning performance.[44] Trying to catch up on memory consolidation by sleeping longer after you've experienced sleep deprivation appears to be as effective as closing the barn door after all the cows have gone.

Research also reveals something else. For efficient learning and memory to occur, you may not want to wake up before your last, longest, and possibly most crucial REM stage is complete. Throughout the night, each REM stage gets progressively longer, with the last stage lasting thirty to forty minutes. Evidence shows that setting the alarm clock so that you awaken before you finish that last stage of REM sleep has very negative consequences on consolidating the information that you need to learn and innovate with.[45]

A study found that medical interns who work more than eighty hours a week make 36 percent more mistakes than their more rested colleagues.

Our Institutional Culture of Sleep Deprivation

Sleep research has significant implications for corporations, the government, the military, and learning institutions that foster a culture of sleep deprivation as an accepted element of the creative, learning, and production process. There's probably a no more effective way to create less institutional learning, impede behavioral change, and decrease innovation than to staff an organization with large numbers of people who haven't completed a full night of slow-wave and REM sleep.

The medical community offers an intriguing example of a group of professionals who are well aware of the research and yet proceed as if they were ignorant of it. A study reported in the *New England Journal of Medicine* over fifteen years ago found that medical interns who work more than eighty hours a week make 36 percent more mistakes than their more rested colleagues.[46] Their mistakes ranged from prescribing overdoses to sticking a needle in the wrong vein. These overworked interns also nodded off more than five times a night during long shifts. It's a persistent wonder to me that this community, which is acutely aware of the research presented here, can still cling to the idea that an eighty-hour workweek is the best way for interns to learn how to practice medicine.

This insane practice not only forces doctors to miss vital brain incubation periods but also puts the people they're trying to heal at high risk. Doctors aren't the only sleep-deprived personnel working in the medical community. Because of staffing shortages, large numbers of nurses and medical technicians are forced to work multiple high-stress shifts on very little sleep.

It's estimated that over 500,000 critical medical errors are made in hospitals each year and that over 30 percent of patients in hospitals are there because of misdiagnoses or medical errors.[47] Although "do no harm" has long been an accepted aim of medical professionals, unleashing large numbers of sleep-deprived, dumbed-down medical personnel on very sick people goes counter to that pledge.

Being burnt-out and sleep-deprived is often worn like a badge of honor in corporations. Unfortunately, this "work hard, play hard" culture has been implicated as a significant factor in four of the most memorable manmade disasters of the last four decades: the explosion of the space shuttle *Challenger*, the *Exxon Valdez* oil spill, and both the Three Mile

Island and Chernobyl nuclear power plant meltdowns.[48] It is believed with too little sleep and too much stress, which resulted in depleted monoamine systems and low creativity, key workers were unable to effectively focus on and implement standard or innovative remedies when confronted with these catastrophes. Anyone who has spent any time in a corporation staffed with overstressed, sleep-deprived, burnt-out workers probably witnessed numerous mistakes. While misjudgment in your personal life may not have as far-reaching consequences as it does for doctors, ships' crews, and power plant workers, failure to pass through your normal slow-wave and REM sleep cycles can still have devastating effects on your academic, business, financial, and social interactions.

An organization that pushes its charges too hard and limits time for the brain to drop into the incubation phases of learning dumbs down not only the individual but the whole institution. Managers, employees, teachers, and students can't hope to achieve the levels of learning, productivity, and creativity they strive for if they're locked in organizations that fail to recognize the inherent value of the *Incubation Stage* of learning.

Cracking the Learning Code

Sleep is a *Super-Incubation Stage* that is much more potent for promoting learning, emotional maturation, and creativity than our waking incubation periods. This is because of three main factors that sleep provides that our waking incubation periods do not: longer slow-wave periods, more widespread slow-wave periods, and REM (Rapid Eye Movement) stage sleep.

38

Stress: The Death of Learning, Innovation, and Emotional Control

The sprouting of the seeds of creativity, intuition and wisdom takes place in a relaxed mind. Only anger, greed and ego require a disturbed mind.

—Shivanshu K. Srivastava, Indian lawyer and writer[1]

For the last 2,000 years, institutions and bosses have, in general, ignored the stress levels of their charges. The persistent belief was, the harder you pressured or whipped that person, the more work you could get out of them. But in a knowledge society, where the key to your success rests in your mental capacity, not your muscles, stress, and its effects on the biology of learning, creativity, and behavioral change must be seriously considered. New research is now uncovering a disturbing fact: the greatest inhibitor of your *Individual Adaptability/Intelligence Factor* and your ability to live a joyful life is the opposite of relaxed states—that is, prolonged stressful states. Considering that most of us feel that we're under vast amounts of stress, no wonder the World Health Organization has declared stress a "worldwide epidemic."[2]

If scholastic and corporate institutions want healthy, happy, engaged, innovative, and compassionate learners instead of unhealthy, disgruntled, disengaged, uninspired individuals, we'll need to develop systems that avoid, as much as possible, the effects excess stress has on our bodies and brains. And if you want to transform your own life, eliminating unnecessary sources of stress is the first place to start

This Is Your Brain and Body on Stress, and It's Not a Pretty Picture!

There are numerous reasons that excessive stress kills learning, innovation, emotional control, and more. Stress can:

- Destroy your brain's ability to effectively drop into the vital *Incubation Stage* of learning
- Atrophy and wither your brain connections[3]
- Wipe out your memory by killing neurons[4]
- Inhibit the creation of new brain cells
- Negatively affect specialized neurons called glial cells, preventing them from transporting nutrients and cleaning up the waste your neurons produce
- Compromise learning by promoting the negative expression of specific genes, resulting in consequences like tangles in the connections between your brain cells, producing neural static and confusion[5]
- Lay down the structural foundations for mental disorders such as

schizophrenia, depression, suicidal ideation, ADHD, and PTSD[6]
- Make you sick, and you can't learn well if you aren't feeling well
- Reduce longevity by shortening special regions of DNA called telomeres[7]

As briefly described earlier in the book, prolonged stress also causes a *neurological downshift*. That is a downward shift in your brain's energy, away from your most evolved flexible upper brain structures toward your lower, more fixed brain structures. The result? Systems thinking, innovation, understanding the future consequences of your present actions, emotional regulation, empathy, and cooperation are all profoundly compromised. Once you understand how the stress response affects your thinking and behaviors, you begin to see stress as a major culprit in the dysfunction of

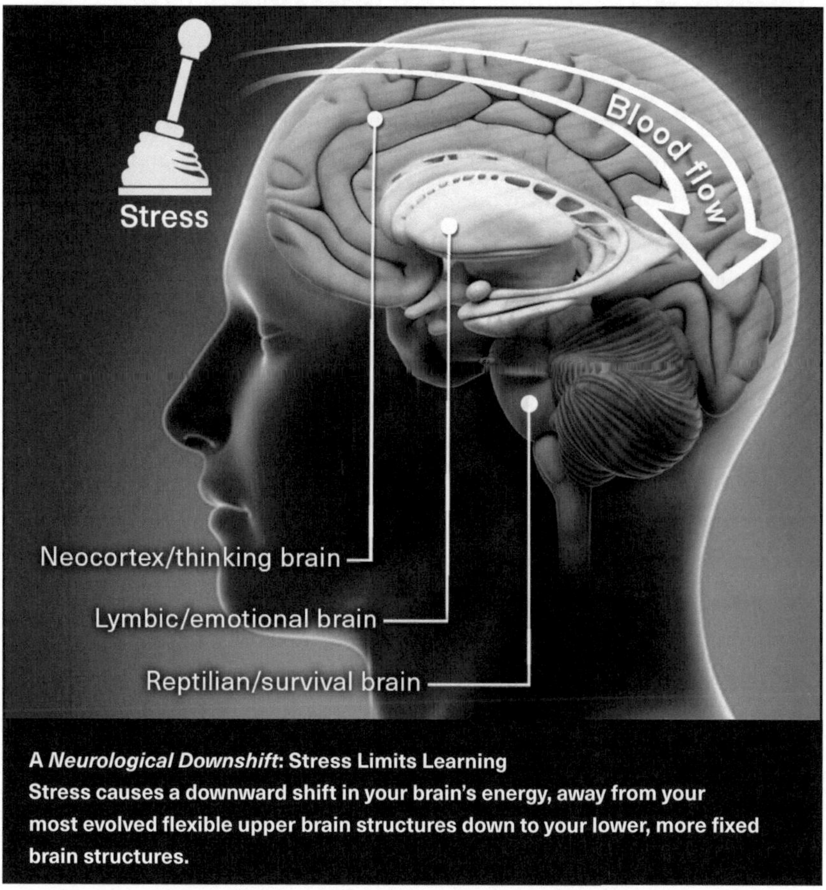

A *Neurological Downshift*: Stress Limits Learning
Stress causes a downward shift in your brain's energy, away from your most evolved flexible upper brain structures down to your lower, more fixed brain structures.

everything from Congress and the medical system to education and the conflicts in the Middle East. Because the stress response can be so devastating to the kinds of thinking we need to transform our turbulent 21st century, developing environments that help us significantly reduce our stress levels may be one of the most important things we can focus on for the success of our nation, our world and ourselves.

But before we can effectively create the learning and behavioral-change environments we demand, we must have a robust understanding of the causes, mechanisms, and consequences of creating organizational and living environments that produce high-stress levels. Knowledge is power. And that's why this chapter and the next delve so deeply into the neurobiology and genetic ramifications of stress on your learning and behaviors.

The Myth: Working under Stress Makes You Smarter

Until recently, it was believed that stress increased your ability to learn. But research into the effects of stress on learning is forcing brain-based educators to conclude that the systems favored by most institutions may in themselves produce so much stress that not only is effective learning and innovation compromised, but so too are our capacities for big-picture thinking, future planning, and empathy. The result is scholastic and corporate institutions filled not with joyful, productive, creative, and compassionate learners but with disgruntled individuals as unhappy and uninspired as Sisyphus, the mythological Greek king who was condemned to spend the afterlife pushing a huge stone to the top of a hill, only to have it roll down again.

The man who gave the world the conditioned response was also one of the first researchers to recognize the profound negative influence that stress has on our learning processes. In 1924, during a major Leningrad flood, water surged high into Ivan Pavlov's laboratory. The terrified dogs that Pavlov had spent weeks and months conditioning were trapped in their cages and had to paddle for their lives with their noses just barely above the high-water mark. After these animals were rescued, Pavlov was shocked to find that the stress of the flood had produced such profound changes in these animals' brains that they had unlearned the conditioned responses he had just worked so hard to implant![8]

Studies Reveal the Devastating Consequences of Stress on Learning

There's a huge body of research that has since confirmed Pavlov's findings of the consequences of stress:

- Over 125 studies of more than 36,000 individuals show conclusively that the more stress people are under, the lower their memory performance, no matter how it's measured—by grade point average, IQ, or other forms of achievement tests.[9, 10, 11]
- Researcher John Hunsley reports that the amount of stress you feel while taking a test directly predicts how poorly you'll do on it.[12]
- A study of 1,790 individuals training to be air traffic controllers showed that even though these individuals had superior IQs, high stress disrupted their abilities to score well on air-traffic tests.[13]
- In a McGill University study, researchers checked stress hormone levels of 130 volunteers and found that high amounts of the stress hormone cortisol correlated with inhibited memory and attention.[14]
- A survey of stressed and maltreated children showed that the rate of having to repeat the first grade was an incredibly sad 65 percent. These stressed-out kids just could not think or learn![15]

I'm a perfect example of the effects of stress on learning. Living with two unpredictable alcoholic and drug-addicted parents, I was under so much stress at home, I flunked not only second grade but also the sixth. I was such a behavioral problem that after seventh grade, I was sent to military school. This was a gift that allowed me to escape my "crazy train" family. The result? My stress levels dropped substantially, which allowed me to become not only the cadet with the highest grade point average in eighth grade but also to be voted cadet of the year in the junior school. The lower stress levels also help transform my view of myself as a troublemaker and screw up. Learning expert Gordon Stokes sums up the research nicely, saying, "Eighty percent of learning difficulties are related to stress. Remove the stress, and you remove the difficulties."[16]

What Is Stress?

The history of the word *stress* goes back to Walter Cannon, who in the 1920s was credited with coming up with this term to describe the fight-or-flight response we experience when we perceive threatening stimuli. Cannon was the first scientist to make the correct hypothesis that when you encounter a threat to life or limb, like a bear in the woods, your sympathetic nervous system is immediately engaged, which in turn sends nerve impulses down the spinal cord and redirects the blood flow away from your internal organs to the larger muscle groups so that you can fight or flee more effectively.[17]

Since Cannon's first efforts, scientists have learned the stress response not only alters blood flow but has other wide-ranging effects, such as increasing your heart rate, upping your energy (glucose) utilization, escalating your breath rate and oxygen consumption, dilating your pupils, raising your blood pressure, increasing your galvanic skin response (sweating), and altering your neurochemical activity so you can only focus your attention on what's threatening you.

Scientists have discovered something else that's imperative to our understanding of how stress affects us: the sympathetic nervous system isn't the only way to turn on the stress response as Cannon discovered. It can also be activated via the stress hormone *cortisol*.[18] Comprehending how the sympathetic and the cortisol-hormone stress routes affect your body and brain is instrumental in grasping how excess stress compromises your ability to learn and remember.

Understanding Your Two Responses to Stress: They Support Each Other

To increase your understanding of the two systems of the stress response—the cortisol-hormone stress route and the sympathetic-nerve stress route. I'm going to generalize greatly because, in truth, these systems are intricate, intertwined, and very complex. But for our purposes, we can say the cortisol-hormone stress route acts to preheat your body and mind, getting you ready for the sympathetic-nerve stress system to launch the physical activity needed to fight or flee. To get a sense of how these two systems of stress support one another, we will revisit what we discovered about the stress response to media violence (see Chapter 19) from a different angle.

Imagine you're home alone and fall asleep while watching a scary movie late at night. After several hours, you're roused from your slumber by a loud creak of the floorboards. It sounds like someone is coming down the hallway! As you lie there trying to figure out what to do next, the hypothalamus in the middle of your brain releases CRF (corticotropic releasing factor) to the pituitary, the pear-shape structure that hangs just beneath it, this primes your brain and body to be ready to fight or flee this unknown stressor. Once stimulated by CRF, the pituitary secretes the hormone ACTH (also known as *corticotropin*) into your bloodstream. When the ACTH flow reaches the outer structure (cortex) of your adrenal glands, which sit atop your kidneys, the adrenal glands are prompted to release another hormone called *glucocorticoid* into the bloodstream. For simplicity, we call the *glucocorticoid* hormone, cortisol.[19]

This cortisol-hormone stress process primes your brain and body to be ready to engage the sympathetic nervous system in case you need to take powerful physical survival action. In the body, the cortisol-hormone system helps you do this by causing your heart to race, your blood pressure to rise, your glucose (sugar) utilization to increase, your breathing to accelerate, your oxygen consumption to increase, your blood flow to shift to the major muscle groups, and sweat to be produced, so your body temperature remains stable while all this internal activation is taking place. As these body changes occur, cortisol is altering its neurochemical activity in your brain, so you can focus your attention more effectively on that scary sound in the hall.

If the creaking sound turns out to be your cat out for a late-night prowl, no problem. Eventually, your cortisol-stress reaction will disengage, the relaxation response will be reactivated, and you'll fall back asleep. But if the

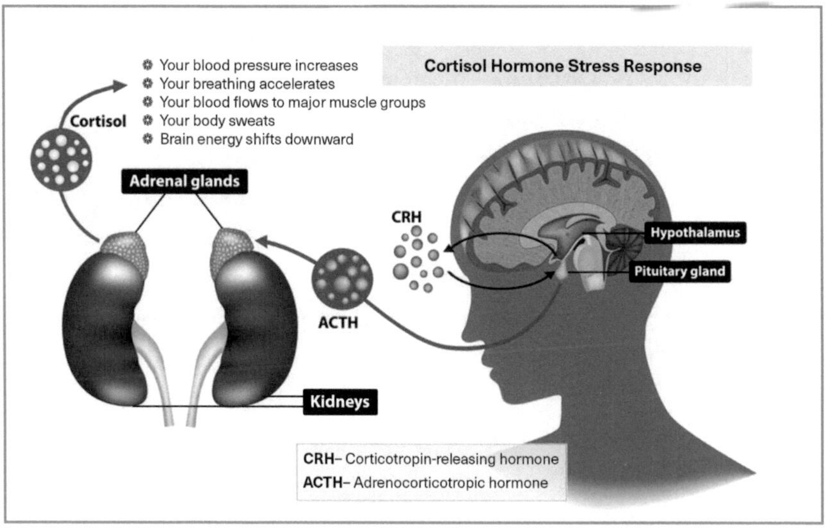

sound is caused by a real threat, like a burglar who pokes his head into the room, your sympathetic nervous system will immediately be activated. Virtually all the brain and body reactions that cortisol has primed and revved up will now be supercharged by the activation of the sympathetic nervous system and the release of adrenaline (epinephrine) that accompanies it. Whether you grab the baseball bat under your bed and try to hit the guy on the head or dive out the window in panicked flight, your sympathetic nervous system's fight-or-flight response will be that much more effective at keeping you alive because your body and brain have been preheated by the cortisol-hormone stress route.

Psychological/Intellectual Stressors Are Your Problem!

Evolution didn't design your cortisol-hormone stress response to be switched on and revving up your body and mind for extended periods. In days gone by, the stress response didn't need to be activated for long periods. This is because physical threats to life and limb were usually resolved in very short periods. This is because when predator and prey met, the prey either got eaten or got away. For both parties, the stress response system was switched on for only minutes to hours before it returned to baseline.

But in our modern world, your human cortisol-stress response can be turned on and revved-up for days, weeks, months, and even years. This isn't only because we live in a world where the threat of terrorism and violence can keep us constantly locked in fear of physical harm, but also because our sophisticated neocortex allows us to switch on our cortisol-stress response because of things that aren't life-threatening that we constantly think about, like the fear of losing your job or your mate.

The problems that cortisol-hormone-stress response causes for learning and memory in modern humans are directly linked to the duration of the response. Short periods of cortisol production actually increase memory formation, but research shows the longer the stress response is activated, the greater the learning and memory disruption. Studies show that, after only twenty to thirty minutes of exposure to cortisol, neural tissue begins to show damage.[20]

At The Institute, we call non-life-threatening, high-anxiety stressors that you think about and obsess upon *Psychological/Intellectual Stressors.* They can keep your cortisol system pumping out stress hormones for days to years. For adults, these stressors include worrying about things that can't be resolved quickly, such as whether you'll get a raise or get fired, have enough money to pay the mortgage, and ever find the perfect mate. *Psychological/Intellectual Stressors* in kids can be activated by concerns like "Will my acne ever go away?" "Will Mom and Dad divorce?" "Will I get a date for the prom?" "Will I be accepted by the 'in' group?" "Will I pass the test next week?" "Will I graduate high enough in my class to get into the right college?" "Will I eventually get a good job?"

Psychological/Intellectual Stressors can keep modern humans trapped in various chronic states of stress almost indefinitely. Without fast resolutions to your *Psychological/Intellectual Stressors*, your cortisol system does something it wasn't designed to do: it floods your body and brain for weeks, months, or years, not only destroying your health but also causing neurological damage that ruins your ability to learn and change your behaviors. As Dr. Pamela Peeke of the University of Maryland says, "We were never meant to deal with prolonged chronic stress. The human body isn't designed to drag around bad memories, anxieties, and frustrations."[21]

The Key to Lowering Stress, Enhancing Learning, and Increasing Health

A major factor that impacts the effects of both physical and *Psychological/Intellectual Stressors* is how much control you feel you have over a situation. Research shows that the less control you feel you have, the more the cortisol stress response will be activated. The main reason brain-based educators don't consider overly authoritarian educational systems to be the most effective learning environments is they take control away from learners for extended periods, thus increasing the learners' stress responses—which in turn negatively affects their learning abilities and their health. That's one reason why so many new educational initiatives encourage students to self-direct their own learning.

Working with several schools that serve children, the majority of whom live in poverty and have parents who are in jail, abusive, or absent, helped us at The Learning Code Institute to develop one of the first educational platforms designed to combat the damage high-stress living environments have on children's biology. Many of these students have PTSD. At these schools, it isn't out of the ordinary to have a child come to school directly from the city jail, where their father may have been arrested for something like smoking crack in the car with his child at two in the morning. (The first time I encountered a child who had just experienced something like this, I was so overwhelmed by the child's pain that I could barely control my own emotions over the damage being done to his delicate brain and health.)

As we are developing this model, we're integrating higher levels of control and self-directed learning by the students as a way to lower their stress hormone levels and up the activation in the neurochemical systems and brain areas that code for calmness, memory formation, and behavioral change. We've also found an added benefit: the more control these children feel they have in their lives, the less they feel the need to act out in the classroom.

Cracking the Learning Code

Excess stress is not your friend. It doesn't increase your productivity, boost your creativity, help you learn, control your emotions, or instigate personal change for the better. In fact, it undermines your work, health, creativity, education, and efforts at self-improvement.

39
How Stress Damages Your Learning, Health, and Your Future

I promise you nothing is as chaotic as it seems. Nothing is worth diminishing your health. Nothing is worth poisoning yourself into stress, anxiety, and fear.

—*Steve Maraboli,* Unapologetically You: Reflections on Life and the Human Experience[1]

In the last chapter, we discussed in general terms why stress isn't a badge of honor. It's holding you back, damaging your brain and health as well as stifling your creativity, fulfillment, and joy. Unfortunately, we live in a society that wrongly lauds stress as a mark of virtue and achievement. In this chapter, I want to guide you into a more in-depth explanation of *how* stress disrupts your ability to learn, innovate, and manage your emotions. This is critically important—one of the most important takeaways from this book. The truths are radical and countercultural, which makes them even more crucial for us to understand if you have any desire to have a more powerful impact on our own learning and the learning and behavior of those you work and live with.

There is another reason why understanding the scientific bases of stress is vital. This is because altered stress-cortisol levels have been found in as one of the primary bases for not only PTSD and high levels of anxiety but what many experts call two of the most dangerous epidemics the world is experiencing today, addiction and depression. Our experience is that few of those treating these disorders have been afforded the opportunity to grasp the science behind what is causing them. Without this knowledge, how are we ever to develop effective programs to reverse their rising trends?

Eight Reasons Why Stress Disrupts, Learning, and Positive Behavioral Change

For simplicity's sake, I've divided this chapter up into eight main sections. They cover the primary reasons stress kills your ability to prosper. Such as how stress makes you think like a lizard, destroys the valuable *Incubation Stage* of learning, creates brain static, kills neurons, destroys glial cells, produces negative gene expression, makes you sick and shortens your life span. In each section, we'll take a deep dive into the science, so you understand more specifically *why* stress is hurting you and limiting our scholastic and corporate organizations from reaching their potential.

FAST FORWARD *or* DEEP DIVE

As in other chapters, you can get the big conceptual picture in the first paragraphs of each section below. If you need more granular understanding of the details, all you need to do is read deeper into each section. Remember, you only need to go as far into each section as your current needs demand.

REASON 1
The Downshift—You Can't Learn If You're Thinking Like a Lizard!

I've described how stress creates *neurological downshift* in an earlier chapter, here let's dive deeper into how this relates to stress.

Excess of the stress hormone cortisol increases blood flow to the muscles and brain, so you're physically and mentally prepared to fight or flee a stressor. In the 1980s, brain-based educational pioneer Leslie Hart theorized that when you're under stress, blood flow and energy usage in your brain aren't evenly distributed.[2] He used the term *downshift* to describe the phenomenon that occurs when you're under excessive stress. The blood

flow and energy in your brain shift from the upper brain centers, in your neocortex and prefrontal lobes, to your lower brain centers, in your limbic system and brain stem.

The upper and rear structures of the neocortex house virtually all the factual and experiential information you've acquired in your lifetime. And the prefrontal lobes are the main center of your working memory and provide you with your human attributes, such as insight, abstraction, self-awareness, association, metaphor, judgment, empathy, compassion, systems thinking, future planning, enhanced social skills, and the mediation of your raw emotions.

As discussed in Chapters 4 and 5, the neocortex and prefrontal lobes sit atop your oldest evolutionary structures, the brain stem and the limbic system. In the 1970s, neuroscientist Paul MacLean labeled the limbic system and brain stem the "mammalian" and "reptilian" brains because these structures control our raw, most basic emotional urges and house the instincts we share with our reptilian and mammalian cousins, such as fighting, fleeing, imitating, hoarding, deceiving, and territory marking.[3] Compared to your upper neocortical brain centers, these lower brain centers are hardwired—that is, the connections aren't very flexible, and therefore, very little new learning can occur there.

Today, neural imaging has shown Hart's theory of stress-induced *downshifting* to be right on. Brain scans clearly illustrate that when you're under the influence of excess cortisol, you think, act, and learn more like your animal cousins because the brain's blood and energy flow shift from the neocortex and prefrontal lobe areas down to the lower mammalian and reptilian brain areas.[4] Educational researcher Eric Jensen comments, "This fact has tremendous implications for learning."[5]

The Downshift Impairs Your Working Memory

As we touched on earlier in the book, the *downshift* of blood and energy flow has a drastic effect on your memory because it affects a major structure of working memory called your lateral prefrontal cortex. In conjunction with the hippocampus, your lateral prefrontal cortex provides the temporary storage area where meaningful information is housed before it's selected into your long-term memory banks. Under stress, when blood flow is redirected away from the lateral prefrontal cortex to your lower brain centers, working memory becomes compromised, which is one reason you might have difficulty trying to learn new information when someone is pressuring you to do so.[6]

This is also the reason that eyewitness accounts of stressful situations are so suspect. Many experts believe witnesses are notoriously inept at identifying details at crime scenes and accidents because watching these events produces high levels of stress, launching the *downshifting* phenomenon, which in turn compromises working memory and, therefore, long-term memory formation.[7]

Impairs Your Ability to Access What You Already Know

Not only does *downshifting* limit your ability to acquire new information, but it also limits your ability to access what you've previously learned and stored in your long-term memory banks. The upper rear areas of your neocortex contain the factual and experiential knowledge that you've acquired in your lifetime. When a stress-induced *downshift* diverts blood flow out of these upper areas, the information held there becomes extremely difficult for you to access.

While *downshifting* doesn't completely cut off blood flow to your upper rear memory networks, it does reduce it to such an extent that neuro-efficiency declines. The end result is that you have a difficult time accessing what you already know. This is why, if you're overly stressed when you give a speech, you can't remember what you've previously memorized. It's also why, when you get into a heated discussion with a coworker or mate, you often can't think of the right thing to say. Yet after the stressful period has passed and blood flow returns to the upper areas of your neocortex, you can easily remember your speech or what you would like to have said.

More on Test Anxiety and the Downshift

Brain-based educators recognize that many students who don't do well on tests (those with test anxiety) or who have been labeled "slow" are indeed quite bright. As we touched on in an earlier chapter, the problem is that their brains *downshift* at the fear of taking a test, limiting access to the very information they've placed in the neocortex and need to pass the test. One of the country's leading experts on intelligence, Yale's Robert J. Sternberg, had such bad test anxiety in the sixth grade that when he took the IQ test with his peers, to the shock of his family, he did quite poorly. When he was allowed to take the test again with fifth graders, he felt confident surrounded by his juniors. Taking this second test and feeling more self-assured created less of a *downshift* in his brain, and therefore he received an outstanding score.[8]

Downshifting Negatively Affects Your Behaviors

Downshifting also negatively affects behavior patterns. Not only do your upper brain areas house systems thinking, future planning, cooperation, and the other evolved traits that make you human, but they also exert control over your lower reptilian and mammalian urges. You don't defecate in public when you really have to go or automatically hit or bite someone who has just taken your parking place (even if you want to!) because your upper brain centers override your more impulsive mammal/reptile instincts. This mediation of the upper over the lower brain centers allows us to cooperate with our fellow humans and to act appropriately in society. When you're under prolonged or intense stress, the control over these lower centers dissolve—which is why some people soil themselves when under threat or fear.

Blame Evolution for the Downshift

Evolutionary biologists believe that we experience *downshifting* because when we're under stress, the brain prefers ancient behavior patterns that have existed for 350 million years.[9] It's as if your brain was saying, "Let's forget what we've learned in this lifetime about socially acceptable behavior and what makes us most human, and revert to the behaviors that our animal and reptile ancestors used to get us out of this stressful jam." *Downshifting* is why, under stress, even the most mild-mannered people can do atrocious things, like commit road rage or carry out an office or school shooting. Hart writes, "Downshifting is always to the more traditional, more familiar, cruder behavior—to what we would do if we had much less brain."

Does Downshifting Create White-Collar Crime?

The cortisol-induced *downshift* may be one of the primary reasons we saw so much deception and self-centeredness in corporate executives when the dot-com bubble burst in the late 1990s and when the stock market crashed in 2008. During these crises, when stock prices plummeted, the huge paychecks, stock options, and luxurious lifestyles of CEOs and top executives were under threat, which increased their cortisol levels. As their blood flow shifted from the upper and frontal neocortical areas to the lower reptilian and mammalian brain centers, the theory is their deception and hoarding behaviors became more pronounced, and their compassion and empathetic qualities became less evident.

While I'm hoping that no defense attorney will try to use *downshifting* as a valid defense for lizard-thinking scoundrels, we must consider the threat of financial loss may have been significant enough to create a neural climate where deception, greed, and lack of compassion can allow executives to betray and deceive their stockholders and loyal coworkers.

Does Education Cause Teachers and Students to Downshift?

As most of us have experienced, educational institutions can be very stressful places. Being prompted by extrinsic motivators to rush from one forty-five-minute study-for-the-test period to the next, having information delivered in a way that doesn't activate one's primary *Biological Preference* (see Chapter 10), competing for grades, being forced to study information that has little personal meaning, and spending long periods locked in small rooms away from the real world can all *downshift* even the most resilient. The result is that many students (especially those who don't have a linguistic dominance) act out or physically or mentally drop out. Why do an astonishing 74 percent of students admit to cheating?[10] Because when students are *downshifted* to their lower brain centers, deception seems reasonable.

Many experts believe the dramatic increase in school violence, such as the Columbine and Sandy Hook massacres, can be attributed to students whose school and life circumstances have *downshifted* them to such an extent that they're able to carry out incredible antisocial and inhuman acts. So many kids feel under stress. The Youth Risk Behavior Survey revealed that 16 percent of students have brought a weapon to school.[11] As Renate and Geoffrey Caine report, "The more threatening and helpless students feel, the more we would expect the behaviors that we would characterize as 'reptilian.'"[12]

What About Violent Conflicts?

In war-torn areas where the threat of death is omnipresent, whole populations can become *downshifted.* Without access to their upper brain centers, large groups of people collectively act in very inhuman ways. Only a *downshifted* society would ever conceptualize that the best resolution to a conflict is to have people strap sticks of dynamite onto their bodies, step on buses, and kill themselves and large numbers of innocent people. Under prolonged stress, *downshifted* societies lack access to the very brain structures that allow them to come up with creative and peaceful solutions to the problems that confront them. To get countries to act more humane and friendly toward one another, the United States and the rest of the world would do better to spend billions not on bullets but on ways to *"upshift"* stressed populations to their higher brain centers so that they can learn new strategies that may help them create peaceful solutions to their plights.

When we work with organizations to transform their cultures from suffering the effects of *downshifting* to more cooperative and innovative cultures, we introduce the concept of the *Noble Cause,* discussed briefly in Chapter 18. We first have them develop a higher mission for their organization—not just bigger profits. Why? Because just like when you are intrinsically motivating yourself when you're pursuing a *Noble Cause* greater than yourself… like ending world hunger, healing cancer, or even something simple like TED Talks and their mission of "spreading ideas," you naturally cause the brain to employ the most evolved and frontal structures. As long as the *Noble Cause* is not just a slogan but ingrained into the fabric of the organization, the result is you start to *"upshift"* the whole culture to more productive brain states.

REASON 2
Stress Inhibits Your Incubation Stage of Learning

Stress can produce such high levels of cortisol that your brain is inhibited from effectively entering the *Incubation Stage* of learning. As you saw in Chapters 31, 35 and 36, before information can be effectively transferred from working memory into your long-term memory banks, and before you can go into highly creative states, the brain must enter the *Incubation Stage.* As discussed, about every ninety minutes during both waking and sleeping phases of incubation, your brain cycles in and out of different neurochemical combinations, which in turn create different EEG brain-wave states—alpha, theta, delta, and REM.[13] When your brain enters these slow-wave and REM states, long-term memory consolidation and creativity are enhanced.

Cortisol primarily inhibits the formation of the EEG waves of incubation by prompting the release of the neurochemical norepinephrine. Under normal circumstances, norepinephrine's job is to keep your brain focused on meaningful stimuli. But when you're under stress for extended periods, cortisol and the accompanying overactivity of norepinephrine have the same effect on your brain as too much caffeine or amphetamine drugs: they lock your mind in anxious, hypervigilant states where physical relaxation and the states of incubation are inhibited. The end result is that your memory formation and creativity suffer greatly.

Stress Ruins Your Sleep

During the day, you may be so focused on your stressors that you're unaware that cortisol and norepinephrine are interrupting your ability to naturally drop into the waking incubation states of learning. But at night, when you can't fall asleep because you're worried about being downsized at work, passing that final exam, or leaving your partner, the effects of excessive cortisol and norepinephrine are all too clear.

At bedtime, if you're relaxed, your cortisol and norepinephrine levels naturally decrease, making it easy for you to drop into alpha, theta, delta, and REM sleep stages. But when your stress thermostat is set on high, the excessive production of norepinephrine can keep you from falling into these stages of sleep. Stress is just like drinking an energy drink or cup of coffee before you go to bed: they're all very effective at keeping norepinephrine levels high.

Sometimes you're so exhausted from being stressed out that you can fall asleep the minute your head hits the pillow, because your anxiety has exhausted your norepinephrine system. But once a few hours of sleep have restored this system, you find yourself wide awake and again fixated on your worries.[14] Drugs like alprazolam and ketoconazole help anxious individuals sleep by inhibiting the adrenal glands' output of excess cortisol.[15]

REASON 3
Stress Causes What Feels Like Static in Your Brain

As we touched on in Chapter 16 on working memory, stress levels can compromise your learning and memory by creating what feels like electrical static in your brain. Like a radio that's not getting good reception, hindering your ability to get clear, concise communication from it, brain static does the same thing to the efficient communication between your brain cells. This effect occurs because high levels of cortisol reduce the activity of one of the brain's central neurotransmitters, serotonin—which plays a vital role in stabilizing cellular communications. Without the calming influence of serotonin, your neurons and their connections fall into a churning ocean of disorganized cross talk.[16]

Every second that you move through complex, sensory-rich environments, billions of bits of sensory information are stimulating billions of neurons, and trillions of neural connections. Serotonin helps regulate this sensory bombardment, providing a tranquil background for neural firing to take place. This allows essential information that has been marked as meaningful by other neurotransmitters to pop out so that it can more easily be selected into your working and long-term memory structures.

The more stress you're under, the harder it is to think clearly and rationally because of your lower serotonin. In addition to the *Downshift,* neural static caused by losing cellular stability is another reason why when you find yourself in an argument, you can't think of what to say next. As we saw, the hallucinogenic drugs LSD and psilocybin cause such wild, unorganized thoughts is because they're excellent serotonin-receptor blockers. And the reason suicide seems like a rational act to certain depressed people is because their high cortisol levels have driven their serotonin levels so low.

Serotonin is also one of the brain's major behavior impulse modulators. This is why low serotonin has been implicated in increased levels of anger and aggression.[17] When serotonin activity is low and neural stability is lost, you can feel anxious and on edge. It can be said our stressful school and corporate environments are perfect serotonin blockers. This is one reason why so many of us in these stressful environments find ourselves acting out of our natural character.

Many researchers believe the majority of prisoners have been so stressed by their circumstances—such as poverty, abusive parents, and threatening environments—that they're plagued by chronically low levels of serotonin, which causes them to habitually act out in socially unacceptable ways.[18] Then they are sent to the threatening environment of prison, and we expect them to learn how to reform. How can they transform in this high cortisol–low serotonin environment which compromises the very learning and behavior change we want them to experience? You can't learn not to be a criminal when your brain is full of static. No wonder the National Institute of Justice reports that the recidivism rate of prisoners five years after release is a mind-boggling 84 percent![19] Clearly, prisons aren't changing behaviors. Low serotonin levels have also been implicated as the cause of the impulsive behaviors in children and adults with ADHD.[20]

REASON 4
Stress Kills Brain Cells, Shrivels Connections, and More

Stress compromises your learning because it literally damages your brain. It does this by withering neural connections, killing existing brain cells, and stopping the production of new ones.[21]

Did you ever feel like you forgot the answers for a test you'd studied for because of stress? Well, you're not alone. In the 1960s, 70s, and 80s, there was much speculation about the role excess cortisol played in the damage to brain tissue, but it wasn't until the 1990s that neuroscientists got a clear picture of just how destructive excess cortisol could be.[22]

Some of the first evidence came from the sufferers of Cushing's syndrome, a condition where various tumors cause the adrenal glands to secrete excessive amounts of cortisol. The sufferers of Cushing's can develop severe memory problems called *cushingoid dementia.* To find the cause of these patients' memory problems, researchers discovered a curious fact: while the size of the rest of the brain remained normal, the hippocampus, which supports working memory by placing information into the correct long-term concept networks, had shrunk. When these patients had operations that removed their adrenal tumors, and cortisol levels returned to normal, the shriveling of the hippocampus stopped.[23]

Further research in the 1990s revealed that not only could diseases like Cushing's cause shrinkage to hippocampus tissues, but so too could *Psychological/Intellectual Stressors.* While working at the Yale University

School of Medicine, J. Douglas Bremner did extensive research with MRIs on different groups of individuals who had experienced high levels of stress. He found that sufferers of post-traumatic stress disorder, an ailment characterized by psychic distress and learning and memory impairment, had hippocampal shrinkage of 8 percent,[24] that sexual abuse victims had a 12 percent reduction in hippocampal size,[25] and that individuals with depression, a mental disorder marked by high levels of cortisol, showed hippocampal shrinkage of nearly 20 percent.[26]

Yvette Sheline at the University of Pennsylvania Perelman School of Medicine found that the longer the depressive bout, the more severe the hippocampal damage.[27] At McGill University, a study that lasted four years revealed that stressed-out elderly women had hippocampi 14 percent smaller than those of their peers who weren't under stress.[28] Research is showing the ugly and disquieting fact that those of us who experience stress for extended periods may never be the same again biologically.[29]

Over the last decade, we've gained great insight into how stress causes the hippocampus to shrivel. The hippocampus is exquisitely sensitive to cortisol because it has more cortisol receptors than any other part of the brain.[30] This intense sensitivity to cortisol is a good thing when you're under stress for short durations. After all, it increases hippocampal function and learning. But when stress goes on for too long, high cortisol levels begin to have a damaging effect on the tissues of the hippocampus and other brain areas.

Stress Destroys the Branches of Your Neurons

High levels of cortisol also attack your dendrites (branches of a neuron), causing them to wither and die. To these delicate branches, which hold the synaptic receptor buds vital to cell-to-cell communication and make up the physical foundations of long-term memory, excess cortisol is like a poison.

Eminent stress researcher Bruce McEwen at Rockefeller University showed that after only a few weeks of exposure to excess cortisol levels, dendrites begin to atrophy, just as if someone had poured acid over them.[31] When these connections wither, they can get pulled apart, which compromises not only your existing memory but also your ability to learn by destroying the physical structures that new information attaches itself to. McEwen found a good news/bad news scenario with his research. The good news is that if your stress is short-lived, the dendrite damage is reversible; the bad news is that if the stressor goes on for too long, the damage becomes permanent.[32]

Excessive Stress Kills Your Brain Cells

Extended exposure to cortisol can literally kill your neurons. As we've discussed, learning initially increases when moderate amounts of cortisol enter the brain for short periods. One of the primary reasons for this is that cortisol can make neurons more excitable, mainly by increasing the activity of the brain's major excitatory neurotransmitter glutamate.

Cortisol enhances glutamate's activity in a roundabout manner, by restraining the activity of the brain's primary chemical of inhibition, GABA. When you're under stress, without GABA to impede glutamate's actions, this excitatory neurotransmitter can bombard neurons without restraint, creating a destructive environment. Too much excitation by glutamate eventually boils neurons in their own overactive chemical stew, actually killing them. Neuroscientists call this overactive, death-imposing neural process *excitotoxicity.*

Excitotoxicity occurs when glutamate stimulates your neurons for too long at specialized receptors sites called NMDA receptors (see Chapter 7). The overstimulation by glutamate keeps these receptor sites jammed open, allowing sodium and calcium to flood unchecked into neurons, which in turn forces them to swell, destroys genetic controls, produces enzymes that devour the neuron's supportive structure, and manufactures corrosive molecules called *free radicals*, which tear away at membrane walls.[33] If this glutamate onslaught goes on for too long, neurons literally die from overstimulation. Excitotoxicity is also the cause of cell death that neuroscientists see with epileptic seizures and strokes.

Stress Stops the Birth of New Neurons

Excess cortisol can also inhibit the birth of new neurons. One of the most persistent neuroscience myths was the brain couldn't create new neurons. Controversial research conducted at the dawn of the 21st century showed that in a process called *neurogenesis,* new brain cells do indeed get birthed when children and adults learn new things. This discovery has given new hope to those who suffer from the neuron loss of Parkinson's, Alzheimer's, strokes, and other nervous system disorders.[34] Unfortunately, when you're under stress for too long, excess cortisol can inhibit the genes that prompt the neurogenesis process in the hippocampus.[35] With stress inhibiting the production of new neurons, your ability to learn and innovate deteriorates.

Stress Reduces Energy to Your Brain

At the same time excess cortisol is withering neural connections, killing neurons, and stopping new ones from being produced, another insidious side effect of cortisol overload is taking place: energy utilization in the brain is being reduced. This reduction of energy to neural tissues exacerbates an already dangerous situation.

Initially, stress increases the delivery of your brain's primary fuel: glucose. This is an evolutionary advantage, giving you the mental energy necessary to pay attention to and remember survival situations. Problems occur when high levels of cortisol production go on for longer than a half-hour. Laboratory experiments show that after only thirty minutes of excess cortisol exposure, glucose utilization by neurons actually begins to decrease.

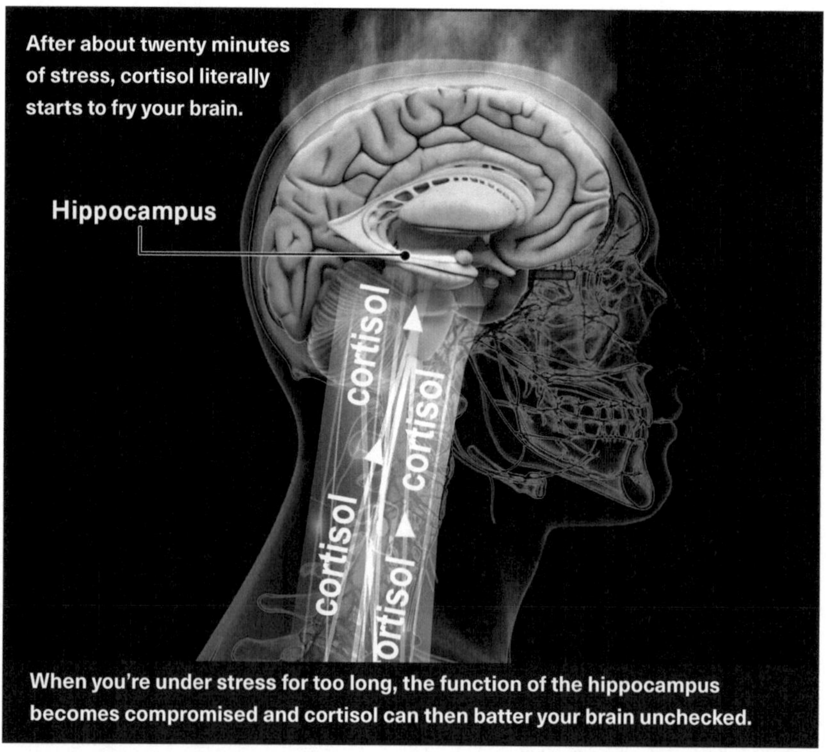

When you're under stress for too long, the function of the hippocampus becomes compromised and cortisol can then batter your brain unchecked.

In a counterintuitive process, as more stress hormones build-up, less glucose gets utilized.[36] The hippocampus is more affected by this reduction in glucose utilization than any other brain region. Here, under long-term stress, glucose utilization can decrease by as much as by 25 percent.[37]

Therefore, under extended periods of stress, the very structure that's imperative to converting working memory to long-term memory and placing new memory into the correct concept networks—and that may already be experiencing withering connections, excitotoxicity, and inhibited cell growth—is thrown into even more significant trauma by the limiting of the very fuel it needs to deal with this destructive onslaught!

The Hippocampus Acts Like a Float in a Toilet, but Stress Disrupts How It Works

As if excitotoxicity combined with a reduction in glucose utilization weren't bad enough for hippocampal cells, cortisol causes another complication in this unique structure. In a bizarre twist of evolutionary fate, the brain structure that's most negatively affected by cortisol is also the brain structure that's most responsible for its regulation!

The hippocampus controls cortisol production through a process called negative feedback. One example of negative feedback is what happens to the water in the toilet tank after you flush. Increasing the amounts of water pour into the tank, decreasing the likelihood that more water will be released. In your toilet, the flotation device controls the water flow; in your brain, the cortisol receptors in the hippocampus control the flow of cortisol. When a sufficient number of your receptors are occupied by cortisol, your hippocampus signals the hypothalamus to stop releasing *corticotropin-releasing factor* (CRF), which in turn reduces the outflow of ACTH from your pituitary, which then slows the release of cortisol from the outer surface of your adrenal gland.[38]

Under normal circumstances, when you're under stress for short periods, this is an exquisite and finely tuned system. But when you're under stress for too long, and the function of hippocampal neurons are compromised by excitotoxicity and a reduction of glucose by 25 percent, the negative feedback system breaks down and fails to operate efficiently. Cortisol can then batter the brain unchecked, like waves battering a ship in a storm. Without the cortisol regulation system in full function, connection shrinkage, and neuron death accelerate.[39] Also, because cortisol is inhibiting neurogenesis, new neurons don't take the place of your dying ones.

Reduction of Your Brain's Fertilizer: BDNF

More research must be done in this area, but initially, it looks like stress affects BDNF, the *Brain-derived Neurotrophic Factor* that acts like a fertilizer to cell growth that we looked at in earlier chapters. This hypothesis is based in part on studies demonstrating that stress negatively affects BDNF such that it reduces new cell growth and causes atrophy of specialized neurons called pyramidal neurons.

Brain imaging and postmortem studies provide additional support, demonstrating shriveling and cell loss in the hippocampus, prefrontal cortex, and amygdala. The reduction of BDNF has also been implicated in a cause of depression. An interesting hypothesis is that antidepressant treatments oppose the effects of stress and depression, by increasing levels of BDNF, boosting cell growth, and reversing or blocking the atrophy and cell loss caused by stress.[40] Additional brain imaging and postmortem studies will be required to further test this hypothesis.

REASON 5
Stress Negatively Affects Your Glial Cells

Your regular neurons (motor neurons, sensory neurons, interneurons, and their hundreds of subtypes) aren't the only brain cells affected by the stress-induced reduction in glucose utilization. Specialized neurons called *glial cells* are also negatively affected. Until quite recently, scientists virtually ignored glial cells when analyzing brain function. Today, neuroscientists recognize that your glial cells play an essential role in supporting regular neurons' ability to perform their learning and memory functions.[41]

Your glial neurons help to process and transport nutrients, act as neural janitors (by cleaning up cell waste that can clog synaptic connections) and produce the fatty substance that wraps around axons called *myelin. Axons* are the cords that transmit information from neuron to neuron. The myelin around an axon acts like insulation around an electrical wire, enhancing the speed and efficiency of signals. When the right amount of myelin is wrapped around your axons, the faster and more efficiently your neurons can communicate with one another, which is another way to increase your learning and behavior change.

When you're under stress, just like other brain cells in the hippocampus, glial neurons use 25 percent less glucose,[42] severely compromising their ability to support your learning and behavior change by transporting nutrients and cleaning cell waste. New research shows that chronic stress also forces glial cells called oligodendrocytes to overproduce myelin, which one would think would be a good thing. It is not! These cells get so hyperactive under stress that they produce too much myelin, thus clogging up communication between your brain cells. These negative changes can become permanent.[43] This overproduction of myelin is believed to be a foundation not only for compromised learning, but for mental disorders such as schizophrenia, autism, depression, suicidal ideation, ADHD, and PTSD.[44]

REASON 6
Stress-Induced Negative Gene Expression

Environmental stressors that create high cortisol levels combined with excitotoxicity and reduced glucose utilization also prompt your genes to express themselves in negative ways. As we've discussed, genes are responsible for producing the proteins and enzymes that provide the foundation of all learning and behavioral change.

How Does Stress Affect Genetic Expression? Epigenetics

Epigenetics is the study of how the environment turns your genes on and off. You can think of DNA as the hard drive of a computer that contains vast amounts of information. The way that information is accessed is via the software, which for our purposes are epigenetic (environmental) factors. While that explanation is simple, the underlying molecular biology is a little more complex.

Environmental stressors turn your genes on and off through a process called *DNA methylation,* which is the addition of a methyl group (one carbon atom bonded to three hydrogen atoms) to a section of DNA. It's like environmental stimuli placing a tag on a segment of your DNA, altering the way your genes express themselves in good or bad ways. Another mechanism of epigenetics is histone modification. *Histones* are proteins that act like spools that the six feet of DNA in each one of your cells wraps itself around. This keeps things nice, neat, and tidy inside the cell. Factors in your environment can affect how tightly or loosely your DNA is wrapped around your histones, which in turn affects which of your genes are turned on and which ones are turned off.

Stress, DNA Methylation, and Neural Tangles

Research reveals that high cortisol levels have an epigenetic effect on genes that can reset your genetic switches in ways that don't serve you well at all. Robert Post, head of Bipolar Collaborative Network of Brain and Behavior Research Foundation, found that cortisol turns on the c-Fos gene. Under typical learning situations, the c-Fos gene prompts normal receptor and dendrite growth. But Post discovered that when we're under stress for too long, the epigenetic effect clicks the c-Fos gene into high gear, causing the production of so much protein that tangles develop in our dendrite branches. These neural tangles not only severely inhibit your learning, innovation, and memory but also may cause and exacerbate depression, manic episodes, and seizures.[45, 46]

At Yale Medical School, Dr. Amy F. T. Arnsten discovered stress activates a gene that releases an enzyme called protein kinase C (PKC) in your brain, which impairs working memory and other high-level prefrontal learning functions.[47] Because efficient prefrontal lobe function is instrumental in the process of rational thinking and the control of impulsive behaviors, many neuroscientists believe the stress-induced expression of PKC is another cause of inappropriate behaviors when we're stressed out, like yelling at or demeaning our loved ones.[48]

Epigenetics and Norepinephrine

Research also shows that high cortisol levels can have a negative epigenetic effect on the genes that produce norepinephrine as well as on the receptor sites that slow its activity, called the *alpha-2 adrenergic receptors*. Studies show that long-term exposure to stress can reset your genes that produce norepinephrine, setting them chronically on high.

Researcher Dennis Charney reports that individuals who were exposed to high-stress levels had 40 percent fewer of these norepinephrine-stopping receptor sites.[49] As discussed in Chapter 32, norepinephrine is one of the monoamines that enhance focus and attention. Long-term overactivity of this neurotransmitter can keep you in a constant state of readiness, also raising your blood pressure and heart rate. These responses make you feel like you're living with your nerves always on edge.[50] Once these epigenetic effects have reset your norepinephrine genes so that more is produced and less is inhibited, it becomes easy for you to lose your temper, and you have a difficult time relaxing, sleeping, focusing, and learning.

Another worrying side effect of the epigenetic mal-expression of the genes that produce and inhibit norepinephrine activity is that under long-term exposure to the excitatory effects of this neurotransmitter, neural structures may become permanently altered so that your reactions are permanently set on a hair-trigger. Dr. Linda Mayes at Yale says, "If you have experiences that are overwhelming, and have them again and again, it changes the structure of the brain." Megan Gunnar at the University of Minnesota notes the research shows, "The kids with the higher cortisol levels score lowest on inhibitory control… Kids from high-stress environments [have] problems in attention regulation and self-control."[51]

Stress Interrupts Genetic Super Windows of Learning Opportunity

While excessive stress can compromise learning and decrease inhibition by resetting genetic switches at any age, this problem is most devastating in a young child's brain. As you saw in Chapter 24, for normal maturation to occur, a child's brain must pass through numerous *super-learning windows of opportunity* at specific times. These windows, which produce dramatic and accelerated neurological development, are controlled by genetic switches. For a child to develop normal motor, tactile, hearing, speech, spatial, bonding, social, and similar abilities, specific genes must be turned on at precise times in a very exact sequence.

Under stress, your brain and body forgo all long-range building projects to deal with immediate stressors. When a child is under stress for long periods, they are at risk of having the genes that can most benefit their immediate survival turned on at the expense of the developmental genes that will help them mature into a complete, loving, and competent adult. Poverty, violence, sexual abuse, drugs, family breakups, and neglect have been shown to have negative consequences on the formation of specific brain areas during windows of genetic learning opportunity.[52] It must be emphasized that there's no evidence to date that once a neurological window of learning opportunity closes, it can ever be effectively opened again. Commenting on the missed opportunity, esteemed developmental researcher Dr. Bruce Perry at Baylor College of Medicine, states, "A piece of the child is lost forever."[53]

Think you're too tightly wound because your mom was? You may be right. Studies show that the epigenetic effects experienced by our parents can be passed down to us. Studies on pregnant women who witnessed the 9/11 attacks on the World Trade Center found that epigenetic tags on their DNA were passed on to their children, resulting in higher levels of cortisol in them.[54]

In another revealing study on epigenetic inheritance, scientists discovered that generations after the Second World War, the children of Holocaust survivors suffered from terrible nightmares of being "chased, persecuted, tortured, or annihilated," even though they weren't even around during the war. Research suggests these epigenetic effects caused by methyl tagging may be able to be passed down for an incredible three generations.[55] This means you could be experiencing stress responses in your life today passed down from your grandmother!

Research on the epigenetic legacy of stress makes it very clear why we need to rescue as many children from war-torn areas or extreme poverty as we can. Nations and politicians need to recognize that actions such as bombing civilian populations have severe consequences that can plague our world three generations after the bombs have been dropped. When you hear of ISIS fighters who behead victims without any signs of remorse, you might want to consider their biological inheritance. The epigenetic effects experienced by their ancestors many decades ago may be what allows them to take those horrifying actions with little feeling today.

Not All Gloom and Doom: Positive Epigenetic Tagging

But epigenetic tagging can cause positive genetic outcomes as well as negative ones. Research indicates you can engage in actions and behaviors (and according to controversial reports, even thoughts) that reverse the effects of negative genetic tagging.[56] This is one of the primary goals of The Institute: to reverse negative biological consequences as well as create more positive ones, not only in brain cells but also in gene expression. It's one of the reasons The Institute is so invested in helping not only Wounded Warriors, addicts, and children with PTSD but also those of us who have been negatively impacted by the environments in which we live and work. Progress made in this area can change people's brains and lives, with the positive influences potentially echoing for generations.

REASON 7
Stress Makes You Sick, and it's Hard to Learn if You Are Sick!

You can't effectively learn, manage your emotions, or be creative if you're sick. And the research is now clear: nothing can make you sicker than having physical and *Psychological/Intellectual Stressors* causing your body and brain to experience high levels of the stress hormone cortisol for extended periods.

Stress Redirecting Blood Flow Causes Disease

While Walter Cannon first recognized that stress redirected the blood flow from your internal organs to your major muscle groups, it was Hans Selye who, in the 1930s, uncovered that this redirection of blood flow, along with the other physiological responses to stress, is a significant contributor to disease.

As a budding endocrinologist, Selye came to his findings while studying the effects of an ovarian extract on rats. After his research, he found a strange phenomenon: both groups of rats developed peptic ulcers, enlarged adrenal glands, and atrophy of the immune system tissues. Selye recognized that something other than the ovarian extract had to be causing the health problems in both groups of rats. He didn't have to look far for his answer. It turned out to be his own inept handling of these wiggly little guys. While Selye had excellent analytical qualities, he had miserable motor dexterity. When he would try to inject his rats, he'd miss them or just nick them with the needle, drop them on the floor, chase them around for a while with a broom, recapture them, and try to inject them again. Selye discovered that the months of maltreatment were causing both groups of rats to get sick.[57]

Selye's discovery gave birth to the field of psychoneuroimmunology—the study of the link between people's stress levels and their vulnerability to disease.[58] Stress-related disorders are now estimated to be the fastest-rising category of illness in the West.

Psychoneuroimmunologists recognize that under the influence of cortisol, your body shifts its resources from long-term survival projects like cell growth, digestion, reproduction, and immune response to shorter-term survival responses that support potential fighting and fleeing actions. Mother Nature saw it made no sense for your body to allocate resources to long-term projects like fighting a virus or cancer if a bear was attacking you. Estimates are that 60 to 90 percent of all primary care doctor visits are driven by stress.[59]

Stress Opens Pandora's Box

A virtual Pandora's Box of disorders has been linked to stress and excess cortisol production, including diabetes, metabolic syndrome, asthma attacks, blood clots, arteriosclerosis,[60] heart disease, strokes, kidney disease, migraine headaches, hypertension, rheumatoid arthritis, backaches, muscle tension, skin ailments, insomnia, chronic fatigue syndrome, addiction, anxiety attacks, depression, and other mental illnesses.[61]

In Lamentations 1:20, the ancient Hebrews wrote, "Behold, O Lord, for I am in distress, my bowels are troubled." They were right: the cortisol-induced shift in blood flow and energy away from the digestive tract creates conditions that aggravate the gastrointestinal tract, causing diarrhea, constipation, ulcers, irritable bowel syndrome, and colitis.[62]

Scientists now have a much better understanding of how the cortisol-stress response negatively affects your immune system. The primary job of your immune system is to fight infectious agents, such as bacteria, viruses, fungi, and parasites, by unleashing lymphocytes, which are helper T-cells and B-cells that identify and rid the body of invaders.[63] Research shows that cortisol actually destroys helper T-cells and B-cells by causing their DNA to produce a suicide protein.[64] One study showed that people under stress for more than four weeks are 250 percent more susceptible to cold viruses because of the stress-induced impairment to their immune systems.[65]

Stress Stops the Immune System from Fighting Cancer and Viruses

High cortisol levels have also been implicated as a cause of cancer. This is because specific cells, with the potential to turn into cancer cells, are continually circulated in the body. When you're feeling relaxed, the immune system identifies and eliminates these rogue cells. With excess cortisol compromising this function, mutant cells take root, causing cancerous tumors.[66] Some viruses, such as herpes simplex, that are always present in the host body, only become active when the immune system is compromised by too much stress.

Stress Can Make You Fat, Short and Negatively Affect Your Sex Life

New research demonstrates that stress is a major factor in the obesity epidemic sweeping the country. Excess cortisol can cause you to gain excessive amounts of fat by making your body resistant to your own insulin.[67] This causes trouble for people who want to lose weight because when you become resistant to your own insulin, your cells can't effectively burn sugars or fats.

Stress can even make people short. Its negative effect of stress on human growth hormone (HGH) has been implicated in the disorder of emotional dwarfism, which the author of *Peter Pan,* J. M. Barrie, endured because of his traumatic childhood.[68] No wonder his famous book was about a child who never grew up, it was his experience.

As the famous soul singer James Brown once sang, man is a "sex machine," but under stress, we are not. Stress can ruin your sex life by redirecting blood flow away from the sex organs. When you're under stress, it makes no sense for your body and brain to waste vital energy on erections and vaginal engorgement.[69]

The Sooner Your Stress Starts, the Worse Things Get

The Adverse Childhood Experiences Study (ACES), a study of over 17,000 people conducted by Kaiser Permanente and the Centers for Disease Control (CDC),[70] reveals a troubling finding: the sooner stress starts in your life, the more negative impact it has on your future health. This well-respected study found that if you experienced childhood stressors from factors such as parental separation/divorce, emotional abuse/neglect, household substance abuse, or physical and/or sexual mistreatment, your susceptibility to physical and mental illness in adulthood would rise dramatically. When risk factors for those with numerous adverse childhood experiences were compared to those of people who'd had zero adverse experiences, the risk for disorders skyrocketed as the number of those experiences increased.

The examples of increased risk are shocking and disheartening. They include a 700 percent increase in alcoholism, a 200 percent increase in cancer, a 400 percent increase in emphysema, a 240 percent increase in sexually transmitted diseases, a 240 percent increase in hepatitis, and a breathtaking 3,000 percent increase in attempted suicides. This study is a testimony to why if we want to lower our healthcare costs in the world, we need to reduce our stress levels.

REASON 8
Stress Reduces Your Life Span

Stress can also limit what you learn by shortening your life. Research shows that a length of DNA segments called *telomeres* on the ends of your chromosomes is a direct index of how well your cells repair and reproduce. As you age, your telomeres shorten, and therefore, cellular reproduction suffers—which is why we get wrinkles as we get older. What's the fastest way to accelerate the shortening of your telomeres and therefore shorten your life span? Live a life of high stress. How does stress shorten your telomeres? They're similar to the plastic caps at the end of a shoelace. The higher the stress on the plastic caps, the more frayed and distressed they become.

The proof of the effect of stress on your telomeres and lifespan comes from studies like the one done at the University of California at San Francisco by Elizabeth Blackburn (who received a Nobel Prize for co-discovering telomeres) and Elissa Epel. They made the alarming discovery that mothers who had high-stress levels due to taking care of children with chronic diseases had substantially shorter telomeres than women who weren't under the same amount of stress. Their study found these overly stressed mothers had telomeres so short that it was equivalent to having lost nine to seventeen years of their lives! Other studies have found similar results in other stressed adult and child populations.[71]

While, to date, this and other research on telomeres has focused on severely stressed groups of adults (those with PTSD) and children (those in orphanages), the consensus among experts is that if you live a stressed life, you'll live a shortened life.[72]

Spawning Salmon Give Us a Graphic Picture of How Stress Kills

A graphic example of how cortisol kills health and life can be seen in salmon after they spawn. You've probably seen documentaries showing beautiful scenes of Pacific salmon leaping over short dams and waterfalls to return to their place of birth to generate billions of eggs. Shortly after spawning, these once-healthy salmon die a grotesque death. Their adrenals become huge, peptic ulcers and kidney lesions develop, their skin peels off, their immune systems collapse, and they're teeming with parasites and infections. Why? This programmed die-off of salmon commences when the genetic program turns on specific genes that prompt the fish's bodies to be overwhelmed with a nonstop flood of cortisol. In effect, these fish boil in their own cortisol stew. Many of us that are under heavy stress walk around feeling sick as these poor dying salmon look.[73]

In light of the above research, consider that during the 2020 Coronavirus pandemic our biggest fear was that the virus could damage our health or kill us. But we may find in the coming years, that the greatest damage to our species health and longevity was not only the virus, but our stress response to it. As President Franklin Delano Roosevelt correctly said in his 1933 inaugural address, referring another upheaval, the Great Depression, "…the only thing we have to fear is fear itself."[74]

So, first of all, let me assert my firm belief that the only thing we have to fear is... fear itself—nameless, unreasoning, unjustified terror which paralyzes needed efforts to convert retreat into advance.

Positive Thinking, Does It Work?

Finally, before I finish, here is a little insight into the role stress plays in understanding the truth of the saying "what we resist… persists."

Science is revealing that the age-old standby, detailed by Norman Vincent Peale in his 1952 book *The Power of Positive Thinking*,[75] may have not had the effect on our biology we want it to have. There is much speculation about how many conscious and subconscious thoughts we have a day. Estimates range from 5,000 to 50,000. It is also estimated of those thoughts, 80 percent are negative! Negative thinking is something evolution gave to us because paying attention to the negative stuff in your life is a better survival strategy than just paying attention to the positive stuff.

But here's the rub, which goes against the existing dogma and is unbeknownst to most motivational professionals: trying to resist this onslaught of negative thoughts with positive thinking can have a very negative impact on your biology. Think about it... when you pit your positive thoughts against your negative thoughts, you create a mental combat zone. If your goal is to transform your thoughts, your behaviors, and your life, this internal conflict can be perilous because it raises your stress hormone levels comprising learning.

Let me explain. You are a closed system—a test tube of complex biochemistry, where everything is connected to everything else. When you start to throw positive thoughts into your neurological environment, they must compete with large existing brain structures that house your negative thoughts. It creates internal conflict. The bottom line is embodied in this quote from Carl Jung, "What you resist not only persists but will grow in size."[76]

Trying to combat and change numerous negative thoughts a day with positive thoughts not only takes a tremendous amount of energy and effort, but the conflict creates high-stress hormone levels. However, mindfulness and the *Incubation Stage* of learning give us a different approach that is less combative and more joyful. We're now discovering that when you don't resist what exists in your head, but instead develop a process to become more aware of your thoughts and emotions, you lower your stress hormone levels, which in turn has a positive effect on many neurotransmitters. For example, when cortisol decreases in your brain, serotonin, a neurotransmitter connected to making order out of chaos, and enhancing your levels of peace and fulfilment, increases. With lower stress levels, you can now also access your more evolved frontal flexible brain structures (upshift), which in turn allows you to begin to recognize that you are not your thoughts or your emotions... but instead, the "real you" is the observer of your thoughts and emotions. The result is your brain climate becomes much more plastic, meaning more favorable to change.

The research on happiness helps validate the research on mindfulness. This shift to *observing* your thoughts and emotions instead of *being* your thoughts and emotions lowers internal conflict, which brings the added benefit of joy. Studies reveal that happy people don't fight their thoughts. Instead, they naturally observe and question their thoughts. What the science is illuminating is that the act of observing is the biological sweet spot that can allow you not only to feel more joyful about your life, but also creates the biological climate in your brain that accelerates the rate at which you can make transformations in your life.[77]

Let me be clear. Just because the science is revealing that there may be a better way of creating changes in your brain than pitting negative thoughts against positive thoughts, we are not saying that this mental conflict cannot produce results. It can. But at what cost? Do you want higher stress and higher biological inefficiency or lower stress and higher biological efficiency?

I'd like to share an anecdote here. When I was sixteen years old, I read Dr. Peale's *The Power of Positive Thinking*. It was the second book that I ever read cover-to-cover. As discussed, my childhood was a little messed up, and I desperately needed a more positive way to look at the world and my life. I credit Dr. Peale and his book for helping me to view the world and my life from a more positive space. It became my bible for over thirty years. In fact, I wrote to Dr. Peale when he was still alive, thanking him for being a mentor in my life. I still deeply treasure his written response. But as I began to better understand the biological differences between positive thinking and being the observer, I shifted my daily practices and now use mindfulness and being the observer of my thoughts almost exclusively. As a result, my ability to change my thoughts and feelings, as well as learn new things, has accelerated dramatically. More importantly because of this *"less stress more observer approach"* I have more energy, joy and feelings of fulfillment than at any time in my life. What a gift.

Cracking the Learning Code

Stress damages your learning, health, and your future by literally destroying your brain tissue. If we want to increase our *Individual* and *Species Adaptability/Intelligence Factors*, we must understand how the stress response is negatively affecting our ability to flourish in the 21st century.

To learn more and see our videos visit our website:

SCAN ME

https://qrco.de/thelearningcode

Conclusion

The Learning Code science reveals the simple fact that we are designed with the biological capacity to quickly and effectively transform ourselves and the world for the better. I hope that by presenting the science of learning in some detail, this book will serve as a tool by which we can accelerate the rate at which we transform our world into a place where all humans can live more joyful, fulfilled, meaningful lives.

Political unrest, deadly viruses, and the realization that our traditional learning strategies prevent us from building a secure, peaceful world have all combined to destabilize our cultures, economies, and collective consciousness. In their book Race Against the Machine MIT economists, Brynjolfsson and McAfee point out what is now apparent: "We are in the early throes of a Great Restructuring."[1] It is our hope at the Learning Institute that the knowledge you have discovered in these pages about the science of how we learn, motivate ourselves, and change our behaviors will allow you to have greater influence during this "Great Restructuring." Not only enriching your own life but also the lives of those you love and work with.

If we are to create a stable, sustainable world, the transformation of human consciousness is no longer a luxury.[2] If you would like to join us on this mission to transform the collective consciousness of the world in relation to learning, take what you have learned here and share it with others in ways that resonate with your personality. Don't try to eat the whole elephant in one bite. Chunk it down. Take the elements of The Learning Code that most resonate with you and share them in ways and places where you can make the most impact. We are here to support you.

To learn more about The Learning Code Institute and how you can support the mission to transform how the world learns, please visit our website at www.thelearningcode.com.

References

Introduction

1 https://www.drucker.institute/perspective/about-peter-drucker/
2 https://www.bfi.org/about-fuller

Chapter 1

1 *It ain't what you don't know that gets you into trouble. It's what you know for sure that just ain't so.* (2018, Nov. 18). https://quoteinvestigator.com/2018/11/18/know-trouble/
2 Berler, N. (2017). College consortia: A cooperative model that offers students greater value. *Forbes*. https://www.forbes.com/sites/noodleeducation/2017/02/01/college-consortia-a-cooperative-model-that-offers-students-greater-value/#53f26d871bae
3 Master's. (n.d.). *About our school.* https://www.masters.ab.ca/about.html
4 https://www.warriorsheart.com/?utm_source=GMB&utm_medium=organic
5 Clayton Christensen Institute. (n.d.) Disruptive innovation. https://www.christenseninstitute.org/disruptive-innovations/
6 Friedman, Z. (2019, Jan. 11). 78% of workers live paycheck to paycheck. *Forbes*. https://www.forbes.com/sites/zackfriedman/2019/01/11/live-paycheck-to-paycheck-government-shutdown/#4ef4e64f10ba
7 Sherter, A. (2019, May 23). Nearly 40% of Americans can't cover a surprise $400 expense. CBS News. https://www.cbsnews.com/news/nearly-40-of-americans-cant-cover-a-surprise-400-expense/
8 Frankel, M. (2017, July 29). 9 baby-boomer statistics that will blow you away. Madison. https://madison.com/business/investment/markets-and-stocks/baby-boomer-statistics-that-will-blow-you-away/article_8f5aed80-4b3d-5586-818a-67466833313b.html
9 Friedberg, B. A. (2019, Sep. 23). Are we in a baby boomer retirement crisis? Investopedia. https://www.investopedia.com/articles/personal-finance/032216/are-we-baby-boomer-retirement-crisis.asp
10 National Center for Health Statistics. (2016). Obesity and overweight. Centers for Disease Control and Prevention. https://www.cdc.gov/nchs/fastats/obesity-overweight.htm

11 Cleveland Clinic. (2018). 80% of Americans don't get enough exercise—and here's how much you actually need. https://health.clevelandclinic.org/80-of-americans-dont-get-enough-exercise-and-heres-how-much-you-actually-need/

12 Rowley, W. R., Bezold, C., Arikan, Y., Byrne, E., & Krohe, S. (2017). Diabetes 2030: Insights from yesterday, today, and future trends. *Population Health Management*, 20(1), 6–12. https://doi.org/10.1089/pop.2015.0181

13 Zhong, W., Maradit-Kremers, H., St. Sauver, J. L., Yawn, B. P., Ebbert, J. O., Roger, V. L., Jacobson, D. J., McGree, M. E., Brue, S. M., & Rocca, W. A. (2013). Age and sex patterns of drug prescribing in a defined American population. *Mayo Clinic Proceedings*, 88(7), 697–707. https://doi.org/10.1016/j.mayocp.2013.04.021

14 Horwitz A. V. (2010). How an age of anxiety became an age of depression. *The Milbank Quarterly*, 88(1), 112–138. https://doi.org/10.1111/j.1468-0009.2010.00591.x

15 Borchard, T. J. (2018, July 8). Why is there more depression in the world? PsychCentral. https://psychcentral.com/blog/why-is-there-more-depression-in-the-world

16 Helliwell, J. F., Layard, R., & Sachs, J. D. (Eds.). (2019). *World happiness report 2019*. https://worldhappiness.report/ed/2019/#read

17 Gallup releases new insights on the state of the global workplace. (2013, Oct. 8). *Gallup*. https://news.gallup.com/opinion/gallup/171632/gallup-releases-new-insights-state-global-workplace.aspx

18 The World Bank (Oct. 2018) Nearly Half the World Lives on Less than $5.50 a Day. Press Release accessed online at: https://www.worldbank.org/en/news/press-release/2018/10/17/nearly-half-the-world-lives-on-less-than-550-a-day

19 Krishnamurti, J. (1989) *Think of these things*. Harper One.

20 Ibid.

21 The Nation's Report Card. (2019). State average scores. https://www.nationsreportcard.gov/mathematics/states/scores/?grade=8

22 National Center for Education Statistics. (2018). *Fast facts: Dropout rates*. https://nces.ed.gov/fastfacts/display.asp?id=16

23 Strauss, V. (2011, Nov. 1). What the new NAEP test results really tell us. *Washington Post*. https://www.washingtonpost.com/blogs/answer-sheet/post/what-the-new-naep-test-results-really-tell-us/2011/11/01/gIQADSOtcM_blog.html

24 Silverman, R. E. (2012, Oct. 26). So much training, so little to show for it. *The Wall Street Journal*. https://www.wsj.com/articles/SB10001424052970204425904578072950518558328

25 Strauss, V. (2015, Oct. 15).Confirmed: Standardized testing has taken over our schools. But who's to blame? https://www.washingtonpost.com/news/answer-sheet/wp/2015/10/24/confirmed-standardized-testing-has-taken-over-our-schools-but-whos-to-blame/

26 The death and life of the great American school system. (n.d.). CRAM. https://www.cram.com/essay/The-Death-And-Life-Of-The-Great/F39KJQXKGZ3Q

27 Forum On Education In America - Bill & Melinda Gates ... (n.d.). https://www.gatesfoundation.org/media-center/speeches/2008/11/bill-gates-forum-o

28 Horsley, M., & Walker, R. (2013). *Reforming homework: practices, learning and policy*. Palgrave Macmillan.

29 Ibid.

30 National Center for Education Statistics. (2019). Trends in international mathematics and science study. https://nces.ed.gov/timss/

31 Ojalvo, H. E. (2011, September. 14) Does your homework help you learn? *The New York Times.* https://learning.blogs.nytimes.com/2011/09/14/does-your-homework-help-you-learn/

32 Kohn, A. (2007). *The homework myth: Why our kids get too much of a bad thing.* Hatchette Books.

33 McFarland, J., Hussar, B., Zhang, J., Wang, X., Wang, K., Hein, S., Diliberti, M., Forrest Cataldi, E., Bullock, Mann, F., & Barmer, A. (2019). *The condition of education 2019* (NCES 2019-144). National Center for Education Statistics. https://nces.ed.gov/pubs2019/2019144.pdf

34 Powell, F., & Kerr, E. (2019, Sep. 9). See the average college tuition in 2019-2020. *US News.* https://www.usnews.com/education/best-colleges/paying-for-college/articles/paying-for-college-infographic

35 Friedman, Z. (2019, Feb. 25). Student loan debtstatistics in 2019: A $1.5 trillion crisis. *Forbes.* https://www.forbes.com/sites/zackfriedman/2019/02/25/student-loan-debt-statistics-2019/#4a42c4bc133f

36 Rainey, M. (2018, Sep. 28). The US government's biggest financial asset: Student debt. *The Fiscal Times.* http://www.thefiscaltimes.com/2018/09/28/US-Government-s-Biggest-Financial-Asset-Student-Debt

37 American Student Assistance. (2015). *Life delayed: The impact of student debt on the daily lives of young Americans.* https://www.canisius.edu/sites/default/files/*/life_delayed_12-2015_1.pdf

38 Reich, J., & Ruipérez-Valiente, J. A. (2019). The MOOC pivot. *Science,* 363(6423),130-131 https://doi.org/10.1126/science.aav7958

39 Ibid

40 Srinivasan, V. (2014, 15 Dec.) The Myth of the Brain Game. *The Atlantic Magazine* https://www.theatlantic.com/health/archive/2014/12/the-myth-of-the-brain-game/383356/

41 Federal Trade Commission. (2016). *Lumosity to pay $2 million to settle FTC deceptive advertising charges for its "brain training" program.* https://www.ftc.gov/news-events/press-releases/2016/01/lumosity-pay-2-million-settle-ftc-deceptive-advertising-charges

42 Wilens ,T. E., Adler, L. A., Adams, J. Sgambati, S., Rotrosen, J., Sawtelle, R., Utzinger, L., & Fusillo, S. (2008). Misuse and diversion of stimulants prescribed for ADHD: a systematic review of the literature. *Journal of the American Academy of Child and Adolescent Psychiatry*, 47(1), 21-31. https://doi.org/10.1097/chi.0b013e31815a56f1.

43 Taylor, K. (2019, Oct. 21). More parents plead guilty in college admissions scandal. *New York Times*. https://www.nytimes.com/2019/10/21/us/college-admissions-scandal.html

Chapter 2

1 Kuhn, T. S. (1962). *The structure of scientific revolutions* (1st ed). University of Chicago Press.

2 Walzer, M. (1977). *Just and unjust wars*. Basic Books.

3 Einstein, A., & Infeld, L. (1938). *The evolution of physics*. Cambridge University Press.

4 Edberg, H. (2019, March 4). Gandhi's top 10 fundamentals for changing the world. *The Positivity Blog*. https://www.positivityblog.com/gandhis-top-10-fundamentals-for-changing-the-world/

5 Otero, C. P. (Ed.). (1994). *Noam Chomsky: From artificial intelligence to theology*. (2 v.). Taylor & Francis.

6 Coveney, P., & Highfield, R. (1996). *Frontiers of complexity: The search for order in a chaotic world*. Ballantine Books.

7 Malthus, T. R. (1998). *An essay on the principle of population*. Electronic Scholarly Publishing Project. (Original work published 1798). http://www.esp.org/books/malthus/population/malthus.pdf

8 Lettvin, J. Y., Maturana, H. R., McCulloch, W. S., & Pitts, W. H. (1959). What the frog's eye tells the frog's brain. *Proceedings of the IRE*, 47(11), 1940-1951. https://hearingbrain.org/docs/letvin_ieee_1959.pdf

Chapter 3

1 Dewey, J. (1902). *The child and the curriculum*. University of Chicago Press.

2 Caine, R. N., & Caine, G. (1991). *Making connections: Teaching and the human brain*. Association for Supervision and Curriculum Development.

3 *It ain't what you don't know*, 2018, loc.cit.

4 Schwartz, B. (2018) Google's search knows about over 130 trillion pages. https://searchengineland.com/googles-search-indexes-hits-130-trillion-pages-documents-263378

5 Ibid

6 https://www.quora.com/How-many-books-are-published-worldwide-every-day

7 Figlio, D., Hamersma,S. & Roth, J. (2011, April) Information Shocks and Social Networks. NBER Working Paper No. 16930

8 https://www.modernworkplacelearning.com/cild/mwl/the-effect-of-information-explosion-and-information-half-life/

9 Deenen, P. (2016, Feb) *How A Generation Lost Its Culture.* Minding The Campus: Reforming Our Universities. Accessed online at https://www.mindingthecampus.org/2016/02/02/how-a-generation-lost-its-common-culture/

10 Geitner, B. (2018, Oct 18) How Do We Learn? The History of Learning Theories. Medium. Accessed online at https://medium.com/@bgietner/how-do-we-learn-the-history-of-learning-theories-413cd1b395ac

11 Snyder, T. D. (1993). *120 years of American education: A statistical portrait.* National Center for Education Statisitcs. https://nces.ed.gov/pubs93/93442.pdf

12 Moore, G. E. (n.d.). *Moore's Law.* http://www.mooreslaw.org

13 Drucker, P. (1993) *Managing for the future.* Routeledge.

14 Perry, M. J. (2017, October 20). Fortune 500 firms 1955 v. 2017: Only 60 remain, thanks to the creative destruction that fuels economic prosperity. *American Enterprise Institute.* https://www.aei.org/carpe-diem/fortune-500-firms-1955-v-2017-only-12-remain-thanks-to-the-creative-destruction-that-fuels-economic-prosperity/

15 de Geus, A. (1998). Planning as learning. *Harvard Business Review.* https://hbr.org/1988/03/planning-as-learning.

16 Connolly, P. J. (n.d.). John Locke (1632-1704). *Internet Encyclopedia of Philosophy.* https://www.iep.utm.edu/locke/

17 *Wilder Graves Penfield.* (n.d.). The Neuro. https://www.mcgill.ca/neuro/about/history/notable-figures/wilder-graves-penfield

18 Calvin, W. H., & Ojemann, G. A. (1994) *Conversations with Neil's brain: The neural nature of thought & language.* http://williamcalvin.com/bk7/bk7.htm

19 Kahneman, D. (2011). *Thinking, fast and slow.* Farrar, Straus and Giroux Publishers.

20 Deming, W. E. (1986). *Out of the crisis.* MIT Press.

21 *In our view: GOP, Improve Health Care.* (2015, July 2). Columbian Publishing Company.

22 https://www.politifact.com/factchecks/2013/apr/26/johnny-isakson/senator-consistently-favor-vote-gun-background-che/

23 Smith, M. K. (2001) Peter Senge and the learning organization. The Encyclopedia of Pedagogy and Informal Education. https://infed.org/mobi/peter-senge-and-the-learning-organization/

24 Deming, 1986, loc.cit.

25 Meadows, D. (2008). *Thinking in systems*: A primer. Earthscan/Dunstan House.

26 Smith, 2001, loc.cit.

Chapter 4

1 Stemberg, R. (1988). *The triarchic mind: A new theory of human intelligence.* Viking Penguin.

2 Augustine of Hippo. (1955). *Confessions.* (A. C. Outler, Trans.). Westminster John Knox Press. (Original work published 400 A.D.). https://www.ling.upenn.edu/courses/hum100/augustinconf.pdf

3 Cherry, K. (2019, Oct. 17). Alfred Binet and the Simon-Binet Intelligence Scale. *Verywell Mind.* https://www.verywellmind.com/alfred-binet-biography-2795503

4 Boring, E. G. (1923). Intelligence as the Tests Test It. *New Republic*, 36, 35-37. https://brocku.ca/MeadProject/sup/Boring_1923.html

5 Gibbs, N. (2001, June 24). The EQ factor. *Time.* http://content.time.com/time/magazine/article/0,9171,133181,00.html

6 Gardner, H. (1983). *Frames of mind: The theory of multiple intelligences.* Basic Books.

7 Ibid

8 Lloyd, P., & Fernyhough, C. (Eds.) (1999). *Lev Vygotsky: Critical Assessments (Critical Assessments of Leading Psychologists).* Routledge.

9 Goleman, D. (1995). *Emotional Intelligence: Why it can matter more than IQ.* Bantam Books.

10 Cherry, K. (2019, Aug. 7). Leta Stetter Hollingworth biography. *Verywell Mind.* https://www.verywellmind.com/leta-stetter-hollingworth-biography-2795499

11 Campbell-Dollaghan, K. (2016, June 8). The long-lost study that tried to quantify creative personalities. *Fast Company.* https://www.fastcompany.com/3060634/the-long-lost-study-that-tried-to-quantify-creative-personalities

12 Fiedler, F. E., & Link, T. G. (1994). Leader intelligence, interpersonal stress, and task performance. In R. J. Sternberg & R. K Wagner (Eds.) *Mind in Context:* Interactionist perspectives on human intelligence (p. 152). Cambridge University Press.

13 Attallah, S., Ahmed, K., & Meisenberg, G. (2014). Factor structure of Wechsler Intelligence Scale for children - Third edition (WISC - III) among gifted students in the Sudan. *Mankind Quarterly,* 55(1/2), 147-170.

14 Leung, R. (2003, Nov. 21). This is 'Jeopardy!' *CBS news.* https://www.cbsnews.com/news/this-is-jeopardy-21-11-2003/

15 Feldman, R, D. (1982). *Whatever happened to the quiz kids? Perils and profits of growing up gifted.* Chicago Review Press.

16 Barrett, S. (1996). It's all in your head: *A guide to understanding your brain and boosting your brain power.* Free Spirit Press.

17 Dryden, G., & Vos, J. (1994). *The learning revolution.* Jalmar Press.

18 Arnold, K. D. (1993). *Academic achievement - A view from the top. TheIllinois Valedictorian Project.* North Central Regional Educational Laboratory. https://files.eric.ed.gov/fulltext/ED368304.pdf

19 Goleman, 1995, loc.cit.

20 Coleman, H., & Argue, M. (2015). Mediating with emotional intelligence: When "IQ" just isn't enough. *Dispute Resolution Journal,* 70(3), 15-24. http://www.onemediator.net/wp-content/uploads/2019/03/interplay-of-emotions-and-rationale-decision-making.pdf

21 Kelly, R., & Caplan, J. (1993). How Bell Labs creates star performers. *Harvard Business Review.* https://hbr.org/1993/07/how-bell-labs-creates-star-performers

22 Stemberg, R. (1988). *The triarchic mind. A new theory of human intelligence.* Viking Penguin.

23 Stanley, T.J. & Robbin, J. (2001). *The millionaire mind.* Andrews McMeel Publishing.

24 Sterman, P. (2005). Meet "Kinko" Paul Orfalea. *Ability Magazine.* https://abilitymagazine.com/past/Natalie_Glebova/Paul_Kinko.html

25 Gazzaniga, M., Irvy, R., & Mangun, G. (2014). *Cognitive neuroscience:* The biology of the mind. Norton.

26 Gardner, 1983, loc.cit.

27 Bickerton, D. (1996). *Language and human behavior.* University of Washington Press.

28 Gardner, 1983, loc.cit.

29 *Brief history of American eugenics and intelligence testing.* (n.d.). LiveLeak. https://www.liveleak.com/view?i=20d_1355781544

30 Leung, E. (n.d.). *Eugenics Education Society founded in Great Britain.* https://eugenicsarchive.ca/discover/timeline/517145a4eed5c60000000004

31 Galton, F. (1904). Eugenics: Its definition, scope, and aims. *The American Journal of Sociology,* 10(1). http://galton.org/essays/1900-1911/galton-1904-am-journ-soc-eugenics-scope-aims.htm

32 *Brief history of American eugenics*, loc.cit.

33 Darrow, C. (2011). *The eugenics cult: The American Mercury, V8, No. 30, June,* 1926. Literary Licensing, LLC. (Original work published 1926).

34 Curtis, T. (2019, Jan. 9). George Bernard Shaw was so enamored with socialism he advocated genocide to advance it. *Foundation for Economic Education.* https://fee.org/articles/george-bernard-shaw-was-so-enamored-with-socialism-he-advocated-genocide-to-advance-it/

35 Eugenics in the Colleges. (1914). *Journal of Heredity*, 5(4), 186. https://academic.oup.com/jhered/article-abstract/5/4/186/772008?redirectedFrom=fulltext

36 Cohen, A. S. (2016). Harvard's eugenics era. *Harvard Magazine.* https://harvardmagazine.com/2016/03/harvards-eugenics-era

37 Micklos, D., & Carlson, E. (2000). Engineering American society: the lesson of eugenics. *Nature Reviews*, 1, 153-158. https://www.bio.davidson.edu/people/kahales/pdfs/pdfsprot/MicklosCarlsoneugenics.pdf

38 Braswell, S. (2019, Oct. 1). Charles Lindbergh's secret life in Germany. *Ozy.* https://www.ozy.com/true-and-stories/charles-lindberghs-secret-double-life-in-germany/96337/

39 Kaelber, L. (2012). *Eugenics: Compulsory sterilization in 50 American states.* http://www.uvm.edu/%7Elkaelber/eugenics/

40 Galton, F. (1869). *Hereditary genius.* Macmillan.

41 Binet, A., & Simon, T. (1916). *The intelligence of the feeble-minded.* Williams & Wilkins company. (Original work published 1909).

42 Downey, M. (1961). *Carl Campbell Brigham: Scientist and educator.* Princeton.

43 Brigham, C. C. (1924). A study of American intelligence. *Nature*, 113, 158. https://doi.org/10.1038/113158e0

44 Troy, G. (2016, June 26). The Racist Origins of the SAT. *The Daily Beast.* http://giltroy.com/the-daily-beast/the-racist-origins-of-the-sat/

45 Galton, F. (1889). *Natural Inheritance.* Macmillan.

46 Burt, C.L. (1912). The inheritance of mental characteristics. *Eugenics Review,* 4, 168-200.

47 Eysenck, H. J., & Kamin, L. (1982). The Intelligence Controversy. *The American Journal of Psychology*, 95(2), 346-349. https://doi.org/10.2307/1422481

48 Dryden & Vos, 1994, loc.cit.

49 Goleman, 1995, loc.cit.

50 Melkikh, A. V. (2013). *Quantum field theory solves the problem of the collapse of the wave function.* https://arxiv.org/pdf/1311.0205.pdf

51 Zukav, G. (1979). *The dancing Wu Li masters: An overview of the new physics.* William Morrow and Company.

52 Rosenthal, R., & Jacobson, L. (1966). Teachers' expectancies: Determinants of pupils' IQ gains. *Psychological Reports,* 19(1), 115–118. https://doi.org/10.2466/pr0.1966.19.1.115

53 Zohar, D. & Marshall, I. (1995) *The Quantum Society: Mind, Physics and a New Social Vision.* William Morrow and Company.

54 Bornedal, P. (2019). *Nietzsche's naturalist deconstruction of truth: A world fragmented in late nineteenth-century epistemology.* Roman & Littlefield.

55 https://quoteinvestigator.com/2015/02/03/you-can/

56 Tate, K. (2016) *Government Gone Wild:* How D.C. Politicians Are Taking You for a Ride -- and What You Can Do About It. Center Street Books.

57 Grade 4 Common Core Mathmatics Assessment Guide. (1987) Boston: Houghton-Mifflin. Accessed online at https://gcsdstaff.org/4thgrade/wp-content/uploads/2014/07/Chapter-5-Study-Guide-Answers.pdf

Chapter 5

1 Winfrey, O. (2011, Oct. 19). The powerful lesson Maya Angelou taught Oprah Winfrey. [Video]. *Oprah's Life Class.* http://www.oprah.com/oprahs-lifeclass/the-powerful-lesson-maya-angelou-taught-oprah-video

2 Darwin, C. (1859). *On the Origin of Species by means of natural selection.* John Murray.

3 Malthus, 1798, loc.cit.

4 Mayr, E. (2000). Darwins influence on modern thought. *Scientific American,* 283(1), 78–83. https://www.scientificamerican.com/article/darwins-influence-on-modern-thought1/

5 Miko, I. (2008) Gregor Mendel and the principles of inheritance. *Nature Education* 1(1), 134. https://www.nature.com/scitable/topicpage/gregor-mendel-and-the-principles-of-inheritance-593/

6 *Thomas Hunt Morgan.* (1866-1945). (n.d.). DNA Learning Center. http://www.dnaftb.org/10/bio.html

7 *Oswald Theodore Avery* (1877-1955). (n.d.). DNA Learning Center. http://www.dnaftb.org/17/bio.html

8 National Human Genome Research Institute. (2019). *Deoxyribonucleic acid* (DNA) fact sheet. https://www.genome.gov/about-genomics/fact-sheets/Deoxyribonucleic-Acid-Fact-Sheet

9 Schrodinger, E. (1944). *What is life?* Cambridge University Press. http://www.whatislife.ie/downloads/What-is-Life.pdf

10 *James Watson, Francis Crick, Maurice Wilkins, and Rosalind Franklin.* (2017). Science History Institute. https://www.sciencehistory.org/historical-profile/james-watson-francis-crick-maurice-wilkins-and-rosalind-franklin

11 Stegmann U. E. (2016). 'Genetic coding' reconsidered: An analysis of actual usage. *The British Journal for the Philosophy of Science*, 67(3), 707–730. https://doi.org/10.1093/bjps/axv007

12 Ayala, F. J., & Fitch, W. M. (1997). Genetics and the origin of species: An introduction. *Proceedings of the National Academy of Sciences,* 94(15), 7691-7697. https://doi.org/10.1073/pnas.94.15.7691

13 Sternberg, R. J. (2019). A theory of adaptive intelligence and its relation to general intelligence. *Journal of Intelligence,* 7(4), 23. https://doi.org/10.3390/jintelligence7040023

14 Dobzhansky, T. (1937). *Genetics and the origin of species.* Columbia University Press.

15 Adams, J. & Rosenweig, F. (2014). Experimental microbial evolution: History and conceptual underpinnings. *Genomics,* 104(6A), 393-398. https://doi.org/10.1016/j.ygeno.2014.10.004

16 Dobzhansky, T. (1962). Mankind evolving: the evolution of the human species. *The Eugenics Review,* 54(3), 168–169.

17 Adams, J. (2004). Microbial evolution in laboratory environments. *Research in Microbiology,* 155(5), 311-318. https://doi.org/10.1016/j.resmic.2004.01.013.

18 Dawkins, R. (1995). *River out of Eden: A Darwinian view of life.* Orion Publishing Group.

19 Darwin, C. (1871). *The descent of man, and selection in relation to sex. John Murray Publisher.*

20 Darwin, C. (1872). *The expression of the emotions in man and animals.* John Murray Publisher.

21 Jantsch, E. (1980). *The Self-Organizing Universe: Scientific and Human Implications of the Emerging Paradigm of Evolution.* Pergamon Press.

22 Plutchik, R., Kellerman, H. (1981). *Emotion: Theory, research, and experience*: Vol. 1. Theories of emotion. Academic Press. https://doi.org/10.1017/S0033291700053769

23 Gardner, 1983, loc.cit.

24 Britten, R., & Davidson, E. H. (1969). Gene regulation for higher cells: a theory. *Science,* 165(3891), 349–357. https://doi.org/10.1126/science.165.3891.349

25 MacLean, P. D. (1977). An evolutionary approach to brain research on prosematic (nonverbal) behavior. In J. S. Rosenblatt & B. R. Komisaruk, *Reproductive Behavior and Evolution* (pp. 137-164). Plenum Press. https://doi.org/10.1007/978-1-4684-2625-0_7.

26 MacLean, P. D. (1982). A triangular brief on the evolution of brain and law. *Journal of Social and Biological Structures,* 5(4), 369-379. https://doi.org/10.1016/S0140-1750(82)92064-4.

27 MacLean, 1977, loc.cit.

28 Goleman, 1995, loc.cit.

Chapter 6

1 Hobson, J. A., & McCarley, R. W. (1977). The brain as a dream state generator: An activation-synthesis hypothesis of the dream process. *The American Journal of Psychiatry*, 134(12), 1335-1348. https://doi.org/10.1176/ajp.134.12.1335

2 Tierney, A. J. (1986). The evolution of learned and innate behavior: Contributions from genetics and neurobiology to a theory of behavioral evolution. *Animal Learning & Behavior*, 14, 339–348. https://doi.org/10.3758/BF03200077

3 Dawkins, 1995, loc.cit.

4 Weiss, P. A. (1968). *Dynamics of development: Experiments and inferences - selected papers on developmental biology. Academic Press.* https://doi.org/10.1016/C2013-0-12166-6

5 Daisley, J. N., Gruss, M., Rose, S. P. R., & Braun, K. (1998). Passive avoidance training and recall are associated with increased glutamate levels in the intermediate medial hyperstriatum yentrale of the day-old chick. *Neural Plasticity*, 6(3), 53-61. https://www.ncbi.nlm.nih.gov/pmc/articles/PMC2565313/pdf/NP-06-03-053.pdf

6 National Research Council (US) Committee on Research Opportunities in Biology. Opportunities in Biology. National Academies Press (US); 1989. 6, The Nervous System and Behavior. Available from: https://www.ncbi.nlm.nih.gov/books/NBK217810/

7 Committee on Research Opportunities in Biology. (1989). The nervous system and behavior. *In Opportunities in Biology* (ch. 6). National Academies Press. https://www.ncbi.nlm.nih.gov/books/NBK217810/

8 Ibid.

9 Lamarck, J. B. (1914). *Zoological Philosophy.* Cambridge University Press.

10 Ibid

11 Guy, B. (2013), Lamarckian evolution explains human brain evolution and psychiatric disorders. *Frontiers in Neuroscience*, 7, 24. https://doi.org/10.3389/fnins.2013.00224

12 Calvin, W. H. (1996). *The cerebral code: Thinking a thought in the mosaics of the mind.* MIT Press.

13 Damasio, A. (1997). *Exploring the minded brain: The Tanner lectures.* University of Michigan Press.

14 Budd G. E. (2015). Early animal evolution and the origins of nervous systems. *Philosophical Transactions of the Royal Society of London. Series B, Biological sciences*, 370(1684), 20150037. https://doi.org/10.1098/rstb.2015.0037

15 MacLean, 1982, loc.cit.

16 Hofman M. A. (2014). Evolution of the human brain: When bigger is better. *Frontiers in Neuroanatomy*, 8, 15. https://doi.org/10.3389/fnana.2014.00015
17 Lartet, L. (1868). Une sépulture des troglodytes du Périgord. *Bulletins de la Société d'Anthropologie de Paris*, 3, 335-349.
18 Ibid.
19 Mithen, S. (1999). *The prehistory of the mind: The cognitive origins of art, religion and science*. Thames Hudson.
20 Panagiotaropoulos, T., Deco, G., Kapoor, V. & Logothetis, N. (2012). Visual consciousness in the lateral prefrontal cortex. *Neuron,* 74(5), 924-935. https://doi.org/10.1016/j.neuron.2012.04.013
21 Heinrich, J. (2011). A cultural species: How culture drove human evolution. *Psychological Science Agenda*. https://www.apa.org/science/about/psa/2011/11/human-evolution
22 Leakey, L. S. B., Tobias, P. V., & Napier, J. R. (1964). A new species of the genus *Homo from Olduvai Gorge*. Nature, 202, 7-9. https://doi.org/10.1038/202007a0
23 Spoor, F., Leakey, M. G., Gathogo, P. N., Brown, F. H., Antón, S. C., McDougall, I., Kiarie, C. Manthi, F. K, & Leakey, L.N. (2007). Implications of new early Homo fossils from Ileret, east of Lake Turkana, Kenya. *Nature*, 448, 688–691. https://doi.org/10.1038/nature05986
24 Domínguez-Rodrigo, M., Pickering, T. R., Semaw, S., & Rogers, M. J. (2005). Cutmarked bones from Pliocene archaeological sites at Gona, Afar, Ethiopia: Implications for the functions of the world's oldest stone tools. *Journal of Human Evolution*, 48, 109-121. https://doi.org/10.1016/j.jhevol.2004.09.004
25 Kotchoubey, B. (2018). Human consciousness: Where is it from and what is it for. *Frontiers in Psychology*, 9, 567. https://doi.org/10.3389/fpsyg.2018.00567

Chapter 7

1 Hobson, J. A., & McCarley, R. W. (1977). The brain as a dream state generator: An activation-synthesis hypothesis of the dream process. *The American Journal of Psychiatry*, 134(12), 1335-1348. https://doi.org/10.1176/ajp.134.12.1335
2 Konorski, J. (1967). *Integrative activity of the brain: An interdisciplinary approach*. University of Chicago Press
3 Hebb, D. O. (1949). *The Organization of behavior: A neuropsychological theory* (1st ed.). Wiley.
4 Crick, F. (1995). *Astonishing hypothesis: The scientific search for the soul.* Scribner.
5 Buzsáki, G. (2010). Neural syntax: Cell assemblies, synapsembles and readers. *Neuron*, 68(3), 362-385. https://doi.org/10.1016/j.neuron.2010.09.023

6 Owens, M. T., & Tanner, K. D. (2017). Teaching as brain changing: Exploring connections between neuroscience and innovative teaching. *CBE Life Sciences Education*, 16(2), fe2. https://doi.org/10.1187/cbe.17-01-0005

7 Buzan, T. (1991). *Use both sides of your brain: New mind-mapping techniques* (3rd ed.). Plume.

8 Caine & Caine, 1991, loc.cit.

9 Building a better mouse brain. (1999, September 9). *Reuters*. https://www.wired.com/1999/09/building-a-better-mouse-brain/

10 Tsien, J. Z. (2000). Building a brainier mouse. Scientific America, 282(4), 62-68. https://www.scientificamerican.com/article/building-a-brainier-mouse/

11 Wang, R., & Reddy, P. H. (2016). Role of glutamate and NMDA receptors in Alzheimer's Disease. *Journal of Alzheimer's Disease*, 57(4), 1-7. https://doi.org/10.3233/JAD-160763

12 Meador, K. J. (2006). The basic science of memory as it applies to epilepsy. *Epilepsia*, 48(Suppl. 9), 23–25. https://doi.org/10.1111/j.1528-1167.2007.01396.x

13 Kandel, E. R. (2012). The molecular biology of memory: cAMP, PKA, CRE, CREB-1, CREB-2, and CPEB. *Molecular Brain*, 5, 14. https://doi.org/10.1186/1756-6606-5-14

14 Kandel, E. R. (1979). *Behavioural bio of Aplysia*: Origin & evolution. W.H. Freeman & Company

15 Yin, J. C. P., Del Vecchio, M., Zhou, H., & Tully, T. (1995). CREB as a memory modulator: induced expression of a dCREB2 activator isoform enhances long-term memory in Drosophila. *Cell*, 81(1), 107-115.

16 Tully, T., Bourtchouladze, R., Scott, R., & Tallman, J. (2003). Targeting the CREB pathway for memory enhancers. *Nature Reviews Drug Discovery*, 2(4), 267–277. https://doi.org/10.1038/nrd1061

17 Joshi, M. (2008, June 23). Drug may reverse mental retardation caused by genetic disorder. *TopNewsHealth*. https://www.topnews.in/health/drug-may-reverse-mental-retardation-caused-genetic-disorder-23169

18 Alberini C. M. (2009). Transcription factors in long-term memory and synaptic plasticity. *Physiological Reviews*, 89(1), 121–145. https://doi.org/10.1152/physrev.00017.2008

19 Ming, G. L., & Song, H. (2011). Adult neurogenesis in the mammalian brain: significant answers and significant questions. *Neuron*, 70(4), 687–702. https://doi.org/10.1016/j.neuron.2011.05.001

20 Griffin E. W., Mulally, S., Foley, C., Warmington, S. A., O'Mara, S. M., & Kelly, A. M. (2011). Aerobic exercise improves hippocampal function and increases BDNF in the serum of young adult males. *Physiology & Behavior*, 104(5), 934-941. https://doi.org/10.1016/j.physbeh.2011.06.005

21 Brink, L. F., Bolandzadeh, N., Nagamatsu, L. S., Hsu, C. H., Davis, J. C., Miran-Khan, K., & Liu-Ambrose, T. (2015). Aerobic exercise increases hippocampal volume in older women with probable mild cognitive impairment: A 6-month randomized controlled trial. *Journal of Sports Medicine*, 49(4), 248-254. https://doi.org/10.1136/bjsports-2013-093184

22 Committee on Physical Activity and Physical Education in the School Environment. (2013). Physical activity, fitness, and physical education: Effects on academic performance. In Educating the student body: *Taking physical activity and physical education to school* (ch. 4). https://www.ncbi.nlm.nih.gov/books/NBK201501/

Chapter 8

1 Gardner, 1983, loc.cit

2 Giedd, J. N. (2008). The teen brain: Insights from neuroimaging. *Journal of Adolescent Health*, 42(4), 335-343. https://doi.org/10.1016/j.jadohealth.2008.01.007

3 Zembrzycki, A., Stocker, A. M., Leingärtner, A., Sahara, S., Chou, S. J., Kalatsky, V., May, S. R., Stryker, M. P., & O'Leary, D. M. (2015). Genetic mechanisms control the linear scaling between related cortical primary and higher order sensory areas. *eLife,* 4, e11416. https://doi.org/10.7554/eLife.11416

4 Manzano, O. D., & Ullen, F. (2018). Same genes, different brains: Neuroanatomical differences between monozygotic twins discordant for musical training. *Cerebral Cortex*, 28(1), 387-394. https://pubmed.ncbi.nlm.nih.gov/29136105/

5 Mackenzie, J. C. (1894). The report of the Committee of Ten. T*he School Review*, 2(3), 146-155. www.jstor.org/stable/1074830

6 Magestretti, P. J. (2013). Food for thought: *What fuels brain cells?* https://www.dana.org/article/food-for-thought-what-fuels-brain-cells/

7 Kidd, J. M., Cooper, G. M., Donahue, W. F., Hayden, H. S., Sampas, N., Graves, T., Hansen, N., Teague, B., Alkan, C., Antonacci, F., Haugen, E., Zerr, T., Yamada, N. A., Tsang, P., Newman, T. L., Tuzun, E., Cheng, Z., Ebling, H. M., Tusneem, N., … Eichler, E. E. (2008). Mapping and sequencing of structural variation from eight human genomes. *Nature*, 453(7191), 56-64. https://doi.org/10.1038/nature06862

8 Alberts, B., Johnson, A., Lewis, J., Raff, M., Roberts, K., & Walter, P. (2002). General recombination. *In Molecular Biology of the Cell* (4th ed.) Garland Science. https://www.ncbi.nlm.nih.gov/books/NBK26898/

9 Cohen, P. (2006, September 4). FAQ: Genetics. *New Scientist*. https://www.newscientist.com/article/dn9965-faq-genetics/

10 Alberts, B., Johnson, A., Lewis, J., Raff, M., Roberts, K., & Walter, P. (2002). Studying gene expression and function. *In Molecular Biology of the Cell* (4th ed.) Garland Science. https://www.ncbi.nlm.nih.gov/books/NBK26818/

11 Darwin, 1859, loc.cit.

12 NIH Curriculum Supplement Series. (2007). *Understanding human genetic variation.* National Institutes of Health. https://www.ncbi.nlm.nih.gov/books/NBK20363/

13 Ewing, A. D., & Kazazian, H. H. (2011). Whole-genome resequencing allows detection of many rare LINE-1 insertion alleles in humans. Genome Research, 21(6), 985-990. https://doi.org/10.1101/gr.114777.110

14 Wills, C. (1991). *Wisdom of the Genes* (1st ed.). Basic Books

15 Dobzhansky, T. (1951). *Genetics and the origin of species* (3rd ed.). Columbia University Press

16 Edelman, G. M. (1998). Building a picture of the brain. *Daedalus*, 127(2), 37-69.

17 Herculano-Houzel, S. The human brain in numbers: A linearly scaled-up primate brain. *Frontiers in Human Neuroscience*, 3(31). https://doi.org/10.3389/neuro.09.031.2009

18 Springer, M. (1999). Learning and memory: *The brain in action.* Association for Supervision & Curriculum.

19 Posner, M. I., & Raichle, M. E. (1999). *Images of mind* (2nd ed.). Scientific American Library

20 Mulligan, L. (2017). The evolution of neurosurgery in the 1940s. *AANS Neurosurgeon*, 28(4). https://aansneurosurgeon.org/features/evolution-neurosurgery-1940s/

21 Wada J (1949). A new method for the determination of the side of cerebral speech dominance. A preliminary report of the intra-carotid injection of sodium amytal in man. *Igaku to Seibutsugaki*, Tokyo, 14, 221–222.

22 Feindel, W., Leblanc, R., & De Almeida, A. N. (2009). Epilepsy surgery: Historical highlights 1909–2009. *Epilepsia*, 50(Suppl. 3),131–151. https://doi.org/10.1111/j.1528-1167.2009.02043.x

23 Mayfield Brain and Spine. (2018). *Epilepsy surgery.* https://mayfieldclinic.com/pe-epilepsysurg.htm

24 Fiez, J. A., Raife, E. A., Balota, D. A., Schwarz, J. P., Raichle, M. E., & Petersen, S. E. (1996). A positron emission tomography study of the short-term maintenance of verbal information. *The Journal of Neuroscience*, 16(2), 808–822. https://doi.org/10.1523/JNEUROSCI.16-02-00808.1996

25 Committee on the Science of Children Birth to Age 8: Deepening and Broadening the Foundation for Success. (2015). *Transforming the workforce for children birth through age 8: A unifying foundation.* National Academies Press

Chapter 9

1 Litchka, P. R. (2019). *Leading schools with unique populations: An international perspective on school leadership.* Rowman & Littlefield.

2 Gardner, H. (1983). *Frames of mind: The theory of multiple intelligences.* Basic Books.

3 Census of Marine Life. (2011, August 24). How many species on Earth? About 8.7 million, new estimate says. *ScienceDaily.* www.sciencedaily.com/releases/2011/08/110823180459.htm

4 Gould, S. J. (1990). Wonderful life: *The Burgess Shale and the nature of history.* W. W. Norton & Company.

5 Fodor, J. A., Fodor, P. J. A., & Fodor, J. (1983). *The modularity of mind: An essay on faculty psychology.* MIT Press.

6 Stavrianos, L. S. (1991). *A global history: From prehistory to the present* (5th ed.). Prentice Hall.

7 Wilson, M. A., & Tonegawa, S. (1997). Synaptic plasticity, place cells and spatial memory: study with second generation knockouts. *Trends in neurosciences,* 20(3), 102-106. https://doi.org/10.1016/s0166-2236(96)01023-5

8 Gopnik, M. (1997). Language deficits and genetic factors. *Trends in Cognitive Sciences,* 1(1), 5–9. https://doi.org/10.1016/S1364-6613(97)01005-X

9 Peng, Y.-R., Shekhar, K., Yan, W., Herrmann, D., Sappington, A., Bryman, G. S., Zyl, T. van, Do, M. T. H., Regev, A., & Sanes, J. R. (2019). Molecular classification and comparative taxonomics of foveal and peripheral cells in primate retina. *Cell,* 176(5), 1222-1237. https://doi.org/10.1016/j.cell.2019.01.004

10 Hoffman, B. U., Baba, Y., Griffith, T. N., Mosharov, E. V., Woo, S.-H., Roybal, D. D., Karsenty, G., Patapoutian, A., Sulzer, D., & Lumpkin, E. A. (2018). Merkel cells activate sensory neural pathways through adrenergic synapses. *Neuron,* 100(6), 1401-1413. https://doi.org/10.1016/j.neuron.2018.10.034

11 Adler, E., Hoon, M. A., Mueller, K. L., Chandrashekar, J., Ryba, N. J. P., & Zuker, C. S. (2000). A novel family of mammalian taste receptors. *Cell,* 100(6), 693–702. https://doi.org/10.1016/S0092-8674(00)80705-9

12 Witelson, S. F., Kigar, D. L., & Harvey, T. (1999). The exceptional brain of Albert Einstein. *The Lancet,* 353(9170), 2149–2153. https://doi.org/10.1016/S0140-6736(98)10327-6

13 Dryden & Vos, 1994, p. 343, loc.cit.

14 National Education Association. (n.d.). *Research spotlight on recruiting & retaining highly qualified teachers.* http://ftp.arizonaea.org/tools/17018.html

15 Gardner, H. (1991). *The unschooled mind: How children think and how schools should teach.* Basic Books.

16 Bogen, J. E. (1975). *Educational aspects of hemispheric specialization.* UCLA Educator.

17 Houston, J. (1979). *Report to the President, U.S. National Commission on the International Year of the Child.* National Education Association.

18 Russell, P. (1979). *The brain book.* Dutton.

19 Pearce, J. C. (1992). *Evolution's end: Claiming the potential of our intelligence.* Harpercollins.

20 Gardner, 1991, loc.cit.

21 Buzan, T., & Buzan, B. (1993). *The mind map book: How to use radiant thinking to maximize your brain's untapped potential.* BBC Books.

22 Gardner, 1991, loc.cit.

23 Einstein, A. (1954). *Ideas and opinions.* Bonanza Books.

24 Isaacson, W. (2008). *Einstein: His life and universe.* Simon & Schuster.

25 Roberts, A. (2019). *Churchill: walking with destiny.* Penguin Books.

26 Sternberg, R. J., & Grigorenko, E. (1997). Intelligence, heredity and environment. Cambridge University Press.

27 Wroten, C. (2015, Aug. 6). Emotional engagement: How positive and negative emotions affect eLearning. *eLearning Industry.* https://elearningindustry.com/emotional-engagement-positive-negative-emotions-af

28 Dunn, R. (1990). Rita Dunn answers questions on learning styles. *Educational Leadership,* 48(2), 15.

29 Gall, F. J., & Lewis, W. (1835). *On the functions of the brain and of each of its parts.* Marsh, Capen & Lyon.

30 Dubuc, B. (n.d.). Broca's area, Wernicke's area, and other language-processing areas in the brain. *The Brain from Top to Bottom.* https://thebrain.mcgill.ca/flash/d/d_10/d_10_cr/d_10_cr_lan/d_10_cr_lan.html

31 Zilles K. (2018). Brodmann: A pioneer of human brain mapping-his impact on concepts of cortical organization. *Brain,* 141(11), 3262–3278. https://academic.oup.com/brain/article/141/11/3262/5144588

32 Feindel, W., Leblanc, R., & Almeida, A. N. D. (2009). Epilepsy surgery: Historical highlights 1909–2009. *Epilepsia,* 50(s3), 131–151. https://doi.org/10.1111/j.1528-1167.2009.02043.x

33 Communications and discussions: Sir Francis Galton. (1911). *Journal of Educational Psychology,* 2(3), 149–150. https://doi.org/10.1037/h0074146

34 Jung, C. (1971). *Psychological Types* (H. G. Baynes, Trans.; R. F. C. Hull, Rev. Trans.). Princeton University Press. (Original work published 1921).

35 Bandler, R., & Grinder, J. (1975). *The structure of magic: A book about language and therapy.* Science and Behavior Books.

36 Gardner, 1983, loc.cit.

37 Niednagel, J. P. (2001). Brain typing: *The amazing new science for understanding yourself and others*. Nightingale Conant.

Chapter 10

1 Kant, I. (1949). What is enlightenment? In C. J. Friedrich, *The philosophy of Kant* (p. 132). The Modern Library.

2 Carbo, M., Dunn, R., & Dunn, K. (1986). *Teaching students to read through their individual learning styles*. Pearson.

3 Gluckman, P. D., Beedle, A., Buklijas, T., Low, F., & Hanson, M. (2016). *Principles of evolutionary medicine*. Oxford University Press.

4 Ford-Jones, P. C. (2015). Misdiagnosis of attention deficit hyperactivity disorder: 'Normal behaviour' and relative maturity. P*aediatrics & Child Health,* 20(4), 200–202.

5 Volkow, N. D., & Swanson, J. M. (2008). Does childhood treatment of ADHD with stimulant medication affect substance abuse in adulthood? *The American Journal of Psychiatry*, 165(5), 553–555. https://doi.org/10.1176/appi.ajp.2008.08020237

6 Dryden & Vos, 1994, loc.cit.

7 Grinder, M. (2010). *Righting the educational conveyor belt*. Metamorphous.

8 Barlow, N. (Ed.) (1958). *The autobiography of Charles Darwin*. Collins Clear-Type Press.

9 Ayala & Fitch, 1997, loc.cit.

10 Bear series, part one: A bear's sense of smell. (2014, Oct. 1). *U.S. National Park Service.* https://www.nps.gov/yose/blogs/bear-series-part-one-a-bears-sense-of-smell.htm

11 Alonso, S., López, S., Izagirre, N., & de la Rúa, C. (2008). Overdominance in the human genome and olfactory receptor activity. *Molecular Biology and Evolution*, 25(5), 997–1001. https://doi.org/10.1093/molbev/msn049

12 Anthoney, T. R. (1994). Neuroanatomy and the neurologic exam: *A thesaurus of synonyms, similar-sounding non-synonyms, and terms of variable meaning*. CRC Press.

13 Courtiol, E., & Wilson, D. A. (2014). Thalamic olfaction: Characterizing odor processing in the mediodorsal thalamus of the rat. *Journal of Neurophysiology*, 111(6), 1274–1285. https://doi.org/10.1152/jn.00741.2013

14 Ibid

15 Roper, S. D., & Chaudhari, N. (2017). Taste buds: Cells, signals and synapses. *Nature Reviews. Neuroscience*, 18(8), 485–497. https://doi.org/10.1038/nrn.2017.68

16 Hazlehurst, K., & Hendry, L. (2016, July 7). Eyes on the prize: The evolution of vision. *Natural History Museum.* https://www.nhm.ac.uk/discover/eyes-on-the-prize-evolution-of-vision.html

17 Jensen, E., & McConchie, L. (2020). *Brain-based learning: Teaching the way students really learn.* Corwin Sage.

18 Mandelbrot, B. (2012). *The fractalist: Memoir of a scientific maverick.* Vintage.

19 Feynman, R. (1949). The theory of positrons. *Physical Review*, 76(6), 749–759. https://doi.org/10.1103/PhysRev.76.749

20 Watson, J. D., & Crick, F. (1953). Molecular structure of nucleic acids: A structure for deoxyribose nucleic acid. *Nature*, 171(4356), 737-738. https://doi.org/10.1038/171737a0

21 Manguel, A. (2014). *A history of reading.* Penguin Books.

22 Gardner, 1983, loc.cit.

23 Peretz, I., Vuvan, D., Lagrois, M.-É., & Armony, J. L. (2015). Neural overlap in processing music and speech. *Philosophical Transactions of the Royal Society B: Biological Sciences*, 370(1664), 20140090. https://doi.org/10.1098/rstb.2014.0090

24 Xu, Y., Gandour, J., Talavage, T., Wong, D., Dzemidzic, M., Tong, Y., Li, X., & Lowe, M. (2006). Activation of the left planum temporale in pitch processing is shaped by language experience. *Human Brain Mapping*, 27(2), 173–183. https://doi.org/10.1002/hbm.20176

25 Brown, S., Martinez, M., & Parsons, L. (2006). Music and language side by side in the brain: A PET study of the generation of melodies and sentences. *The European Journal of Neuroscience*, 23, 2791-2803. https://doi.org/10.1111/j.1460-9568.2006.04785.x.

26 Itoh, K., Fujii, Y., Suzuki, K., & Nakada, T. (2001). Asymmetry of parietal lobe activation during piano performance: A high field functional magnetic resonance imaging study. *Neuroscience Letters,* 309(1), 41–44. https://doi.org/10.1016/S0304-3940(01)02024-9

27 Dartmouth College. (2002, Dec. 13). Melodies in your mind: Researchers map brain areas that process tunes. *ScienceDaily.* www.sciencedaily.com/releases/2002/12/021213062832.htm

28 Platel, H., Price, C., Baron, J. C., Wise, R., Jany, L., Frackowiak, R., Lechevalier, B., & Eustache, F. (1997). The structural components of music perception. A functional anatomical study. *Brain,* 120(Pt 2), 229-243. https://doi.org/10.1093/brain/120.2.229.

29 Blood, A. J., & Zatorre, R. J. (2001). Intensely pleasurable responses to music correlate with activity in brain regions implicated in reward and emotion. *Proceedings of the National Academy of Sciences,* 98(20), 11818–11823. https://doi.org/10.1073/pnas.191355898

30 Hyde, K. L., Lerch, J., Norton, A., Forgeard, M., Winner, E., Evans, A. C., & Schlaug, G. (2009). Musical training shapes structural brain development. *The Journal of Neuroscience*, 29(10), 3019–3025. https://doi.org/10.1523/JNEUROSCI.5118-08.2009

31 Rauscher, F., Shaw, G., Levine, L., Wright, E., Dennis, W., & Newcomb, R. (1997). Music training causes long-term enhancement of preschool children's spatial–temporal reasoning. *Neurological Research*, 19(1), 2-8, https://doi.org/10.1080/01616412.1997.11740765

32 College Board. (2012). 2012 college-bound seniors: *Total group profile report*. https://secure-media.collegeboard.org/digitalServices/pdf/research/TotalGroup-2012.pdf

33 Ibid.

34 Sridharan, D., Levitin, D. J., Chafe, C. H., Berger, J., & Menon, V. (2007). Neural dynamics of event segmentation in music: converging evidence for dissociable ventral and dorsal networks. *Neuron*, 55(3), 521-532. https://doi.org/10.1016/j.neuron.2007.07.003

35 Civil, M. D. (2007). Using music to improve learning in mathematics. *Education and Human Development Master's Theses*. https://digitalcommons.brockport.edu/cgi/viewcontent.cgi?article=1276&context=ehd_theses

36 Leaver, R. (2015). More than simple psalm-singing in English: sacred music in early colonial America. *Yale Journal of Music & Religion*, 1(1). https://doi.org/10.17132/2377-231X.1012

37 Souster, T. (1983). Voice of God [Review of *Beethoven and the Voice of God*]. *London Review of Books,* 5(7). https://www.lrb.co.uk/the-paper/v05/n07/tim-souster/voice-of-god

38 Loewy, J. (1999). The use of music psychotherapy in the treatment of pediatric pain. Music therapy and medicine: *Theoretical and clinical approaches.* 189-206. https://www.researchgate.net/publication/292411212_The_use_of_music_psychotherapy_in_the_treatment_of_pediatric_pain

39 Nichols, J. (1998). The origins and dispersal of languages: linguistic evidence. https://www.researchgate.net/publication/238347292_The_origins_and_dispersal_of_languages_linguistic_evidence Academy of Sciences.

40 Prince, M. (2016, 9 Dec) Why Monkeys Can't Talk and What They Would Sound Like If They Could. *Science Magazine*. Accessed online at https://www.sciencemag.org/news/2016/12/why-monkeys-can-t-talk-and-what-they-would-sound-if-they-could

41 Chomsky, N. (1965). *Aspects of the theory of syntax*. M.I.T. Press.

42 Démonet, J. F., Chollet, F., Ramsay, S., Cardebat, D., Nespoulous, J. L., Wise, R., Rascol, A., & Frackowiak, R. (1992). The anatomy of phonological and semantic processing in normal subjects. *Brain*, 115(Pt 6), 1753–1768. https://doi.org/10.1093/brain/115.6.1753

43 Butterworth, B. (1999). What counts: *How every brain is hardwired for math and the mathematical brain*. Free Press.

44 Amalric, M., & Dehaene, S. (2016). Origins of the brain networks for advanced mathematics in expert mathematicians. Proceedings of the National Academy of Sciences, 113(18), 4909. https://doi.org/10.1073/pnas.1603205113

45 Dehaene, S. (2002). Précis of "the number sense" *Mind and Language,* 16(1), 16-36. https://doi.org/10.1111/1468-0017.00154

46 Higgs, M. (2001). Is there a relationship between the Myers-Briggs type indicator and emotional intelligence? Journal of Managerial Psychology, 16(7), 509–533. https://doi.org/10.1108/EUM0000000006165

47 Aron, E. N. (2002). *The highly sensitive child: Helping our children thrive when the world overwhelms them.* Harmony.

48 Goleman, D., & Boyatzis, R. (2008). Social intelligence and the biology of leadership. *Harvard Business Review.* https://hbr.org/2008/09/social-intelligence-and-the-biology-of-leadership

49 Gardner, 1983, loc.cit.

50 Goleman, 1995, loc.cit.

51 Humphrey, N. (1984). *Consciousness regained: Chapters in the development of mind.* Oxford University Press.

52 Charvet, C. J., & Finlay, B. L. (2012). Embracing covariation in brain evolution: large brains, extended development, and flexible primate social systems. *Progress in Brain Research*, 195, 71–87. https://doi.org/10.1016/B978-0-444-53860-4.00004-0

53 Brent, L., Heilbronner, S., Horvath, J. Gonzalez-Martinez, J., Ruiz-Lambides, A., Robinson, A. G., Skene, J. H. P., & Platt, M. L. (2013). Genetic origins of social networks in rhesus macaques. *Scientific Reports*, 3, 1042. https://doi.org/10.1038/srep01042

Chapter 11

1 Jerne, N. K. (1967). [Various basic problems of current immunology]. *Der Landarzt,* 43, 1526-1530.

2 Piattelli-Palmarini, M. Evolution, selection and cognition: From "learning" to parameter setting in biology and in the study of language. *Cognition.* Volume 31, Issue 1, February 1989, Pages 1-44. Cambridge, MA: MIT

3 Locke, J. (1690). *An essay concerning humane understanding.* Scolar Press.

4 Watson, J. B., & Rayner, R. (1920). Conditioned emotional reactions. *Journal of Experimental Child Psychology*, 3(1), 1–14. https://doi.org/10.1037/h0069608

5 Kant, I. (1929). *Critique of pure reason*. Bedford. (Original work published 1781).

6 Darwin, 1859, loc.cit.

7 Weiner, J. (2014). *The beak of the finch: A story of evolution in our time.* Vintage Books.

8 Bateson, W. (1913). *Problems of genetics*. Cambridge University Press.

9 Darwin, 1871, loc.cit.

10 Darwin, 1872, loc.cit.

11 James, W. (2012). *The Letters of William James, vol.1. The Atlantic Monthly Press.* (Original work published 1920).

12 James, W. (1890). *Principles of psychology.* Henry Holt and Company.

13 Spencer, H. (1864). *The Principles of biology*. Williams and Norgate.

14 Darwin, 1859, loc.cit.

15 Spencer, 1864, loc.cit.

16 Spencer H. (1897). Social statics; *or the conditions essential to happiness specified,* and the first of them developed. John Chapman.

17 Todd, D. (2014, Nov. 29). Abusing Darwin and biology. *Vancouver Sun*. https://vancouversun.com/news/staff-blogs/abusing-darwin-and-biology

18 Galbraith, J. K. (1998). *The Affluent society*. Houghton Mifflin.

19 Fisher, R., Haldane, J.B.S., & Wright, S. (2020). Random mutations and evolutionary change. *Understanding Evolution*. University of California Museum of Paleontology. https://evolution.berkeley.edu/evolibrary/article/0_0_0/history_19

20 Hockley, W. E., & Lewandowsky, S. (Eds.). (2009). *Relating theory and data: Essays on human memory in honor of Bennet B. Murdock*. Psychology Press.

21 Antibody and antigen. (n.d.). *Science clarified.* http://www.scienceclarified.com/Al-As/Antibody-and-Antigen.html

22 Nicholson, L. B. (2016). The immune system. *Essays in Biochemistry,* 60(3), 275-301. https://doi.org/10.1042/EBC20160017

23 Jerne, N. K. (1974). Towards a network theory of the immune system. *Annales D'Immunologie,*125, 373-89.

24 Jerne, 1974, loc.cit.

25 Sporns, O., & Tononi, G. (1994). *Selectionism and the brain*, volume 37. Academic Press.

26 Pinker, S. (1997). How the mind works. W. W. Norton & Company.

27 Calvin, 1996, loc.cit.

28 Gazzaniga, M. S., Ivry, R. B., & Mangun, G. R. (2019). *Cognitive neuroscience: The biology of the mind.* W.W. Norton & Company.

29 Sporns O. (2013). Structure and function of complex brain networks. *Dialogues in Clinical Neuroscience,* 15(3), 247–262.

30 Vormberg, A., Effenberger, F., Muellerleile, J., & Cuntz, H. (2017). Universal features of dendrites through centripetal branch ordering. *PLoS Computational Biology*, 13(7), e1005615. https://doi.org/10.1371/journal.pcbi.1005615

31 LeDoux, 2003, loc.cit.

32 Kagan, J. (1978). *The growth of the child: Reflections on human development.* W. W. Norton & Company.

33 Askonas, B.A., & Howard, J. G. (1997). Niels Kaj Jerne. 23 December 1911-7 October 1994. *Biographical Memoirs of Fellows of the Royal Society*, 43, 236-251.

Chapter 12

1 *They asked me what I wanted to be when I grew up. I said 'happy.'* (2013, May 29). Quote Investigator. Retrieved May 28, 2020, from https://quoteinvestigator.com/2013/05/29/grow-up-happy/

2 Frankl, V. (1946). *Man's search for meaning.* Beacon Press.

3 Levesque, C., Copeland, K. J., & Pattie, M. D. (2010). Intrinsic and Extrinsic Motivation. *International Encyclopedia of Education* (3rd ed.), 618-623. https://doi.org/10.1016/B978-0-08-044894-7.00612-6

4 Langer, S. (1942) *Philosophy in a new key: A study in the symbolism of reason, rite, and art.* Harvard University Press.

5 Gardner, H. (1984). *Art, mind, and brain: A cognitive approach to creativity.* Basic Books.

6 Postman, N. & Weingartner, C. (1969). *Teaching as a subversive activity.* Delacorte Press.

7 Hiley, B., & Peat, F. (1987). *Quantum implications. Essays in honour of David Bohm.* Routledge & Kegan Paul.

8 Wheatley, M. (2006). *Leadership and the new science: Discovering order in a chaotic world.* Berrett-Koehler Publishers.

9 Sacks, O. (1994). A new vision of the mind. In O. Sporns & G. Tononi's (Eds.) *Selectionism and the Brain: International review of neurobiology volume 37* (p. 365). Academic Press.

10 Caine, R., & Caine, G. (1999). *Making connections: Teaching and the human brain.* Addison-Wesley Pub. Co.

11 Piaget, J. (1945). *Play, dreams and imitation in Childhood.* Heinemann.

12 Senge, P. M. (1990). *The fifth discipline: The art and practice of the learning organization.* Currency

13 Harrison, S. A., & Stolurow, L. M. (Eds.). (1973). *Improving instructional productivity in higher education.* Educational Technology Publications.

14 Harrison, S. A., & Stolurow, L. M. (Eds.). (1973). *Improving instructional productivity in higher education.* Educational Technology Publications.
15 Yang, Y., Lu, J., & Zuo, Y. (2018). Changes of synaptic structures associated with learning, memory and diseases. *Brain Science Advances,* 4(2), 99–117. https://doi.org/10.26599%2FBSA.2018.2018.9050012
16 Vormberg, A., Effenberger, F., Muellerleile, J., & Cuntz, H. (2017). Universal features of dendrites through centripetal branch ordering. *PLoS computational biology,* 13(7), e1005615. https://doi.org/10.1371/journal.pcbi.1005615
17 Scheibel, A., Conrad, T., Perdue, S., Tomiyasu, U., & Wechsler, A. (1990). A quantitative study of dendrite complexity in selected areas of the human cerebral cortex. *Brain and Cognition,* 12, 85-101. https://doi.org/10.1016/0278-2626(90)90006-A
18 Bartlett, F. (1932). *Remembering: A study in experimental and social psychology.* Cambridge University Press.
19 Edelman, G. (1989). *The remembered present: A biological theory of consciousness.* Basic Books.
20 Kringelbach, M. L., & Berridge, K. C. (2010). The neuroscience of happiness and pleasure. *Social research,* 77(2), 659–678.
21 Csikszentmihalyi, M. (1990). *Flow: The psychology of optimal experience.* Harper & Row.
22 Selye, H. (1964). *From dream to discovery: On being a scientist.* McGraw-Hill.
23 Caine & Caine, 1999, loc.cit

Chapter 13

1 Kojouri, K. (2016, November 21). *Kamand Kojouri, writer, poet.* Retrieved May 28, 2020, from https://kamandkojouri.com/2016/11/21/658/
2 Villanueva, J. C. (2010, Apr. 8). How Many Atoms Are There in the Universe? *Universe Today.* https://www.universetoday.com/36302/atoms-in-the-universe/
3 Ordaz, S. J., Lenroot, R. K., Wallace, G. L., Clasen, L. S., Blumenthal, J. D., Schmitt, J. E., & Giedd, J. N. (2010). Are there differences in brain morphometry between twins and unrelated singletons? A pediatric MRI study. *Genes, Brain, and Behavior,* 9(3), 288–295. https://doi.org/10.1111/j.1601-183X.2009.00558.x
4 Harrison, S. A., & Stolurow, L. M. (Eds.). (1973). *Improving instructional productivity in higher education.* Educational Technology Publications.
5 Baddeley, A. (2014). *Essentials of human memory.* Psychology Press.
6 Pardede, P. (2008). *A review on reading theories and its implication to the teaching of reading.* https://parlindunganpardede.wordpress.com/articles/language-teaching/a-review-on-reading-theories-and-its-implication-to-the-teaching-of-reading/

7 Ashton-Warner, S. (1963). *Teacher.* Bantam Books.

8 Blumenfeld, S. L. (2002, Sep. 1). The Victims of Dick and Jane. *Chalcedon.* https://chalcedon.edu/magazine/the-victims-of-dick-and-jane

9 Freire, P. (1993). *Pedagogy of the oppressesed.* Continuum.

10 Harste, J., Woodward, V., & Burke, C. (1984). *Language Stories & Literacy Lessons.* Heinemann Educational Books.

11 Richards, T. (2002). Reproducibility of proton MR spectroscopic imaging (PEPSI): comparison of dyslexic and normal-reading children and effects of treatment on brain lactate levels during language tasks. *American Journal of Neuroradiology*, 23(10), 1678-1685.

Chapter 14

1 Jensen, E. (1998). *Teaching with the brain in mind.* Association for Supervision and Curriculum Development.

2 Jouen, A. L., Ellmore, T. M., Madden, C. J., Pallier, C., Dominey, P. F., & Ventre-Dominey, J. (2015). Beyond the word and image: Characteristics of a common meaning system for language and vision revealed by functional and structural imaging. *NeuroImage,* 106, 72–85. https://doi.org/10.1016/j.neuroimage.2014.11.024.

3 Crebbin, W. (2000). Revisioning learning-contributions of postmodernism, constructivism and neurological research. *Australian Association for Research in Education Annual Conference.* https://www.aare.edu.au/data/publications/2000/cre00462.pdf

4 Tyng, C. M., Amin, H. U., Saad, M. N. M., & Malik, A. S. (2017). The influences of emotion on learning and memory. *Frontiers in Psychology, 8*, 1454. https://doi.org/10.3389/fpsyg.2017.01454

5 Winkelman, M. J. (2017). The mechanisms of psychedelic visionary experiences: Hypotheses from evolutionary psychology. *Frontiers in Neuroscience*, 11. https://doi.org/10.3389/fnins.2017.00539

6 Sylwester, R. (1994). How emotions affect learning. *Educational Leadership*, 7. http://www.ascd.org/publications/educational-leadership/oct94/vol52/num02/How-Emotions-Affect-Learning.aspx

7 Davy, D. (2014). *Emotional life - managing your feelings to make the most of your precious time on earth: How to gain mastery over your feelings.* John Hunt Publishing.

8 Knuuttila, P. of T. E. and the P. of R. S., & Knuuttila, S. (2004). *Emotions in ancient and medieval philosophy*. Oxford University Press. https://doi.org/10.1093/0199266387.001.0001

9 Descartes, R. (2007). *Discourse on the method of rightly conducting the reason, and seeking truth in the sciences* (J. Veitch, Trans.). Jonathan Bennett. (Original work published 1637). https://www.earlymoderntexts.com/assets/pdfs/descartes1637.pdf

10 Karren, K. J., Smith, L., Gordon, K. J., & Frandsen, K. J. (2013). *Mind/body health: The effects of attitudes, emotions, and relationships* (5th ed.). Pearson.

11 LeDoux, 2003, loc.cit.

12 Ekman, P. (1999). Basic emotions. In T. Dalgleish & M. Power (Eds.), *Handbook of cognition and emotion* (ch. 3). John Wiley & Sons. https://pdfs.semanticscholar.org/305a/0ff88a70b49f745e95bd5ef004a33f5afc4a.pdf

13 Sacks, O. (1999). Making up the mind. In P. Murphy (Ed.), *Learners, learning and assessment* (pp. 58–68). Sage.

14 Rogan, M. T., & LeDoux, J. E. (1996). Emotion: Systems, cells, synaptic plasticity. *Cell,* 85(4), 469–475. https://doi.org/10.1016/s0092-8674(00)81247-7

15 Shariff, A. F., & Tracy, J. L. (2011). What are emotion expressions for? *Current Directions in Psychological Science*, 20(6), 395–399. https://doi.org/10.1177/0963721411424739

16 MacLean, P. D. (1990). *The triune brain in evolution: Role in paleocerebral functions.* Springer Science & Business Media.

17 Damasio, 1994, loc.cit.

18 Reisberg, D., & Hertel, P. (Eds.). (2004). Memory and emotion: *Series in affective science.* Oxford University Press. https://doi.org/10.1093/acprof:oso/9780195158564.001.0001

19 Roddenberry, G., & Daniels, M. (Directors). (1966). *Star Trek: The original series* [Television Series]. NBC.

20 Goleman, 1995, loc.cit.

21 Greenspan, S. I. (1997). *The growth of the mind: And the endangered origins of intelligence.* Addison-Wesley.

22 Greaves, J. (n.d.). Emotional intelligence (EQ) tests, training, certification & coaching. *TalentSmart.* https://www.talentsmart.com//articles/How-Do-You-Stack-Up--EQ-Trends-by-Industry-41897549-p-1.html

23 Evans, D. (2002). *Emotion: The science of sentiment.* Oxford University Press.

24 Beecher-Monas, E., & Garcia-Rill, E. (2019). *Fundamentals of neuroscience and the law: Square Peg, Round Hole.* Cambridge Scholars Publishing.

25 Hiser, J., & Koenigs, M. (2018). The multifaceted role of the ventromedial prefrontal cortex in emotion, decision making, social cognition, and psychopathology. *Biological Psychiatry,* 83(8), 638–647. https://doi.org/10.1016/j.biopsych.2017.10.030

26 Sylwester, 1994, loc.cit.
27 Tyng et al., 2017, loc.cit.
28 Hiser & Koenigs, 2018, loc.cit.
29 Shamay-Tsoory, S. G. (2011). The neural bases for empathy. *The Neuroscientist,* 17(1), 18-24. https://doi.org/10.1177/1073858410379268.
30 U.S. Congress. (2014). *Congressional record: Proceedings and debates of the Congress.* Government Printing Office.
31 Wagner, U., Handke, L., & Walter, H. (2015). The relationship between trait empathy and memory formation for social vs. non-social information. *BMC psychology,* 3(1), 2. https://doi.org/10.1186/s40359-015-0058-3
32 De Waal, F. B. (2012). The antiquity of empathy. *Science,* 336(6083), 874-876. https://doi.org/10.1126/science.1220999.
33 MacLean, P. D. (1978). A Mind of Three Minds: Educating the triune brain. In J. Chall & A. Mirsky (Eds.), *Education and the Brain* (pp. 308–342). University of Chicago Press.
34 King, A. J., Johnson, D. D., & Van Vugt, M. (2009). The origins and evolution of leadership. *Current Biology,* 19(19), R911-R916. https://doi.org/10.1016/j.cub.2009.07.027
35 Radzvilavicius, A. L., Stewart, A. J., & Plotkin, J. B. (2019). Evolution of empathetic moral evaluation. *eLife,* 8, e44269. https://doi.org/10.7554/eLife.44269.
36 Decety, J., & Moriguchi, Y. (2007). The empathic brain and its dysfunction in psychiatric populations: Implications for intervention across different clinical conditions. *BioPsychoSocial medicine,* 1(1), 22. https://doi.org/10.1186/1751-0759-1-22
37 Blair, R. J. R. (2007). The amygdala and ventromedial prefrontal cortex in morality and psychopathy. *Trends in Cognitive Sciences,* 11(9), 387-392. https://doi.org/10.1016/j.tics.2007.07.003
38 Soyka, M. (2011). Neurobiology of aggression and violence in schizophrenia. *Schizophrenia Bulletin,* 37(5), 913-920. https://www.ncbi.nlm.nih.gov/pmc/articles/PMC3160223/
39 Tan, S. Y., & Yip, A. (2014). António Egas Moniz (1874-1955): Lobotomy pioneer and Nobel laureate. *Singapore Medical Journal,* 55(4), 175–176. https://doi.org/10.11622/smedj.2014048
40 Sacks, O. (1995). *An anthropologist on Mars: Seven paradoxical tales.* Alfred A. Knopf.
41 Weiten, W. (2008). P*sychology: Themes and variations* (8th ed.). Cengage Learning.

42 Luciana, M. (2013). Adolescent brain development in normality and psychopathology. *Development and Psychopathology*, 25(4pt2), 1325-1345. https://doi.org/10.1017/S0954579413000643.

43 LeDoux, J. (2000). Emotion circuits in the brain. *Annual Review of Neuroscience*, 23, 155-184. https://doi.org/10.1146/annurev.neuro.23.1.155

44 Buckley, M. A., & Zimmermann, S. H. (2003). *Mentoring children and adolescents: A guide to the issues.* Greenwood Publishing Group.

45 Common Sense Media, & Rideout, V. (2011). *Zero to eight: Children's media use in America.* Common Sense Media

46 Rideout, V., & Robb, M. B. (2019). *The common sense census: Media use by tweens and teens, 2019.* https://www.commonsensemedia.org/sites/default/files/uploads/research/2019-census-8-to-18-key-findings-updated.pdf

47 LeDoux, J. (1996). *The emotional brain: The mysterious underpinnings of emotional life.* Simon and Schuster.

48 Lykken, D. T. (1995). *The antisocial personalities.* Lawrence Erlbaum Associates.

Chapter 15

1 Pert, C. B. (1997). *Molecules of emotion: The science behind mind-body medicine.* Scriber.

2 Cummings, E. (1926). 'Since feeling is first'. In *is 5*. Boni and Liveright.

3 Tyng, C. M., Amin, H. U., Saad, M. N. M., & Malik, A. S. (2017). The influences of emotion on learning and memory. *Frontiers in Psychology*, 8, 1454. https://doi.org/10.3389/fpsyg.2017.01454.

4 Cacioppo, J. T., & Freberg, L. (2012). *Discovering psychology: The science of mind.* Cengage Learning.

5 Lees, C., & Hopkins, J. (2013). Effect of aerobic exercise on cognition, academic achievement, and psychosocial function in children: a systematic review of randomized control trials. *Preventing Chronic Disease,* 10, E174. https://doi.org/10.5888/pcd10.130010

6 Hofman M. A. (2014). Evolution of the human brain: when bigger is better. *Frontiers in Neuroanatomy*, 8, 15. https://doi.org/10.3389/fnana.2014.00015

7 James, W. (1890). *The principles of psychology.* Henry Holt and Company. http://infomotions.com/sandbox/great-books-redux/corpus/html/principles.html.

8 LeDoux, 1996, loc.cit.

9 Green, E., Green, A., & Walters, E. D. (1999). Voluntary control of internal states: Psychological and physiological. *Subtle Energies and Energy Medicine,* 10, 71-88. https://pdfs.semanticscholar.org/e7c5/d791df607359cda5c327e086ad92a2b4303d.pdf

10 Descartes, 2007, loc.cit.

11 James, 1890, loc.cit.

12 Damasio, 1994, loc.cit.

13 Pert, 1997, loc.cit.

14 Green et al., 1999, loc.cit.

15 Bergland, R. (1989). *The fabric of mind.* Penguin Books.

16 LeDoux, 2003, loc.cit.

17 Sacks, O. (1999). Making up the mind. In P. Murphy (Ed.), *Learners, Learning and Assessment* (pp. 58–68). Sage.

18 Catton, W. R. (1996). The problem of denial. *Human Ecology Review,* 3(1), 53-62. http://www.zo.utexas.edu/courses/THOC/Denial.source.html. President Wilson suffers a stroke, 1919. (2002). *EyeWitness to History*. www.eyewitnesstohistory.com

19 Robertson, S. L. (n.d.). William Douglas. *The First Amendment Encyclopedia.* https://www.mtsu.edu/first-amendment/article/1328/william-douglas

20 Bechara, A., Tranel, D., Damasio, H., & Damasio, A. R. (1996). Failure to respond autonomically to anticipated future outcomes following damage to prefrontal cortex. *Cerebral Cortex*, 6(2), 215-225. https://academic.oup.com/cercor/article/6/2/215/338675

21 Bechara, A., & Damasio, A. R. (2005). The somatic marker hypothesis: A neural theory of economic decision. *Games and Economic Behavior*, 52(2), 336-372. https://doi.org/10.1016/j.geb.2004.06.010

22 Hines, L. A., Sundin, J., Rona, R. J., Wessely, S., & Fear, N. T. (2014). Posttraumatic stress disorder post Iraq and Afghanistan: Prevalence among military subgroups. *Canadian Journal of Psychiatry,* 59(9), 468–479. https://doi.org/10.1177/070674371405900903

23 Kemp, J., & Bossarte, R. (2013). *Suicide data report: 2012.* Department of Veterans Affairs, Mental Health Services, Suicide Prevention Program. https://www.va.gov/opa/docs/Suicide-Data-Report-2012-final.pdf

24 The Learning Code Institute. (2020, March 8). *Our founder—Learning Code Institute.* https://www.thelearningcode.com/our-founder

25 Kuiper, J., van Netten, S. M., Keijzer, F. A. (2011). *Explicit memory as a framework for the neural correlate of consciousness* (Doctoral dissertation, Faculty of Science and Engineering). fse.studenttheses.ub.rug.nl/9666/1/MindMatters.pdf

Chapter 16

1 Leary, T., Metzner, R., & Alpert, R. (1964). *The psychedelic experience: A manual based on the Tibetan Book of the Dead.* Citadel Press.

2 McTighe, J., & Willis, J. (2019). *Upgrade your teaching: Understanding by design meets neuroscience.* ASCD.

3 Alloway, T. P., & Alloway, R. G. (2015). *Understanding working memory.* SAGE Publications.

4 Herbert, N. (1993). *Elemental mind: Human consciousness and the new physics.* Dutton.

5 Tyng, C. M., Amin, H. U., Saad, M., & Malik, A. S. (2017). The Influences of Emotion on Learning and Memory. *Frontiers in psychology,* 8, 1454. https://doi.org/10.3389/fpsyg.2017.01454

6 Miller, G. A. (1956). The magical number seven, plus or minus two: Some limits on our capacity for processing information. *Psychological Review*, 63(2), 81-97.

7 Baddeley, A. D., & Hitch, G. (1974). Working memory. *Psychology of Learning and Motivation,* 8, 47-89. https://doi.org/10.1016/S0079-7421(08)60452-1

8 Reisberg, D. (2012). The acquisition of memories and the working-memory system. *In Cognition: Exploring the science of the mind* (5th ed.) (ch. 5). W. W. Norton & Company.

9 Cowan N. (2014). Working Memory Underpins Cognitive Development, Learning, and Education. *Educational Psychology Review*, 26(2), 197–223. https://doi.org/10.1007/s10648-013-9246-y

10 Damasio, 1994, loc.cit.

11 Kosslyn, S. M., & Koenig, O. (2010). *Wet mind: The new cognitive neuroscience.* Free Press.

12 Kosslyn, S. M., & Koenig, O. (2010). loc.cit.

13 Weissman, D. (2004). Dorsal anterior cingulate cortex resolves conflict from distracting stimuli by boosting attention toward relevant events. *Cerebral Cortex*, 15(2), 229–237. https://pubmed.ncbi.nlm.nih.gov/15238434/

14 Chiao, J. Y., Harada, T., Komeda, H., Li, Z., Mano, Y., Saito, D., et al. (2009). Neural basis of individualistic and collectivistic views of self. *Human Brain Mapping*, 30(9), 2813-2820. https://doi.org/10.1002/hbm.20707.

15 Reevy, G., Ozer, Y. M., & Ito, Y. (2010). *Encyclopedia of emotion*. Greenwood.

16 Dichter, G. S., Felder, J. N., & Bodfish, J. W. (2009). Autism is characterized by dorsal anterior cingulate hyperactivation during social target detection. *Social Cognitive and Affective Neuroscience,* 4(3), 215–226. https://www.ncbi.nlm.nih.gov/pmc/articles/PMC2728636/

17 Huang, A. X., Hughes, T. L., Sutton, L. R., Lawrence, M., Chen, X., Ji, Z., & Zeleke, W. (2017). Understanding the self in individuals with Autism Spectrum Disorders (ASD): A review of literature. *Frontiers in Psychology*, 8, 1422. https://doi.org/10.3389/fpsyg.2017.01422

18 Ledoux, J. E. (2012). Evolution of human emotion. Evolution of the primate brain *Progress in Brain Research,* 431–442. https://doi.org/10.1016/b978-0-444-53860-4.00021-0

19 Szczepanski, S. M., & Knight, R. T. (2014). Insights into human behavior from lesions to the prefrontal cortex. *Neuron*, 83(5), 1002–1018. https://doi.org/10.1016/j.neuron.2014.08.011

20 Catafu, A. M., Parellada, E., Lomena, F. J., Bernardo, M., Pavia, J., Ros, D., Setoain, J., & Gonzalez-Moncluz, E. (1994). Prefrontal and Temporal Blood Flow in Schizophrenia: Resting and Activation Technetium-99m-.HMPAO SPECT Patterns in Young Neuroleptic-Naive Patients with Acute Disease. *The Journal of Nuclear Medicine*, 35(6), 935–941.

21 Esherick, J. E. (2015). *Brain injury*. Mason Crest.

22 Squire L. R. (2009). The legacy of patient H.M. for neuroscience. *Neuron*, 61(1), 6–9. https://doi.org/10.1016/j.neuron.2008.12.023

23 Herrmann, N., Lanctôt, K. L., & Khan, L. R. (2004). The role of norepinephrine in the behavioral and psychological symptoms of dementia. *The Journal of Neuropsychiatry and Clinical Neurosciences,* 16(3), 261–276. https://doi.org/10.1176/jnp.16.3.261

24 Xing, B., Li, Y.-C., & Gao, W.-J. (2016). Norepinephrine versus dopamine and their interaction in modulating synaptic function in the prefrontal cortex. *Brain Research*, 1641, 217–233. https://doi.org/10.1016/j.brainres.2016.01.005

25 Berger, M., Gray, J. A., & Roth, B. L. (2009). The expanded biology of serotonin. *Annual review of medicine*, 60, 355–366. https://doi.org/10.1146/annurev.med.60.042307.110802

26 Yaribeygi, H., Panahi, Y., Sahraei, H., Johnston, T. P., & Sahebkar, A. (2017). The impact of stress on body function: A review. *EXCLI Journal,* 16, 1057–1072. https://doi.org/10.17179/excli2017-480

27 Gupta, M., Jin, L. & Homa, N. (2003). *Static and dynamic neural networks: From fundamentals to advanced theory.* Wiley.

Chapter 17

1 Interview with Saurabh Sharma. (2014, January 11). *Smashwords*. Retrieved May 28, 2020, from https://www.smashwords.com/interview/SaurabhSharma

2 Olds, J. (1956). Runway and maze behavior controlled by basomedial forebrain stimulation in the rat. *Journal of Comparative and Physiological Psychology*, 49(5), 507–512. https://doi.org/10.1037/h0047544

3 ibid

4 Ikemoto, S. (2007). Dopamine reward circuitry: Two projection systems from the ventral midbrain to the nucleus accumbens-olfactory tubercle complex. *Brain Research Reviews*, 56(1), 27–78. https://doi.org/10.1016/j.brainresrev.2007.05.004

5 Berridge, K. C., & Kringelbach, M. L. (2015). Pleasure systems in the brain. *Neuron*, 86(3), 646–664. https://doi.org/10.1016/j.neuron.2015.02.018

6 Teicher, M. H., Andersen, S. L., & Hostetter, J. C. (1995). Evidence for dopamine receptor pruning between adolescence and adulthood in striatum but not nucleus accumbens. Developmental Brain Research, 89(2), 167–172. https://doi.org/10.1016/0165-3806(95)00109-q

7 Bhatia, A., & Saadabadi, A. (2019). Biochemistry, *dopamine receptors.* StatPearls Publishing. https://www.ncbi.nlm.nih.gov/books/NBK538242/

8 Bromberg-Martin, E. S., Matsumoto, M., & Hikosaka, O. (2010). Dopamine in motivational control: Rewarding, aversive, and alerting. Neuron, 68(5), 815–834. https://doi.org/10.1016/j.neuron.2010.11.022

9 Herd, S., Mingus, B., & O'Reilly, R. (2010). Dopamine and self-directed learning. *In Proceedings of the 2010 conference on Biologically Inspired Cognitive Architectures 2010* (pp. 58–63). https://grey.colorado.edu/mediawiki/sites/mingus/images/c/c0/HerdMingusOReilly10.pdf

10 Robins, L. N., Davis, D. H., & Nurco, D. N. (1974). How permanent was Vietnam drug addiction? *American Journal of Public Health, 64* (12 Suppl), 38–43. https://doi.org/10.2105/ajph.64.12_suppl.38

11 Robins, L. N. (1993). Vietnam veterans rapid recovery from heroin addiction: A fluke or normal expectation? *Addiction*, 88(8), 1041–1054. https://doi.org/10.1111/j.1360-0443.1993.tb02123.x

12 Compton, P. A., Anglin, M. D., Khalsa-Denison, M. E., & Paredes, A. (1996). The D2 dopamine receptor gene, addiction, and personality: Clinical correlates in cocaine abusers. *Biological Psychiatry*, 39(4), 302–304. https://doi.org/10.1016/0006-3223(95)00476-9

13 Comings, D. E., Muhleman, D., Ahn, C., Gysin, R., & Flanagan, S. D. (1994). The dopamine D2 receptor gene: a genetic risk factor in substance abuse. *Drug and Alcohol Dependence*, 34(3), 175-180. https://doi.org/10.1016/0376-8716(94)90154-6

14 *Co-occurring and dual diagnosis treatment guide.* (2020). American Addiction Centers. https://americanaddictioncenters.org/co-occurring-disorders

15 *Why willpower doesn't work, and why more meaning does.* (2016, Apr. 27). Warriors Heart. https://www.warriorsheart.com/why-willpower-doesnt-work/

16 Blum, K., Thompson, B., Demotrovics, Z., Femino, J., Giordano, J., Oscar-Berman, M., Teitelbaum, S., Smith, D. E., Roy, A. K., Agan, G., Fratantonio, J., Badgaiyan, R. D., & Gold, M. S. (2015). The molecular neurobiology of twelve steps program & fellowship: Connecting the dots for recovery. *Journal of Reward Deficiency Syndrome*, 1(1), 46–64. https://doi.org/10.17756/jrds.2015-008

17 Why Willpower Doesn't Work - Warriors Heart. (n.d.). Retrieved from https://www.warriorsheart.com/why-willpower-doesnt-work/

Chapter 18

1 Senge, P. M. (1990). The leader's new work: Building learning organizations. *Sloan Management Review*, 7-23.

2 Pink, D. H. (2009). *Drive: The surprising truth about what motivates us.* Riverhead Books.

3 Di Domenico, S. I., & Ryan, R. M. (2017). The emerging neuroscience of intrinsic motivation: A new frontier in self-determination research. *Frontiers in Human Neuroscience*, 11, 145. https://doi.org/10.3389/fnhum.2017.00145

4 Giorgi, G., Fiz Perez, J. & Mancuso, S. (2016). The negative effects of extrinsic motivation on organizational emotional intelligence. *General Management,* 17(154).

5 Pounds, J. (2006, March 13). The great motivational myth. *Management Issues.* https://www.management-issues.com/opinion/3076/the-great-motivational-myth/

6 Root, R., & Haudan, J. (1998). Creating a company of business people. *Organization Development Journal.*

7 Rabi, I. I. (1960). *My Life and Times as a Physicist.* Claremont College.

8 Ginsberg, M. B., & Wlodkowski, R. (2015). *Motivation and Culture,* 634-637. SAGE.

9 Kohn, A. (1994). The risks of rewards. *ERIC Digest.* https://www.alfiekohn.org/article/risks-rewards/

10 Kohn, A. (1993, Sep.). Why incentive plans cannot work. *Harvard Business Review.* https://hbr.org/1993/09/why-incentive-plans-cannot-work

11 Alper, M., Durose, M. R., & Markman, J. (2018, May). *2018 update on prisoner recidivism: A 9-year follow-up period (2005-2014).* U.S. Department of Justice, Bureau of Justice Statistics. https://www.bjs.gov/content/pub/pdf/18upr9yfup0514.pdf

12 Kohn, A. (1999). *Punished by rewards: The trouble with gold stars, incentive plans, A's, praise, and other bribes* (25th ed.). Mariner Books.

13 Committee on Ways and Means. (1985, May 7). *Teenage pregnancy issues: Hearing before the subcommittee on public assistance and unemployment compensation of the committee on ways and means, House of Representatives, Ninety-ninth Congress, First Session.* U.S. Government Printing Office.

14 Hamilton, S. F., Hamilton, M. A., & Pittman, K. (2004). Principles for youth development. In S. F. In Hamilton & M. A. Hamilton (Eds.), *The youth development handbook: Coming of age in American communities.* Thousand Oaks: Sage Publications, Inc. http://ecommons.library.cornell.edu/bitstream/1813/21945/2/PrinciplesYD.pdf

15 Jones, D. (2001, May 10). Firms spend billions to fire up workers—with little luck. *USA Today*. https://worldbusiness.org/wp-content/uploads/2013/05/sp060701.pdf

16 Caine & Caine, 1991, loc.cit.

17 Kohn, 1999, loc.cit.

18 Fischer, C., Malycha, C. P., & Schafmann, E. (2019). The influence of intrinsic motivation and synergistic extrinsic motivators on creativity and innovation. *Frontiers in Psychology*, 10, 137. https://doi.org/10.3389/fpsyg.2019.00137

19 Kohn, 1999, loc.cit.

20 Brown, B., & Rosenbaum, L. (1983). Stress and competence. In J. H. Humphrey, *From Stress in Childhood* (p. 47). AMS Press.

21 Boggiano, A. K., Main, D., Flink, C., Barrett, M., Silvern, L., & Phyllis, K. (1989). A model of achievement in children: The role of controlling strategies in helplessness and affect. *Advances in Text Anxiety Research*, 13-26.

22 Stiller, J. D., & Ryan, R. M. (1992). *Teachers, parents, and student motivation: The effects of involvement and autonomy support* [conference paper].

23 de Montaigne, M. & Hazlitt, W. (1845). *The works of Michael de Montaigne: Comprising his essays, letters, journey through Germany and Italy.* C. Templemon

24 Kratcoski, P. C., & Kratcoski, L. D. (1990). *Juvenile delinquency*, (3rd ed.). Prentice Hall Publications.

25 Hermann, H. (2003). *Beneath the wheel* (M. Roloff, Trans.). Picador. Originally published in 1906.

26 Crowther, J. G. (1995). *Six great scientists: Copernicus, Galileo, Newton, Darwin, Marie Curie, Einstein. Barnes & Noble.*

27 Einstein, A., & Schilpp, P. A. (1979). *Autobiographical notes*. Open Court.

28 Jensen, E. (2006). *Enriching the brain: How to maximize every learner's potential.* John Wiley & Sons

29 Rogers, C. R. (1969). *Freedom to learn*. Merrill.

30 Brophy, J., and Kher, N. (1986). Teacher socialization as a mechanism for developing student motivation to learn. In R. S. Feldman (ed.), *The Social Psychology of Education* (pp. 257–288). Cambridge University Press.

31 Florida, R. (2004, Oct.). America's looming creativity crisis. *Harvard Business Review*. https://hbr.org/2004/10/americas-looming-creativity-crisis

32 Land, G., & Jarman, B. (1993). *Breakpoint and beyond: Mastering the future today.* HarperBusiness.

33 Amabile, T. M. (1985). Motivation and creativity: Effects of motivational orientation on creative writers. *Journal of Personality and Social Psychology*, 48(2), 393-399. https://doi.org/10.1037/0022-3514.48.2.393

34 Deci, E. L. (1975). *Intrinsic motivation.* Plenum Press. https://doi.org/10.1007/978-1-4613-4446-9

35 Caine & Caine, 1991, loc.cit.

36 Kohn, A. (1999). *Punished by rewards: The trouble with gold stars, incentive plans,* A's, praise, and other bribes (25th ed.). Mariner Books.

37 Condry, J. (1977). *Enemies of exploration: Self-initiated versus other-initiated learning. Journal of Personality and Social Psychology*, 35(7), 459–477. https://doi.org/10.1037/0022-3514.35.7.459

38 Chhangur, R., Overbeek, G., Verhagen, M., Weeland, J., Matthys, W., & Engels, R. (2015). DRD4 and DRD2 genes, parenting, and adolescent delinquency: Longitudinal evidence for a gene by environment interaction. *Journal of Abnormal Psychology,* 124, 791-802. https://doi.org/10.1037/abn0000091.

39 Caine & Caine, 1991, loc.cit.

40 Garbarino, J. (1975). The impact of anticipated reward upon cross-age tutoring. *Journal of Personality and Social Psychology,* 32(3), 421–428. https://doi.org/10.1037/h0077087

41 Pittman, T. S., Emery, J., & Boggiano, A. K. (1982). Intrinsic and extrinsic motivational orientations: Reward-induced changes in preferences for complexity. *Journal of Personality and Social Psychology*, 42(5), 789-797. https://doi.org/10.1037/0022-3514.42.5.789

42 Piaget, J. (1932). *The moral judgment of the child.* Kegan Paul, Trench, Trübner and Co.

43 Csikszentmihalyi, 1990, loc.cit.

44 Collins, S. (2015). *Neuroscience for learning and development: How to apply neuroscience and psychology for improved learning and training.* Kogan Page Publishers.

45 Anderson, R., Manoogian, S. T., & Reznick, J. S. (1976). The undermining and enhancing of intrinsic motivation in preschool children. *Journal of Personality and Social Psychology,* 34(5), 915–922. https://doi.org/10.1037/0022-3514.34.5.915

46 Roemer, J. (1992). Stars and bribes forever. *Parenting,* 58, 61.

47 Senge, 1990, loc.cit.

48 Deming, W. E. (1993). T*he new economics for industry, government, and education.* MIT Press.

49 Deci, E. L., Koestner, R., & Ryan, R. M. (1999). A meta-analytic review of experiments examining the effects of extrinsic rewards on intrinsic motivation. *Psychological Bulletin* 125, 627-668. https://doi.org/10.1037/0033-2909.125.6.627

50 Mursell, J. L. (1952). *Psychology for modern education.* W W Norton & Co.

51 Skinner, B. F., & Ferster, C. B. (1957). *Schedules of reinforcement.* Appleton-Century-Crofts. https://doi.org/10.1037/10627-000

52 Berns, G. (2008, Dec. 6). In hard times, fear can impair decision-making. *The New York Times.* https://www.nytimes.com/2008/12/07/jobs/07pre.html

53 Sylwester, R. (1995). *An educator's guide to the human brain.* Association for Supervision and Curriculum Development.

54 Gad, T. (2016). *Customer experience branding: Driving engagement through surprise and innovation.* Kogan Page.

55 Kotter, J. P., & Heskett, J. L. (1992). *Corporate culture and performance.* Free Press.

56 de Mello, A. (1984). *The song of the bird.* Image Books.

Chapter 19

1 Deci, E. L. (1978). Application of research on the effects of rewards. In M. R. Lepper & D. Greene (Eds.), *The hidden costs of rewards: New perspectives on the psychology of human motivation* (pp. 193-203). Lawrence Erlbaum & Association.

2 Miele, J. (2014, Nov. 4). Family life matters: Combating codependency. *U.S. Army.* https://www.army.mil/article/137572/family_life_matters_combating_codependency

3 Caine & Caine, 1991, loc.cit.

4 Boldt, L. G. (2004). *How to find the work you love.* Penguin Books

5 Ferguson, M. (1980). *The Aquarian conspiracy: Personal and social transformation in the 1980s.* J. P. Tarcher Inc.

6 Anderson, S. C. (1994). A critical analysis of the concept of codependency. *Social Work*, 39(6), 677-685.

7 Treadway, D. (1990). Codependency: Disease, metaphor, or fad? *Family Therapy Networker*, 14(1), 39-42.

8 Schultz W. (2010). Dopamine signals for reward value and risk: Basic and recent data. *Behavioral and Brain Functions*, 6, 24. https://doi.org/10.1186/1744-9081-6-24

9 Deci, E., & Ryan, R. M. (1985). *Perspectives in social psychology: Intrinsic motivation and self-determination in human behavior.* Plenum Press. https://doi.org/10.1007/978-1-4899-2271-7

10 Jensen, E. (1995). *Brain-based learning: The new paradigm of teaching.* Turning Point.

11 Center for Substance Abuse Treatment. (1999). How stimulants affect the brain and behavior. *Treatment Improvement Protocol (TIP) Series*, 33. https://www.ncbi.nlm.nih.gov/books/NBK64328/

12 Seligman, M. E. P. (1972). Learned helplessness. *Annual Review of Medicine,* 23(1), 407–412. https://doi.org/10.1146/annurev.me.23.020172.002203

13 Maier, S. F., & Seligman, M. E. (1976). *Learned helplessness: Theory and evidence.* Journal of Experimental Psychology: General, 105(1), 3-46. https://doi.org/10.1037/0096-3445.105.1.3

14 Chua, A. (2011). *Battle hymn of the tiger mother.* Penguin Press

15 Anwar, Y. (2013, June 18). The verdict on tiger-parenting? Studies point to poor mental health. *Berkeley News.* https://news.berkeley.edu/2013/06/18/chinese-parenting/

16 Dewar, G. (2017, June). Authoritarian parenting: What happens to the kids? *Parenting Science.* https://www.parentingscience.com/authoritarian-parenting.html

17 Baumrind, D. (2013). Authoritative parenting revisited: History and current status. In R. E. Larzelere, A. S. Morris, & A. W. Harrist (Eds.), *Authoritative parenting: Synthesizing nurturance and discipline for optimal child development* (p. 11–34). American Psychological Association. https://doi.org/10.1037/13948-002

18 Rothrauff, T. C., Cooney, T. M., & An, J. S. (2009). Remembered parenting styles and adjustment in middle and late adulthood. *The Journals of Gerontology: Series B,* 64(1), 137-146. https://www.ncbi.nlm.nih.gov/pmc/articles/PMC2654985/

19 Griffiths, J. (2014). *A country called childhood: Children and the exuberant world.* Catapult.

20 Miller, A. (1985). *For your own good: Hidden cruelty in child-rearing and the roots of violence.* Farrar, Straus and Giroux.

21 Waldersee, V. (2019, Apr. 15). Don't become a teacher, warn most teachers. *YouGov.* https://yougov.co.uk/topics/education/articles-reports/2019/04/14/dont-become-teacher-warn-most-teachers

22 Illinois New Teacher Collaborative. (n.d.). *Support new teacher mentoring and induction.* http://intc-static.education.illinois.edu/sites/default/files/NewTeacherInduction_IL.pdf

23 Valli, L. & Buese, D. The Changing Roles of Teachers in an Era of High-Stakes Accountability. *American Educational Research Journal.* Vol. 44, No. 3 (Sep., 2007), pp. 519-558

24 Hart, L. A. (1969). *The classroom disaster: How the outworn classworn classroom system cripples our schools and cheats our children, and how to replace it.* Teachers College Press.

25 Gordon, T. (1989). *Teaching children self-discipline . . . at home and at school.* Times Books.

26 Lozanov, G. (1978). *Suggestology and outlines of suggestopedy.* Gordon and Breach.

27 Watts, A. (2011). Foreword. In C. Al Huang, *Embrace tiger, return to mountain* (pp. 1-3) [Foreword]. Singing Dragon.

28 Zukav, G. (2009). *Dancing Wu Li masters: An overview of the new physics.* Harper Collins.

29 Kruse, K. (2017, March 17). This Yale psychologist says carrots and sticks don't motivate. *Forbes.* https://www.forbes.com/sites/kevinkruse/2017/03/27/this-yale-psychologist-says-carrots-and-sticks-dont-motivate/#5a0ee8b222db

30 Gordon, I., Martin, C., Feldman, R., & Leckman, J. F. (2011). Oxytocin and social motivation. *Developmental Cognitive Ceuroscience,* 1(4), 471–493. https://doi.org/10.1016/j.dcn.2011.07.007

Chapter 20

1 SANTAYANA, G., & Gouinlock, J. (2014). *The Life of Reason or The Phases of Human Progress: Reason in Religion, Volume VII, Book Three* (Wokeck M. & Coleman M., Eds.). MIT Press. doi:10.2307/j.ctt9qf8s2

2 Deming, W. E. (1986). *Out of the crisis.* MIT Press.

3 "The Divine Comedy, Inferno - "The Master Of Those Who Know"" Magill's Quotations in Context Ed. Frank N. Magill. eNotes.com, Inc. 1965 eNotes.com 4 May, 2020

4 Darwin on Aristotle. Gotthelf, A. *Journal of the History of Biology.* Vol. 32, No. 1 (Spring, 1999), pp. 3-30. Vol. 32, No. 1 (Spring, 1999), pp. 3-30

5 Garner, E. & Buchanan, R. eds. (2000) *Pluralism in theory and practice: Richard McKeon and American Philosophy.* Vanderbilt University Press.

6 Machiavelli, N. (1964). *The prince.* (M. Musa, Trans.). St. Martin's Press (Originally published in 1532)

7 Ibid.

8 Newton, I. (2010) *Mathematical principes of natural philsophy.* (A. Motte, Trans.). Snowball. (Originally published in 1686).

9 Spencer, H. (1864). *The Principles of biology.* Williams and Norgate.

10 Zohar, D. & Marshall, I. (1995) *The quantum society: Mind, physics and a new social vision.* William Morrow and Company.

11 Ibid

12 Van Horn Melton, J. (1988) *Absolutism and the eighteenth-century origins of compulsory schooling in Prussia and Austria.* Cambridge University Press.

13 Watson, J. B. (1913). Psychology as the behaviorist views it. *Psychological Review,* 20(2), 158–177.

14 Watson, J. B. (1928). *The Ways of Behaviorism.* Harper & Brothers.

15 Old Testament, Genesis 3:5

16 Skinner, B.F. (1938). *Behavior of organisms.* Appleton-Century-Crofts.

17 Sylwester, R. (1995) *A celebration of neurons: a teacher's guide to the human brain.* ASCD.

18 Pavlov, I. (1903). The Experimental Psychology and Psychopathology of Animals. *The 14th International Medical Congress,* Madrid, Spain, April 23-30, 1903.

19 Ebbinghaus, H. (1885). *Memory: A contribution to experimental psychology.* Dover.

20 Russell, P. (2010) *The brain book: Know your own mind and how to use it.* Routledge.

21 Frankl, V. (1946). *Man's search for meaning.* Beacon Press.

Chapter 21

1 Huntington, S. P. (1993). If not civilizations, what? Samuel Huntington responds to his critics. *Foreign Affairs.* https://www.foreignaffairs.com/print/node/1069597

2 Hart, 1969, loc.cit.

3 Anderson, R. C., & Pearson, P. D. (1984). *Schema-theoretic view of basic processes in reading comprehension* (Technical report No. 306). Longman's Inc.

4 Dooling, D., & Lachman, R. (1971). Effects of comprehension on retention of prose. *Journal Of Experimental Psychology,* 88(2), 216-222. https://doi.org/10.1037/h0030904

5 Fisher, K., Wandersee, J., & Moody, D. (2005). *Mapping biology knowledge.* Kluwer Academic Publishers.

6 Baird, K., & Phillips, K. (2011). *Memory connection.* Massey University

7 Rolls, E. T. (2013). The mechanisms for pattern completion and pattern separation in the hippocampus. Frontiers in Systems Neuroscience. https://doi.org/10.3389/fnsys.2013.00074

8 Schalkwijk, F. (2018). A New Conceptualization of the Conscience. *Frontiers In Psychology,* 9. doi: 10.3389/fpsyg.2018.01863

9 Bower, G., & Morrow, D. (1990). Mental models in narrative comprehension. *Science,* 247(4938), 44-48. https://doi.org/10.1126/science.2403694

10 Caine & Caine, 1991, loc.cit.

11 Caine, G. N., & Caine, R. (1997). *Unleashing the power of perceptual change: The potential of brain-based teaching.* Association for Supervision & Curriculum.

12 Dryden & Vos, 1994, loc.cit.

13 American Association for the Advancement of Sciences. (2000). AAAS's Project 2061 report: *Big biology books fail to convey big ideas.* https://www.eurekalert.org/pub_releases/2000-06/AAft-AP2r-2606100.php

14 Gardner, 1991, loc.cit.

15 Hart, L. A. (1983). *Human brain and human learning.* Longman Publishing Group.

16 Bybee, R. (1997). *Achieving scientific literacy.* Heinemann.

17 Bartlett, F. C. (1932). *Remembering: A study in experimental and social psychology.* Cambridge University Press.

18 Bodamer, J. (1947). *Die prosop-agnosie.* Arch. f. Psychiatr. u. Z. Neur., 179, 6–53. https://doi.org/10.1007/BF00352849

19 Ellis, H. D., & Florence, M. (1989). Bodamer's (1947) paper on prosopagnosia. *Cognitive Neuropsychology*, 7(2), 81-105. https://doi.org/10.1080/02643299008253437

20 Cytowic, R. E. (1993). *The man who tasted shapes.* Tarcher.

21 Hermelin, B. (2001). *Bright splinters of the mind: A personal story of research with autistic savants.* Jessica Kingsley Publishers.

22 de Groot, A.D. (1966). Perception and memory versus thought: Some old ideas and recent findings. In B. Kleinmuntz (Ed.), *Problem solving: Research, methods, and theory.* Wiley.

23 Calvin, W. H. (1998). Competing for consciousness: A Darwinian mechanism at an appropriate level of explanation. *Journal of Consciousness Studies,* 5(4), 389–404.

24 Just, M. A., Carpenter, P. A., Keller, T. A., Emery, L., Zajac, H., & Thulborn, K. R. (2001). Interdependence of nonoverlapping cortical systems in dual cognitive tasks. *Neuroimage,* 14(2), 417-426. https://doi.org/10.1006/nimg.2001.0826

25 Stigler, J. W., & Hiebert, J. (1997). Understanding and improving classroom mathematics instruction: An overview of the TIMSS video study. *Phi Delta Kappan,* 79, 14-21. https://lsc-net.terc.edu/do/paper/8037/show/use_set-math_ref.html

26 Jacobs, J., Yoshida, M., Stigler, J.W., & Fernandez, C. (1997). Japanese and American teachers' evaluations of mathematics lessons: A new technique for exploring beliefs. *The Journal of Mathmatical Behavior*, 16(1), 7.24. https://doi.org/10.1016/s0732-3123(97)90004-3

27 Jensen, 2008, loc.cit.

28 Over 70% of students don't like homework, survey says. (2017, January 16). *Ejinsight.* https://www.ejinsight.com/eji/article/id/1474553/20170116-over-70-percent-of-students-don-t-like-homework-survey-says

29 Math relevance to U.S. middle school students: *A survey commissioned by Raytheon Company.* (2012, February). https://www.raytheon.com/sites/default/files/news/rtnwcm/groups/corporate/documents/content/rtn12_studentsmth_results.pdf

30 *Advance organisers: Giving your students the big picture.* (n.d.). Evidence Based Teachers Network. https://ebtn.org.uk/big-picture-2/

31 Murphy, J., Weil, M., Hallinger, P., & Mitman, A. (1985). School effectiveness: *A conceptual framework.* The Educational Forum, 49(3), 361-374. https://doi.org/10.1080/00131728509335856

32 Donchin E. (1979). Event-related brain potentials: A tool in the study of human information processing. In H. Begleiter (Ed.), *Evoked brain potentials and behavior* (pp. 13-75). Plenum Press. https://doi.org/10.1007/978-1-4684-3462-0_2

33 Gazzaniga, M. (Ed.). (2009). *The cognitive neurosciences* (4th ed.). MIT Press.

34 Buzan, T. (2004) *The mind map handbook.* HarperCollins.

35 Shatalov, *Victor Fedorovich.* (n.d.). Clever Geek. https://clever-geek.github.io/articles/1112619/index.html

36 Jensen & Jensen, 2008, loc.cit.

Chapter 22

1 Grothe, M. (2008). *I never metaphor I didn't like: A comprehensive compilation of history's greatest analogies, metaphors, and similes.* HarperCollins US.

2 Michalowicz, M. (2012, Februrary 24). Master the ultimate sales tool: Perfect recall. *American Express.* https://www.americanexpress.com/en-us/business/trends-and-insights/articles/master-the-ultimate-sales-tool-perfect-recall/

3 Hebb, 1949, loc.cit.

4 Shatz, C. J. (1992). The developing brain. *Scientific America*, 267, 60–67. https://science.sciencemag.org/content/274/5290/1133.abstract

5 Coveney, P., & Highfield, R. (1996). *Frontiers of complexity: The search for order in a chaotic world.* Fawcett.

6 Pavlov, I. (1903). The experimental psychology and psychopathology of animals. *Animals*, 1, 47-60,

7 O'Brien, D. (2000). *Learn to remember.* Duncan Baird Publishers.

8 Sherin, J. E., & Nemeroff, C. B. (2011). Post-traumatic stress disorder: the neurobiological impact of psychological trauma. *Dialogues in Clinical Neuroscience*, 13(3), 263–278.

9 Ohashi, K., Anderson, C. M., Bolger, E. A., Khan, A., McGreenery, C. E., & Teicher, M. H. (2017). Childhood maltreatment is associated with alteration in global network fiber-tract architecture independent of history of depression and anxiety. *NeuroImage*, 150, 50–59. https://doi.org/10.1016/j.neuroimage.2017.02.037

10 Minton, K., Ogden, P., & Pain, C. (2006). *Trauma and the body: A sensorimotor approach to psychotherapy.* W. W. Norton & Company.

11 Edelman, G. (1987). *Neural Darwinism.* Basic Books.

12 Altrogge, M. (2013, July 29). The passion of Gutenberg. *The Blazing Center.* https://theblazingcenter.com/2013/07/give-me-more-of-gutenbergs-passion.html

13 Clancey, W. (1993). The strange, familiar, and forgotten: An anatomy of consciousness. *Artificial Intelligence,* 60(2), 313-356. https://doi.org/10.1016/0004-3702(93)90007-x

14 Steward, A. M. (1999). Aristotle's theory of genius examined through Reid's theory of natural signs. *Episteme,* 10, 1-9. https://digitalcommons.denison.edu/cgi/viewcontent.cgi?article=1074&context=episteme

15 Kumar, N. (2003, November). Buddha and Christ - Two Gods on the Path to Humanity. *Exotic India.* https://www.exoticindiaart.com/article/buddhaandchrist/

16 Kant, 1929, loc.cit.

17 Pagels, H. (1989). *The dreams of reason.* Bantam Books.

18 Calvin, 1996, loc.cit.

19 Wilson, J. W. (2018, March 8). *Cracking the addiction code* [Video file]. https://www.youtube.com/watch?v=LIOQRgEzXcc

20 Gardner, H., & Winner, E. (1978). The development of metaphoric competence: Implications for humanistic disciplines. *Critical Inquiry,* 5(1), 123-141.

21 Ferguson, M. (1981). *The Aquarian conspiracy: Personal growth and social transformation in the 1980s.* J. P Tarcher.

22 Caine & Caine, 1991, loc.cit.

23 Nolan, P. C. J., & McKinnon, D. H. (2003). Enhancing the middle in a New Zealand secondary school: Integration, experiential learning, and computer use. *International Journal of Educational Reform,* 12(3), 230-243. https://doi.org/10.1177/105678790301200304

24 Buzan & Buzan, 1993, loc.cit.

25 Hart, 1983, loc.cit.

Chapter 23

1 Branson, R. (n.d.). You learn by doing and by falling over. *Virgin.* https://www.virgin.com/richard-branson/you-learn-doing-and-falling-over

2 Goldin, P. R. (2018). Xunzi. *The Stanford Encyclopedia of Philosophy.* Retrieved March 24, 2020, from https://plato.stanford.edu/entries/xunzi/

3 Carrol, L. (1865). *Alice's adventures in wonderland.* Macmillan.

4 Einstein, 1954, loc.cit.

5 Wittrock, M. C. (1974). Learning as a generative process. *Educational Psychologist,* 11(2), 87-95.

6 Russell, 1979, loc.cit.

7 Sylwester, R. (2004). *How to explain a brain: An educator's handbook of brain terms and cognitive processes.* Corwin.

8 Dewey, J. (1902). *The child and the curriculum.* University of Chicago Press.

9 Crick, F. (1995). *Astonishing hypothesis: The scientific search for the soul.* Scribner.

10 Ornstein, R., & Ehrlich, P. (1989). *New world, new mind: Moving toward conscious evoluation.* Doubleday.

11 Begley, S. (1996, February 18). Your child's brain. *Newsweek.* https://www.newsweek.com/your-childs-brain-179930

12 Pearce, J. C. (1977). *Magical child: Rediscovering nature's plan for our children.* Dutton.

13 Hart, 1983, loc.cit.

14 Allen, C. G. (1894). *Post-prandial philosophy.* Chatto & Windus.

15 Jackendoff, R. (2006). FAQ: *How did language begin?* https://www.linguisticsociety.org/resource/faq-how-did-language-begin

16 Urban, T. (2013). *Putting time in perspective - updated.* https://waitbutwhy.com/2013/08/putting-time-in-perspective.html

17 Kolb, D. A., Boyatzis, R. E., & Mainemelis, C. (2001). Experiential learning theory: Previous research and new directions. In R. J. Sternberg & L.-f. Zhang (Eds.), *The educational psychology series. Perspectives on thinking, learning, and cognitive styles* (p. 227–247). Lawrence Erlbaum Associates Publishers.

Chapter 24

1 Whitman, W. (1892). 'There was a child went forth.' In *Leaves of Grass* (Deathbed ed.). David McKay. (Original work published 1855).

2 Edelman, G. M. (1992). *Bright air, brilliant fire.* BasicBooks.

3 Cherkis, J. (2015, January 28). There's a treatment for heroin addiction that actually works. *Why aren't we using it?* Huffington Post. http://projects.huffingtonpost.com/dying-to-be-free-heroin-treatment

4 Committee on Integrating the Science of Early Childhood Development. (2000). The developing brain. In J. P. Shonkoff & D. A. Phillips, (Eds.) *From neurons to neighborhoods: The science of early childhood development* (ch. 8). National Academies Press. https://www.ncbi.nlm.nih.gov/books/NBK225562/

5 Tierney, A. L., & Nelson, C. A. (2009). Brain development and the role of experience in the early years. *Zero to Three*, 30(2), 9–13.

6 Damasio, 1994, loc.cit.

7 Márquez, G. G. (2007). *Love in the Time of Cholera.* (A. Grossman, Trans.) Vintage. (Original work published 1985).

8 Ornstein, R. (1992). *The evolution of consciousness: The origins of the way we think.* Simon and Schuster.

9 Lubis, P. H., & Halim, B. (2018). *Human blastocyst formation and development.* IntechOpen. https://doi.org/10.5772/intechopen.82095

10 Ross, M. G., & Ervin, M. G. (2017). Fetal development and physiology. In S. G. Gabbe, J. R. Niebyl, J. L. Simpson, et al. (Eds), *Obstetrics: Normal and problem pregnancies* (7th ed.). Elsevier.

11 Walsh, C., & Cepko, C. L. (1992). Widespread dispersion of neuronal clones across functional regions of the cerebral cortex. *Science*, 255(5043), 434-440. https://doi.org/10.1126/science.1734520

12 von Melchner, L., Pallas, S. L., & Sur, M. (2000). Visual behaviour mediated by retinal projections directed to the auditory pathway. *Nature,* 404(6780), 871-876. https://doi.org/10.1038/35009102

13 Johannes, M., Ackeren, V., Barbero, F. M., Mattioni, S., Bottini, R., & Collignon, O. (2018). Neuronal populations in the occipital cortex of the blind synchronize to the temporal dynamics of speech. *eLife*, 7, e31640. https://www.ncbi.nlm.nih.gov/pmc/articles/PMC5790372/

14 Benettia, S., van Ackerena, M. J., Rabinia, G., Zoncaa, J., Foaa, V., Baruffaldib, F., Rezka, M., Pavania, F., Rossione, F., & Collignona, O. (2017). Functional selectivity for face processing in the temporal voice area of early deaf individuals. *Proceedings of the National Academy of Sciences of the United States of America,* 114(31), E6437-E6446. https://doi.org/10.1073/pnas.1618287114

15 Li, Z., McKercher, S. R., Cui, J., Nie, Z., Soussou, W., Roberts, A. J., Sallmen, T., Lipton, J. H., Talantova, M., Okamoto, S., & Lipton, S. A. (2008). Myocyte enhancer factor 2C as a neurogenic and antiapoptotic transcription factor in murine embryonic stem cells. *Journal of Neuroscience*, 28(26), 6557-6568. https://www.ncbi.nlm.nih.gov/pmc/articles/PMC2679693/

16 Blakensee, S. (2000, April 25) Rewired Ferrets Overturn Theories of Brain Growth. *The New York Times*. Accessed online at https://www.nytimes.com/2000/04/25/science/rewired-ferrets-overturn-theories-of-brain-growth.html

17 Ackerman, C. E. (2019). *What is neuroplasticity? A psychologist explains.* https://positivepsychology.com/neuroplasticity/

18 Schwartz, J. M., & Begley, S. (2003). Birth of the brain. In The mind and the brain: *Neuroplasticity and the power of mental force.* Harper Perennial. https://publicism.info/psychology/mind/4.html

19 Eliot, L. (2000). *What's going on in there? : How the brain and mind develop in the first five years of life.* Bantam.

20 The New York Academy of Sciences. (2014, November 11-13). *Shaping the developing brain: Prenatal through early childhood.* The Fifth Annual Aspen Brain Forum, New York Academy of Sciences.

21 Greenough, W. T., Black, J. E., & Wallace, C. S. (1987). Experience and brain development. *Child Development*, 58(3), 539-559. https://csds.qld.edu.au/sdc/Provectus/GAPP/Neuroplasticity/files/Greenough%201987%20Experience%20and%20Brain%20Development.pdf

22 Perry, B., & Pollard, M. (1997). *Altered brain development following global neglect in early childhood.* Proceedings from the annual meeting of the Society for Neuroscience, New Orleans.

23 Jacobs, B., Schall, M., & Scheibel, A. B. (1993). A quantitative dendritic analysis of wernicke's area in humans. II. Gender, hemispheric, and environmental factors. *The Journal of Comparative Neurology*, 327(1), 97-111. https://doi.org/10.1002/cne.903270108.

24 Stern Y. (2012). Cognitive reserve in ageing and Alzheimer's disease. *The Lancet. Neurology,* 11(11), 1006–1012. https://doi.org/10.1016/S1474-4422(12)70191-6

25 Cancer, A., Stievano, G., Pace, G., Colombo, A., & Antonietti, A. (2019). Cognitive processes underlying reading improvement during a rhythm-based intervention. A small-scale investigation of italian children with dyslexia. *Children (Basel, Switzerland)*, 6(8), 91. https://doi.org/10.3390/children6080091

26 Eden, G. F., Jones, K. M., Cappell, K., Gareau, L., & Wood, F. B. (2004). Neural changes following remediation in adult developmental dyslexia. *Neuron*, 44(3), 411-422. https://doi.org/10.1016/j.neuron.2004.10.019

27 Taub, E., Uswatte, G., & Pidikiti, R. (1999). Constraint-induced movement therapy: A new family of techniques with broad application to physical rehabilitation--A clinical review. *Journal of Rehabilitation Research & Development*, 36(3), 237-251. https://www.uab.edu/citherapy/images/pdf_files/citreview_jrrd99.pdf

28 Snowdon, D. A. (1997). Aging and Alzheimer's disease: Lessons from the nun study. *The Gerontologist*, 37(2), 150-156. https://doi.org/10.1093/geront/37.2.150

29 Edelman, 1989, loc.cit.

30 Dahl, G. B., Kostol, A. R., & Mogstad, M. (2014). Family welfare cultures. *The Quarterly Journal of Economics*, 129(4), 1711-1752. https://academic.oup.com/qje/article-abstract/129/4/1711/1852847?redirectedFrom=fulltext

31 Fantuzzo, J. W., & Mohr, W. K. (1999). Prevalence and effects of child exposure to domestic violence, *The Future of Children—Domestic Violence and Children*, 9(3), 21-32. https://doi.org/10.2307/1602779

32 Alper, M., Durose, M. R., & Markman, J. (2018). *2018 update on prisoner recidivism: A 9-year follow-up period (2005-2014).* Bureau of Justice Statistics. https://www.bjs.gov/content/pub/pdf/18upr9yfup0514.pdf

33 Rueve, M. E., & Welton, R. S. (2008). Violence and mental illness. *Psychiatry*, 5(5), 34-48. Retrieved from https://www.ncbi.nlm.nih.gov/pmc/articles/PMC2686644/#B27

34 Kruesi, M. J. P., Rapoport, J. L., Hamburger, S., Hibbs, E., Potter, W. Z., Lenane, M., & Brown, G. L. (1990). Cerebrospinal fluid monoamine metabolites, aggression, and impulsivity in disruptive behavior disorders of children and adolescents. *Archives of General Psychiatry*, 47(5), 419-426. https://doi.org/10.1001/archpsyc.1990.01810170019003

35 Bremner J. D. (2006). Traumatic stress: Effects on the brain. *Dialogues in Clinical Neuroscience*, 8(4), 445-461. https://www.ncbi.nlm.nih.gov/pmc/articles/PMC3181836/

36 Kotulak, R. (1997). *Inside the brain: Revolutionary discoveries of how the mind works.* Andrews McMeel Publishing.

37 Perry, B., & Szalavitz, M. (2007). Stairway to heaven: Treating children in the crosshairs of trauma. *Psychotherapy Networker*. https://www.psychotherapynetworker.org/magazine/article/667/stairway-to-heaven

38 Committee on Child Maltreatment Research. (2014). Consequences of child abuse and neglect. *In New directions in child abuse and neglect research.* National Academies Press. https://www.ncbi.nlm.nih.gov/books/NBK195987/

39 Huttenlocher, J., Haight, W., Bryk, A., Seltzer, M., & Lyons, T. (1991). Early vocabulary growth: Relation to language input and gender. *Developmental psychology*, 27(2), 236-248. https://doi.org/10.1037/0012-1649.27.2.236

40 Bugaj, K., & Brenner, B. (2011). The effects of music instruction on cognitive Development and reading skills-an overview. *Bulletin of the Council for Research in Music Education,* (189), 89-104. https://doi.org/10.5406/bulcouresmusedu.189.0089

41 Woollett, K., & Maguire, E. A. (2011). Acquiring "the Knowledge" of London's layout drives structural brain changes. *Current Biology*, 21(24), 2109–2114. https://doi.org/10.1016/j.cub.2011.11.018

42 Campbell, F., & Ramey, C. (1994). Effects of early intervention on intellectual and academic achievement: A follow-up study of children from low-income families. *Child Development*, 65(2), 684-698. https://doi.org/10.2307/1131410

43 Nielsen Company. (2018). *The Nielsen total audience report.* https://www.nielsen.com/wp-content/uploads/sites/3/2019/04/q3-2018-total-audience-report.pdf

44 Kaiser Family Foundation. (2010). *Generation M2: Media in the lives of 8- to 18-year olds.* https://www.kff.org/wp-content/uploads/2013/04/8010.pdf

Chapter 25

1 Hunter, M. (1989). Madeline Hunter in the English Classroom. *The English Journal*, 78(5), 16-18.

2 Piaget, J. (1936). *The origin of intelligence in children.* Routledge & Kegan Paul.

3 Bueno, D. (2019). Genetics and learning: How the genes influence educational attainment. *Frontiers in Psychology.* https://doi.org/10.3389/fpsyg.2019.01622

4 Gilmore, J. H., Knickmeyer, R. C., & Gao, W. (2018). Imaging structural and functional brain development in early childhood. Nature reviews. *Neuroscience,* 19(3), 123–137. https://doi.org/10.1038/nrn.2018.1

5 Romberg, A. R., & Saffran, J. R. (2010). *Statistical learning and language acquisition. Wiley interdisciplinary reviews. Cognitive Science,* 1(6), 906–914. https://doi.org/10.1002/wcs.78

6 Hofer, M. A. (1994). Hidden regulators in attachment, separation, and loss. *Monographs of the Society for Research in Child Development,* 59(2-3), 192-207. https://doi.org/10.2307/1166146

7 Rowe, M. L., Raudenbush, S. W., & Goldin-Meadow, S. (2012). The pace of vocabulary growth helps predict later vocabulary skill. *Child development,* 83(2), 508–525. https://doi.org/10.1111/j.1467-8624.2011.01710.x

8 Blakeslee, S. (1995, August 29). Behind the veil of thought: Advances in brain research; In Brain's early growth, timetable may be crucial. *The New York Times.* https://www.nytimes.com/1995/08/29/science/behind-veil-thought-advances-brain-research-brain-s-early-growthtimetable-may-be.html

9 Hofer, 1994, loc.cit.

10 Coulton, G. G. (1906). *St. Francis to Dante.* David Nutt.

11 Spitz, R.A. (1949). *The role of ecological factors in emotional development in infancy. Child Development,* 20(3), 145-155. https://doi.org/10.2307/1125870

12 Chisholm, K., Carter, M. C., Ames, E. W., & Morison, S. J. (1995). Attachment security and indiscriminately friendly behavior in children adopted from Romanian orphanages. *Development and Psychopathology,* 7(2), 283–294. https://doi.org/10.1017/s0954579400006507

13 Hofer, 1994, loc.cit.

14 Champagne F. A. (2008). Epigenetic mechanisms and the transgenerational effects of maternal care. *Frontiers in neuroendocrinology,* 29(3), 386–397. https://doi.org/10.1016/j.yfrne.2008.03.003

15 Ardiel, E. L., & Rankin, C. H. (2010). The importance of touch in development. *Paediatrics & Child Health,* 15(3), 153–156. https://www.ncbi.nlm.nih.gov/pmc/articles/PMC2865952/

16 Perry, B. (2002). Childhood experience and the expression of genetic potential: What childhood neglect tells us about nature and nurture. *Brain and Mind,* 3, 79–100. https://doi.org/10.1023/A:1016557824657

17 Haley, D. W., & Stansbury, K. (2003). Infant stress and parent responsiveness: Regulation of physiology and behavior during still-face and reunion. *Child Development,* 74(5), 1534–1546. https://doi.org/10.1111/1467-8624.00621

18 *Most learning happens in the first 3 years.* (n.d.). Centre for Educational Neuroscience. http://www.educationalneuroscience.org.uk/resources/neuromyth-or-neurofact/most-learning-happens-in-the-first-3-years/

19 Nevills, P., & Wolfe, P. (2009). *Building the reading brain*, preK-3. Corwin Press.

20 Mcmurray, B. (2007). Defusing the childhood vocabulary explosion. *Science,* 317(5838), 631. https://doi.org/10.1126/science.1144073

21 Klarowska, B. (2011, July 25). Speech and language development. *Virtual Speech Center.* https://www.virtualspeechcenter.com/blog/37/speech-and-language-development-milestones

22 Mayo Clinic Staff (2019, March) Language Development: Speech Milestones for Babies. Accessed online at: https://www.mayoclinic.org/healthy-lifestyle/infant-and-toddler-health/in-depth/language-development/art-20045163

23 Broca's Area , Wernicke's Area, and Other Language-Processing Areas in the Brain. Accessed online at https://thebrain.mcgill.ca/flash/d/d_10/d_10_cr/d_10_cr_lan/d_10_cr_lan.html

24 Ibid.

25 Hart, B., & Risley, T. R. (1992). American parenting of language-learning children: Persisting differences in family-child interactions observed in natural home environments. *Developmental Psychology,* 28(6), 1096–1105. https://doi.org/10.1037/0012-1649.28.6.1096

26 Weisleder, A., & Fernald, A. (2013). Talking to children matters: Early language experience strengthens processing and builds vocabulary. *Psychological Science,* 4(11), 2143-2152. https://doi.org/10.1177/0956797613488145

27 Shattuck, R. (1994). The forbidden experiment: *The story of the Wild Boy of Aveyron.* Kodansha International.

28 Vyshedskiy, A., Mahapatra, S., Dunn, R. (2017). Linguistically deprived children: meta-analysis of published research underlines the importance of early syntactic language use for normal brain development. *Research Ideas and Outcomes,* 3. https://doi.org/10.1101/166538

29 Fromkin, V., Krashen, S., Curtiss, S., Rigler, D., & Rigler, M. (1974). The development of language in Genie: A case of language acquisition beyond the "critical period." *Brain and Language*, 1, 81-107. https://doi.org/10.1016/0093-934X(74)90027-3

Chapter 26

1 Cajal, S. R. (1999). *Advice for a young investigator.* (N. Swanson & L. W. Swanson, Trans). MIT Press. Original work published 1897.

2 Greenough, W. T. (1986). What's special about development? Thoughts on the bases of experience-sensitive synaptic plasticity. In W. T. Greenough, & J. M. Juraska, *Developmental neuropsychobiology* (pp. 387-407). Academic Press. https://doi.org/10.1016/C2013-0-10765-9

3 Rakic, P. (1972). Mode of cell migration to the superficial layers of fetal monkey neocortex. *Journal of Comparative Neurololgy*, 145 (1), 61–83. https://doi.org/10.1002/cne.901450105.

4 Petanjek, Z., Judaš, M., Šimić, G., Rašin, M. R., Uylings, H. B. M., Rakic, P., & Kostović, I. (2011). Extraordinary neoteny of synaptic spines in the human prefrontal cortex. *Proceedings of the National Academy of Sciences,* 108(32), 13281-13286. https://doi.org/10.1073/pnas.1105108108

5 Reinert, H., & Reinert, E. S. (2006). Creative destruction in economics: Nietzsche, Sombart, Schumpeter. In Backhaus J.G., Drechsler W. (Eds.), *Friedrich Nietzsche* (1844–1900) (pp. 55-85). *The European Heritage in Economics and the Social Sciences,* 3. https://doi.org/10.1007/978-0-387-32980-2_4

6 Eliot, L. (2000). *Whats going on in there?: How the brain and mind develop in the first five years of life.* Bantam Books.

7 ibid.

8 Ackerman S. (1992). The development and shaping of the brain. In *Discovering the brain* (ch. 6). National Academies Press. https://www.ncbi.nlm.nih.gov/books/NBK234146/

9 Paus, T., Keshavan, M., & Giedd, J. N. (2008). *Why do many psychiatric disorders emerge during adolescence?* Nature reviews. Neuroscience, 9(12), 947–957. https://doi.org/10.1038/nrn2513

10 Tang, G., Gudsnuk, K., Kuo, S. H., Cotrina, M. L., Rosoklija, G., Sosunov, A., Sonders, M. S., Kanter, E., Castagna, C., Yamamoto, A., Yue, Z., Arancio, O., Peterson, B. S., Champagne, F., Dwork, A. J., Goldman, J., & Sulzer, D. (2014). Loss of mTOR-dependent macroautophagy causes autistic-like synaptic pruning deficits. *Neuron*, 83(5), 1131-1143. https://doi.org/10.1016/j.neuron.2014.07.040

11 Edelman, 1987, loc.cit.

12 Lin H, Yang Y, Chen J, Zhong X, Liu Z, et al. (2014) Congenital cataract: Prevalence and surgery age at Zhongshan Ophthalmic Center (ZOC). *PLOS ONE* 9(7), e101781. https://doi.org/10.1371/journal.pone.0101781

13 Choyce, D. P. (1959). Correction of uniocular aphakia by means of all-acrylic anteriorchamber implants: Reivew of 100 cases. *British Medical Journal*, 2(5152), 609-612. https://doi.org/10.1136/bmj.2.5152.609

14 Hubel, D. H., & Wiesel, T. N. (1959). Receptive fields of single neurones in the cat's striate cortex. *The Journal of Physiology,* 148(3), 574–591. https://doi.org/10.1113/jphysiol.1959.sp006308

15 Slater, A., & Lewis, M. (2007). Development of the senses. *In Introduction to infant development* (2nd ed.) (p. 88). Oxford University Press.

16 Lambert S. R. (2016). The timing of surgery for congenital cataracts: Minimizing the risk of glaucoma following cataract surgery while optimizing the visual outcome. *Journal of the American Association for Pediatric Ophthalmology and Strabismus*, 20(3), 191–192. https://doi.org/10.1016/j.jaapos.2016.04.003

17 Taub, E., & Wolf, S.L. (1997). Constraint-Induced (CI) Movement techniques to facilitate upper extremity use in stroke patients. *Topics in Stroke Rehabilitation*, 3, 38-61. https://www.tandfonline.com/doi/abs/10.1080/10749357.1997.11754128

18 *Edward Taub*. (n.d.). University of Alabama, Department of Psychology. Retrieved April 7, 2020 from https://www.uab.edu/cas/psychology/people/faculty/edward-taub

19 Best, C. C., & McRoberts, G. W. (2003). Infant perception of non-native consonant contrasts that adults assimilate in different ways. *Language and Speech*, 46(Pt 2-3), 183–216. https://doi.org/10.1177/00238309030460020701

20 Smith, D. G. (2018, May 4). At what age does our ability to learn a new language like a native speaker disappear? *Scientific America*. https://www.scientificamerican.com/article/at-what-age-does-our-ability-to-learn-a-new-language-like-a-native-speaker-disappear/

21 Blakeslee, S. (1986, June 24). Rapid Changes Seen in Young Brains. *The New York Times*. https://www.nytimes.com/1986/06/24/science/rapid-changes-seen-in-young-brain.html

22 Kuhl, P. K. (1993). Innate predispositions and the effects of experience in speech perception: The Native Language Magnet theory. In B. de Boysson-Bardies, S. de Schonen, P. W. Jusczyk, P. McNeilage, & J. Morton (Eds.), NATO ASI series D: Behavioural and social sciences, 69. *Developmental neurocognition: Speech and face processing in the first year of life* (p. 259–274). Kluwer Academic/Plenum Publishers. https://doi.org/10.1007/978-94-015-8234-6_22

23 Freed, J., & Parsons, L. (1997). Is there any such thing as ADD? *In Right-brained children in a left-brained world* (p. 36). Simon & Schuster.

24 Klawans, H. L. (1996). *Why Michael couldnt hit: and other tales of the neurology of sports.* W.H. Freeman.

25 Alkon, D. L. (1994). *Memory's voice: Deciphering the mind-brain code.* HarperPerennial.

26 American Academy of Child and Adolescent Psychiatry. (2016, April). *Obesity in children and teens.* https://www.aacap.org/AACAP/Families_and_Youth/Facts_for_Families/FFF-Guide/Obesity-In-Children-And-Teens-079.aspx

27 Zmigrod, L., Rentfrow, P. J., & Robbins, T. W. (2019). The partisan mind: Is extreme political partisanship related to cognitive inflexibility? *Journal of Experimental Psychology: General.* https://doi.org/10.1037/xge0000661
28 Kiyosaki, R., & Lechter, S. (1997). *Rich dad, poor dad.* TechPress.
29 Benes, F. M. (1988). Post mortem structural analyses of schizophrenic brain: Study designs and the interpretation of data. *Psychiatric Developments,* 3, 213–226.
30 Benes, F. M., Turtle, M., Khan, Y., & Farol, P. (1994). Myelination of a key relay zone in the hippocampal formation occurs in the human brain during childhood, adolescence, and adulthood. *Archives of General Psychiatry,* 51(6), 477–484. https://jamanetwork.com/journals/jamapsychiatry/article-abstract/496649
31 Schaie, K. W., Willis, S. L., & Caskie, G. I. (2004). The Seattle Longitudinal Study: Relationship between personality and cognition. *Aging, Neuropsychology, and Cognition,* 11(2-3), 304–324. https://doi.org/10.1080/13825580490511134
32 Madigan, E. (2014, October 23). Old masters at the top of their game. *New York Times.* https://www.nytimes.com/interactive/2014/10/23/magazine/old-masters-at-top-of-their-game.html

Chapter 27

1 Criswell, W. (n.d.). *Warren Criswell.* https://www.warrencriswell.com/
2 Gray, P. (2015, May 5). Early academic training produces long-term harm. *Psychology Today.* https://www.psychologytoday.com/us/blog/freedom-learn/201505/early-academic-training-produces-long-term-harm
3 *What grown-ups understand about child development: A national benchmark survey.* (1997). https://static.virtuallabschool.org/activities/physical/IT.Phys_2.Milestones_A2.GrwnupsUnderstandChildDev.pdf
4 Blakeslee, S. (1997, April 17). Studies show talking with infants shapes basis of ability to think. *The New York Times.* https://www.nytimes.com/1997/04/17/us/studies-show-talking-with-infants-shapes-basis-of-ability-to-think.html
5 Doman, G., & Doman, J. (1994). *How to multiply your baby's intelligence.* Avery Publishing Group.
6 Park, M. (2009). Study: Want a smart baby? TV's not going to help. *CNN.* https://edition.cnn.com/2009/HEALTH/03/03/babies.watch.TV/index.html
7 Strauss, V. (2015, August 17). Why pushing kids to learn too much too soon is counterproductive. *The Washington Post.* https://www.washingtonpost.com/news/answer-sheet/wp/2015/08/17/why-pushing-kids-to-learn-too-much-too-soon-is-counterproductive/
8 Piaget, 1936, loc.cit.
9 Miller, B. L., & Cummings, J. L. (1999). *The human frontal lobes: Functions and disorders.* Guilford Publications

10 Jarvis, M., & Chandler, E. (2001). *Angles on child psychology.* Nelson Thornes.

11 Armor, D. J., & Sousa, S. (2014). The dubious promise of universal preschool. *National Affairs.* https://www.nationalaffairs.com/publications/detail/the-dubious-promise-of-universal-preschool

12 Barnett, W. S., & Hustedt, J. T. (2005). Head Start's lasting benefits. *Infants & Young Children,* 18(1), 16–24. https://doi.org/10.1097/00001163-200501000-00003.

13 Protzko, J. (2015). The environment in raising early intelligence: A meta-analysis of the fadeout effect. *Intelligence*, 53, 202-210. https://doi.org/10.1016/j.intell.2015.10.006

14 Herrnstein, R. J., & Murray, C. (1994). *The bell curve: Intelligence and class structure in American life.* Free Press.

15 Zigler, E. (1992). *Head Start: The inside story of America's most successful educational experiment.* Harper Collins.

16 Rosenberg, D. (1997, February 28). Pots, blocks, and socks. *Newsweek.* https://www.newsweek.com/pots-blocks-amp-socks-174966

17 Cullen, F. T., & Jonson, C. L. (2011). *Correctional theory: Context and consequences.* SAGE.

18 Hirsh Pasek, K. (1991), Pressure or challenge in preschool? How academic environments affect children. *New Directions for Child and Adolescent Development,* 1991(53), 39-46. https://doi.org/10.1002/cd.23219915307

19 Elkind, D. (1981). *The hurried child: Growing up too fast too soon.* Addison-Wesley Publishing Company.

20 Elkind, D. (1987). *Miseducation: Preschoolers at risk.* Knopf.

21 Diamond, M, & Hopson, J. (1998). *Magic trees of the mind: How to nuture your child's intelligence, creativity, and healthy emotions from birth through adolescence.* Penguin Putnam, Inc.

22 Fischer, K. W., & Bullock, D. (1984). Cognitive development in school-age children: Conclusions and new directions. In W. A. Collins (Ed.), *Development during middle childhood: The years from six to twelve* (ch. 3). https://www.ncbi.nlm.nih.gov/books/NBK216774/

23 Quart, A. (2006). Hothouse kids: *How the pressure to succeed threatens childhood.* Penguin Random House.

24 White, R. E. (2013). *The power of play: A research summary on play and learning.* Minnesota Children's Museum. https://www.childrensmuseums.org/images/MCMResearchSummary.pdf

25 Developing balance, coordination and fine motor skills on the playground. (2018, December 28). *Playworld.* https://playworld.com/blog/playgrounds-develop-balance-coordination-motor-skills/

26 Marzollo, J., & Lloyd, J. (1977). *Learning through play*. Pengiun.

27 Pearce, 1992, loc.cit.

28 Pearce, 1977, loc.cit.

29 Bruer, J. (1999). *The myth of the first three years: A new understanding of early brain development and lifelong learning*. Free Press.

30 Coulter, D. (2014) *Original Mind: Uncovering Your Natural Brilliance.* Sounds True Publishing.

31 Bruer, 1999, loc.cit.

32 Jensen, E. (1998). *Teaching with the brain in mind.* Association for Supervision and Curriculum Development.

33 Bueno, D. (2019). Genetics and learning: How the genes influence educational attainment. *Frontiers in Psychology*, 10, 1622. https://doi.org/10.3389/fpsyg.2019.01622

34 Einstein, A., & Harris, A. (1934). *The world as I see it.* Covici, Friede.

35 Jensen, 1995, loc.cit.

36 Ramsden, S., Richardson, F. M., Josse, G., Thomas, M. S. C., Ellis, C., Shakeshaft, C., Seghier, M. L., & Price, C. J. (2011). Verbal and non-verbal intelligence changes in the teenage brain. *Nature,* 479, 113–116. https://doi.org/10.1038/nature10514

37 Medina, J. (2010). *Brain rules for baby: How to raise a smart and happy child from zero to five.* Pear Press.

38 Zimmerman, F. J., Christakis, D. A., & Meltzoff, A. N. (2007). Associations between media viewing and language development in children under age 2 years. *The Journal of Pediatrics*, 151(4), 364 - 368. https://doi.org/10.1016/j.jpeds.2007.04.071

39 Richert, R. A., Robb, M. B., Fender, J. G., & Wartella, E. (2010). Word learning from baby videos. *Archives of Pediatrics & Adolescent Medicine*, 164(5), 432-437. https://doi.org/10.1001/archpediatrics.2010.24

40 Brown, A. (2011). Media use by children younger than 2 years. *Pediatrics,*128(5). https://doi.org/10.1542/peds.2011-1753

41 Ward, V. (2013, April 21). Toddlers becoming so addicted to iPads they require therapy. *The Telegraph.* https://www.telegraph.co.uk/technology/10008707/Toddlers-becoming-so-addicted-to-iPads-they-require-therapy.html

42 Brown, 2011, loc.cit.

43 Lewin, T. (2009, October 23). No Einstein in your crib? Get a refund. *The New York Times.* https://www.nytimes.com/2009/10/24/education/24baby.html

44 Ward, 2013, loc.cit.

45 Parker, C. B. (2014, October 20). Scientific evidence does not support the brain game claims, Stanford scholars say. *Stanford University News.* https://news.stanford.edu/news/2014/october/brain-games-carstensen-102014.html

46 Smith, C. (2020, February 2). Interesting Luminosity statistics and facts (2020) | By the numbers. *DMR.* https://expandedramblings.com/index.php/lumosity-statistics-and-facts/

47 Caporale, N., & Dan, Y. (2008). Spike timing–dependent plasticity: A Hebbian learning rule. *The Annual Review of Neuroscience*, 31, 25–46. https://doi.org/10.1146/annurev.neuro.31.060407.125639

48 Plagens, P. (1997, February 28). Little artists and athletes. *Newsweek.* https://www.newsweek.com/little-artists-and-athletes-174958

Chapter 28

1 Louv, R. (2005). *Last child in the woods: Saving our children from nature-deficit disorder.* Algonquin Books.

2 Schmall, T. (2018, August 13).Americans spend half their lives in front of screens. *New York Post.* https://nypost.com/2018/08/13/americans-spend-half-their-lives-in-front-of-screens/

3 Fisher, N. (2019, January 24). How much time Americans spend in front of screens will terrify you. *Forbes.* https://www.forbes.com/sites/nicolefisher/2019/01/24/how-much-time-americans-spend-in-front-of-screens-will-terrify-you/#5fda7a391c67

4 Rideout, v. (2010) Generation M2: Media in the Lives of 8- to 18-Year-Olds. Oakland, CA: Kaiser Family Foundation. Accessed online at https://www.kff.org/other/report/generation-m2-media-in-the-lives-of-8-to-18-year-olds/

5 Rideout, V. (2017). *The Common Sense census: Media use by kids age zero to eight.* https://www.commonsensemedia.org/sites/default/files/uploads/research/csm_zerotoeight_fullreport_release_2.pdf

6 https://www.commonsensemedia.org/sites/default/files/uploads/research/csm_zerotoeight_fullreport_release_2.pdf

7 Gentry, J., & Eron, L. D. (1993). American Psychological Association Commission on violence and youth. *American Psychologist*, 48(2), 89. https://doi.org/10.1037/0003-066X.48.2.89

8 Strasburger, V. C., & Wilson, B. J. (2014). Television violence: Sixty years of research. In D. A. Gentile (Ed.), *Media violence and children: A complete guide for parents and professionals* (p. 135–177). Praeger/ABC-CLIO.

9 Children and the media. (1999). *Paediatrics & Child Health*, 4(5), 350–362. https://academic.oup.com/pch/article/4/5/350/2666579

10 Committee on Public Education. (2001). Children, adolescents, and television. *Pediatrics,* 107(2) 423-426. https://doi.org/10.1542/peds.107.2.423

11 *Americans don't want to unplug from phones while on vacation, despite latest digital detox trend.* (2018, May 17). Asurion. https://www.asurion.com/about/press-releases/americans-dont-want-to-unplug-from-phones-while-on-vacation-despite-latest-digital-detox-trend/

12 Ophir, E., Nass, C., & Wagner, A. D. (2009). Cognitive control in media multitaskers. *Proceedings of the National Academy of Sciences,*106(37) 15583-15587. https://www.pnas.org/content/106/37/15583

13 Foerde, K., Knowlton, B. J., & Poldrack, R. A. (2006). Modulation of competing memory systems by distraction. *Proceedings of the National Academy of Sciences of the United States of America,* 103(31), 11778–11783. https://doi.org/10.1073/pnas.0602659103

14 Siegel, M., Warden, M. R., & Miller, E, K. (2009). Phase-dependent neuronal coding of objects in short-term memory. *Proceedings of the Nationa Academy of Sciences,* 106(50) 21341-21346. https://doi.org/10.1073/pnas.0908193106

15 Medina, J. (2008). *Brain rules: 12 principles for surviving and thriving at work, home, and school.* Pear Press.

16 Loh, K. K., & Kanai, R. (2014). Higher media multi-tasking activity is associated with smaller gray-matter density in the anterior cingulate cortex. *PLOS ONE,* 9(9), e106698. https://doi.org/10.1371/journal.pone.0106698

17 Jason, L. A., Danielewicz, J., & Mesina, A. (2005). Reducing media viewing: Implications for behaviorists. *Journal of Early and Intensive Behavior Intervention,* 2(3), 194-206. http://dx.doi.org/10.1037/h0100313

18 Anderson, J. (2016, November). The impact of media use and screen time on children, adolescents, and families. *American College of Pediatricians.* Retrieved April 6, 2020, from https://acpeds.org/position-statements/the-impact-of-media-use-and-screen-time-on-children-adolescents-and-families

19 Science and Technology Committee. (2019, January 29). *Impact of social media and screen-use on young people's health: Fourteenth report of session* 2017–19. Parliamentary Copyright House of Commons. https://publications.parliament.uk/pa/cm201719/cmselect/cmsctech/822/822.pdf

20 Poll finds Americans critical of television. (1996, April 15). *The Christian Science Monitor.* https://www.csmonitor.com/1996/0415/15132.html

21 Ziglar, Z. (2010). See you at the top: *Twenty-fifth anniversary edition.* Pelican Publishing. (Original work published 1974).

22 Centerwall, B. S. (1992). Television and violence: The scale of the problem and where to go from here. *The Journal of the American Medical Association.* https://www.congress.gov/congressional-record/1999/05/13/senate-section/article/S5195-1

23 Kolata, G. (2001, February 20). Doubters fault theory finding earlier puberty. *The New York Times*. https://www.nytimes.com/2001/02/20/us/doubters-fault-theory-finding-earlier-puberty.html

24 Herman-Giddens, M. E., Slora, E. J., Wasserman, R. C., Bourdony, C. J., Bhapkar, M. V., Koch, G. G., & Hasemeier, C. M. (1997). Secondary sexual characteristics and menses in young girls seen in office practice: A study from the Pediatric Research in Office Settings Network. *Pediatrics*, 99(4), 505-512. https://doi.org/10.1542/peds.99.4.505

25 Scutti, S. (2015, January 26). Puberty comes earlier and earlier for girls. *Newsweek*. https://www.newsweek.com/2015/02/06/puberty-comes-earlier-and-earlier-girls-301920.html

26 Euling, S. Y., Herman-Giddens, M. E., Lee, P. A., Selevan, S. G., Juul A, Sørensen, T. I., Dunkel, L., Himes, J. H., Teilmann, G., & Swan, S. H. (2008). Examination of US puberty-timing data from 1940 to 1994 for secular trends: panel findings. *Pediatrics,* 121(suppl 3), S172–S191. https://doi.org/10.1542/peds.2007-1813D

27 Finer, L. B. (2007). Trends in premarital sex in the United States, 1954-2003. *Public Health Reports*, 122(1), 73–78. https://doi.org/10.1177/003335490712200110

28 Centers for Disease Control and Prevention. (2018). Youth risk behavior surveillance—United States, 2017. *MMWR Surveillance Summaries*, 67(8). https://www.cdc.gov/healthyyouth/data/yrbs/pdf/2017/ss6708.pdf

29 Song, A. V., & Halpern-Felsher, B. L. (2011). Predictave relationship between adolescent oral and vaginal sex: Results from a prospective, longitudinal study. *Archives of Pediatrics and Adolescent Medicine*, 165(3), 243-249. https://jamanetwork.com/article.aspx?doi=10.1001/archpediatrics.2010.214

30 Herman-Giddens et al, 1997, loc.cit.

31 Salti, R., Tarquini, R., Stagi, S., Perfetto, F., Cornélissen, G., Laffi, G., Mazzoccoli, G., & Halberg, F. (2006). Age-dependent association of exposure to television screen with children's urinary melatonin excretion? *Neuroendocrinology Letters*, 27(1-2), 73-80. https://citeseerx.ist.psu.edu/viewdoc/download?doi=10.1.1.424.5941&rep=rep1&type=pdf

32 Steimer T. (2002). The biology of fear- and anxiety-related behaviors. *Dialogues in Clinical Neuroscience,* 4(3), 231–249. https://www.ncbi.nlm.nih.gov/pmc/articles/PMC3181681/

33 Mrug, S., Madan, A., Cook, E. W., & Wright, R. A. (2015). Emotional and physiological desensitization to real-life and movie violence. *Journal of Youth and Adolescence,* 44(5), 1092–1108. https://doi.org/10.1007/s10964-014-0202-z

34 Love, T., Laier, C., Brand, M., Hatch, L., & Hajela, R. (2015). Neuroscience of internet pornography addiction: A review and update. *Behavioral sciences,* 5(3), 388–433. https://doi.org/10.3390/bs5030388

35 Smith, S. M., & Vale, W. W. (2006). The role of the hypothalamic-pituitary-adrenal axis in neuroendocrine responses to stress. *Dialogues in Clinical Neuroscience,* 8(4), 383–395. https://www.ncbi.nlm.nih.gov/pmc/articles/PMC3181830/

36 Facts for Families. (2014, December). *TV violence and children.* American Academy of Child and Adolescent Psychiatry. https://www.aacap.org/AACAP/Families_and_Youth/Facts_for_Families/FFF-Guide/Children-And-TV-Violence-013.aspx

37 Laor, N., Wolmer, L., Mayes, L. C., Golomb, A., Silverberg, D. S., Weizman, R., & Cohen, D. J. (1996). Israeli preschoolers under scud missile attacks: A developmental perspective on risk-modifying factors. *Archives of General Pscyhiatry,* 53(5), 416-423. https://doi.org/10.1001/archpsyc.1996.01830050052008

38 Gunnar, M. R., & Donzella, B. (2002). Social regulation of the cortisol levels in early human development. *Psychoneuroendocrinology,* 27(1-2), 199-220. https://doi.org/10.1016/s0306-4530(01)00045-2

39 Anderson, C. A., Berkowitz, L., Donnerstein, E., Huesmann, L. R., Johnson, J. D., Linz, D., Malamuth, N. M., & Wartella, E. (2003). The influence of media violence on youth. *Psychological Science in the Public Interest*, 4(3), 81-110. https://doi.org/10.1111/j.1529-1006.2003.pspi_1433.x.

40 Eron, L. D., Huesmann, L. R., Lefkowitz, M. M., & Walder, L. O. (1972). Does television violence cause aggression? *American Psychologist*, 27(4), 253–263. https://doi.org/10.1037/h0033721

41 Singer, D. G., & Singer, J. L. (1995). *Readying children for school through brief exposure to television segments from Barney and Friends, Mr. Rogers' Neighborhood and Sesame Street: A report on workshops for day care teachers and family home care providers.* Connecticut Public Broadcasting.

42 Christakis, D. A., Garrison, M. M., Herrenkohl, T., Haggerty, K., Rivara, F. P., Zhou, C., & Liekweg, K. (2013). Modifying media content for preschool children: a randomized controlled trial. *Pediatrics*, 131(3), 431-438. https://doi.org/10.1542/peds.2012-1493

43 Sincero, S. M. (2011, August 9). *Habituation.* https://explorable.com/habituation

44 Rankin, C. H., Abrams, T., Barry, R. J., Bhatnagar, S., Clayton, D. F., Colombo, J., Coppola, G., Geyer, M. A., Glanzman, D. L., Marsland, S., McSweeney, F. K., Wilson, D. A., Wu, C. F., & Thompson, R. F. (2009). Habituation revisited: an updated and revised description of the behavioral characteristics of habituation. *Neurobiology of Learning and Memory*, 92(2), 135–138. https://doi.org/10.1016/j.nlm.2008.09.012

45 Turk-Browne, N. B., Scholl, B. J., & Chun, M. M. (2008). Babies and brains: habituation in infant cognition and functional neuroimaging. *Frontiers in Human Neuroscience,* 2, 16. https://doi.org/10.3389/neuro.09.016.2008

46 Howard, P. J. (1994). *The owner's manual for the brain: Everyday applications from mind-brain research.* Leornian Press.

47 Huesmann, L. & Kirwil, L. (2011). Why observing violence increases the risk of violent behavior by the observer. *The Cambridge Handbook of Violent Behavior and Aggression.* https://doi.org/10.1017/CBO9780511816840.029.

48 Seeman, P., Bzowej, N. H., Guan, H. C., Bergeron, C., Becker, L. E., Reynolds, G. P., Bird, E. D., Riederer, P., Jellinger, K., Watanabe, S., Wallace, W., & Tourtellotte, W. W. (1987). Human brain dopamine receptors in children and aging adults. *Synapse,* 1(5), 399-404. https://doi.org/10.1002/syn.890010503

49 ibid.

50 Smith, N. (2018, April 6). Social media looks like the new opiate of the masses. *The Japan Times*. https://www.japantimes.co.jp/opinion/2018/04/06/commentary/world-commentary/social-media-looks-like-new-opiate-masses/#.Xmu4jqj0nIU

51 Huesmann L. R. (2007). The impact of electronic media violence: scientific theory and research. *The Journal of Adolescent Health,* 41(6 Suppl 1), S6–S13. https://doi.org/10.1016/j.jadohealth.2007.09.005

52 Brockmyer, J. F. (2015). Playing violent video games and desensitization to violence. *Child & Adolescence Psychiatric Clinics of North America,* 24(1), 65-77. https://doi.org/10.1016/j.chc.2014.08.001

53 Zahn, P. (1999, April 25). *Who's to blame?/Dunblane/Audra* [60 Minutes]. CBS.

54 Flannery, D. J., & Singer, M. I. (2016, March 28). Here's how witnessing violence harms children's mental health. *The Conversation*. http://theconversation.com/heres-how-witnessing-violence-harms-childrens-mental-health-53321

55 Cantor, J., & Sparks, G. G. (1984). Children's fear responses to mass media: Testing some Piagetian predictions. *Journal of Communication*, 34(2), 90–103. https://academic.oup.com/joc/article-abstract/34/2/90/4282748?redirectedFrom=fulltext

56 Cairns, E. (1990) Impact of Television News Exposure on Children's Perceptions of Violence in Northern Ireland, *The Journal of Social Psychology*, 130:4, 447-452, DOI: 10.1080/00224545.1990.9924605

57 Ginsburg, K. R. (2007). The importance of play in promoting healthy child development and maintaining strong parent-child bonds. *Pediatrics*, 119(1), 182-191. https://doi.org/10.1542/peds.2006-2697

58 Chazan, S. E. (2002). *Profiles of play: Assessing and observing structure and process in play therapy*. Jessica Kingsley Publishers.

59 Santer, J., & Griffiths, C. (2007). Free play in early childhood: *A literature review.* National Children's Bureau. https://www.bl.uk/collection-items/free-play-in-early-childhood-a-literature-review

60 Panksepp, J. (2004). *Affective neuroscience: The foundations of human and animal emotions.* Oxford University Press.

61 Caine, & Caine, 1991, loc.cit.

62 Singer & Singer, 1995, loc.cit.

63 Singer, J. L., & Singer, D. G. (1979). Television viewing, family style and aggressive behavior. *Education Resources Information Center.* Paper presented at the Annual Meeting of the American Association for the Advancement of Science, Houston, Texas.

64 Mikulak, M. (2015). Childhood unmasked - *The agency of Brazil's street and working children.* Cognella Academic Publishing.

65 Marzollo, J., & Lloyd, J. (1977). Learning through play. Pengiun.

66 Pearce, 1992, loc.cit.

67 Wilson, F. R. (1999). *The hand: How its use shapes the brain, language, and human culture.* Vintage.

68 Gabbatt, A. (2010, February 3). Excessive internet use linked to depression, research shows. *The Guardian*. https://www.theguardian.com/technology/2010/feb/03/excessive-internet-use-depression

69 Stoll, C. (1996). *Silicon snake oil: Second thoughts on the information highway.* Anchor

70 Dickson, E. (2019, October 18). Teen suicide is on the rise and no one knows why. *Rolling Stone*. https://www.rollingstone.com/culture/culture-news/teen-suicide-study-increase-why-900711/

71 Curtin, S. C., & Heron, M. (2019, October). *Death rates due to suicide and homicide among persons aged 10–24: United States, 2000–2017* [NCHS Data Brief No. 352]. https://www.cdc.gov/nchs/products/databriefs/db352.htm?deliveryName=USCDC_171-DM11162

72 Sedgwick, R., Epstein, S., Dutta, R., & Ougrin, D. (2019). Social media, internet use and suicide attempts in adolescents. *Current Opinion in Psychiatry,* 32(6), 534–541. https://doi.org/10.1097/YCO.0000000000000547

73 Christakis, D. A., Zimmerman, F. J., DiGiuseppe, D. L., & McCarty, C. A. (2004). Early television exposure and subsequent attentional problems in children. *Pediatrics*, 113(4) 708-713. https://doi.org/10.1542/peds.113.4.708

74 Wamsley, E. J., & Stickgold, R. (2011). Memory, sleep and dreaming: Experiencing consolidation. *Sleep Medicine Clinics*, 6(1), 97–108. https://doi.org/10.1016/j.jsmc.2010.12.008

75 Harris, G. (2005, Oct 19) Sleeping Pill Use by Youths Soars, Study Says. *New York Times.*

76 Peracchia, S., & Curcio, G. (2018). Exposure to video games: effects on sleep and on post-sleep cognitive abilities. A sistematic review of experimental evidences. *Sleep Science*, 11(4), 302–314. https://doi.org/10.5935/1984-0063.20180046

Chapter 29

1 George Bernard Shaw quotes. (n.d.). *Quote Park.* https://quotepark.com/authors/george-bernard-shaw/

2 Young, E. (1759). *Conjectures on original composition.* London. https://case.edu/affil/sce/authorship/Young.pdf

3 Dugatkin, L. (2000). *The imitation factor.* Free Press.

4 Aristotle. (1997). *Poetics* (M. Health, Trans.). Penguin Classics. (Original work published ca. 335 B.C.).

5 Ramachandran, V. S., & Altschuler, E. L. (2009). The use of visual feedback, in particular mirror visual feedback, in restoring brain function. *Brain,* 132(Pt 7):1693-710. https://academic.oup.com/brain/article/132/7/1693/328686

6 Jordan, M. A. (2014, February 4). What's in a meme? *Richard Dawkins Foundation.* https://www.richarddawkins.net/2014/02/whats-in-a-meme/

7 Dawkins, R. (1976). *The selfish gene.* Oxford University Press.

8 Rizzolatti, G., Fadiga, L., Gallese, V., & Fogassi, L. (1995). Premotor cortex and the recognition of motor actions. *Cognitive Brain Research*, 3(1996) 131-141. https://doi.org/10.1016/0926-6410(95)00038-0

9 Ramachandran, V. (2000). Mirror neurons and imitation learning as the driving force behind "the great leap forward" in human evolution. https://www.edge.org/3rd_culture/ramachandran/ramachandran_index.html

10 Heiser, M., Iacoboni, M., Maeda, F., Marcus, J., & Mazziotta, J. (2003). The essential role of Broca's area in imitation. *European Journal Of Neuroscience,* 17(5),1123-1128. https://doi.org/10.1046/j.1460-9568.2003.02530.x

11 *AAAS Symposium to take place on mirror neurons.* (2005, February 19). American Psychoanalytic Association. https://www.eurekalert.org/pub_releases/2005-02/apa-ast021405.php

12 Oberman, L., Hubbard, E., Mccleery, J., Altschuler, E., Ramachandran, V., & Pineda, J. (2005). EEG Evidence for Mirror Neuron Dysfunction in Autism Spectrum Disorders. *Cognitive Brain Research*, 24, 190-198. https://doi.org/10.1016/j.cogbrainres.2005.01.014.

13 Ramachandran, V., & Rogers-Ramachandran, D. (1996). Denial of disabilities in anosognosia. *Nature*, 382, 501. https://doi.org/10.1038/382501a0

14 Eliot, L. (1999). *What's going on in there?* Bantam Books.

15 Ostlund, B. D., Measelle, J. R., Laurent, H. K., Conradt, E., & Ablow, J. C. (2017). Shaping emotion regulation: Attunement, symptomatology, and stress recovery within mother-infant dyads. *Developmental Psychobiology*, 59(1), 15–25. https://doi.org/10.1002/dev.21448

16 Pearce, 1977, loc.cit.

17 Chisholm et al., 1995, loc.cit.

18 Perry & Pollard, 1997, loc.cit.

19 *Raise happy and well-behaved kids with positive, loving tools.* (n.d.). Love and Logic. https://www.loveandlogic.com/a/info/love-and-logic-parenting

20 Radke-Yarrow, M., & Klimes-Dougan, B. (2002). Parental depression and offspring disorders: A developmental perspective. In S. H. Goodman & I. H. Gotlib (Eds.), *Children of depressed parents: Mechanisms of risk and implications for treatment* (pp. 155-173). American Psychological Association. https://doi.org/10.1037/10449-006

21 Dimberg, U., Thunberg, M., & Elmehed, K. (2000). Unconscious facial reactions to emotional facial expressions. *Psychological Science*, 11(1), 86-89. https://doi.org/10.1111/1467-9280.00221

22 Dawson, G., Frey, K., Panagiotides, H., Yamada, E., Hessl, D., & Osterling, J. (1999). Infants of depressed mothers exhibit atypical frontal electrical brain activity during interactions with mother and with a familiar, nondepressed adult. *Child Development*, 70(5), 1058-1066. https://doi.org/10.1111/1467-8624.00078

23 Aloni, M., & Bernieri, F. (2004). Is love blind? The effects of experience and infatuation on the perception of love. *Journal Of Nonverbal Behavior*, 28(4), 287-296. https://doi.org/10.1007/s10919-004-4160-0

24 Koenig, K. (2015). *Outsmarting overeating.* New World Library.

25 Deming, 1986, loc.cit.

26 Mennella, J., Jagnow, C., & Beauchamp, G. (2001). Prenatal and postnatal flavor learning by human infants. *Pediatrics*, 107(6), e88-e88. https://doi.org/10.1542/peds.107.6.e88

27 Farroni, T., Menon, E., Rigato, S., & Johnson, M. (2007). The perception of facial expressions in newborns. *European Journal Of Developmental Psychology*, 4(1), 2-13. https://doi.org/10.1080/17405620601046832

28 Farroni, T., Csibra, G., Simion, F., & Johnson, M. (2002). Eye contact detection in humans from birth. Proceedings Of The National Academy Of Sciences, 99(14), 9602-9605. doi: 10.1073/pnas.152159999

29 Nordqvist, E., Rudner, M., Johansson, M., Lindgren, M., & Heimann, M. (2015). The relationship between deferred imitation, associative memory, and communication in 14-months-old children. Behavioral and electrophysiological indices. *Frontiers In Psychology*, 6. https://doi.org/10.3389/fpsyg.2015.00260

30 Meltzoff A. N. (1988). Infant imitation and memory: nine-month-olds in immediate and deferred tests. *Child Development*, 59(1), 217–225. https://doi.org/10.1111/j.1467-8624.1988.tb03210.x

31 Snyder, 1993, loc.cit.

32 Gardner,1991, loc.cit.

33 Bureau of Labor Statistics. (2019). *Employmen characteristics of families - 2018* [News release]. https://www.bls.gov/news.release/pdf/famee.pdf

34 Gardner, 1991, loc.cit.

35 Gardner, 1991, loc.cit.

36 Gee, J. (2014). Decontextualized language: A problem, not a solution. *International Multilingual Research Journal,* 8(1), 9-23. https://doi.org/10.1080/19313152.2014.852424

37 Parker, K., Horowitz, J. M., Brown, A., Fry, R., Cohn, D., & Igielnik, R. (2018, May 22). *How urban, suburban and rural residents interact with their neighbors.* Pew Research Center. https://www.pewsocialtrends.org/2018/05/22/how-urban-suburban-and-rural-residents-interact-with-their-neighbors/

38 *Oral sex behaviors among teens.* (2015). Child Trends Data Bank. https://www.childtrends.org/wp-content/uploads/2013/12/95_Oral_Sex_1.pdf

39 *Drug use among youth: Facts and statistics.* (2017). National Center for Drug Abuse Statistics. https://drugabusestatistics.org/teen-drug-use/

40 Shore, M. F., Massimo, J. L., & Mack, R. (1965). Changes in the perception of interpersonal relationships in successfully treated adolescent delinquent boys. *Journal of Consulting Psychology*, 29(3), 213–217. https://doi.org/10.1037/h0022127

41 Shore, M. F., & Massimo, J. L. (1979). Fifteen years after treatment: A follow-up study of comprehensive vocationally-oriented psychotherapy. *American Journal of Orthopsychiatry*, 49(2), 240–245. https://doi.org/10.1111/j.1939-0025.1979.tb02605.x

42 Califano, J. (2009). *How to raise a drug-free kid.* Fireside.

43 Light, R. (2001). *Making the most of college*. Harvard University Press.

44 Goldberg, M. F. (1999). The most important education in America: An interview with Shirley Strum Kenny. *Education Resources Information Center,* 80(7), 526-528. https://eric.ed.gov/?id=EJ581590

45 Hermann, E. (1981). *Shinichi Suzuki*. Summy-Birchard Music.

46 Szymanski, M., & Whalen, J. (2011). *Making work visible.* Cambridge University Press.

47 Polanyi, M. (1962). *Personal knowledge*. The University of Chicago Press.

Chapter 30

1 Toffler, A. (1984). *Future Shock*. Bantam.

2 Hart, 1983, loc.cit.

3 Schmidt, M., Bailey, S., Goodwin, T., Jones, J., Killian, J., Legato, M., Limoli, C., Moussa, S., & Ploutz-Snyder, L. (2017). Men, women, and space travel: Gene-linked molecular networks, human countermeasures, and legal and ethical considerations. *Gender and the Genome*, 1, 54-67. https://doi.org/10.1089/gg.2017.29004.rtl.

4 Smith, A. (2000). *Accelerated learning in practice: Brain-based methods for accelerating motivation and achievement.* Network Educational Press.

5 Dubuc, loc.cit.

6 Hart, 1983, loc.cit.

7 Robson, D. (2011, September 21). A brief history of the brain. *New Scientist.* https://www.newscientist.com/article/mg21128311-800-a-brief-history-of-the-brain/

8 Hart, 1983, loc.cit.

9 Miettinen, R. (2000). The concept of experiential learning and John Dewey's theory of reflective thought and action. *International Journal Of Lifelong Education*, 19(1), 54-72. https://doi.org/10.1080/026013700293458

10 Dewey, 1902, loc.cit.

11 *Association of experiential education (AEE).* (n.d.). The Association for Experiential Education. https://www.aee.org/about

12 *Achieving environmental literacy in a standards-based world.* (2014, January 12). State Education and Environment Roundtable. Retrieved on April 9, 2020, from http://seer.org/blog/

13 Raywid, M. A., Schmerler, G., Phillips, S. E., & Small, G. A. (2003). Not so easy going: The policy environments of small urban schools and schools-within-schools. ERIC Clearinghouse on Rural Education and Small Schools.

14 Harr, J. J., Parrish, T., Chambers, J., Levin, J., & Segarra, M. (2006, December 17). *Considering special education adequacy in California.* American Institutes for Research. https://cepa.stanford.edu/sites/default/files/21-AIR-Special-Ed%283-07%29.pdf

15 Grahn, P., Martensson, F., Lindblad, B., Nilsson, P., & Ekman, A. (1997). Ute på Dagis Stad and Land. *Swedish University of Agriculture.*

16 Nolan & McKinnon, 2003, loc.cit.

17 National Center for Education Statistics. (2003). *Highlights from the Program for International Student Assessment (PISA).* https://nces.ed.gov/pubs2002/2002116.

pdf
18 Sobel, D. (2004). *Place-based education: Connecting classrooms & communities.* Orion Society.
19 Heschong Mahone Group. (1999). Daylighting in schools: *An investigation into the relationship between daylighting and human performance.* Detailed Report. https://eric.ed.gov/?id=ED444337
20 Coyle, K. J. (2004, May). *Understanding environmental literacy in America: And making it a reality.* National Environmental Education & Training Foundation. https://eric.ed.gov/?id=ED522820
21 *Achieving environmental literacy,* 2014, loc.cit.
22 Lieberman, G. A., & Hoody, L. L. (1998). *Cosing the achievement gap: Using the environment as an integrating context for learning.* State Education and Environment Roundtable. http://www.seer.org/extras/execsum.pdf
23 Ibid.

Chapter 31

1 Krishnamurti, J. (n.d.). Krishnamurti on education. J. *Krishnamurti Online.* http://legacy.jkrishnamurti.org/it/krishnamurti-teachings/view-context.php?tid=43&chid=297&w=education&s=Context
2 National Institute of Diabetes and Digestive and Kidney Diseases. (2017, August). *Overweight & obesity statistics.* Retrieved March 11, 2020, from https://www.niddk.nih.gov/health-information/health-statistics/overweight-obesity
3 von Neumann, J. (Burks, A.W., Ed.). (1966). *Theory of self-reproducing automata theory of self-reproducing automata.* University of Illinois Press
4 Burks, A. W. (1970). *Essays on cellular automata.* University of Illinois Press.
5 Jensen, 1998, loc.cit.
6 Kircher, T., David, A. S., & David, A. (2003). *The self in neuroscience and psychiatry.* Cambridge University Press.
7 Sousa, D. A. (2006). *How the brain learns.* Corwin Press.
8 Prep, K. T. (2019). *MCAT behavioral sciences review 2020-2021.* Simon and Schuster.
9 Bermudez-Rattoni, F. (2007). *Neural plasticity and memory: From genes to brain imaging.* CRC Press.
10 Organization for Economic Co-operation and Development. (2007). *Understanding the brain: The birth of a learning science.* Learning in the 21st Century: Research, innovation and policy [OECD/CERI International Conference]. https://www.oecd.org/site/educeri21st/40554190.pdf

11 Tiwari, P., Dwivedi, S., Singh, M. P., Mishra, R., & Chandy, A. (2013). Basic and modern concepts on cholinergic receptor: A review. *Asian Pacific Journal of Tropical Disease,* 3(5), 413–420. https://doi.org/10.1016/S2222-1808(13)60094-8

12 Shen, H. (2018, September 18). Does the adult brain really grow new neurons? *Scientific American.* https://www.scientificamerican.com/article/does-the-adult-brain-really-grow-new-neurons/

13 Obernier, K., Cebrian-Silla, A., Thomson, M., Parraguez, J. I., Anderson, R., Guinto, C., Rodriguez, J. R., Garcia-Verdugo, J. M., & Alvarez-Buylla, A. (2018). Adult neurogenesis is sustained by symmetric self-renewal and differentiation. *Cell Stem Cell,* 22(2), 221-234.e8. https://doi.org/10.1016/j.stem.2018.01.003

14 Kempermann, G., Kuhn, H. G., & Gage, F. H. (1998). Experience-induced neurogenesis in the senescent dentate gyrus. *The Journal of Neuroscience,* 18(9), 3206–3212. https://doi.org/10.1523/JNEUROSCI.18-09-03206.1998

15 Gage, F. H. (2019). Adult neurogenesis in mammals. *Science,* 364(6443), 827-828. https://doi.org/10.1126/science.aav6885

16 Fratiglioni, L., Paillard-Borg, S., & Winblad, B. (2004). An active and socially integrated lifestyle in late life might protect against dementia. *The Lancet. Neurology,* 3(6), 343–353. https://doi.org/10.1016/S1474-4422(04)00767-7

17 Bathina, S., & Das, U. N. (2015). Brain-derived neurotrophic factor and its clinical implications. *Archives of Medical Science,* 6, 1164–1178. https://doi.org/10.5114/aoms.2015.56342

18 Willis, J., & Willis, M. (2020). *Research-based strategies to ignite student learning: Insights from neuroscience and the classroom.* ASCD.

Chapter 32

1 Blanchard, K. (n.d.). Feedback is the breakfast of champions. *Ken Blanchard.* https://www.kenblanchardbooks.com/feedback-is-the-breakfast-of-champions/

2 van Gerven, M. (2017). Computational foundations of natural intelligence. *Frontiers in Computational Neuroscience,* 11. https://doi.org/10.3389/fncom.2017.00112

3 Lőrincz, M. L., & Adamantidis, A. R. (2017). Monoaminergic control of brain states and sensory processing: Existing knowledge and recent insights obtained with optogenetics. *Progress in Neurobiology,* 151, 237–253. https://doi.org/10.1016/j.pneurobio.2016.09.003

4 Jacob, S. N., & Nienborg, H. (2018). Monoaminergic Neuromodulation of Sensory Processing. *Frontiers in Neural Circuits,* 12. https://doi.org/10.3389/fncir.2018.00051

5 Austin, J. H. (1999). *Zen and the brain: Toward an understanding of meditation and consciousness.* MIT Press.

6 Lin, T. W., & Kuo, Y. M. (2013). *Exercise benefits brain function: The monoamine connection. Brain Sciences,* 3(1), 39–53. https://doi.org/10.3390/brainsci3010039

7 Schmeichel, B. E., & Berridge, C. W. (2013). Neurocircuitry underlying the preferential sensitivity of prefrontal catecholamines to low-dose psychostimulants. Neuropsychopharmacology, 38(6), 1078–1084. https://doi.org/10.1038/npp.2013.6

8 Sommi, R. W., Crismon, M. L., & Bowden, C. L. (1987). Fluoxetine: A serotonin-specific, second-generation antidepressant. *Pharmacotherapy,* 7(1), 1–15. https://doi.org/10.1002/j.1875-9114.1987.tb03496.x

9 Chen, Y.-H., Huang, E. Y.-K., Kuo, T.-T., Miller, J., Chiang, Y.-H., & Hoffer, B. J. (2017). Impact of traumatic brain injury on dopaminergic transmission. *Cell Transplantation*, 26(7), 1156–1168. https://doi.org/10.1177/0963689717714105

10 Cain, C. K., Maynard, G. D., & Kehne, J. H. (2012). Targeting memory processes with drugs to prevent or cure PTSD. *Expert Opinion on Investigational Drugs,* 21(9), 1323–1350. https://doi.org/10.1517/13543784.2012.704020

11 Csikszentmihalyi, 1990, loc.cit.

12 Sechenov, I. (1965). *Reflexes of the brain* (G. Gibbons, Ed.; S. Belsky, Trans.). MIT Press. (Original work published 1863.)

13 Datta, P. (2014). *It's all in your mind.* Leadstart.

14 Barros, S. P., & Offenbacher, S. (2009). Epigenetics. *Journal of Dental Research*, 88(5), 400–408. https://doi.org/10.1177/0022034509335868

15 Senge, 1990, loc.cit.

16 Csikszentmihalyi, 1990, loc.cit.

17 Jacob & Nienborg, 2018, loc.cit.

18 Chatterjee, H. J., & Hannan, L. (2016). *Engaging the senses: Object-based learning in higher education.* Routledge.

19 Kuiper, J. (2011). *Explicit memory as a framework for the neural correlate of consciousness* (Doctoral dissertation, Faculty of Science and Engineering). fse.studenttheses.ub.rug.nl/9666/1/MindMatters.pdf

20 Hickok, G. (2014). *The myth of mirror neurons: The real neuroscience of communication and cognition.* W. W. Norton & Company.

21 Serafim, A. de P., Barros, D. M. de, Valim, A., & Gorenstein, C. (2009). Cardiac response and anxiety levels in psychopathic murderers. *Brazilian Journal of Psychiatry*, 31(3), 214–218. https://doi.org/10.1590/S1516-44462009000300006

22 Center for Substance Abuse Treatment. (1999). How stimulants affect the brain and behavior. *In Treatment for stimulant use disorders*. Substance Abuse and Mental Health Services Administration. https://www.ncbi.nlm.nih.gov/sites/books/NBK64328/

23 Austrian, S. G. (2005). *Mental disorders, medications, and clinical social work.* Columbia University Press.

Chapter 33

1 Breuning, L. G. (2012). *Meet your happy chemicals: Dopamine, endorphin, oxytocin, serotonin.* CreateSpace Independent Publishing Platform.

2 Frazer, A., & Hensler, J G.. (1999). Serotonin receptors. In Siegel, G. J., Agranoff, B. W., Albers. R. W., et al., (eds.) *Basic neurochemistry: Molecular, cellular and medical aspects (6th ed).* Lippincott-Raven. https://www.ncbi.nlm.nih.gov/books/NBK28234/

3 Purves, D., Augustine, G. J., Fitzpatrick, D. Hall, W. C., Lamantia, A. S., McNamara, J. O., & Williams, S. M. (Eds.). (2004). *Neuroscience (3rd ed.).* Sinauer Associates, Inc.

4 ibid

5 Schwarz, L. A., & Luo, L. (2015). Organization of the locus coeruleus-norepinephrine system. *Current Biology,* 25(21), R1051-R1056 https://www.cell.com/current-biology/fulltext/S0960-9822(15)01150-1?_returnURL=https%3A%2F%2Flinkinghub.elsevier.com%2Fretrieve%2Fpii%2FS0960982215011501%3Fshowall%3Dtrue

6 Chmielewski, W.X., Mückschel, M., Ziemssen, T., & Beste, C. (2017). The norepinephrine system affects specific neurophysiological subprocesses in the modulation of inhibitory control by working memory demands. *Human Brain Mapping,* 38, 68-81. https://doi.org/10.1002/hbm.23344

7 Somers, D. C., & Sheremata, S. L. (2013). Attention maps in the brain. *Wiley Interdisciplinary Reviews: Cognitive Science,* 4(4), 327–340. https://doi.org/10.1002/wcs.1230

8 Spinelli, M. G. (2000). Effects of steroids on mood/depression. *In Menopause: Biology and pathology* (563-582). https://doi.org/10.1016/B978-012453790-3/50040-8

9 Lin, T. W., & Kuo, Y. M. (2013). Exercise benefits brain function: The monoamine connection. *Brain Dciences,* 3(1), 39–53. https://doi.org/10.3390/brainsci3010039

10 Arain, M., Haque, M., Johal, L., Mathur, P., Nel, W., Rais, A., Sandhu, R., & Sharma, S. (2013). Maturation of the adolescent brain. *Neuropsychiatric Disease and Treatment,* 9, 449–461. https://www.ncbi.nlm.nih.gov/pmc/articles/PMC3621648/

11 Gantz, S. C., Ford, C. P., Morikawa, H., & Williams, J. T. (2018). The evolving understanding of dopamine neurons in the substantia nigra and ventral tegmental area. *Annual Review of Physiology*, 80(1), 219-241. https://doi.org/10.1146/annurev-physiol-021317-121615

12 Kamiński, J., Mamelak, A. N., Birch, K., Mosher, C. P., Tagliati, M., & Rutishauser, U. (2018). Novelty-sensitive dopaminergic neurons in the human substantia nigra predict success of declarative memory formation. *Current Biology*, 28(9), 1333-1343. https://doi.org/10.1016/j.cub.2018.03.024

13 Heijnen, S., Hommel, B., Kibele, A., & Colzato, L. S. (2016). Neuromodulation of aerobic wxercise-A review. *Frontiers in Psychology*, 6, 1890. https://doi.org/10.3389/fpsyg.2015.01890

14 Kurniawan, I. T., Guitart-Masip, M., & Dolan, R. J. (2011). Dopamine and effort-based decision making. *Frontiers in Neuroscience*, 5, 81. https://doi.org/10.3389/fnins.2011.00081

15 Nall, R. (2019). What to know about Parkinson's dementia. *Medical News Today*. https://www.medicalnewstoday.com/articles/314486

16 Sacks, O. (1973). *Awakenings*. Duckworth & Co.

17 Gründer, G., & Cumming, P. (2016). The dopamine hypothesis of schizophrenia. *The Neurobiology of Schizophrenia*, 109-124. https://doi.org/10.1016/B978-0-12-801829-3.00015-X

18 Anscombe, R. (1987). The disorder of consciousness in schizophrenia. *Schizophrenia Bulletin*, 13(2), 241-60. https://pdfs.semanticscholar.org/015f/5ce864f939cebb0d9df06f8fa5d5319e1830.pdf?_ga=2.116670351.887179552.1605910649-110028425.1605910649

19 Owens, D. C., & Johnstone, E. C. (2018). The development of antipsychotic drugs. *Sage Journals*. https://doi.org/10.1177/2398212818817498

20 Li, P., Snyder, G. L., & Vanover, K. E. (2016). Dopamine targeting drugs for the treatment of schizophrenia: Past, present and future. *Current Topics in Medicinal Chemistry*, 16(29), 3385–3403. https://doi.org/10.2174/1568026616666160608084834

21 Lally, J., & MacCabe, J. H. (2015). Antipsychotic medication in schizophrenia: A review. *British Medical Bulletin*, 114(1), 169–179. https://academic.oup.com/bmb/article/114/1/169/246291

22 Bancos, I. (Ed.). (2020). *What is serotonin?* Hormone Health Network, Endocrine Society. https://www.hormone.org/your-health-and-hormones/glands-and-hormones-a-to-z/hormones/serotonin

23 Carew, T. J. (2004). *Behavioral neurobiology: The cellular organization of natural behavior*. Sinauer Associates.

24 Hornung, J. P. (2003). The human raphe nuclei and the serotonergic system. *Journal of Chemical Neuroanatomy*, 26(4), 331-43. https://doi.org/10.1016/j.jchemneu.2003.10.002

25 The Human Memory. (2019). *Brain neurons and synapses.* https://human-memory.net/brain-neurons-synapses/

26 Jenkins, T. A., Nguyen, J. C., Polglaze, K. E., & Bertrand, P. P. (2016). Influence of Tryptophan and Serotonin on Mood and Cognition with a Possible Role of the Gut-Brain Axis. *Nutrients,* 8(1), 56. https://doi.org/10.3390/nu8010056

27 Chantranupong, L. & Sabatini, B. L. (2018). Sunlight brightens learning and memory. *Cell,* 173(7), 1570-1572. https://doi.org/10.1016/j.cell.2018.05.044

28 The National Optical Astronomy Observatory. (n.d.). *Recommended light levels.* https://www.noao.edu/education/QLTkit/ACTIVITY_Documents/Safety/LightLevels_outdoor+indoor.pdf

29 Tafet, G. E., & Nemeroff, C. B. (2016). The links between stress and depression: Psychoneuroendocrinological, genetic, and environmental interactions. *The Journal of Neuropsychiatry and Clinical Neurosciences,* 28(2), 77-88. https://doi.org/10.1176/appi.neuropsych.15030053

30 Albizu, L., Holloway, T., González-Maeso, J., & Sealfon, S. C. (2011). Functional crosstalk and heteromerization of serotonin 5-HT2A and dopamine D2 receptors. *Neuropharmacology*, 61(4), 770–777. https://doi.org/10.1016/j.neuropharm.2011.05.023

31 National Center for Health Statistics. (2017). *Antidepressant use among persons aged 12 and over: United States, 2011–2014* (NCHS data brief, no 283). Centers for Disease Control and Prevention. https://www.cdc.gov/nchs/data/databriefs/db283.pdf

32 Baumann, M. H., Wang, X., & Rothman, R. B. (2007). 3,4-Methylenedioxymethamphetamine (MDMA) neurotoxicity in rats: A reappraisal of past and present findings. *Psychopharmacology*, 189(4), 407–424. https://doi.org/10.1007/s00213-006-0322-6

33 Krishnakumar, D., Hamblin, M. R., & Lakshmanan, S. (2015). Meditation and yoga can modulate brain mechanisms that affect behavior and anxiety-A modern scientific perspective. *Ancient Science*, 2(1), 13–19. https://doi.org/10.14259/as.v2i1.171

34 Seo, D., Patrick, C. J., & Kennealy, P. J. (2008). Role of serotonin and dopamine system interactions in the neurobiology of impulsive aggression and its comorbidity with other clinical disorders. *Aggression and Violent Behavior,* 13(5), 383–395. https://doi.org/10.1016/j.avb.2008.06.003

35 Meijers, J., Harte, J. M., Jonker, F. A., & Meynen, G. (2015). Prison brain? Executive dysfunction in prisoners. *Frontiers in Psychology*. https://doi.org/10.3389/fpsyg.2015.00043

36 Lanctôt, K. L., Herrmann, N., & Mazzotta, P. (2001). Role of serotonin in the behavioral and psychological symptoms of dementia. *The Journal of Neuropsychiatry and Clinical Neurosciences,* 13(1), 5-21. https://neuro.psychiatryonline.org/doi/full/10.1176/jnp.13.1.5

37 Sylwester, R. (1997). The neurobiology of self-esteem and aggression. *Education for Democratic Life,* 54(5), 75-79. http://www.ascd.org/publications/educational_leadership/feb97/vol54/num05/The_Neurobiology_of_Self-Esteem_and_Aggression.aspx

38 Krakowski, M. (2003). Violence and serotonin: Influence of impulse control, affect regulation, and social functioning. *The Journal of Neuropsychiatry and Clinical Neurosciences,* 15(3), 294-305. https://neuro.psychiatryonline.org/doi/full/10.1176/jnp.15.3.294

Chapter 34

1 Kahneman, D. (2011, October 19). Don't blink! The hazards of confidence. *The New York Times.* https://www.nytimes.com/2011/10/23/magazine/dont-blink-the-hazards-of-confidence.html

2 O'Keefe, J., & Nadel, L. (1978). *The hippocampus as a cognitive map.* Clarendon Press.

3 Lisman, J., Buzsáki, G., Eichenbaum, H., Nadel, L., Ranganath, C., & Redish, A. D. (2017). Viewpoints: How the hippocampus contributes to memory, navigation and cognition. *Nature Neuroscience,* 20(11), 1434–1447. https://doi.org/10.1038/nn.4661

4 Epstein, R. A., Patai, E. Z., Julian, J. B., & Spiers, H. J. (2017). The cognitive map in humans: Spatial navigation and beyond. *Nature Neuroscience,* 20(11), 1504–1513. https://doi.org/10.1038/nn.4656

5 Caine & Caine, 1991, loc.cit.

6 Center for Microeconomic Data. (2020). *Quarterly report on household debt and credit: 2019 Q4.* Federal Reserve Bank of New York. https://www.newyorkfed.org/medialibrary/interactives/householdcredit/data/pdf/hhdc_2019q4.pdf

7 Wittman, J. (n.d.). *The forgetting curve.* https://www.csustan.edu/sites/default/files/groups/Writing%20Program/forgetting_curve.pdf

8 National Physical Activity Plan Alliance. (2018). *The 2018 United States report card on physical activity for children and youth.* http://www.physicalactivityplan.org/projects/PA/2018/2018%20US%20Report%20Card%20Full%20Version_WEB.PDF?pdf=page-link

9 Delate, T., Gelenberg, A. J., Simmons, V. A., Brenda, M. S., & Motheral, R. (2004). Trends in the use of antidepressants in a national sample of commercially insured pediatric patients, 1998 to 2002. *Psychiatric Services,* 55(4), 387-391. https://ps.psychiatryonline.org/doi/pdf/10.1176/appi.ps.55.4.387

10 Medco Health Solutions. (2004). *Study reveals pediatric spending spike on drugs to treat behavioral problems.* https://www.express-scripts.com/corporate/

11 Whitney, D. G., & Peterson, M. D. (2019). US national and state-level prevalence of mental health disorders and disparities of mental health care use in children. *JAMA Pediatrics*, 173(4), 389–391. https://doi.org/10.1001/jamapediatrics.2018.5399

12 Children's Hospital And Regional Medical Center. (2004). Study finds link between television viewing and attention problems in children. *ScienceDaily*. https://www.sciencedaily.com/releases/2004/04/040406090140.htm

13 Kahn, P. H. (1999). *The human relationship with nature: Development and culture.* MIT Press.

14 Human-Environment Research Laboratory. (2004). *Kids who spend more time outside end up paying more attention inside.* http://www.state.sc.us/forest/kidscon.pdf

15 Wells, N. M., & Evans, G. W. (2003). *Nearby nature: A buffer of life stress among rural children.* Environment and Behavior, 35(3), 311–330. https://doi.org/10.1177/0013916503035003001

16 Louv, 2005, loc.cit.

Chapter 35

1 Inayat Khan. (1996). *The mysticism of sound and music: The Sufi teaching of Hazrat Inayat Khan* (Rev. ed). Shambhala. (Original work published 1923).

2 Ritter, S. M., & Dijksterhuis, A. (2014). Creativity - the unconscious foundations of the incubation period. *Frontiers in Human Neuroscience*, 8, 215. https://doi.org/10.3389/fnhum.2014.00215

3 Henok, N., Vallée-Tourangeau, F. & Vallée-Tourangeau, G. (2020). Incubation and interactivity in insight problem solving. *Psychological Research,* 84, 128–139. https://link.springer.com/article/10.1007/s00426-018-0992-9

4 DePorter, B., & Hernacki, M. (1992). *Quantum learning: Unleashing the genius in you.* Dell.

5 Nails, D. (2020). Socrates. *The Stanford Encyclopedia of Philosophy* (E. N. Zalta, Ed.). Retrieved April 10, 2020, from https://plato.stanford.edu/entries/socrates/

6 Holt, L. (n.d.). *Archimedes 287-212 BCE.* http://www.math.wichita.edu/history/Men/archimedes.html

7 Altrogge, 2013, loc.cit.

8 Descartes, 2007, loc.cit.

9 Suzuki, J. (2009). *Mathematics in historical context.* Mathematical Association of America.

10 Ibid.

11 Borkan, A. (2014, June 24). Invented in bed: Elias Howe's sewing machine. *Casper the Blog.* https://blog.casper.com/invented-in-bed-elias-howes-sewing-machine/

12 Atkins, H. J. B. (1974). *Down, the home of the Darwins: The story of a house and the people who lived there.* Royal College of Surgeons of England.

13 Nietzsche, F. W., Ludovici, A. M., & Sweet, D. (2008). *Twilight of the idols: Or, how to philosophize with a hammer.* Barnes & Noble. (Original work published 1889).

14 Shelley, M. W. (1823). *Frankenstein:* Or, the modern Prometheus (Rev. ed.) H. Colburn and R. Bentley.

15 Balfour, G. (1912). *The life of Robert Louis Stevenson.* Charles Scribner's Sons

16 Dyer, F. L., & Martin, T. C. (2018). *Edison: His life and inventions.* SNova.

17 Heinzmann, G., & Stump, D. (2017). Henri Poincaré. *The Stanford Encyclopedia of Philosophy* (E. N. Zalta, Ed.). Retrieved April 10, 2020, https://plato.stanford.edu/entries/poincare/

18 Rock, A. J. (2019, September 3). August Kekule von Stradonitz. *Encyclopedia Brittanica.* Retrieved April 10, 2020, from https://www.britannica.com/biography/August-Kekule-von-Stradonitz

19 Webb, W. B., & Cartwright, R. D. (2020, January 9). Dream. *Encyclopedia Brittanica.* Retrieved April 10, 2020, from August Kekule von Stradonitz

20 Taylor, N. (2000). *Laser: The inventor, the Nobel laureate, and the thirty-year patent war.* Simon & Schuster.

21 Hawking, S. (2013). *My brief history.* Bantam Books.

22 Einstein & Harris, 1934, loc.cit.

23 Borkan, A. (2014, July 1). Invented in bed: Einstein's theory of general relativity. *Casper the Blog.* https://blog.casper.com/einsteins-theory/

24 Damasio, 1994, loc.cit.

25 Peever, J., & Fuller, P. M. (2016). The biology of REM sleep. *Current Biology,* 26(1), R34–R35. https://doi.org/10.1016/j.cub.2015.11.011

26 Ibid

27 Rasch, B., & Born, J. (2013). About sleep's role in memory. *Physiological reviews,* 93(2), 681–766. https://doi.org/10.1152/physrev.00032.2012

28 Gold P. E. (2008). Protein synthesis inhibition and memory: Formation vs amnesia. *Neurobiology of Learning and Memory,* 89(3), 201–211. https://doi.org/10.1016/j.nlm.2007.10.006

29 Rasch & Born, 2013, loc.cit.

30 Koyama, Y. (2012). [Regulation of sleep and wakefulness through the monoaminergic and cholinergic systems]. *Brain and Nerve = Shinkei Kenkyu No Shinpo,* 64(6), 601–610.

31 Incubate. (n.d.). *Your Dictionary.* Retrieved April 10, 2020, from https://www.yourdictionary.com/incubate

32 *Brain balance guide: The neurochemistry of sleep.* (n.d.). Power On, Power Off. https://poweronpoweroff.com/blogs/longform/the-neurochemistry-of-sleep

33 Crow, T. J., & Grove-White, I. G. (1973). An analysis of the learning deficit following hyoscine administration to man. *British Journal of Pharmacology,* 49(2), 322–327. https://doi.org/10.1111/j.1476-5381.1973.tb08379.x

34 Tiwari, P., Dwivedi, S., Singh, M. P., Mishra, R., & Chandy, A. (2013). Basic and modern concepts on cholinergic receptor: A review. *Asian Pacific Journal of Tropical Disease,* 3(5), 413–420. https://doi.org/10.1016/S2222-1808(13)60094-8

35 Luppi, P.-H., & Fort, P. (2019). Neuroanatomical and neurochemical bases of vigilance states. *Handbook of Experimental Pharmacology*, 253, 35–58. https://doi.org/10.1007/164_2017_84

36 *Brain basics: Understanding sleep.* (2019, August 13). National Institute of Neurological Disorders and Stroke. Retrieved March 13, 2020, from https://www.ninds.nih.gov/Disorders/Patient-Caregiver-Education/Understanding-sleep

37 Brodal, P. (2004). *The central nervous system: Structure and function* (3rd ed). Oxford University Press.

38 McCorry L. K. (2007). Physiology of the autonomic nervous system. *American Journal of Pharmaceutical Education*, 71(4), 78. https://doi.org/10.5688/aj710478

39 Ibid.

40 Purves, D., Augustine, G. J., & Fitzpatrick, D. (Eds.). (2001). Acetylcholine. *In Neuroscience* (2nd ed.). Sinauer Associates. https://www.ncbi.nlm.nih.gov/books/NBK11143/

41 Austin, 1999, loc.cit.

42 Ferreira-Vieira, T. H., Guimaraes, I. M., Silva, F. R., & Ribeiro, F. M. (2016). Alzheimer's disease: Targeting the Cholinergic System. *Current Neuropharmacology,* 14(1), 101–115. https://doi.org/10.2174/157015 9x13666150716165726

43 Zhang, L., Zhou, F.-M., & Dani, J. A. (2004). Cholinergic drugs for Alzheimer's disease enhance in vitro dopamine release. *Molecular Pharmacology,* 66(3), 538–544. https://doi.org/10.1124/mol.104.000299

44 McCoy, A. N., & Tan, S. Y. (2014). Otto Loewi (1873-1961): Dreamer and Nobel laureate. *Singapore Medical Journal*, 55(1), 3–4. https://doi.org/10.11622/smedj.2014002

45 Caine & Caine, 1991, loc.cit.

46 Vaughan, J. (1988). The potential of reflective practice: Rainbow or reality? In H. Waxman, H. Freiberg, J. Vaughan, & M. Weil (Eds.), *Images of reflection in teacher education: Summaries of papers presented at the National Conference on Reflective Inquiry in Teacher Education* (pp. 47-49). Association of Teacher Educators.

47 Schön, D. A. (1983). *The reflective practitioner: How professionals think in action.* Basic Books.

48 Jensen, E. P. (1988). *Super teaching: Master strategies for building student success.* Kendall/Hunt Publishing Company.

49 Ibid.

Chapter 36

1 Harrison, S. (2020). *Creative zing! Spark your creativity and powerfully present your ideas.* MacHillock Publishing

2 *Brain basics: Genes at work in the brain.* (2019, August 13). National Institute of Neurological Disorders and Stroke. Retrieved April 3, 2020, from https://www.ninds.nih.gov/Disorders/Patient-Caregiver-Education/Genes-Work-Brain

3 Berry, J. (2019). *What are neurotransmitters?* https://www.medicalnewstoday.com/articles/326649#key-types-of-neurotransmitters

4 Nakatsuka, N., Yang, K. A., Abendroth, J. M., Cheung, K. M., Xu, X., Yang, H., Zhao, C., Zhu, B., Rim, Y. S., Yang, Y., Weiss, P. S., Stojanović, M. N., & Andrews, A. M. (2018). Aptamer-field-effect transistors overcome Debye length limitations for small-molecule sensing. *Science*, 362(6412), 319-324. https://doi.org/10.1126/science.aao6750

5 Amin, H. U., Malik, A. S., Badruddin, N., & Chooi, W. T. (2014). Brain behavior in learning and memory recall process: A high-resolution EEG analysis. In: Goh J. (eds.) *The 15th International Conference on Biomedical Engineering.* IFMBE Proceedings, 43, 683-686.

6 Brainworks. (n.d.). *What are brainwaves?* https://brainworksneurotherapy.com/what-are-brainwaves

7 Rasch & Born, 2013, loc.cit.

8 Squire, L. R., Genzel, L., Wixted, J. T., & Morris, R. G. (2015). Memory consolidation. *Cold Spring Harbor Perspectives in Biology*, 7(8), a021766. https://doi.org/10.1101/cshperspect.a021766

9 Evans, C., & Evans, P. (1983). *Landscapes of the night: How and why we dream.* Pearson Ptr.

10 Ferrara, M., Moroni, F., De Gennaro, L., & Nobili, L. (2012). Hippocampal sleep features: Relations to human memory function. *Frontiers in Neurology,* 3, 57. https://doi.org/10.3389/fneur.2012.00057

11 Howard, P. (2014). The owner's manual for the brain (4th ed.): *The ultimate guide to peak mental performance at all ages.* William Morrow Paperbacks.

12 Diekelmann, S., & Born, J. (2010). Slow-wave sleep takes the leading role in memory reorganization. *Nature Reviews Neuroscience,* 11, 218. https://doi.org/10.1038/nrn2762-c2

13 Rasch & Born, 2013, loc.cit.

14 Lilienfeld, S. O., & Arkowitz, H. (2014). The truth about shock therapy: Electroconvulsive therapy is a reasonably safe solution for some severe mental illnesses. *Scienttific America Mind,* 25, 3, 70-71. https://www.scientificamerican.com/article/the-truth-about-shock-therapy/

15 McGaugh J.L. (1966). Time-dependent processes in memory storage. *Science,* 153(3742), 1351-13588. https://doi.org/10.1126/science.153.3742.1351

16 Verleger, R., Rose, M., Wagner, U., Yordanova, J., & Kolev, V. (2013). Insights into sleep's role for insight: Studies with the number reduction task. *Advances in Cognitive Psychology,* 9(4), 160–172. https://www.ncbi.nlm.nih.gov/pmc/articles/PMC3902672/

17 Biletzki, A., & Matar, A. (2020). Ludwig Wittgenstein. *The Stanford Encyclopedia of Philosophy* (E. N. Zalta, Ed.). Retrieved March 24, 2020, from https://plato.stanford.edu/entries/wittgenstein/#PhilInve

18 Panksepp, J. (1998). Affective neuroscience: T*he foundations of human and animal emotions.* Oxford University Press.

19 Howard, 2014, loc.cit.

20 Gendlin, E. T. (1978). *Focusing.* Everest House.

21 Wilson, J. (Ed.). (2013). *Schumann resonance.* National Aeronautics and Space Administration. https://www.nasa.gov/mission_pages/sunearth/news/gallery/schumann-resonance.html

22 Meyers, B. (2011). *Introduction to frequency (the key to understanding earth energy).* Magnetic Pulsers. http://www.magneticpulsers.com/frequency.html

23 Bentov, I. (1988). *Stalking the wild pendulum: On the mechanics of consciousness.* Destiny Books.

24 Pritchard, A. (2005). *Ways of learning: Learning theories and learning styles in the classroom.* Routledge.

25 Jensen, 1998, loc.cit.

26 Senge, 1990, loc.cit.

27 Webb, T. W., & Webb, D. (1990). *Accelerated learning with music: A trainer's manual.* Accelerated Learning Systems.

28 Waldorf Education. (n.d.). *Waldorf education: An introduction.* https://www.waldorfeducation.org/waldorf-education

29 Blunt, R. J. (1982). *A study of the educational thought of Rudolf Steiner.* https://core.ac.uk/download/pdf/145045623.pdf

30 Daube, W. C., & Jakobsche, C. E. (2015). Biochemical effects of meditation: A literature review. *Scholarly Undergraduate Research Journal at Clark*, 1, 10. https://commons.clarku.edu/cgi/viewcontent.cgi?article=1017&context=surj

31 National Center for Complementary and Integrative Health. (2019). *Meditation: In depth*. U.S. Department of Health and Human Services, National Institutes of Health. https://nccih.nih.gov/health/meditation/overview.htm

32 Klemp, N. (2019, 20 Aug) Want to Improve Your Memory? A Decade-Long Stanford Study Suggests You Should Stop Doing This 1 Thing. *Inc. Magazine.* Accessed online at https://www.inc.com/nate-klemp/want-to-improve-your-memory-a-decade-long-stanford-study-suggests-you-should-stop-doing-this-1-thing.html

33 Abrams, A. I. (1972). Paired-associate learning and recall: A pilot study of the transcendental meditation program. *Scientific Research on the Transcendental Meditation Program,* 1. https://pdfs.semanticscholar.org/1131/a64e594c820dea456396bbe7af60567f2f65.pdf

34 Wallace, R. K., Orme-Johnson, D. W., Mills, P. J., & Dillbeck, M. C. (1984) Academic achievement and the paired hoffman reflex in students practicing meditation. *International Journal of Neuroscience,* 24(3-4), 261-266. https://doi.org/10.3109/00207458409089814

35 van Leeuwen, S., Singer, W., & Melloni, L. (2012). Meditation increases the depth of information processing and improves the allocation of attention in space. *Frontiers in Human Neuroscience,* 6, 133. https://doi.org/10.3389/fnhum.2012.00133

36 Baijal, S., & Srinivasan, N. (2009). Theta activity and meditative states: Spectral changes during concentrative meditation. *Cognitive Processing,* 11, 31-38. https://doi.org/10.1007/s10339-009-0272-0.

37 Kaur, C., & Singh, P. (2015). EEG derived neuronal dynamics during meditation: Progress and challenges. *Advances in Preventive Medicine,* 1-10. https://doi.org/10.1155/2015/614723.

38 Stroebel, C.F., & Glueck, B.C. (1978). Passive meditation: Subjective, clinical, and electrographic comparison with biofeedback. In: Schwartz G.E., Shapiro D. (Eds.) *Consciousness and self-regulation.* Springer. https://doi.org/10.1007/978-1-4684-2571-0_10

39 Hafenbrack, A. C., Kinias, Z., & Barsade, S. G. (2013). Debiasing the mind through meditation: Mindfulness and the sunk-cost bias. *Psychological Science,* 25(2), 369-376. https://doi.org/10.1177/0956797613503853

40 Lucid Forever. (2018). *How Thomas Edison used the hypnagogic state.* https://www.lucidforever.com/hypnagogia/

41 Jabr, F. (2013). Why your brain needs more downtime. *Scientific American.* https://www.scientificamerican.com/article/mental-downtime/

42 Jensen, 1998, loc.cit.
43 Salk Institute for Biological Studies. (n.d.). *History of Salk: About Jonas Salk.* https://www.salk.edu/about/history-of-salk/jonas-salk/
44 Pert, C. (1997). *Molecules of emotion: The science behind mind-body medicine.* Simon & Schuster.
45 Kiyosaki, R., & Lechter, S. (1997). *Rich dad poor dad.* Warner Books.

Chapter 37

1 Recent Hemingway-related publications. (2020). The Hemingway Society. https://www.hemingwaysociety.org/recent-publications
2 Committee on Sleep Medicine and Research. (2006). Sleep physiology. In Colten H. R., Altevogt B. M. (Eds). Sleep disorders and sleep deprivation: An unmet public health problem (ch. 2). Institute of Medicine, National Academies Press. https://www.ncbi.nlm.nih.gov/books/NBK19956/
3 Pigarev, I. V., & Pigareva, M. L. (2019). Historical view on the attempts to understand the function of sleep in the school of Ivan Pavlov and his Russian forerunners and followers. Clinical and Translational Neuroscience, 1-6. https://doi.org/10.1177/2514183X19834764
4 Aserinsky, E., & Kleitman, N. (1953). Regularly occurring periods of eye motility, and concomitant phenomena, during sleep. Science, 118(3062), 273-274 https://doi.org/10.1126/science.118.3062.273
5 Dement, W., & Kleitman, N. (1957). Cyclic variations in EEG during sleep and their relation to eye movements, body motility, and dreaming. Electroencephalography and Clinical Neurophysiology, 9, 673–690. https://doi.org/10.1016/0013-4694(57)90088-3
6 Peever, J., & Fuller, P. M. (2016). Neuroscience: A distributed neural network controls REM sleep. Current Biology, 26(1), R34–R35. https://doi.org/10.1016/j.cub.2015.11.011
7 Klingelhöfer, J., Hajak, G., Matzander, G., Schulz-Varszegib, M., Sander, D., Rüther, E., & Conrad, B. (1995). Dynamics of cerebral blood flow velocities during normal human sleep. Clinical Neurology and Neurosurgery, 97(2), 142-148. https://doi.org/10.1016/0303-8467(95)00030-N
8 Austin, J. (1999) Zen and the brain: Toward an understanding of meditation and consciousness. MIT Press.
9 Hall, R, H. (1998). Neurotransmitters and sleep. http://web.mst.edu/~rhall/neuroscience/03_sleep/sleepneuro.pdf
10 Gutchess, G. (2001). The Homeric songs: City of dreams. http://englishare.net/literature/POL-HS-City-Dreams.htm

11 Descartes, R. (1641). Meditations on first philosophy. Cambridge. https://www.earlymoderntexts.com/assets/pdfs/descartes1641.pdf

12 Hobson, J. A. (1994). The chemistry of conscious states: How the brain changes its mind. Little Brown & Co.

13 Lewis, P. A. (2014, July 18). What is dreaming and what does it tell us about memory? Scientific America. https://www.scientificamerican.com/article/what-is-dreaming-and-what-does-it-tell-us-about-memory-excerpt/

14 Perogamvrosa, L., & Schwartz, S. (2012). The roles of the reward system in sleep and dreaming. Neuroscience & Biobehavioral Reviews, 36(8), 1934-1951. https://doi.org/10.1016/j.neubiorev.2012.05.010

15 McCoy, A. N., & Tan, S. Y. (2014). Otto Loewi (1873-1961): Dreamer and nobel laureate. Singapore Medical Journal, 55(1), 3–4. https://doi.org/10.11622/smedj.2014002

16 Brain basics: Understanding sleep, 2019, loc.cit.

17 Li, W., Ma, L., Yang, G., & Gan, W. B. (2017). REM sleep selectively prunes and maintains new synapses in development and learning. Nature Neuroscience, 20(3), 427–437. https://doi.org/10.1038/nn.4479

18 Cerri, M., Luppi, M., Tupone, D., Zamboni, G., & Amici, R. (2017). REM sleep and endothermy: Potential sites and mechanism of a reciprocal interference. Frontiers in Physiology. https://doi.org/10.3389/fphys.2017.00624

19 Issa, E. B. (2008). The processing of sounds in auditory cortex during sleep. John Hopkins University. http://web.mit.edu/eissa/www/IssaE_2008_thesis.pdf

20 Sharma, S., & Kavuru, M. (2010). Sleep and metabolism: an overview. International Journal of Endocrinology, 270832. https://doi.org/10.1155/2010/270832

21 Costa, R. M. (2019). Sleep and sexual arousal: A complex relation. The Journal of Sexual Medicine, 16(6), 946. https://doi.org/10.1016/j.jsxm.2019.03.267

22 Corsi Cabrera, M., Velasco, F., del Río Portilla, Y., Armony, J.L., Trejo Martínez, D., Guevara, M.A. and Velasco, A.L. (2016). Human amygdala activation during rapid eye movements of rapid eye movement sleep: An intracranial study. Journal of Sleep Research, 25, 576-582. https://doi.org/10.1111/jsr.12415

23 Squeglia, L. M., Jacobus, J., & Tapert, S. F. (2009). The influence of substance use on adolescent brain development. Clinical EEG and Neuroscience, 40(1), 31–38. https://doi.org/10.1177/155005940904000110

24 Walker, M. P., & van der Helm, E. (2009). Overnight therapy? The role of sleep in emotional brain processing. Psychological Bulletin, 135(5), 731–748. https://doi.org/10.1037/a0016570

25 Colrain, I. M., Nicholas, C. L., & Baker, F. C. (2014). Alcohol and the sleeping brain. Handbook of Clinical Neurology, 125, 415–431. https://doi.org/10.1016/B978-0-444-62619-6.00024-0

26 Brower, J. K. (2001). Alcohol's effects on sleep in alcoholics. The Journal of the National Institute on Alcohol Abuse and Alcoholism, 25(2), 110-125. https://pubs.niaaa.nih.gov/publications/arh25-2/110-125.htm
27 Yang, G., Lai, C. S. W., Cichon, J., Ma, L., Li, W., & Gan, W. (2014). Sleep promotes branch-specific formation of dendritic spines after learning. Science, 344(6188), 1173-1178. https://doi.org/10.1126/science.1249098
28 Rotenberg, V. S. (2006). REM sleep function and brain monoamine regulation. In Sleep and sleep disorders (pp. 27-35). Springer. https://doi.org/10.1007/0-387-27682-3_4
29 Siegel, J. M., & Rogawski, M. A. (1988). A function for REM sleep: Regulation of noradrenergic receptor sensitivity. Brain Research Reviews, 13, 213-233. https://doi.org/10.1016/0165-0173(88)90007-0
30 Hoenselaar, A. (2015). The structure of hippocampal activity during REM sleep. California Institute of Technology. https://doi.org/10.7907/Z98W3B8H.
31 Miller, R. (1991). Cortico-hippocampal interplay and the representation of context in the brain. Springer.
32 Ólafsdóttir, H. F., Bush, D., & Barry, C. (2018). The role of hippocampal replay in memory and planning. Current Biology, 28(1), R37–R50. https://doi.org/10.1016/j.cub.2017.10.073
33 Division of Sleep Medicine. (2007). Changes in sleep with age. Harvard Medical School. http://healthysleep.med.harvard.edu/healthy/science/variations/changes-in-sleep-with-age
34 Committee on Sleep Medicine and Research, 2006, loc.cit.
35 Grigg-Damberger, M. M., & Wolfe, K. M. (2017). Infants sleep for brain. Journal of Clinical Sleep Medicine, 13(11), 1233–1234. https://doi.org/10.5664/jcsm.6786
36 Revonsuo, A. (2000). The reinterpretation of dreams: An evolutionary hypothesis of the function of dreaming. Behavioral and Brain Sciences, 23, 877–1121. https://www.unil.ch/files/live/sites/ln/files/shared/Revonsuo_1.pdf
37 Media Violence Council on Communications and Media. (2009). Pediatrics, 124(5), 1495-1503. https://doi.org/10.1542/peds.2009-2146
38 Harris, G. (2005, Oct 19)Sleeping Pill Use by Youths Soars, Study Says. New York Times. Accessed online at: https://www.nytimes.com/2005/10/19/health/sleeping-pill-use-by-youths-soars-study-says.html
39 Carskadon, M. A., Wolfson, A. R., Acebo, C., Tzischinsky, O., & Seifer, R. (1998). Adolescent sleep patterns, circadian timing, and sleepiness at a transition to early school days. Sleep, 21(8), 871–881. https://academic.oup.com/sleep/article/21/8/871/2726004

40 Semenza, D. & Meldrum, R. & Jackson, D. & Vaughn, M. & Piquero, A. (2020). School Start Times, Delinquency, and Substance Use: A Criminological Perspective. Crime & Delinquency. 66. 163-193. 10.1177/0011128719845147.

41 Carskadon, M. A., Wolfson, A. R., Acebo, C., Tzischinsky, O., & Seifer, R. 1998. loc.cit.

42 Sheldon, S. (2015). Sleep education in schools: Where do we stand? Journal of Clinical Sleep Medicine, 11(6), 595–596. https://doi.org/10.5664/jcsm.4762.

43 Stickgold, R., Whidbee, D., Schirmer, B., Patel, V., & Hobson, J. A. (2000). Visual discrimination task improvement: A multi-step process occurring during sleep. Journal of Cognitive Neuroscience, 12(2), 246–254. https://doi.org/10.1162/089892900562075

44 Robert Stickgold, LaTanya James and J. Allan Hobson. (2000). Visual discrimination learning requires sleep after training. Nature Neuroscience, 3(12), 1237-1238. https://doi.org/10.1038/81756

45 Siegel, J. M. (2001). The REM sleep–memory consolidation hypothesis. Sleep, Dreams, and Memory, 294(5544), 1058-1063. https://doi.org/10.1126/science.1063049

46 Landrigan, C. P., Rothschild, J. M., Cronin, J. W., Kaushal, R., Burdick, E., Katz, J. T., Lilly, C. M., Stone, P. H., Lockley, S. W., Bates, D. W., & Czeisler, C. A. (2004). Effect of reducing interns' work hours on serious medical errors in intensive care units. New England Journal of Medicine, 351(18),1838-1848. https://doi.org/10.1056/NEJMoa041406

47 Wittich, C. M., Burkle, C. M., & Lanier, W. L. (2006). Medication Errors: An overview for clinicians. Mayo Clinical Proceedings, 89(8), 1116-1125. https://doi.org/10.1016/j.mayocp.2014.05.007

48 Mitler, M. M., Carskadon, M. A., Czeisler, C. A., Dement, W. C., Dinges, D. F., & Graeber, R. C. (1988). Catastrophes, sleep, and public policy: consensus report. Sleep, 11(1), 100–109. https://www.ncbi.nlm.nih.gov/pmc/articles/PMC2517096/

Chapter 38

1 Shivanshu K. Srivastava. (n.d.). http://porterfolio.net/shivanshu

2 Scott, C. (2009). Optimal stress. John Wiley & Sons.

3 McEwen, B. S., Nasca, C., & Gray, J. D. (2016). Stress effects on neuronal structure: Hippocampus, amygdala, and prefrontal cortex. Neuropsychopharmacology, 41(1), 3–23. https://doi.org/10.1038/npp.2015.171

4 Branan, N. (2007, June). Stress kills brain cells off. Scientific American. https://www.scientificamerican.com/article/stress-kills-brain-cells/

5 Dubois, A. L., & Mistretta, M. A. (2019). Overcoming burnout and compassion fatigue in schools: A guide for counselors, administrators, and educators. Routledge.

6 Bremner, J. D. (2002). Does stress damage the brain? Understanding trauma-related disorders from a mind-body perspective. WW Norton & Company.

7 Shalev I. (2012). Early life stress and telomere length: Investigating the connection and possible mechanisms: a critical survey of the evidence base, research methodology and basic biology. BioEssays, 34(11), 943–952. https://doi.org/10.1002/bies.201200084

8 Pavlov, I. P. (1960). Conditioned reflexes: An investigation of the physiological activity of the cerebral cortex (G. V. Anrep, Trans.). Oxford University Press. (Original published work 1927).

9 Sun, S. H., & Zoriah, A. (2015). Assessing stress among undergraduate pharmacy students in University of Malaya. Indian Journal of Pharmaceutical Education, 49(2), 99-105. https://doi.org/10.5530/ijper.49.2.4

10 Friedel, E., Schlagenhauf, F., Beck, A., Dolan, R. J., Huys, Q. J., Rapp, M. A., & Heinz, A. (2015). The effects of life stress and neural learning signals on fluid intelligence. European Archives of Psychiatry and Clinical Neuroscience, 265(1), 35–43. https://doi.org/10.1007/s00406-014-0519-3

11 Heissel, J. A., Levy, D. J., & Adam, E. K. (2017). Stress, sleep, and performance on standardized tests: Understudied pathways to the achievement gap. AERA Open. https://doi.org/10.1177/2332858417713488

12 Hunsley, J. (1985). Test anxiety, academic performance, and cognitive appraisals. Journal of Educational Psychology, 77(6), 678. https://doi.org/10.1037/0022-0663.77.6.678

13 Collins, W.E., Schroeder, D.J., & Nye, L. N. (1991). Relationships of anxiety scores to screening and training status of air traffic controllers. Aviation, Space and Environmental Medicine, 62(3), 236-240. https://doi.org/10.1016/0003-6870(92)90250-y

14 Lupien, S. J., Buss, C., Schramek, T. E., Maheu, F., & Pruessner, J. (2005). Hormetic influence of glucocorticoids on human memory. Nonlinearity in Biology, Toxicology, Medicine, 3(1), 23–56. https://doi.org/10.2201/nonlin.003.01.003

15 Goleman, 1995, loc.cit.

16 Stokes, G., & Whiteside, D. (1987). One brain: Dyslexic learning correction and brain integration. Three in One Concepts

17 Cherry, K. (2019, August 18). How the fight or flight response works. VerywellMind. https://www.verywellmind.com/what-is-the-fight-or-flight-response-2795194

18 Lowrance, S. A., Ionadi, A., McKay, E., Douglas, X., & Johnson, J. D. (2016). Sympathetic nervous system contributes to enhanced corticosterone levels following chronic stress. Psychoneuroendocrinology, 68, 163–170. https://doi.org/10.1016/j.psyneuen.2016.02.027

19 Understanding the stress response. (2018, May 1). Harvard Health Publishing. https://www.health.harvard.edu/staying-healthy/understanding-the-stress-response

20 Chetty, S., Friedman, A. R., Taravosh-Lahn, K., Kirby, E. D., Mirescu, C., Guo, F., Krupik, D., Nicholas, A., Geraghty, A., Krishnamurthy, A., Tsai, M. K., Covarrubias, D., Wong, A., Francis, D., Sapolsky, R. M., Palmer, T. D., Pleasure, D., & Kaufer, D. (2014). Stress and glucocorticoids promote oligodendrogenesis in the adult hippocampus. Molecular Psychiatry, 19(12), 1275–1283. https://doi.org/10.1038/mp.2013.190

21 Adler, J. (1999, June 14). Stress and how it attacks your body. Newsweek. https://www.newsweek.com/stress-167544

Chapter 39

1 Maraboli, S. (2013). *Unapologetically you: Reflections on life and the human experience.* Better Today.

2 Hart, 1983, loc.cit.

3 do Amaral, J. R., & de Oliveira, J. M. (1999, October 24). The Triune Brain. https://www.yumpu.com/en/document/read/18990587/the-triune-brain

4 Arnsten A. F. (2009). Stress signalling pathways that impair prefrontal cortex structure and function. *Nature reviews.* Neuroscience, 10(6), 410–422.

5 Jensen, 1998, loc.cit.

6 Shwartz, M. (2007, March 7) *Robert Sapolsky Discusses Physiological Effects of Stress.* Stanford Report. Accessed online at: https://news.stanford.edu/news/2007/march7/sapolskysr-030707.html

7 Aharonian, A. A., & Bornstein, B. H. (2008). Stress and eyewitness memory. *Faculty Publications*, Department of Psychology, 485. https://digitalcommons.unl.edu/cgi/viewcontent.cgi?article=1482&context=psychfacpub

8 Goode, E. (2001, April 3). Scientist at work: Robert Sternberg; his goal: Making intelligence tests smarter. *The New York Times.* https://www.nytimes.com/2001/04/03/science/scientist-at-work-robert-sternberg-his-goal-making-intelligence-tests-smarter.html

9 Steimer, 2002, loc.cit.

10 McCabe, D. L., Treviño, L. K., & Butterfield, K. D. (2001). Cheating in academic institutions: A decade of research. *Ethics & Behavior*, 11(3), 219-232. https://doi.org/10.1207/S15327019EB1103_2

11 Centers for Disease Control and Prevention, 2018, loc.cit.

12 Caine & Caine, 1991, loc.cit.

13 Scammell, T. E., Arrigoni, E., & Lipton, J. O. (2017). Neural circuitry of wakefulness and sleep. *Neuron*, 93(4), 747–765. https://doi.org/10.1016/j.neuron.2017.01.014

14 Wolfson, E. (2013, December 26). Why some people respond to stress by falling asleep. *The Atlantic*. https://www.theatlantic.com/health/archive/2013/12/why-some-people-respond-to-stress-by-falling-asleep/282422/

15 Judd, S. J., Wong, J. A. C. Q. U. I., Saloniklis, S., Maiden, M., Yeap, B., Filmer, S., & Michailov, L. (1995). The effect of alprazolam on serum cortisol and luteinizing hormone pulsatility in normal women and in women with stress-related anovulation. *The Journal of Clinical Endocrinology & Metabolism*, 80(3), 818-823. https://doi.org/10.1210/jcem.80.3.7883836

16 Cowen, P. J. (2002). Cortisol, serotonin and depression: All stressed out? *The British Journal of Psychiatry,* 180(2), 99-100. https://doi.org/10.1192/bjp.180.2.99

17 Olivier, B. (2015). Serotonin: a never-ending story. *European Journal of Pharmacology,* 753, 2-18. https://doi.org/10.1016/j.ejphar.2014.10.031

18 Raine, A. (2002). The biological basis of crime. In J. Q. Wilson, & J. Petersilia (Eds.), *Crime: Public policies for crime control* (pp. 43-74). ICS Press. https://pdfs.semanticscholar.org/6ade/3d8978b1b17d68fcab42e0070120527e20fa.pdf

19 National Institute of Justice. (2016). *Programs and legislation targeting justice-involved young adults* [White paper]. https://nij.ojp.gov/redirect-legacy/topics/crime/pages/programs-andlegislation-targeting-justice-involved-young-adults.aspx

20 Oades, R. D., Lasky-Su, J., Christiansen, H., Faraone, S. V., Sonuga-Barke, E. J., Banaschewski, T., ... & Franke, B. (2008). The influence of serotonin-and other genes on impulsive behavioral aggression and cognitive impulsivity in children with attention-deficit/hyperactivity disorder (ADHD): Findings from a family-based association test (FBAT) analysis. *Behavioral and Brain Functions,* 4(1), 48. https://doi.org/10.1186/1744-9081-4-48

21 Branan, 2007, loc.cit.

22 de Souza-Talarico, J. N., Marin, M. F., Sindi, S., & Lupien, S. J. (2011). Effects of stress hormones on the brain and cognition: Evidence from normal to pathological aging. *Dementia & Neuropsychologia*, 5(1), 8–16. https://doi.org/10.1590/S1980-57642011DN05010003

23 Mauri, M., Sinforiani, E., Bono, G., Vignati, F., Berselli, M. E., Attanasio, R., & Nappi, G. (1993). Memory impairment in Cushing's disease. *Acta Neurologica Scandinavica*, 87(1), 52-55. https://doi.org/10.1111/j.1600-0404.1993.tb04075.x

24 Bremner, 2006, loc.cit.

25 Bremner, J. D., Randall, P., Vermetten, E., Staib, L., Bronen, R. A., Mazure, C., Capelli, S., McCarthy, G., Innis, R. B., & Charney, D. S. (1997). Magnetic resonance imaging-based measurement of hippocampal volume in posttraumatic stress disorder related to childhood physical and sexual abuse--a preliminary report. *Biological Psychiatry*, 41(1), 23–32. https://doi.org/10.1016/s0006-3223(96)00162-x

26 Bremner, J. D., Narayan, M., Anderson, E. R., Staib, L. H., Miller, H. L., & Charney, D. S. (2000). Hippocampal volume reduction in major depression. *American Journal of Psychiatry*, 157(1), 115-118. https://doi.org/10.1176/ajp.157.1.115

27 Sheline, Y. I., Wang, P. W., Gado, M. H., Csernansky, J. G., & Vannier, M. W. (1996). Hippocampal atrophy in recurrent major depression. *Proceedings of the National Academy of Sciences*, 93(9), 3908-3913. https://doi.org/10.1073/pnas.93.9.3908

28 Lupien, S. J., & Lepage, M. (2001). *Stress, memory, and the hippocampus: Can't live with it, can't live without it.* Behavioural Brain Research, 127(1-2), 137-158. https://doi.org/10.1016/S0166-4328(01)00361-8

29 Moffitt, T. (2010, November 16). Childhood stress and lifelong health. *Psychology Today.* https://www.psychologytoday.com/us/blog/genes-and-the-environment/201011/childhood-stress-and-lifelong-health

30 Hart, S. (2018). *Brain, attachment, personality: An introduction to neuroaffective development.* Routledge.

31 Watanabe, Y., Gould, E., & McEwen, B. S. (1992). Stress induces atrophy of apical dendrites of hippocampal CA3 pyramidal neurons. *Brain Research*, 588(2), 341-345. https://doi.org/10.1016/0006-8993(92)91597-8

32 McEwen B. S. (2008). Central effects of stress hormones in health and disease: Understanding the protective and damaging effects of stress and stress mediators. *European Journal of Pharmacology*, 583(2-3), 174–185. https://doi.org/10.1016/j.ejphar.2007.11.071

33 Wang, Y., & Qin, Z. H. (2010). *Molecular and cellular mechanisms of excitotoxic neuronal death.* Apoptosis, 15(11), 1382-1402. https://doi.org/10.1007/s10495-010-0481-0

34 Barry, S. R. (2011, January 16). How to grow new neurons in your brain. *Psychology Today.* https://www.psychologytoday.com/us/blog/eyes-the-brain/201101/how-grow-new-neurons-in-your-brain

35 Saaltink, D. J., & Vreugdenhil, E. (2014). Stress, glucocorticoid receptors, and adult neurogenesis: a balance between excitation and inhibition? *Cellular and Molecular Life Sciences,* 71(13), 2499–2515. https://doi.org/10.1007/s00018-014-1568-5

36 McEwen, B. S., Nasca, C., & Gray, J. D. (2016). Stress Effects on Neuronal Structure: Hippocampus, Amygdala, and Prefrontal Cortex. *Neuropsychopharmacology : official publication of the American College of Neuropsychopharmacology*, 41(1), 3–23. https://doi.org/10.1038/npp.2015.171

37 Torres, I. L., Gamaro, G. D., Silveira-Cucco, S. N., Michalowski, M. B., Corrêa, J. B., Perry, M. L., & Dalmaz C. (2001). Effect of acute and repeated restraint stress on glucose oxidation to CO2 in hippocampal and cerebral cortex slices. *Brazilian Journal of Medical and Biological Research*, 34(1), 111-116. https://doi.org/10.1590/s0100-879x2001000100013

38 Gjerstad, J. K., Lightman, S. L., & Spiga, F. (2018). Role of glucocorticoid negative feedback in the regulation of HPA axis pulsatility. *Stress,* 21(5), 403–416. https://doi.org/10.1080/10253890.2018.1470238

39 Vyas, S., Rodrigues, A. J., Silva, J. M., Tronche, F., Almeida, O. F., Sousa, N., & Sotiropoulos, I. (2016). Chronic stress and glucocorticoids: From neuronal plasticity to neurodegeneration. *Neural Plasticity, 2016.* https://doi.org/10.1155/2016/6391686

40 Duman R. S. (2004). Neural plasticity: consequences of stress and actions of antidepressant treatment. *Dialogues in Clinical Neuroscience*, 6(2), 157–169.

41 University of California - Riverside. (2018, June 18). Often overlooked glial cell is key to learning and memory: Biomedical scientists offer simple advice: Keep the brain active. *ScienceDaily*. Retrieved from www.sciencedaily.com/releases/2018/06/180618222427.htm

42 Torres et al., 2001, loc.cit.

43 Jauregui-Huerta, F., Ruvalcaba-Delgadillo, Y., Gonzalez-Castañeda, R., Garcia-Estrada, J., Gonzalez-Perez, O., & Luquin, S. (2010). Responses of glial cells to stress and glucocorticoids. *Current Immunology Reviews*, 6(3), 195–204. https://doi.org/10.2174/157339510791823790

44 Sanders, R. (2014, February 11). New evidence that chronic stress predisposes brain to mental illness. *Berkeley News*. https://news.berkeley.edu/2014/02/11/chronic-stress-predisposes-brain-to-mental-illness/

45 Post, R., Weiss, S., & Leverich, G. (1994). Recurrent affective disorder: Roots in developmental neurobiology and illness progression based on changes in gene expression. *Development and Psychopathology*, 6(4), 781-813. https://doi.org/10.1017/S0954579400004788

46 Kotulak, 1997, loc.cit.

47 Arnsten A. F. (2009). Stress signalling pathways that impair prefrontal cortex structure and function. *Nature reviews. Neuroscience,* 10(6), 410–422. https://doi.org/10.1038/nrn2648

48 Kim, S., & Lee, D. (2011). Prefrontal cortex and impulsive decision making. *Biological Psychiatry*, 69(12), 1140–1146. https://doi.org/10.1016/j.biopsych.2010.07.005

49 Southwick, S. M., Vythilingam, M., & Charney, D. S. (2005). The psychobiology of depression and resilience to stress: implications for prevention and treatment. *Annual Review of Clinical Psychology*, 1, 255-291. https://doi.org/10.1146/annurev.clinpsy.1.102803.143948

50 Carnahan, J. (2011, August 16). *Balancing your brain chemistry: Treating neurotransmitter imbalances*. https://www.jillcarnahan.com/2011/08/16/balancing-your-brain-chemistry-testing-and-treating-neurotransmitter-imbalances/

51 Begley, Sh. (1997, February 28). How to build a baby's brain. *Newsweek*. https://www.newsweek.com/how-build-babys-brain-174940

52 Petersen, A. C., Joseph, J., Feit, M., on Law, C., & National Research Council. (2014). Consequences of child abuse and neglect. In *New directions in child abuse and neglect research*. National Academies Press (US). https://www.ncbi.nlm.nih.gov/books/NBK195987/

53 Perry, B. D., Pollard, R. A., Blakley, T. L., Baker, W. L., & Vigilante, D. (1995). Childhood trauma, the neurobiology of adaptation, and "use-dependent" development of the brain: How "states" become "traits." *Infant Mental Health Journal*, 16(4), 271-291. https://onlinelibrary.wiley.com/doi/10.1002/1097-0355(199524)16:4%3C271::AID-IMHJ2280160404%3E3.0.CO;2-B

54 Yehuda, R., Engel, S. M., Brand, S. R., Seckl, J., Marcus, S. M., & Berkowitz, G. S. (2005). Transgenerational effects of posttraumatic stress disorder in babies of mothers exposed to the World Trade Center attacks during pregnancy. *The Journal of Clinical Endocrinology & Metabolism*, 90(7), 4115-4118. https://academic.oup.com/jcem/article/90/7/4115/2837310

55 Kellermann, N. P. (2013). Epigenetic transmission of holocaust trauma: can nightmares be inherited. *The Israel Journal of Psychiatry and Related Sciences*, 50(1), 33-39. http://doctorsonly.co.il/wp-content/uploads/2013/07/08_Epigenetic-Transmission.pdf

56 Kanherkar, R. R., Bhatia-Dey, N., & Csoka, A. B. (2014). Epigenetics across the human lifespan. *Frontiers in Cell and Developmental Biology*, 2, 49. https://doi.org/10.3389/fcell.2014.00049

57 Jackson, M. (2014). Evaluating the role of Hans Selye in the modern history of stress. In *Stress, shock, and adaptation in the twentieth century*. University of Rochester Press. https://www.ncbi.nlm.nih.gov/books/NBK349158/

58 Berczi I. (1994) Stress and disease: The contribution of Hans Selye to psychoneuroimmunology. A personal reminiscence. In I Berczi, & J. Szélenyi (Eds.) *Advances in psychoneuroimmunology.* Hans Selye Symposia on Neuroendocrinology and Stress, 3. Springer. https://doi.org/10.1007/978-1-4757-9104-4_1

59 Avey, H., Matheny, K. B., Robbins, A., & Jacobson, T. A. (2003). Health care providers' training, perceptions, and practices regarding stress and health outcomes. *Journal of the National Medical Association*, 95(9), 836-845.

60 Yousaf, S., Saleem, M., & Naseer, F. (2019). Stress: A key player for the induction of many diseases. *Journal of Psychiatry and Psychiatric Disorders*, 3 (2), 37-56. https://www.academia.edu/39386696/Stress_A_Key_Player_for_the_Induction_of_Many_Diseases

61 Nerurkar, A., Bitton, A., Davis, R. B., Phillips, R. S., & Yeh, G. (2013). When physicians counsel about stress: Results of a national study. *JAMA Internal Medicine,* 173(1), 76-77. https://doi.org/10.1001/2013.jamainternmed.480

62 Mayer, E. A. (2000). The neurobiology of stress and gastrointestinal disease. *Gut,* 47(6), 861-869. http://dx.doi.org/10.1136/gut.47.6.861

63 Institute for Quality and Efficiency in Health Care. (2010, December 14). *The defense mechanisms of the adaptive immune system.* https://www.ncbi.nlm.nih.gov/books/NBK279397/

64 Moreno-Smith, M., Lutgendorf, S. K., & Sood, A. K. (2010). Impact of stress on cancer metastasis. *Future oncology (London, England)*, 6(12), 1863–1881. https://doi.org/10.2217/fon.10.142

65 Bonneau, R. H., Kiecolt-Glaser, J. K., & Glaser, R. (1990). Stress-induced modulation of the immune response. *Annals of the New York Academy of Sciences,* 594(1), 253-269. http://pni.osumc.edu/KG%20Publications%20(pdf)/044.pdf

66 Zembroski, R. (2017, January 12). How stress caused my cancer. *The ZRT Laboratory Blog.* https://www.zrtlab.com/blog/archive/stress-response-causes-cancer/

67 Shomon, M. (2020, January 05). The cortisol weight-loss controversy. *VeryWell Health.* https://www.verywellhealth.com/the-cortisol-weight-loss-controversy-3233036

68 Bystander, P. J. (2010, May 6). Was J. M. Barrie really Peter Pan? *The Pharmaceutical Journal.* https://www.pharmaceutical-journal.com/news-and-analysis/opinion/blogs/was-j-m-barrie-really-peter-pan/11010109.blog?firstPass=false

69 Miller, K. (2017, January 13). How stress affects your sex life—And what to do about it. *Self.* https://www.self.com/story/how-stress-affects-sex-life

70 Edwards, V. J., Anda, R. F., Dube, S. R., Dong, M., Chapman, D. F., & Felitti,

V. J. (2005). The wide-ranging health consequences of adverse childhood experiences. *Child Victimization: Maltreatment, Bullying, and Dating Violence Prevention and Intervention, Civic Research Institute,* 8-1. https://web.archive.org/web/20151226092010/http://www.cdc.gov/violenceprevention/acestudy/outcomes.html#Over

71 Blackburn, E., & Epel, E. (2017). *The telomere effect: A revolutionary approach to living younger, healthier, longer.* Orion Spring.

72 Haiken, M. (2012, July 5). How stress and sleep loss are shortening your life. *Forbes.* https://www.forbes.com/sites/melaniehaiken/2012/07/05/how-stress-and-sleep-loss-are-shortening-your-life/#6001778a5935

73 Carruth, L. L., Jones, R. E., & Norris, D. O. (2002). Cortisol and Pacific salmon: a new look at the role of stress hormones in olfaction and home-stream migration. *Integrative and Comparative Biology*, 42(3), 574-581. https://academic.oup.com/icb/article/42/3/574/723971

74 Leuchtenburg, William. (1963) *Franklin D. Roosevelt and the New Deal*, 1932-1940. Harper and Row.

75 Peale, N. V. (2003). *The power of positive thinking.* Simon & Schuster.

76 Seltzer, L. F. (2016, June 15). You only get more of what you resist—Why? *Psychology Today.* https://www.psychologytoday.com/us/blog/evolution-the-self/201606/you-only-get-more-what-you-resist-why

77 Obrien, M. (2016). *New research shows that mindfulness is the secret to happiness.* https://mrsmindfulness.com/ted-talk-happiness-is-mindfulness/

Conclusion

1 Race Against the Machine MIT economists Brynjolfsson and McAfee (Digital Frontier Press, 2011); http://b1ca250e5ed661ccf2f1-da4c182123f5956a3d22aa43eb816232.r10.cf1.rackcdn.com/contentItem-5422867-40675649-ew37tmdujwhnj-or.pdf

2 https://eckharttolle.com/stillness-speaks-excerpt/

Glossary

Adaptability/Intelligence Factor
The attributes, skills, and behaviors that allow our species to better adapt, survive, and thrive in the real world.

Amygdalaizing
Exploring meanings deeply enough to effectively stimulate the amygdala, so you experience an emotional reaction.

Ancient Learning Theory
A concept that intelligence is represented not by how happy and successful you are in life, but by scores on pen-and-paper tests—devaluing emotions and what is meaningful to you while treating you like machine parts on a conveyer belt.

Biological Preference
A primary learning method preference. The eleven main biological preferences are Auditory, Internal-Emotional, Kinesthetic, Linguistic, Mathematical, Musical, Social-Emotional, Spatial/3D, Tactile, Taste/Smell, and Visual.

Completion
The phenomenon in which only a small number of neurons in a concept network need to be activated to stimulate the whole network.

Cycle of Transformation
The four stages of the learning process—Information, Action, Feedback, and Incubation—that allow your brain to create meaning.

Dendrites
Branches extending from neurons in the brain that hold synaptic buds like the leaves on a tree. These grow in length to hold new synapses—newly learned information—and split to hold even more synapses once they reach a certain length.

Disruptive Innovation
Innovation that displaces previously held beliefs about what is possible. Because of the Learning Code science's ability to quickly transform the status quo, it has been called disruptive innovation.

Downregulation
Neurons literally change their shapes from repeated stimulation; at the synapse, neurons have receptor sites on which neurochemicals from other neurons dock and transmit information. If too many neurochemical messengers are produced, neurons are reconfigured so that fewer receptor sites are available to these incoming messengers.

Fast Thinking
Reaching quick conclusions under pressure, fast thinking is the least flexible brain structure and prevents us from accurately predicting consequences to actions.

Habituation
The loss of a neuron's sensitivity resulting from a prolonged pattern of stimulation. After a new stimulus, such as violence, becomes routine, the brain uses habituation to quell its reaction.

Hot Networks
Pre-existing networks that code meaningful information; large numbers of neurons fire simultaneously and cause the chemical cascade to produce synaptic growth that holds long-term memory.

Imitation System
All the genetic and neurological elements that drive us to imitate others.

Jumping Genes
Genes that create mutations in stretches of DNA by spontaneously jumping from one segment of the double helix and reinserting themselves into another, jumping from their initial position to knock out another section, or even jumping up, spinning around 180 degrees, and reinserting themselves backward.

Logotherapy
An approach to psychotherapy that sees humans' search for meaning as our primary motivational force.

Meaning Network
A group of neurological structures that must be stimulated before new information can be selected into its long-term memory banks; otherwise, not only is very little remembered, but levels of motivation, satisfaction, and joy plummet.

Meme
Unconscious imitation that impacts biology so that your behaviors change for extended periods.

Mind Mapping
Drawing a literal map showing how details connect to create the big picture and support memory acquisition of details. Mind mapping was used in developing the outline of this book.

Mutation
The unexpected and chance rearranging of genetic material caused by the adding, dropping, or rearranging one (point mutations) or more (chromosome mutations) base pairs on a sequence of DNA, which changes the meaning of the information conveyed. Mutations are also the source of new characteristics and behaviors that support individual and species options for surviving and thriving.

Myelin
White, fatty substance coating axons, this neural material enhances the speed and efficiency of neuron communication. Like insulation on a telephone line, myelin helps accelerate signals between neurons. It also prevents "cross talk" caused when a signal from one neuron interferes with a nearby one.

Neural Knockout
What happens when trying to cram too much information into your working memory without allowing the brain the time and space it needs to drop into slow-wave (alpha, theta, and delta) states. If trapped in inactive, awake beta states too long, the information stored in your working memory can easily get knocked out before it has been set up for inclusion in long-term memory.

Neural Darwinism
Neurons select the structures and functions supporting the signals that most stimulate them.

Neural Pruning
The natural process by which neurons and synapses are efficiently cut away to shape your brain to fit its environment.

Neurogenesis
The process of birthing neurons. For four months shortly after conception, cells explode from the neural tube at the rate of 500,000 per minute. It's during this short eighteen-week period that the basic building blocks of your brain are formed and put in place.

Neurogenetics of Authenticity
A Learning Code program to help you understand not only the neurobiology of authenticity but also the positive effects it has on other people's behavior because of more biological impact.

Neurogenetics of Cultural Transformation
A Learning Code program encouraging companies to develop a Noble Cause to activate the same neurological upshift as personal meaning.

Neurological Downshift
Excessive levels of stress prompt the overproduction of cortisol, shifting the distribution of glucose and oxygen away from areas of working memory, greatly reducing their efficiency.

Neurological Upshift
When engaged in meaningful activities, stress levels decrease, resulting in an upshift in neural energy to the upper and frontal parts of your brain. When these structures are engaged, emotional control, innovation, and compliance go up, while impulsivity and behavioral problems go down.

Neuromodulators
The neurons dopamine, serotonin, and norepinephrine that modify the way other neurons act. Like a volume dial, the monoamine system gains control, either amplifying or muting the activity of other neurons.

Noble Cause
A cause bigger than any one person or the company; focusing the company's attention on such a cause activates the same neurological upshift as personal meaning.

Orchestrated Incubation Time
Time purposefully taken away from activities that demand a high level of concentration, which creates beta waves, to engage in activities that produce slow-wave, alpha, theta, and delta activity such as listening to music, gardening, zoning out, or engaging in sports and hobbies.

Preheating
The idea that concepts must be presented at the beginning of the learning process.

Psychoneuroimmunology
A study of the link between stress levels and vulnerability to disease.

Selection
The natural process of physiological change within an organism when it selects new information into its biology that helps it adapt more effectively. The human brain can only learn through a selective process, and all new learning is dependent upon what already exists in your brain.

Slow Thinking
A contemplative approach to solving problems allowing "big-picture" systems thinking and more accurately predicting the consequences of your actions. Virtually all great innovations were achieved using a contemplative, slow-thinking process.

Somatic Feedback
All changes in the body (soma) and organs, represented by changes in heart rate, perspiration, respiration, pupil dilation, blood pressure, muscle tone, digestion, and elimination, transmitted back to the brain via the nervous system, as well as changes in concentrations of peptides, glucose, oxygen, carbon dioxide, ions, pH levels, and hormones, like BDNF (Brain-derived Neurotrophic Factors) ferried back to the brain via the bloodstream.

Trophic Factors
Chemical modulators that support the growth, survival, migration, and function of neurons, such as those expressed in infants' brains when the infants are loved and touched by their primary caregivers.

Working Memory
A generalized workspace for conscious and unconscious mental processing. The three main factors of working memory: it prioritizes information based on the intensity of emotional and body feedback; it is limited in both the time frame and storage capacity; and inclusion into its space depends on information already held in your brain as meaningful.

Index

C

D

O

P

Made in the USA
Thornton, CO
08/21/23 19:56:59

714e2e78-afc1-4ad4-bfcc-b8abc3656ce6R02